Advanced Studies in Theoretical and Applied Econometrics

Volume 57

This book series aims at addressing the most important and relevant current issues in theoretical and applied econometrics. It focuses on how the current data revolution has affected econometric modeling, analysis and forecasting, and how applied work has benefited from this newly emerging data-rich environment. The series deals with all aspects of macro-, micro-, financial-, and econometric methods and related disciplines, like, for example, program evaluation or spatial analysis.

The volumes in the series are either monographs or edited volumes, mainly targeting researchers, policymakers, and graduate students.

This book series is listed in Scopus.

Badi H. Baltagi · László Mátyás
Editors

Seven Decades of Econometrics and Beyond

A Tribute to the Life and Work of Marc Nerlove

Springer

Editors
Badi H. Baltagi
Center for Policy Research
Syracuse University
Syracuse, NY, USA

László Mátyás
Department of Economics
Central European University
Vienna, Austria

ISSN 1570-5811 ISSN 2214-7977 (electronic)
Advanced Studies in Theoretical and Applied Econometrics
ISBN 978-3-031-92698-3 ISBN 978-3-031-92699-0 (eBook)
https://doi.org/10.1007/978-3-031-92699-0

This Springer imprint is published by the registered company Springer Nature Switzerland AG
The registered company address is: Gewerbestrasse 11, 6330 Cham, Switzerland

Foreword

In 1959, I enrolled in Marc Nerlove's graduate econometrics course at the University of Minnesota. At the time, I was a graduate student in social psychology with no background in economics, but I had a keen interest in the statistical analysis of social science data.

Least-squares regression methods had been introduced to economics decades earlier, notably in Ragnar Frisch and Henry Schultz's studies of sugar demand in the 1920s, and in Charles Cobb and Paul Douglas's analysis of production. These advances, along with the founding of the Econometric Society in 1930, laid the groundwork for the field. The study of demand systems by Richard Stone in 1945, the pioneering work of the Cowles Commission under Tjalling Koopmans, and the development of two-stage least squares by Robert Basmann and Hans Theil in the 1950s, marked the beginnings of modern econometrics. These topics formed the core of Marc's course.

At the time, digital computers were rudimentary, requiring machine language programming, and computing a regression was an arduous task done on a four-function Friden calculator. Despite this, Marc assigned us to design and estimate a simultaneous equations model with double precision. My project—a seven-variable model of the petroleum market—took two weeks to estimate. This challenging experience captivated me enough to switch to the economics Ph.D. program, focusing on the econometric analysis of individual behavior.

Marc moved to Stanford in 1960, but in the summer of 1961, I worked with him, Kenneth Arrow, and Hirofumi Uzawa on convex analysis and the theory of production, which became the basis of my thesis. When I entered the job market in 1962, my knowledge of economics was still uneven. However, Marc instilled in me the importance of bridging economic theory and empirical data. Although his name did not appear on my thesis, he was in essence my advisor—and I was his first Ph.D. student.

Our research paths diverged over the years: I gravitated toward behavioral economics and individual choice prediction, while Marc delved into economic dynamics and market and economy-wide applications. Nevertheless, the lessons I learned from him remained pivotal, and parallels between our approaches to research are evident.

Throughout his long and storied career, Marc continued to refine methods for economic data analysis. From innovations in time-series and panel data to applied research on business expectations and the impacts of population growth, his contributions profoundly shaped the field of econometrics. The breadth and power of Marc's contributions are evident from his citations in econometrics textbooks, and from the significant surveys and studies included in this volume.

University of California, Berkeley, March 26, 2025 *Daniel McFadden*

Preface

Marc Nerlove was a truly renaissance man in the twentieth century. Cultivated, speaking multiple languages, freely and happily interacting with anybody with an open mind, he was an intellectual giant. His natural curiosity, deep knowledge of a wide range and aspects of social sciences, and the ease to present difficult problems, led him to have a lasting impact on economics, econometrics and policy. This volume pays tribute to his life, personality, and long-lasting influence on the profession.

The book contains high quality original research work in different areas of theoretical and applied econometrics. Some survey type chapters are also featured which help better understand some important areas of econometrics research. When appropriate and relevant, a few personal paragraphs and insights into his research are added. The chapters present cutting edge results in panel data, machine learning, agricultural economics, spatial economics, and income inequality among others. The variety of topics discussed reflect some of the wide range of contributions Marc Nerlove made to the profession.

Although hard to believe these days, five to six decades ago, empirical economics and econometrics were overwhelmingly dominated by macroeconomics. Since then, the use of firm level and other types of micro-level data for econometric analysis have been widely adopted, bearing results unimaginable earlier. The first chapter of the volume provides a very personal, first-hand historical insight into Marc Nerlove's pioneering role in this process. It is fascinating to learn how the vision of a few researchers and institutes helped re-shape economic practice.

The second chapter deals with the 'evergreen' issue of sustainable development. Although Malthus' bleak predictions were repeatedly proven wrong, the future is far from certain. Marc Nerlove's approach was that we can live beyond our means for a time only by depleting our environmental capital stock. Following his steps, the chapter gives a thorough review of this relevant and heated discussion and, unfortunately, its conclusions are not rosy.

Marc Nerlove in his seminal work published almost seven decades ago[1] explored the relevance of price expectations in agricultural production and how these may affect supply elasticities. Chapter 3 extends this 'Nerlovian model' by taking into account some recent developments in econometrics, machine learning and data availability. These new models are then estimated and tested using some FAO data sets. It shows that Nerlove's approach is still relevant these days.

Chapter 4 presents, within a historical perspective, an interesting interaction between game theory and econometrics. A more personal approach helps to get an insight on how Marc Nerlove operated simultaneously as an educator and a researcher and how he inspired his students.

Chapter 5 provides a detailed study of the state of knowledge on measurement and analysis of inequality of outcomes like income and earning. This includes the state of art techniques for identifying the distribution of outcomes and interesting functions of it, such as inequality measures, poverty, and mobility indices. The main aim of the chapter is to facilitate the adoption of the latest developments in this area at a time of heightened interest in evidence based policy analysis.

The Opportunity Zones (OZ), the largest ongoing place-based development program in the U.S., were intended to stimulate investment and drive economic growth in low-income areas by lowering capital gains tax rates. Chapter 6 investigates the spatial spillover effects of the OZ due to their interconnections with high-income neighbouring areas. The empirical results indicate that census tracts located near more developed regions exhibit a stronger response to the OZ program due to the presence of spillover effects. The driving factor of these policies is the number of high-income neighbors. The chapter shows, however, that they play the role of a double-edged sword.

Chapter 7 deals with rationality tests and the estimation of asymmetric loss functions by using information in density forecasts. This chapter shows that often forecasters treat underestimation of real output more dearly than over-prediction, while the opposite is true for inflation.

In economics there are frequently cases when agents make discrete choices depending on past outcomes and on the ones of other agents. This may create network interdependencies. The order of the dymanics and that of the network pattern usually is not known a prori. To deal with this, Chapter 8 parametrizes the higher order time lag and network lag structures to estimate a response function. This chapter suggests panel probit estimation based on control functions and studies their suitability through detailed simulation experiments.

Chapter 9 argues that the measurement of treatment effects using panel data is essentially an issue of predictions. In the literature there are several ways to construct conterfactuals based on hypothetical data generating processes. The chapter proposes a unifying framework for these based on a factor approach to conterfactuals.

Chapter 10 proposes a semiparametric method for the estimation of nonparametric panel data models with correlated random-effects, where both the nonparametric function and a finite-dimensional parameter associated with (potentially) observed

[1] Marc Nerlove: Estimates of the Elasticities of Supply of Selected Agricultural Commodities, *Journal of Farm Economics*, 1956, pp. 496-509.

time-invariant regressors can be identified. Analytical and simulation results are put forward together with an illustrative application on the relationship between firms' research and development expenditures, current assets, and regulatory restrictions across different industries.

Dynamic panel data models have been widely used in the literature for decades, and Marc Nerlove has lucidly examined their problems and applications to the study of growth convergence. One of the workhorses for their estimation is the Generalised Method of Moments (GMM) estimator, which is mainly applied to the first-difference model. Chapter 11 investigates the effect of maintaining the model in levels, combined with a Correlated Random Effects (CRE) specification. Some analytical and simulation results are provided, along with applications to R&D, production and wage functions, to illustrate the advantages of this approach.

Chapter 12 is a tribute to Marc Nerlove's contributions to panel data and spectral analysis in time series. In his book titled *Essays in Panel Data Econometrics*,[2] Marc reviews serial correlation in panel data error components models. This chapter proposes a new estimation method based on Whittle's approximate maximum likelihood method[3] for an ARMA(p,q) in the remainder disturbances and performs Monte Carlo simulations to examine its performance in small samples. The results demonstrate that this spectral method allows one to consider complex structures of the variance-covariance matrix to account for serial correlation in the remainder error.

Chapter 13 investigates three-level dynamic panel data models with mixed coefficient structure, where one level is a group (or stratum) and the time component is short. Identification and estimation issues are discussed in the presence of diverse forms of heterogeneity and cases with unbalanced or missing data.

Chapter 14 reviews basic elements of the Mundlak and Chamberlain projections which are well known classic results in panel data showing that the random effects estimator reduces to the fixed effects estimator when the regressors are correlated with the individual effects. This is done following a simple transformation proposed by Manuel Arellano.[4] Topics discussed include the augmented regression model, the Hausman test, minimum-distance estimation and its link to GMM, unbalanced data, and higher-dimensional data.

Chapter 15 generalizes the Mundlak panel data model to allow for a subset of the variables to have heterogeneous coefficients. These are estimated using fitted values from unit-specific regressions. An important application of this is allowing for heterogenous trends. A simple specification test is proposed to determine whether the usual two-way fixed effects are sufficient or whether unit-specific trends should be added. This approach has also implications for relaxing parallel trends in a difference-in-differences settings with staggered interventions. This is illustrated with an empirical application.

---

[2] Marc Nerlove: Essays in Panel Data Econometrics, *Cambridge University Press*, 2005.

[3] Peter Whittle: Estimation and Information in Stationary Time Series, *Arkiv för Matematik*, 1953, 2(5), 423–434.

[4] Manuel Arellano: On Testing of Correlated Effects with Panel Data, *Journal of Econometrics*, 1993, 59, 87-97.

The inadequacy of standard asymptotic tests in finite samples is well known in a simultaneous equations context. Chapter 16 proposes alternative exact and bound procedures and shows their feasibility. Particular attention is given to identification issues. Simulation results show that relaxing the structure under the alternative hypothesis pays off power wise. While structures hold information, this comes at an important cost: imposing it introduces nuisance parameters that are influenced by the model's identification status.

Chapter 17, the last chapter of the volume, generalizes the static log-linear probability model originally introduced by Nerlove and Press,[5] to the dynamic analysis of qualitative processes with two or three alternatives. This modelling approach has relevant applications in predicting future financial returns.

Syracuse, April 2025 *Badi H. Baltagi*
Budapest and Vienna, April 2025 *László Mátyás*

[5] Nerlove, M. and Press, S.: Univariate and Multivariate Log-linear and Logistic Models, *Rand Corporation*, 1973.

Acknowledgements

We address our thanks to all those who have facilitated the birth of this volume: the contributors who produced quality work, despite tight deadlines, frequent requests, and onerous instructions from the editors, and Anke Meyer for her continuous encouraging and stimulating support.

Comments and remarks received during the writing and editing process acknowledged at the end of each chapter have substantially improved the quality of the book. Financial assistance by the Department of Economics of the Central European University is also kindly acknowledged.

English language editing by Eszter Timár made the text of some chapters easier and more enjoyable to read.

The final camera-ready copy of the volume has been prepared with LaTeX and Overleaf by the authors, the editors, and the help of Ramzi Chariag.

Contents

List of Contributors

Volume Editors:
Badi H. Baltagi
Syracuse University, Syracuse, New York, USA, e-mail: bbaltagi@syr.edu

László Mátyás
Central European University, Budapest, Hungary and Vienna, Austria, e-mail: matyas@ceu.edu

Contributors:
Monika Avila Márquez
University of Bristol, Bristol, UK, e-mail: monika.avilamarquez@gmail.com

Badi H. Baltagi
Syracuse University, Syracuse, New York, USA, e-mail: bbaltagi@syr.edu

Paul A. Bjorn
UH Parma, Parma, Ohio, USA, e-mail: pabjorn@hotmail.com

Maria Elena Bontempi
University of Bologna, Bologna, Italy, e-mail: mariaelena.bontempi@unibo.it

Georges Bresson
Université Paris Panthéon-Assas, Paris, France, e-mail: georges.bresson@u-paris2.fr

Yisroel Cahn
New York University, New York, USA, e-mail: yisroel.cahn@nyu.edu

Felix Chan
Curtin University, Perth, Australia, e-mail: felix.chan@cbs.curtin.edu.au

Jan Ditzen
Free University of Bozen-Bolzano, Bolzano, Italy, e-mail: jan.ditzen@unibz.it

Jean-Marie Dufour
McGill University, Montréal, Canada, e-mail: jean-marie.dufour@mcgill.ca

Richard Dwumfour
Curtin University, Perth, Australia, e-mail: richard.dwumfour@curtin.edu.au

Peter H. Egger
ETH Zurich, Zurich, Switzerland, e-mail: pegger@ethz.ch

Jean-Michel Etienne
Université Paris-Saclay, Sceaux, France, e-mail: jean-michel.etienne@u-psud.fr

Christian Gouriéroux
University of Toronto, Canada; Toulouse School of Economics and CREST, France, e-mail: christian.gourieroux@ensae.fr

Daniel J. Henderson
University of Alabama, Tuscaloosa, USA, e-mail: daniel.henderson@ua.edu

Emma Kate Henry
University of Alabama, Tuscaloosa, USA, e-mail: ekhenry@crimson.ua.edu

Cheng Hsiao
University of Southern California, Los Angeles, USA, and Paula and Gregory Chow Center for Studies in Economics, Xiamen University, China, e-mail: chsiao@usc.edu

Elizabeth Jackson
Curtin University, Perth, Australia, e-mail: Elizabeth.Jackson@curtin.edu.au

Michaela Kesina
University of Groningen, Groningen, The Netherlands, e-mail: m.kesina@rug.nl

Lynda Khalaf
Carleton University, Ottawa, Canada, e-mail: Lynda_Khalaf@carleton.ca

Jing Kong
University of Southern California, Los Angeles, USA, e-mail: jingkong@usc.edu

Jaya Krishnakumar
University of Geneva, Geneva, Switzerland, e-mail: jaya.krishnakumar@unige.ch

Kajal Lahiri
University at Albany, Albany, New York, USA, e-mail: klahiri@albany.edu

Fushang Liu
Massachusetts Department of Revenue, Boston, Massachusetts, USA, e-mail: fushangl@gmail.com

Esfandiar Maasoumi
Emory University, Atlanta, USA, e-mail: emaasou@emory.edu

László Mátyás
Central European University, Budapest, Hungary and Vienna, Austria, e-mail: matyas@ceu.edu

Nour Meddahi

Toulouse School of Economics, Toulouse, France, e-mail: nour.meddahi@tse-fr.eu

Dibya Deepta Mishra
Rice University, Houston, USA, e-mail: ddm5@rice.edu

Isabelle Perrigne
Rice University, Houston, Texas, USA, e-mail: iperrigne@gmail.com

John Rust
Georgetown University, Washington DC, USA, e-mail: jrust@editorialexpress.com

Robin C. Sickles
Rice University, Houston, Texas, USA, e-mail: rsickles@rice.edu

Alexandra Soberon
University of Cantabria, Santander, Spain, e-mail: alexandra.soberon@unican.es

Yanfei Sun
Toronto Metropolitan University, Toronto, Canada, e-mail: yanfei.sun@torontomu.ca

Quang Vuong
New York University, New York, USA, e-mail: qvuong@nyu.edu

Wuwei Wang
Southwestern University of Finance and Economics, Chengdu, China, e-mail: wangwuwei@swufe.edu.cn

Tom Wansbeek
University of Groningen, The Netherlands, e-mail: t.j.wansbeek@rug.nl

Jeffrey M. Wooldridge
Michigan State University, East Lansing, Michigan, USA, e-mail: wooldri1@msu.edu

Yimeng Xie
Xiamen University, Xiamen, China, e-mail: mengxie@xmu.edu.cn

Qiankun Zhou
Louisiana State University, Baton Rouge, USA, e-mail: qzhou@lsu.edu

Klaus F. Zimmermann
Bonn University, Bonn, Germany, e-mail: klaus.f.zimmermann@gmail.com

Chapter 1
Analysis of Business Surveys: The Mannheim Years

Klaus F. Zimmermann

Abstract In 1979, Marc Nerlove and my doctoral advisor, Heinz König, launched a groundbreaking joint project on 'Business Survey Data Analysis', which continued for about 16 years. This project began during a period of transition in the economics profession, marked by a shift from macro theory to applied microeconomics, and from macro-econometrics to the study of qualitative micro data. Under the leadership of Nerlove and König, an international team pioneered the use of firm-level data for microeconometric analyses. This paper documents the team's work, the challenges they faced, their ambitions, and their academic achievements. It also highlights Nerlove's leadership, working style, and personality, as reflected in the project and beyond. As a member of the Mannheim research team, I also had the opportunity to become Nerlove's academic guest at the University of Pennsylvania in 1987.

1.1 Introduction

In the late 1970s, economics as a scientific discipline was still dominated by theoretical approaches, with macroeconomics shaping much of econometric research. Large-scale econometric models sought to model entire national economies, driven by the expectation that they could serve as tools for economic control. In Europe, academic research remained underdeveloped compared to the United States. Only a few European institutions, such as the London School of Economics (LSE) and CORE in Louvain-la-Neuve, had achieved significant international visibility. At that time, the University of Mannheim was emerging as a leading center for economic research in Germany. Among the country's foremost macroeconomists and macro-econometricians was Heinz König of the University of Mannheim, who, alongside Wilhelm Krelle at the University of Bonn, played a central role in shaping Germany's

Klaus F. Zimmermann ✉
Bonn University, Bonn, Germany, e-mail: klaus.f.zimmermann@gmail.com

B. H. Baltagi and L. Mátyás (eds.), *Seven Decades of Econometrics and Beyond*, Advanced Studies in Theoretical and Applied Econometrics 57,
https://doi.org/10.1007/978-3-031-92699-0_1

econometric landscape. König's influence extended beyond research, as he also served as the Rector of the University of Mannheim from 1979 to 1982.

Marc Nerlove shared König's interest in time-series econometrics, particularly in spectral analysis. Despite their primary focus on macroeconomic research, they secured funding from NATO to establish a transatlantic collaboration aimed at creating and analyzing a new dataset based on firm-level surveys. Their efforts were rooted in the rich survey tradition of the Ifo Institute in Munich, which had conducted extensive monthly business surveys since the 1950s. However, the Ifo Institute only used this data in aggregated form for macroeconomic monitoring rather than for micro-level analysis. König's established connections with the Ifo Institute were instrumental in accessing this resource. He maintained regular exchanges with its leadership, particularly at the legendary annual Ottobeurer Seminar, where Germany's leading economists convened to discuss pressing economic issues.

The challenge was that the Ifo business survey data existed only in paper form, making individual firm-level records inaccessible for systematic analysis. Moreover, the data was qualitative and discrete rather than continuous, which posed methodological obstacles. At the time, economists largely dismissed qualitative data, favoring direct observations of economic behavior over subjective assessments. Additionally, statistical methods for analyzing qualitative data were underdeveloped. A breakthrough came through Marc Nerlove's methodological contributions to contingency table analysis, which he had already advanced in 1973 but had yet to apply extensively. The project also benefited from its connection to CIRET, an international research network focused on business cycle survey data, in which the Ifo Institute played a key role. This network facilitated the dissemination of new methods and research approaches, creating an environment for advancing micro-econometric applications.

Through this collaboration, Marc Nerlove, an academic entrepreneur with a global perspective, introduced Heinz König, a leading macroeconomist, to micro-econometric research. Over the years, Nerlove spent frequent research periods in Mannheim, contributing to the methodological and empirical development of firm-level data analysis. Despite the significant contributions of König and Nerlove, little research has focused on the methodological innovations and challenges of their collaboration, particularly in leveraging firm-level data for econometric analysis.

This chapter examines the transformative impact of Heinz König and Marc Nerlove's collaboration on the development of micro-econometric methods, focusing on their innovative use of firm-level data. It will outline the methodological innovations introduced, assess the impact of these innovations on econometric research, and explore the challenges and successes of their collaboration. The following sections detail the evolution of their research endeavor. Section 1.2 introduces Marc Nerlove and outlines my own involvement in the project. Section 1.3 provides an overview of the Ifo data and describes the working process of the Mannheim research team. Section 1.4 presents the methodological foundations and key research findings. Section 1.5 discusses subsequent research developments within the broader network. Finally, Section 1.6 summarizes and evaluates the overall contributions of this long-term collaboration.

1.2 Marc Nerlove – Visionary, Leader, Globalist, Generalist

1.2.1 Marc Nerlove

Marc Nerlove was a towering figure in economics and econometrics, whose methodological innovations and empirical investigations left a lasting imprint on the discipline. After earning his Ph.D. from Johns Hopkins University in 1956, he embarked on an academic career that spanned more than six decades, influencing generations of scholars across multiple domains.

Nerlove was a pioneer in microeconometrics, particularly in the estimation of dynamic models using panel data. His groundbreaking research on adaptive expectations and supply responses in agriculture remains a cornerstone of empirical work on producer behavior. His 1958 book, *The Dynamics of Supply: Estimation of Farmers' Response to Price*, was a pioneering effort to apply econometric techniques to agricultural data, setting a precedent for the integration of economic theory with empirical analysis (Nerlove, 1958b). His work laid the foundation for modern empirical studies on agricultural supply and demand, influencing policies on agricultural markets and price stabilization.

His contributions to time-series econometrics and macroeconomics are also significant. His 1964 Econometrica paper *Spectral Analysis of Seasonal Adjustment Procedures* introduced spectral methods to study economic fluctuations, demonstrating their application in evaluating seasonal adjustment techniques (Nerlove, 1964). His later work, particularly his book *Analysis of Economic Time Series: A Synthesis*, provided an extensive and rigorous framework for time-series modeling (Nerlove, Grether & Carvalho, 1979). This work synthesized approaches to time-series econometrics, bridging traditional econometric methods with modern spectral and state-space models. His research advanced the understanding of economic cycles, particularly how firms and individuals form expectations over time, and influenced the broader study of macroeconomic fluctuations.

Nerlove was also engaged in macroeconomic research. His 1962 American Economic Review paper, *A Quarterly Econometric Model for the U.K.: A Review Article*, was an important contribution to the growing field of macroeconometric modeling (Nerlove, 1962). His 1966 International Economic Review paper, *A Tabular Survey of Macro-Econometric Models*, provided one of the first comprehensive reviews of macroeconometric models, helping to systematize research in this field (Nerlove, 1966).

Beyond macroeconomics, Nerlove was a significant contributor to population economics. Together with Assaf Razin and Efraim Sadka, he explored the interplay between household decisions, demographic trends, and economic welfare using economic micro theory. Their joint book, *Household and Economy: Welfare Economics of Endogenous Fertility*, offered a formalized economic analysis of fertility decisions, treating fertility as an endogenous choice influenced by economic conditions, based on many top publications (Nerlove, Razin & Sadka, 1987). This study provides a theoretical foundation for understanding how economic incentives shape demographic transitions, contributing to debates on population growth, pension systems, and intergenerational transfers. His work challenged traditional Malthusian perspectives by demonstrating that population growth could be optimally managed through economic incentives rather than coercive policies.

A further distinctive aspect of Nerlove's research is his pioneering use of the log-linear probability model for the analysis of categorical economic data. His collaboration with S. James Press on *Univariate and Multivariate Log-Linear and Logistic Models* (Nerlove & Press, 1973 and Nerlove & Press, 1976) laid a foundational framework for applying these models in economics. This work provided the methodology for analyzing contingency tables and categorical survey data, and provided a crucial tool in the study of business test data as will be the focus in this chapter (König, Nerlove & Oudiz, 1981, and Nerlove, 1983). The importance of this line of research was underscored when Nerlove chose to focus on expectations, plans, and realizations of business firms for his Presidential Address to the Econometric Society, published as *Expectations, Plans and Realizations in Theory and Practice* (Nerlove, 1983).

Throughout his career, Nerlove was a visionary, leader, globalist and generalist. He was visionary in the sense that his methodological advances anticipated and shaped the trajectory of modern econometrics. His emphasis on dynamic models, expectation formation, and panel data econometrics prefigured many contemporary approaches in applied economics. As a leader, he trained and mentored numerous students, many of whom became leading economists and econometricians in their own right. His work earned him numerous accolades, including the election as a Fellow of the Econometric Society, later on even the president, and the prestigious John Bates Clark Medal, awarded to the most promising American economist under 40.

Nerlove was globalist in both his research and academic engagement. His work spanned multiple countries and economic contexts, from U.S. agricultural markets to European business surveys to developing economies in Latin America and Asia. He collaborated extensively with international researchers, reflecting on his belief that economic knowledge should transcend national boundaries. His visiting appointments at leading institutions across Europe, Latin America, and Asia underscored his role as a bridge between different traditions and cultures of economic thought.

Finally, Nerlove was a generalist in the best sense. While many scientists specialize narrowly in methodology, theory, or applied work, he has moved seamlessly between theoretical economics and econometrics, empirical analysis, and economic policy. His research encompassed agriculture, macroeconomics, population, expectation formation, time-series analysis, and microeconometrics, reflecting a rare breadth of expertise.

Even in his later years, Nerlove remained intellectually engaged and continued to contribute to econometric methodology and applied economic research. His legacy endures not only in the methodologies he developed, but also in the scholars he trained and the empirical insights he provided. He passed away in 2023 at the age of 90, leaving behind a vast intellectual legacy that continues to shape his fields of analysis.

1.2.2 Background Reflections

The long-time research partner of Marc Nerlove in Germany was Heinz König (1927–2002), a leading figure in post-war German economics, a pioneer of empirical economic research, and econometrics. König began as a macroeconomist and, competing with Wilhelm Krelle from Bonn University, developed the first large-scale macroeconometric models in Germany. In 1958–1959, he was a Rockefeller Fellow at the Massachusetts Institute of Technology (MIT), Harvard University, and Stanford University. He became a Full Professor at the University of Mannheim in 1962, where he remained despite receiving numerous prestigious offers from other universities. He served as Rector of the University of Mannheim from 1979 to 1982, chaired the Verein für Socialpolitik (the German Economic Association) from 1987 to 1988, and was the founding director of the Centre for European Economic Research (ZEW) from 1991 to 1997. indexCentre for European Economic Research (ZEW)

Nerlove and König were both distinguished figures in their respective fields, each commanding a strong national reputation and possessing distinct yet equally formidable personalities. While König, whose name fittingly means 'king' in German, wielded his authority in the hierarchical chair-system of German universities at the time with an almost autocratic style, Nerlove's influence was more understated and diplomatic. Nevertheless, he too mentored a devoted group of PhD students and maintained an extensive global network of established research collaborators.

Both were natural leaders, earning huge respect through their intellectual rigor and visionary contributions. Their research interests overlapped in macroeconomic modeling and time-series econometrics. However, Nerlove's expertise extended into agricultural and population economics, while König also made significant contributions to labor economics. During what I refer to as *The Mannheim Years* (detailed more below), they collaborated on a project initially funded by NATO (research grant no. 1180, 1976–1979) and later by the National Science Foundation (USA, Grant SOC 74-21194), and Deutsche Forschungsgemeinschaft (Grant 219/10) focused on the creation and analysis of categorical business survey data to examine firm-level behavior. Through this collaboration, both evolved into microeconometricians.

Given their shared background, it is unsurprising that the central theme of their joint research was the formation of business expectations. Nerlove had been engaged with adaptive and other expectation-formation models since his doctoral work in agricultural economics in the late 1950s, later expanding this focus within time-series econometrics. König, in turn, explored adaptive and rational expectations in the context of the Phillips curve, a topic that was the subject of intense international debate at the time.

I studied economics and statistics at the University of Mannheim, earning my master's degree (*Diplom-Volkswirt*) in the fall of 1978. My diploma thesis examined the macroeconomic debate on the effectiveness of monetary and fiscal policies in the presence of rational expectations, including an empirical analysis of the Phillips curve in Germany. König awarded my diploma thesis the highest distinction and offered me a full-time position as a research assistant at his chair. This role encompassed not only teaching and grading assistance but also, early on, involvement in the business survey project led by Nerlove and König. Alongside Gebhard Flaig, who had graduated from Mannheim two years earlier, I quickly became a key figure in König's chair system, helping to manage and direct a substantial portion of the research and teaching activities. Writing a dissertation was an after-hours task by university regulation anyway, and I found all these challenges inspiring and rewarding. These experiences later allowed me to conduct my own research with efficiency and the highest academic rigor. The chair system also had the advantage of providing a constant presence of colleagues who were available for guidance when needed. This system provided also more time and support at a later stage to prepare for the academic market.

I served as a research associate until 1984 and earned my doctoral degree in 1985, subsequently becoming a *Hochschulassistent* (Assistant Professor) at the University of Mannheim. In 1986, I was a Research Fellow at CORE, Université Catholique de Louvain in Louvain-la-Neuve, followed by a position as a Senior Research Fellow at the Wissenschaftszentrum Berlin (Social Science Research Center, WZB). I then held a Visiting Associate Professorship at the University of Pennsylvania in Philadelphia. Upon returning to Mannheim in 1988, I was awarded a Heisenberg Fellowship from the Deutsche Forschungsgemeinschaft (German Research Foundation, DFG), before moving to the University of Munich as a Full Professor of Economic Theory and director of the newly established Seminar for Labor and Population Economics. At Munich, I was also responsible for liaising with the Ifo Institute and served as a member of its supervisory board. In 1998, I declined an initiative of the Bavarian government to become President of the Ifo Institute, opting instead to move to the University of Bonn to establish the Institute for the Study of Labor (IZA).

This early success story owes much to Marc Nerlove and the dynamic research environment fostered by the Faculty of Economics at the University of Mannheim, particularly under Heinz König's leadership. For me, the project on the analysis of business survey data played a crucial role in this intellectual climate. Based on early publication successes related to the project (see section 1.5.1 for more details), Jacques Drèze invited me to join CORE, and Edmond Malinvaud to speak in his research seminar in Paris.

Many faculty members and their doctoral students later pursued highly successful careers in academia and beyond. Among them were Hans-Werner Sinn, who later became a professor at the University of Munich and President of the Ifo Institute, and Wolfgang Franz, who went on to serve as President of the ZEW following Heinz König. Gebhard Flaig was also appointed to a faculty position in Munich, and he eventually moved to the Ifo Institute to take over the business survey department and joined Ifo's executive board. Unlike Franz and Flaig, Sinn was not a student of König, although this is sometimes claimed in the social media.

Christoph Schmidt who was a master student and student helper at the König chair, completed his Ph.D. at Princeton University after moving the US on our advice, and got his habilitation with me at the University of Munich. Like Franz he later became a member and then the chair of the German Council of Economic Experts.

Other colleagues in Mannheim included my wife, Astrid Zimmermann-Trapp. A rising star in the faculty was Horst Siebert, an environmental economist, who led a large research center of the faculty before he moved to the University of Konstanz.

Siebert later became President of the Kiel Institute for the World Economy, a position that led to our renewed professional interactions when I served as President of the German Institute for Economic Research (DIW Berlin).

Marc Nerlove was relaxed, inquisitive, and highly sociable. He was genuinely interested in people and engaged with their work. It became my routine task to pick him up from the airport during his annual research visits and take him to his hotel, which was usually the *Goldene Gans* near Mannheim's central station. This location was also a frequent gathering place for König's team, where we would often meet in the restaurant after seminars over a glass of wine. Nerlove was a welcome participant in these informal discussions. Small gestures of his remain in my memory: although he somehow knew of my wife, they had not yet met. One day, when they encountered each other in the elevator of the university building, he walked up to her and introduced himself with the words, *You must be Astrid.*

Nerlove shared my interest in population economics, which I intended to make the focus of my doctoral research. Initially, Heinz König was not particularly enthusiastic about my idea of bringing Gary Becker's family economics to Germany. However, he soon changed his mind, particularly with Nerlove's support. This openness to new ideas was a defining trait of my doctoral advisor. König's understandable concern that I might be overburdened thematically dealing with household and firm decisions at the same time ultimately proved unwarranted, as I was able to apply the econometric techniques I had learned through the business survey project to my research in population economics (Zimmermann, 1985a).

What impressed me about Marc Nerlove was not only his diverse academic interests but also his exceptional ability to build and sustain research networks. For instance, he often combined his visits to Mannheim with research meetings on population economics with Assaf Razin and Efraim Sadka, who traveled from Israel. This early exposure allowed me to establish professional connections with both, and later I maintained frequent contact with Sadka. Nerlove also supported me in founding the *European Society for Population Economics (ESPE)* and delivered an invited lecture at its inaugural conference in Rotterdam. This lecture was later published in the *Journal of Population Economics* (Nerlove, 1988), which I had founded and which quickly became the leading journal in the field. Nerlove, Razin, and Sadka also contributed to an edited volume I published, *Economic Theory of Optimal Population* (Nerlove, Razin & Sadka, 1989).

A defining experience for me was the opportunity, initiated by Nerlove, to serve as a *Visiting Associate Professor* at the University of Pennsylvania in the calendar year 1987. This appointment allowed me to teach introductory courses in microeconomics and macroeconomics, as well as a lecture course on population economics. It also provided a strong foundation for successfully launching the *Journal of Population Economics* and for collaborating on research papers with his doctoral students, including David Ross and Lorenzo Pupillo. His research infrastructure supported me in numerous ways, and I fondly remember both professional discussions and private gatherings with him and my family. Even later, he remained genuinely interested in my daughter's development.

During my time in Philadelphia, I also met (among many other long-lasting connections) Lars-Hendrik Röller, who was completing his doctorate there, and Manfred Deistler, a leading scholar in time-series econometrics, who was on a research visit. Over the years, I maintained regular contact with both. With Röller, in his capacity as Chief Economic Advisor to Chancellor Angela Merkel, we engaged in discussions on labor market reforms and migration policies. With Deistler, we have frequently debated strategic questions of science policy and ways to strengthen research in our respective countries, drawing on insights from our experiences in the United States.

1.3 Business Test Data and the Mannheim Years

1.3.1 The Ifo Business Test

The Ifo Institute in Munich, Germany, a prominent publicly funded economic research institution in the country, has consistently conducted business surveys since 1949, establishing a foundation for systematic data-based economic analysis in Germany. Analogous questionnaires were subsequently developed for Italy (1949), France and Japan (1951), Austria (1953), Belgium, the Netherlands, Sweden, and South Africa (1954), Switzerland (1955), Denmark (1956), Finland (1957), and the United Kingdom (1958), and by 1995 were already available for 56 countries (Zimmermann, 1997).

The Ifo data collected in Germany initially encompassed manufacturing companies from 1949 onward. In 1950, the monthly survey was extended to include the retail trade sector, and in 1951, it incorporated the wholesale trade sector. The construction industry was integrated in 1956, while the service sector was not included until 2001.

The Ifo Business Climate Index for Germany, established through surveys conducted in the 1950s, gained recognition since the 1970s as one of the most significant indicators of economic activity in the country. This index is derived from approximately 7,000 monthly responses from businesses (Becker & Wohlrabe, 2008), and these responses were only recently stored as microdata within the Ifo Business Survey files. Although time series data for various industries and sectors have long been accessible through the Ifo macro database, access to the underlying microdata was historically first impossible and later limited for research purposes only.

Several scholars have provided a comprehensive review of the history of Ifo Business data (formerly referred to as Ifo Business Test or Ifo Konjunkturtest), including Oppenländer and Poser (1989); Zimmermann (1997); Becker and Wohlrabe (2008), and most recently Sauer, Schasching and Wohlrabe (2023).

Since 2004, the Ifo Institute had systematically converted its microdata inventory into Stata format, facilitating access to these data through the Ifo Data-Pool. This development enabled external researchers to conduct scientific analyses at the Ifo Institute utilizing anonymized microdata from four standard Ifo surveys: the Ifo Business Survey, the Ifo Investment Survey, the Ifo Innovation Survey, and the Ifo

World Economic Survey. To maintain confidentiality for participating companies, the dataset is anonymized and was accessible only under stringent criteria at a designated Ifo-based single-user computer.

Economic tendency surveys constitute systematic instruments designed to capture qualitative information regarding the current economic situation and future expectations from businesses and consumers. In contrast to traditional quantitative economic indicators that rely on empirical data such as output, employment, or sales figures, these surveys collect subjective assessments and anticipations, thereby providing timely insights into economic trends. The European Union's Joint Harmonised EU Programme of Business and Consumer Surveys exemplifies this methodological approach, conducting monthly surveys across various sectors—including manufacturing, construction, retail trade, services, financial services, and among consumers—to generate harmonized economic indicators.

The standard questions posed monthly in the Ifo Business Survey pertain to both the current and anticipated economic circumstances of firms, differentiated across several segments. The participating firms provide at the establishment rather than the firm level categorical variables that can be classified into three groups: (i) *ex ante* variables measuring plans or expectations; (ii) *ex post* variables reporting realizations; and (iii) variables reflecting *evaluations* of factors like order backlogs or inventories. Reported categories are typically trichotomous, responses are increase (+), no change (=), or decrease (-); or greater than normal (+), normal (=), or less than normal (-); or too large (+), about right (=), or too small (-). The +, =, - categories can also be coded as 1, 2, 3.

The aggregated indicators derived from such data are instrumental in short-term forecasting and identifying turning points in business cycles, thereby complementing official statistical data that often become available only after significant delays and are subject to subsequent revisions. Due to the categorical nature of micro-level data, the application of regression analysis at the firm level has long been unclear.

The initial documented scientific utilization of Ifo data was carried out by Anderson (1952). He employed time-series data (January 1950 – February 1952) to investigate the correlation between Ifo Business Survey data and official statistics. Through correlation analysis, he demonstrated that partial aggregates of the Business Survey, such as those pertaining to nutrition, closely approximated official statistics. Anderson proposed and illustrated the utility of balances calculated as the difference between the percentage of positive responses minus the percentage of negative responses at a specific point in time. He successfully utilized such data to forecast macroeconomic time-series.

Theil (1955) subsequently expanded this approach, focusing particularly on the use of balances as an aggregation method and pioneering the application of microdata analysis for manufacturing, specifically in the leather and shoe industry. Thonstad and Jochems (1961) further advanced the field by modeling production plans based on company expectations and assessments of the business climate, continuing the research initiated by Theil and applying similar methodologies to data from the leather and shoe industry (1956–1958).

The Centre for International Research on Economic Tendency Surveys (CIRET) emerged as the academic entity within the business survey movement, facilitating conferences and exchanges to promote the collection of such data globally. CIRET's origins can be traced to 1952, when an informal group of economists from institutions such as the Ifo Institute (Germany), the Institut National de la Statistique et des Études Économiques (INSEE, France), and the Association of Italian Chambers of Commerce collaborated under the designation *Comité International pour l'Étude des Méthodes Conjuncturelles* (CIMCO). This informal cooperation was formalized in 1960 with the establishment of the 'Contact International des Recherches Economiques Tendancielles' (CIRET). Initially affiliated with a research group directed by Theil at the Econometrisch Instituut in Rotterdam and later led by Anderson since 1966 at the University of Mannheim, CIRET also maintained a documentation center at the Ifo Institute (see also Knoche, 2025).

In 1971, CIRET and its documentation center merged and were fully integrated into the Ifo Institute, adopting the designation 'Centre for Economic Tendency Surveys'. By 1999, CIRET established a new legal foundation under Belgian law and relocated its headquarters to the KOF Swiss Economic Institute at ETH Zurich, adopting its current designation to reflect its international scope. A study by Abberger et al. (2022) developing a composite monthly indicator for the world business cycle (the Global Economic Barometers) utilizes business survey data from over 50 countries worldwide (Abberger, Graff, Müller & Sturm, 2022).

1.3.2 Marc Nerlove and the Mannheim Team

The *Mannheim Years* refer to the period during which our team at the University of Mannheim was actively engaged in a research project on expectations, plans, and realizations in economic decision-making of business firms. This project was initially funded by NATO from 1976 to 1979. The first publication by a team member appeared in 1979, authored by Heinz König, while the final publication co-authored by Marc Nerlove was in 1995. This marks a span of 16 years, which can be considered the primary project period. However, an alternative perspective extends this timeline from the start of funding in 1976 to the publication of my handbook article in 1997, making it a 21-year period.

The core members of the Mannheim support team included Gebhardt Flaig, Seiichi Kawasaki, and Klaus F. Zimmermann. Flaig was involved from 1976 to 1983, while Kawasaki joined in 1980 after completing his dissertation at Northwestern University under Marc Nerlove in 1979. Kawasaki remained in Mannheim until 1985, constrained by the maximum duration of temporary university contracts. I was at the chair from 1978 to 1985, took leave from 1986 to 1987, returned to Mannheim in 1988 to direct an independent research team, and moved to the University of Munich in 1989.

During the key Mannheim years, the presence of Flaig and Zimmermann defined the team's core period from 1978 to 1983 (five years). If the period is broadened to include years when at least one of them was present, it extends from 1976 to 1985, covering nine years.

Within the team, roles varied. Kawasaki, already holding a Ph.D., focused on complex theoretical and technical challenges, often involving programming or statistical problems. His perseverance was remarkable, and he frequently returned with solutions to problems that others could not resolve. He also contributed a core Fortran program, already developed at Northwestern, which was integral for analyzing data and running regressions for the project. He named this program *Tornado*, signifying speed, though the team humorously dubbed it *Snail*.

At that time, computational work relied on the University of Mannheim's mainframe system. Programs were input via punch cards, which had to be manually loaded in the cellar of our building, since the computing center was far away. The process was cumbersome and prone to errors—cards could be misplaced or damaged, leading to significant setbacks. Each researcher handled their own jobs, as dropping the card decks could be disastrous. Computation times were long, sometimes taking a full week, rapidly exhausting our annual computing quotas. Fortunately, Heinz König, who also served as university rector, ensured that we received additional capacity when needed.

Operational tasks fell primarily to Gebhard Flaig and me. Flaig was a highly skilled econometrician with deep statistical expertise and programming experience. When Marc Nerlove visited, research discussions often led to new ideas requiring additional programming. Occasionally, this meant working overnight to ensure results were ready before Nerlove's departure at the end of the week.

Both König and Nerlove were demanding scholars, always pushing for the best possible results while recognizing the challenges involved. Working with them was intellectually stimulating and rewarding.

Despite intense work periods, there was also space for independent research. The University of Mannheim maintained an exchange program with the University of Western Ontario, allowing us to collaborate with visiting scholars. Through this, John McMillan contributed significantly to our work on business survey data by providing the right framing of the articles (Kawasaki, McMillan & Zimmermann, 1982 and Kawasaki, McMillan & Zimmermann, 1983). Additionally, I pursued research on correlation measures for qualitative data, leading to ideas for pseudo-R^2 measures, which I later developed into publications with Mike Veall (Veall & Zimmermann, 1996). These methodological papers remain among my most highly cited works, surpassing even my publications in top-tier economics journals.

In business surveys, variables are typically categorized as increase (+), no change (=), or decrease (-). The challenge arises in calculating how these variables change over time or differ from one another. Specifically, how is a change defined? For instance, how can one effectively compare a change in price or a shift in production between consecutive periods? Additionally, how can plans or expectations be evaluated against actual outcomes, which is essential for assessing forecast errors, unmet plans, or unexpected results?

After extensive internal discussions, a straightforward solution was identified in the team by utilizing the ordered nature of the variable categories (see Nerlove, 1983,

1259-1260), which has gained broader acceptance in the literature. This is further elaborated upon in Figures 1.1 and 1.2.

Figure 1.1 presents a comparison between the expected or planned value (Y*) and the actual realization (Y). In addition to conducting a regression analysis of Y* on Y, it is pertinent to examine the difference Y-Y*, which represents the forecast error, insufficient plan fulfillment, or unexpected outcomes. The difference Y-Y* can be interpreted as no change (=) when situated on the main diagonal of the figure. It is considered a decrease (-) in the upper right section of the figure and an increase (+) in the lower left section. A Y-Y* value denoted as '+' signifies a positive surprise, an underestimation, or a development exceeding the plan, whereas a Y-Y* value denoted as '–' indicates a negative surprise, an overestimation, or a development falling short of the plan.

		Y_t		
		+	=	–
	+	=	–	–
Y^*_{t-1}	=	+	=	–
	–	+	+	=

Fig. 1.1: Realizations Y_t given expectations or plans Y^*_{t-1} and definition of forecast error, insufficient plan fulfillment or surprise

Simple differences between variables can be categorized in a manner similar to the method suggested in Figure 1.1, as illustrated in Figure 1.2. Beyond regressing a variable on its previous value, it may be interesting to examine changes in the direction of change. In Figure 1.2, no change (=) represents situations along the main diagonal. An increased (+) value indicates an upward trend over time, whereas a decreased (-) value indicates a downward trend.

Although it was possible to define the (3,1) cell of the figures as +, + and the (1,3) cell as –, –, this approach was not adopted due to considerations of simplicity and computational efficiency. The construction of such five-category variables was avoided, particularly considering the substantial computation times required on the mainframe computer, as reported above. The introduction of additional categories would have increased computing time and significantly raised the likelihood of encountering empty cells, thereby rendering the applied models inapplicable.

		Y_t		
		+	=	−
Y_{t-1}	+	=	−	−
	=	+	=	−
	−	+	+	=

Fig. 1.2: Realizations Y_t given past values Y_{t-1} and definition of categorical change

1.4 Business Survey Data Analysis

1.4.1 The Log-linear Probability Model

In the contemporary statistical literature, the log-linear probability (LLP) model is highly valued for its capacity to examine categorical data within an explorative research framework. This approach allows researchers to explore and comprehend complex relationships within contingency tables, thereby shedding light on the interplay between multiple categorical variables. The LLP model is particularly adept at detecting and measuring dependencies, offering a thorough understanding of how various categories affect each other. Researchers from diverse fields such as economics, sociology, demography, psychology, epidemiology, and marketing have shown considerable interest in this method. Typically, LLP models are employed to investigate associations among categorical variables. LLP models can also be expressed as multinomial logit models. This section explains the core econometric methodology of the Mannheim business survey data analysis project.

Drawing on Nerlove and Press (1973) and Nerlove and Press (1976), LLP models emerged as a prominent technique for analyzing business survey data in the 1970s and 1980s. As of March 9, 2025, the former report had garnered 668 Google Scholar citations, while the latter had received 73, demonstrating significant interest from the academic community.

In business surveys, the majority of variables are categorical, and the data can be analyzed using contingency tables. Consequently, it is useful to examine the nature of associations between these variables, or to what extent these associations deviate from a model of statistical independence. Typically, this method assumes a nominal scale for the variables, thereby disregarding the ordinal nature of some data. In addition to the work of Nerlove and Press, key references for the subsequent analysis include Bishop, Fienberg and Holland (1988), Kawasaki and Zimmermann (1981), and Zimmermann (1997).

Assume two categorical variables A and B with categories $i = 1,2,\ldots,I$; $j = 1,2,\ldots,J$. Let $\{\pi_{ij}\}$ be the contingency table of the probabilities involving these variables, where π_{ij} are the probabilities. The statistical model of independence implies

$$\pi_{ij} = \pi_{i+}\pi_{+j},$$

where π_{i+} and π_{+j} are the row and column marginals. The Pearson χ^2 statistic can examine this specification.

To allow for non-independence, the model can be generalized by

$$\pi_{ij} = \bar{\mu}\pi(i)\pi(j)\pi(i,j)$$

with

$$\sum_i \pi(i) = \sum_j \pi(j) = \sum_{i,j} \pi(i,j) = \sum_{i,j} \pi_{ij} = 1,$$

where $\pi(i)$, $\pi(j)$ and $\pi(i,j)$ are component probabilities and $\bar{\mu}$ is a normalization constant. Model (1.2) nests model (1.1) if the departure from independence has equal probability, $\pi(i,j) = 1/IJ$ for all i,j, and one obtains $\bar{\mu} = IJ$, $\pi(i) = \pi_{i+}$, and $\pi(j) = \pi_{+j}$. A logarithmic transformation of (1.2) leads to the log-linear probability model

$$\log \pi_{ij} = \mu + u_i + u_j + u_{ij} \tag{1.1}$$

with restrictions

$$\sum_i u_i = \sum_j u_j = \sum_i u_{ij} = \sum_j u_{ij} = 0. \tag{1.2}$$

Equations (1.2) are the so-called analysis of variance (ANOVA) restrictions. μ ($= \log \bar{\mu}$) is a constant, while u_i and u_j represent the main effects of variables A and B, respectively. The parameters u_{ij} denote the bivariate interaction terms, which quantify the association between categories i and j of both variables. A positive association is indicated by $u_{ij} > 0$, whereas a negative association is indicated by $u_{ij} < 0$. Through straightforward algebraic manipulation of equations (1.1) and (1.2), it can be demonstrated that u_{ij} represents the deviation of $\log \pi_{ij}$ from the arithmetic means of the respective column and row logged probabilities, in addition to the overall mean of the logged probabilities.

Consider now three categorical variables A, B, C with categories $i = 1,2,\ldots,I$; $j = 1,2,\ldots,J$; $k = 1,2,\ldots,K$ with contingency table $\{\pi_{ijk}\}$. Then the corresponding LLP model is

$$\log \pi_{ijk} = \mu + u_i + u_j + u_k + u_{ik} + u_{jk} + u_{ijk}, \tag{1.3}$$

where restrictions similar to (1.2) hold. Restrictions $u_{ijk} = 0$ for all i,j,k impose independence of association. If $u_{ijk} = 0$ and $u_{ij} = 0$ for all i,j,k, variables A and B are conditionally independent. Equation (1.3) (like equation (1.1) in the two-variable case before) is nothing more than a re-parameterization of the underlying three-way contingency table. It is therefore also called a 'saturated' model specification.

Equation (1.3) considers *joint dependence* of variables A, B, and C. A conditional probability model $Pr(A|B,C)$, where A is endogenous and B,C are exogenous, is provided by

$$\log \pi_{ijk} = \mu_{jk} + u_i + u_{ij} + u_{ik}. \tag{1.4}$$

This presumes the independence of association, a common assumption in econometrics. The conditional probabilities of the categories of one or more dependent variables, given one or more independent variables, are determined solely by the main effects of the dependent variables, the interactions among the dependent variables, and the interactions between the dependent and independent variables, excluding the main effects of and the interactions among the independent variables.

Parameter estimates u for (1.4) are obtained by assuming product multinomial sampling and maximizing the concentrated log-likelihood function

$$L(m_{ijk}|u) = \sum_{i,j,k} m_{ijk} \log \pi_{i|jk},$$

using standard techniques. An asymptotically valid covariance matrix Ω of the estimates allows for the usual testing procedures. Estimation details are provided in Nerlove and Press (1973), Kawasaki and Zimmermann (1981) and Bishop et al. (1988).

The LLP model provides detailed category-wise associations between categorical variables; however, it lacks an overall measure that summarizes the effects, such as a correlation coefficient for continuous variables. (Of course, a straightforward likelihood-ratio test can be employed to assess the significance of the entire set of bivariate interaction parameters, as compared to a model that omits these parameters.) Conversely, numerous nominal and ordinal association measures have been employed in traditional contingency table analysis, independent of the LLP approach (for references see Bishop et al., 1988). Despite this, no dominant index for discrete data has emerged. While most variables in the business survey are ordinal, some are nominal. The Mannheim project conducted an intensive examination of this literature and attempted to integrate contingency table association measures into the LLP analysis.

Following Kawasaki and Zimmermann (1981), two association measures are examined within the framework of the LLP model. Numerous applications in the business survey literature have used this research approach (see, for instance, Nerlove, 1983 and Kawasaki et al., 1983). It is noteworthy that the LLP model does not impose any ordering. Thus, the detailed effect parameters capture associations solely on a nominal scale. By connecting these parameters with association measures, the information contained within the various parameters can be consolidated into a single index, which can then be interpreted ordinally.

The bivariate component probabilities $\pi(i, j)$ and $\pi(i, k)$ are directly related to the estimated interaction parameters for equation (1.7), e.g., for $\pi(i, j)$:

$$\pi(i,j) = \frac{\exp(u_{ij})}{\sum_{i'}\sum_{j'}\exp(u_{i'j'})}, \quad i,i' = 1,2,\ldots,I; j,j' = 1,2,\ldots,J.$$

The core idea is now to apply association measures to those tables: Following Kawasaki and Zimmermann (1981), the two measures suggested here are γ and Φ^2. The first is an ordinal measure, while the second is a nominal measure of association. γ was initially introduced by Goodman and Kruskal (1979) for standard contingency table analysis and is highly regarded in that literature.

The first measure is defined as

$$\gamma = \frac{PS - PD}{PS + PD},$$

where

$$PS = 2\sum_i\sum_j \pi(i,j)\left[\sum_{i'>i}\sum_{j'>j}\pi(i',j')\right]$$

$$PD = 2\sum_i\sum_j \pi(i,j)\left[\sum_{i'>i}\sum_{j'<j}\pi(i',j')\right].$$

PS (PD) is the probability of a positive (negative) association between both variables based on the orders of the categories for both variables. Hence, γ is positive (negative) if it is more probable to obtain a positive (negative) than a negative (positive) association if one selects individual observations.

Φ^2 quantifies the difference between a set of probabilities $\pi(i,j)$ and the expected values derived from a specific probability model. When the equal probability model ($\pi(i,j) = 1/IJ$) is used as the reference, the result obtained is:

$$\Phi^2 = \sum_i\sum_j \frac{[\pi(i,j) - \hat{\pi}(i,j)]^2}{\hat{\pi}(i,j)} = \frac{1}{IJ}\sum_i\sum_j [IJ\pi(i,j) - 1]^2.$$

Φ^2 measures how different the association for a given model specification is from a reference model of zero bivariate interaction parameters.

Let $\mathbf{u}_{AB}$ represent the vector of the bivariate interaction parameters u_{ij} between variables A and B, and Ω_{uu} denote the corresponding covariance matrix. The asymptotic distributions of the estimated association measures can then be derived using the delta method. For instance, one obtains for γ the variance formula $\gamma_u'\Omega_{uu}\gamma_u$, where γ_u is the gradient of $\gamma(\mathbf{u}_{AB})$. Kawasaki and Zimmermann (1981) provide detailed formulas.

It is important to note that the LLP model primarily identifies correlations or associations rather than establishing causality. While it provides valuable insights into the relationships between variables, it does not inherently determine causal links. Therefore, researchers must employ additional methods and frameworks, such as experimental designs or causal inference techniques, to establish causality with greater confidence. LLP models nevertheless remain an important instrument for explorative data analysis.

1.4.2 Formation of Price Expectations, Output Plans, and Subsequent Realizations

The Mannheim business survey data project has resulted in a substantial number of published research papers, which are too numerous to comprehensively review and evaluate within this chapter, although some work will be discussed later on. Consequently, this section concentrates on the two flagship publications of the project, examining their efforts to reveal the microdata-based evidence concerning the formation of price expectations, output plans, and their subsequent realizations by business firms. The two key studies are: Marc Nerlove's 1983 paper, *Expectations, Plans, and Realizations in Theory and Practice*, published in *Econometrica*, and the 1981 study co-authored by Heinz König, Marc Nerlove, and Gilles Oudiz, *On the Formation of Price Expectations. An Analysis of Business Test Data by Log-Linear Probability Models*, published in the *European Economic Review* (König, Nerlove & Oudiz, 1981 and Nerlove, 1983).

The paper by König et al. (1981) was presented at the prestigious *International Seminar on Macroeconomics* (ISoM), held on June 23-24, 1980, in Oxford, UK. The inclusion of a business survey paper in a macroeconomic conference underscored the growing significance of microdata analyses in addressing macroeconomic questions.

The ISoM was initiated in 1978 as a joint venture between the *National Bureau of Economic Research* (NBER) and the French *École des Hautes Études en Sciences Sociales* (EHESS). At its inception, it was co-directed by Georges de Ménil, Robert J. Gordon, and Jean Waelbroeck, who were instrumental in guiding its academic focus. The seminar evolved into a crucial forum for the exchange of innovative macroeconomic research, promoting collaboration among economists from Europe and the United States. With the exception of its first year, the seminar's proceedings were consistently published in the *European Economic Review*, facilitating broad distribution of the research presented. Although EHESS was instrumental in ISoM's establishment, the leadership has since 1993 become more globally inclusive, with leading economists from various institutions assuming control. The latest ISoM event was held on June 4–5, 2024, and was hosted by the *Bank for International Settlements* in Basel, Switzerland.

Marc Nerlove delivered Nerlove (1983) as the Presidential Address at the 1981 *European Meeting of the Econometric Society*, which took place in Amsterdam from August 31 to September 4, 1981. The fact that Marc selected this subject for his address as the President of the *Econometric Society* indicates that, among the diverse research areas he engaged in, he considered the outcomes of the Mannheim Business Survey project to be of significant importance. The paper not only reviews previous studies of the project but also considerably expands on the research questions and findings. In the following, I will first summarize and examine the key findings of Nerlove (1983), and then highlight the differences and additions with respect to König et al. (1981).

Marc Nerlove's Presidential Address to the Econometric Society

In his 1983 research, Marc Nerlove explores the complex link between the expectations or plans of firms regarding prices and output and the actual outcomes they experience. Utilizing comprehensive business survey data from manufacturing companies in France (INSEE) and Germany (Ifo data), the study examines how accurately firms predict outcomes, the consistent biases in their forecasts, and the processes that shape expectation formation. A major conclusion of the study is that firms often underestimate the probability of change, with their expectations frequently centering around the 'no change' category, while actual results show more variability. Additionally, the research highlights notable differences between countries, with German firms demonstrating more stability in their expectation-formation processes compared to French firms.

Expectations and plans are crucial in the economic decision-making processes of firms, yet modeling these empirically had been challenging at the time of the research work. The paper examines several straightforward models of expectation formation, such as extrapolative expectations, adaptive expectations, and error-learning mechanisms, to assess their ability to explain firm behavior. The findings indicate that firms mainly apply error-learning models, where expectations or plans are adjusted based on previous forecasting errors, rather than solely on extrapolative models that simply project past trends into the future. A significant finding is that, although price and output expectations show some persistence, firms tend to be systematically conservative in their forecasts about future conditions. This conservatism is evident in a strong tendency to predict 'no change', a pattern observed in both French and German firms. However, the data suggest that this conservative approach is more evident among German firms, while French firms exhibit more variability in their expectations and plans.

The paper further explores the systematic biases present in the expectations of firms. German companies consistently underestimate the extent of changes in demand, production, and prices. Although they predict changes less often than they actually occur, their forecasting errors remain relatively stable over time. This consistency indicates that German firms use fairly uniform rules for forming expectations, making their biases foreseeable. In contrast, French companies show significant variability in how they form expectations. The study reveals that the connection between planned and actual changes in production, demand, and prices is much more erratic among French firms, suggesting that their forecasting rules are less consistent or that they operate in a more unpredictable economic environment. The instability of conditional distributions in the French data suggests that economy-wide factors, such as macroeconomic shocks or policy changes, may affect firms' expectation errors in an inconsistent way.

How closely are firms' price expectations linked to their production plans? If companies determine prices based on forecasted demand and anticipated production limitations, one would anticipate a strong connection between changes in price expectations and adjustments in production plans. Yet, the findings in the paper indicate a surprising level of independence between these two processes. A joint model estimated for price expectations and production plans shows that changes in price expectations and production plans occur almost independently. This observation is consistent among both French and German firms, challenging standard economic models that suggest firms adjust prices and output simultaneously in response to demand shocks. The observed independence might be due to rigidities in price-setting behavior. German firms, in particular, seem to modify their production plans in response to unexpected demand changes but do not necessarily alter their pricing strategies accordingly. This implies that supply-side constraints or competitive pressures might restrict firms from freely adjusting prices in response to actual shocks.

What is the role of demand shocks in plan fulfillment? The study also identifies the elements that influence whether companies stick to their original plans. A central hypothesis examined is that unforeseen shifts in demand significantly impact whether companies alter their production strategies and pricing forecasts. The findings reveal a strong link between unexpected demand changes and the inability to meet production plans. For both French and German firms, when actual demand diverges considerably from what was expected, they are much more inclined to modify their production strategies. However, there are differences in how these companies adjust their pricing strategies. German companies are more likely to change their price forecasts in response to production deficits, whereas French companies do not show a consistent pattern between unexpected demand and changes in price expectations. This indicates that price-setting in France might be more inflexible, potentially due to regulatory limitations, labor market challenges, or institutional factors that restrict firms' ability to adjust prices in response to demand changes.

The paper further explores an economically rich conditional probability model that connects firms' production strategies to crucial economic factors like demand expectations, inventory appraisals, and recent demand fluctuations. The empirical findings indicate that firms are more inclined to plan production increases when (i) they have recently observed a rise in demand, (ii) they perceive their inventory levels as insufficient, and (iii) they anticipate an increase in future demand. These results strongly support the idea that firms' production planning is influenced not just by extrapolative trends but by a combination of demand conditions and inventory assessments. Additionally, the empirical estimates for both French and German firms are strikingly similar, implying that the fundamental economic mechanisms driving production planning are largely consistent across different institutional settings.

In conclusion, Nerlove (1983) enhances the understanding of how expectations are formed and their influence on the decision-making processes of firms. The research emphasizes the systematic biases present in firms' predictions, which often lean towards anticipating stability in prices and output, even though actual outcomes show significant fluctuations. While error-learning models effectively explain price and demand expectations, production plans seem to be more closely linked to economic fundamentals like demand expectations and inventory levels. The apparent disconnect between price expectations and production plans indicates that firms' pricing strategies might be constrained, limiting their adaptability. This has significant implications for economic modeling, especially regarding monetary and fiscal policy, as it implies that firms might not react to demand shocks as standard equilibrium models would predict. The differences observed between French and German firms highlight the impact of institutional factors on expectation formation and the execution of plans. The more stable expectation processes of German firms suggest they operate in more predictable market conditions, whereas the instability in the French data indicates a more volatile economic environment.

Comparing Nerlove, 1983, with König, Nerlove and Oudiz, 1981

Marc Nerlove's 1983 paper and the earlier 1981 study co-authored by Heinz König, Marc Nerlove, and Gilles Oudiz analyze business survey data from German and French firms. Both articles employ data from the Ifo Institute (Germany) and INSEE (France) to examine how firms form expectations, revise their plans, and ultimately adjust their business decisions in light of realized outcomes. However, while the 1981 article focuses exclusively on price expectations, the 1983 study expands the scope to include production plans and demand forecasts, providing a broader view of firm behavior. This comparative analysis highlights the methodological advancements, empirical findings, and theoretical contributions of both works, while also considering their implications for economic modeling and firm decision-making.

Methodological foundations and innovations. Both articles share a *methodological commitment* to using log-linear probability models to analyze categorical business survey data. The 1981 study introduces this approach as an alternative to traditional time-series analysis, arguing that direct survey data on firms' expectations provide richer insights into the expectation formation process than conventional econometric models that rely on observed outcomes alone. The 1983 article builds upon this foundation, maintaining the log-linear probability framework while further extending it with recursive conditional probability models. This additional methodological layer allows the later study to examine how different business expectations—such as price anticipation, production plans, and demand forecasts—interact with one another and evolve over time.

A significant *methodological difference* is how expectations are modeled. While Nerlove (1958a) laid the groundwork with the adaptive expectations model, emphasizing how expectations adjust in response to forecast errors, this early work relied on time-series macro data estimation rather than directly observed micro expectation data. The 1981 study now focuses on price expectations using qualitative micro data, examining them through adaptive and extrapolative models. It investigates whether firms rely more on past realizations or on adjustments based on recent forecast errors. The 1983 study broadens this approach, applying similar models not only to price expectations but also to production planning and demand forecasting. In doing so, it tests whether firms treat these different expectations as interconnected or if they develop them in isolation from one another. The 1983 study also provides a more refined assessment of expectation stability, comparing how German and French firms revise their forecasts in response to past realizations.

A notable *methodological advancement* in the 1983 paper is its application of recursive models to capture the sequential nature of business decision-making. By structuring the analysis to acknowledge the interdependencies among various decision variables, the 1983 study offers a more nuanced view of firm behavior. This is evident in its treatment of production plans, where the paper investigates whether firms adjust their planned output in response to unexpected demand fluctuations.

Empirical findings. The two articles arrive at different conclusions regarding how companies develop and adjust their expectations. The 1981 study reveals a strong link between price expectations and past outcomes, indicating that firms often base their future price forecasts on recent pricing patterns. However, it also highlights notable differences in expectation formation between German and French firms. German firms' price expectations exhibit greater stability over time, whereas French firms' expectations fluctuate more widely. This implies that the process of forming expectations is shaped not only by economic fundamentals but also by institutional and behavioral influences.

The 1983 study builds on these findings by demonstrating that the stability of expectations varies depending on the type of business decision. German firms show consistency in their price and demand expectations but display more variability in production planning, suggesting that they treat pricing and production decisions as somewhat separate. Conversely, French firms exhibit more volatility in their expectations for prices, demand, and production, indicating a less structured approach to business planning.

One of the most striking findings in the 1983 paper is that production plans and price expectations are nearly independent of one another. This contradicts conventional economic models that assume firms jointly determine pricing and output strategies in response to market conditions. Instead, the study finds that firms often revise their price expectations based on past price trends, while production plans are adjusted primarily in response to demand fluctuations. This suggests that firms may not always coordinate their pricing and output decisions optimally, either due to rigidities in pricing strategies or constraints in adjusting production capacity.

The differences between German and French firms are especially insightful in this context. The 1983 paper indicates that German firms typically adjust production in response to demand changes, whereas French firms show greater uncertainty in revising their expectations. This instability might be attributed to macroeconomic factors such as inflationary pressures, labor market rigidities, or variations in industrial policy. The greater stability in German firms' production plans suggests a reliance on structured forecasting methods or long-term strategic planning.

Challenges of rational expectations. Both studies have added to the prevailing debate at the time on rational expectations, a theory suggesting that economic agents form their expectations using all available information in an unbiased statistical manner. The 1981 study already reveals that firms' price expectations do not entirely align with rational expectations; instead, they are shaped by a combination of extrapolative and adaptive processes. Firms adjust their expectations based on past outcomes but also display systematic biases in their predictions. This finding contradicts the rational expectations hypothesis, which assumes that economic agents will eventually eliminate systematic forecast errors.

The 1983 study supports this conclusion and broadens it to include other business decisions beyond price expectations. By demonstrating that firms' production plans and price expectations are largely independent, the later study indicates that firms do not always optimize their decisions in a fully coordinated way. This challenges standard economic models that assume firms maximize profits by jointly determining prices and output levels. Instead, it suggests a more fragmented decision-making process, where pricing and production planning function as separate mechanisms influenced by different sets of expectations.

An additional significant contribution of the 1983 study, beyond the earlier work, is its examination of the stability of expectations over time. While rational expectations theory posits that firms should gradually refine their forecasts as they gather more information, the study finds that expectation formation remains highly variable, particularly among French firms. This implies that firms may encounter constraints in processing information efficiently or that they rely on heuristics rather than formal predictive models.

1.5 Research Impact

1.5.1 Firm Price and Output Changes and Rational Expectations

Marc Nerlove inspired numerous research papers involving him and/or other members of the Mannheim group. In relation to the key papers examined in section 3.2, Nerlove (1983) and König et al. (1981), this section highlights four papers that expand on these themes, authored by junior team members, specifically Kawasaki et al. (1982) on *Disequilibrium dynamics: An empirical study* and Kawasaki et al. (1983), *Inventories and price inflexibility*, on the development of firm price and output changes, as well as Kawasaki and Zimmermann (1986), *Testing the rationality of price expectations for manufacturing firms*, and Zimmermann (1986), *On rationality of business expectations: A micro analysis of qualitative responses*, on rational expectations. The fact that we were able to undertake this work independently was a remarkable acknowledgment of our strong support for the general project.

Output and price flexibility

Kawasaki et al. (1982) primarily examines how firms adjust their prices and output levels in response to disequilibrium situations. It focuses on whether these adjustments move firms closer to or further away from equilibrium. The paper defines disequilibrium based on firms' assessments of their inventory levels and unfilled orders. It finds that firms often experience disequilibrium, with around 60 percent of observations indicating misalignment in either inventories or order backlogs. The study also finds that firms respond to stock disequilibrium within one month, using both price and output adjustments, but with a notable difference in flexibility: output adjustments are more frequent than price changes. Contrary to conventional expectations, the study finds no significant evidence that prices are less flexible downward than upward. The authors also highlight that flexibility in price and quantity adjustments varies significantly across industries.

Kawasaki et al. (1983) extends this analysis by providing a more nuanced explanation of why prices appear less flexible than quantities. Developing a theoretical model following Kirman and Sobel (1974) for orientation, it introduces a distinction between firms' responses to transitory versus permanent changes in demand. The study argues that firms react differently depending on whether demand fluctuations are perceived as short-term or long-term. Using changes in incoming orders from the previous month as a measure of short-run demand shifts, and expected changes in business conditions over the next six months as a proxy for long-run demand shifts, the study demonstrates that firms adjust both price and output when responding to permanent demand changes. In contrast, firms primarily adjust output, rather than prices, in response to transitory changes in demand. This theoretical refinement helps explain why price changes are observed less frequently than output adjustments in the short run.

Overall, while Kawasaki et al. (1982) focuses on the general disequilibrium behavior of firms and their tendency to favor output over price adjustments, Kawasaki et al. (1983) deepens the analysis by distinguishing between different types of demand shocks and showing that price changes are more likely to accompany long-term shifts in demand. The latter study thus provides an explanation for the empirical finding that price flexibility appears lower than quantity flexibility. Together, these papers contribute to a better understanding of firm behavior in disequilibrium situations by clarifying the role of demand expectations in shaping firms' pricing and production decisions.

How are Kawasaki et al. (1982) and Kawasaki et al. (1983), in the following KMZ, related to Nerlove (1983)? Beyond common data and similar methods, a common interest is to understand how firms adjust prices and output in response to economic conditions, though they approach these questions with different emphases.

The 1982 finding of KMZ that firms more frequently adjust output than prices in response to inventory imbalances and unfilled orders aligns with Nerlove's broader theme that expectations and realizations often diverge due to structural constraints and uncertainties in firms' decision-making processes. The 1983 extension by KMZ refines this analysis by distinguishing between permanent and transitory demand shocks, showing that price adjustments primarily occur when demand changes are perceived as long-term, whereas short-term fluctuations tend to induce output changes instead. This finding intersects with Nerlove's work, which examines how firms' expectations about future conditions shape their planning and decision-making.

Nerlove (1983) while explicitly modeling the process by which firms develop price and production plans based on past realizations and expected future demand demonstrates that firms systematically underestimate the volatility of their environment. Their expectations disproportionately concentrated in the 'no-change' category compared to actual realizations. This tendency is consistent with the findings of KMZ

1983, who also observe that firms exhibit inertia in their pricing behavior, preferring to adjust output rather than prices unless they perceive demand shifts as permanent.

The findings of KMZ contributed significantly to the macroeconomic debates of the 1980s, particularly in the discourse surrounding Keynesian and neoclassical perspectives on price and output flexibility. In the Keynesian tradition, particularly in the emerging *New Keynesian* framework, price and wage stickiness were central tenets, implying that firms tend to adjust output rather than prices in response to demand fluctuations. The 1982 study reinforced this view, demonstrating that firms predominantly altered quantities rather than prices when reacting to disequilibrium. This evidence supported Keynesian models emphasizing nominal rigidities, which explain persistent unemployment and output fluctuations. The observation that output is more flexible than prices bolstered the argument that aggregate demand shocks have tangible effects on employment and production rather than being quickly neutralized through price adjustments.

However, their 1983 study introduced a nuanced perspective, complicating the Keynesian interpretation. By differentiating between permanent and transitory demand shocks, the authors found that firms adjusted prices when demand shifts were perceived as permanent but changed output levels when shifts were seen as temporary. This behavior aligned with rational expectations theory, a core component of neoclassical economics, which also gained prominence in the 1980s (see below). The evidence suggested that firms acted with foresight, adjusting prices strategically based on their expectations of future demand rather than being universally constrained by price rigidity.

These findings also had implications for *Real Business Cycle* (RBC) theory, developed by Kydland and Prescott (1982), which posited that business cycles stem primarily from real supply-side shocks rather than demand fluctuations. The tendency of firms to adjust output more than prices in response to short-term shocks was consistent with RBC models, which downplayed price distortions as a driver of economic fluctuations. However, the fact that firms adjusted prices in response to long-term demand shifts indicated that price flexibility was conditional rather than absolute, contradicting the RBC assumption of continuously clearing markets.

Ultimately, KMZ bridged the divide between Keynesian and neoclassical perspectives. The 1982 study reaffirmed the Keynesian argument for output flexibility and price stickiness, justifying fiscal and monetary interventions to stabilize demand. Their 1983 research, however, highlighted the role of expectations and selective price adjustments, incorporating elements of rational expectations into the analysis of market behavior. By distinguishing between short- and long-term adjustments, these studies helped refine macroeconomic modeling, influencing the evolution of New Keynesian economics, which sought to integrate rational expectations into traditional Keynesian frameworks.

Their work also resonated with the broader RBC literature by acknowledging that while short-run price rigidity exists, firms adjust strategically when they anticipate permanent shifts in demand. This insight challenged the pure RBC view that markets always clear efficiently but suggested that elements of RBC modeling could be reconciled with observed price-setting behavior.

In sum, their findings provided empirical support for both Keynesian and neoclassical theories, demonstrating that firm behavior is more complex than either paradigm alone suggests. By illustrating how firms navigate disequilibrium through both output and price adjustments based on expectations, their work contributed to the ongoing development of macroeconomic thought in the 1980s and beyond.

Rational expectations

Kawasaki and Zimmermann (1986) analyze the rationality of price expectations among German manufacturing firms using data from the Ifo Business Survey. Their study examines the biases in firms' prediction-realization tables for prices, production, and demand, testing whether these expectations align with the rational expectations hypothesis. Their findings suggest that firms exhibit systematic biases with a tendency to overestimate their prices and predict price changes more conservatively than actual realizations.

One key finding is that German firms are more likely to overestimate rather than underestimate their future selling prices. This means that firms systematically predict price levels to be higher than they turn out to be. This pattern contradicts the rational expectations hypothesis, which assumes that forecasting errors should be random rather than displaying a systematic bias. The authors quantify this bias using measures of forecast accuracy and consistency and find that firms exhibit a clear tendency toward over-prediction.

Another crucial result relates to firms' expectations regarding price changes. Firms tend to be conservative in their predictions, meaning that they systematically underestimate the magnitude of their price fluctuations. Instead of forecasting large shifts in prices, firms expect smaller and more gradual changes. This finding suggests that firms may relate their expectations too heavily to recent past price movements rather than efficiently incorporating all available information, which is another violation of the rational expectations hypothesis.

To formally test for rationality, the study employs an efficiency test to examine whether price forecast errors are systematically related to past price changes. If firms were forming rational expectations, forecast errors should be uncorrelated with past information. However, the study finds a strong and persistent relationship between price surprises and one-period lagged price changes. This result indicates that firms' price expectations are influenced by past trends in a way that makes their errors predictable, another departure from rationality.

Beyond price expectations, the study also investigates production and demand forecasts. Similar to their findings on prices, the authors observe that firms' expectations for production and demand also exhibit systematic biases, with firms tending to overpredict levels of demand and underpredict variability in production levels. These biases further support the conclusion that firms do not form expectations in a fully rational manner.

The implications of these findings extend to broader economic modeling and policymaking. Many macroeconomic models assume that firm and individual expectations are rational, meaning that systematic forecasting errors should not persist over time. However, Kawasaki and Zimmermann's results suggest that firms' price expectations are neither unbiased nor efficient. This challenges the assumptions underlying many economic models and suggests that firms' price-setting behavior may not fully account for all available information, possibly because of adjustment costs, informational constraints, or behavioral tendencies.

The rational expectations hypothesis was originally formulated by Muth (1961) in his seminal paper. He argues that economic agents form their expectations in a way that is consistent with the true underlying economic model, meaning that, on average, their forecasts do not systematically deviate from the predictions that would be made using all available information. This concept became central to macroeconomics, particularly through the work of Robert E. Lucas Jr. in the 1970s, who integrated it into macroeconomic models (Lucas, 1976 and Lucas, 1972). His application of rational expectations laid the foundation for the *New Classical approach*, which fundamentally challenged Keynesian economics by arguing that systematic monetary policy interventions would be largely ineffective in influencing real economic variables. This perspective was reinforced by Sargent and Wallace (1975), who introduced the policy ineffectiveness proposition, arguing that only unexpected policy changes could affect output and employment.

The findings of Kawasaki and Zimmermann (1986) and Zimmermann (1986) are consistent with the research results by Nerlove (1983) and König et al. (1981) as summarized in section 1.4.2. They had significant implications for these macroeconomic debates. As the rational expectations framework underpinned the policy ineffectiveness proposition, the empirical rejection of unbiased and efficient expectations suggests that government policy could still have real effects, even if anticipated. This provides empirical support for the emerging *New Keynesian* critique of the *New Classical* approach. If expectations were not fully rational and exhibited systematic biases, this implied that price and wage rigidities, as modeled in *New Keynesian* frameworks, could have real economic consequences.

1.5.2 Development of the Research Field

The research output from project-related scholars and beyond experienced a significant surge, expanding in multiple directions. Reviews of this evolution can be found in Zimmermann (1997) and Becker and Wohlrabe (2008). Zimmermann (1997) examines various topics, including 'predictive performance,' 'the formation of anticipations,' 'rational expectations,' 'output and price responses,' 'determinants of labor demand,' 'innovations, patent activity, and trade,' as well as 'seasonality in business surveys'. Meanwhile, Becker and Wohlrabe (2008) focus on 'studies on expectation formation,' 'special survey questions on innovation,' and 'business cycle analysis'.

For a long time, German and French datasets dominated publications in this field. However, research soon expanded to other countries. Notable examples include Nerlove and Zepeda Payeras (1986) for Mexico, Ghysels and Nerlove (1988) for

Belgium, Pupillo and Zimmermann (1991) for Italy, and Nerlove and Schuermann (1995) for Switzerland and the United Kingdom.

The project's earliest publications include König (1979) written in German, and Koenig and Oudiz (1979) written in French. An important milestone was König and Nerlove (1980), initially presented at the CIRET conference in Lisbon and later published in the conference proceedings. These early contributions laid the groundwork for later studies such as König et al. (1981) and Nerlove (1983).

Over a span of 16 years, Marc Nerlove maintained a strong research focus on business cycle-related topics. Of the 23 papers he published on the topic between 1979 and 1995, nine appeared in CIRET conference volumes—representing approximately 39 percent of his output in this area. This translates to an average of 1.4 papers per year, alongside numerous other contributions across diverse fields.

In the following discussion, I highlight several key studies carried out or inspired by the work of Marc Nerlove and his team. These studies examine various topics, including expectation formation, labor demand, innovation, international trade, and seasonality. Beyond expanding the range of topics, researchers have also introduced different econometric methods, enriching the analytical approaches applied in this field.

Expectation formation. The debate on expectation formation remains unresolved, with findings varying depending on the measurement approach and data source. Using a latent variable model and business survey data, Ivaldi (1992) finds that the rational expectations hypothesis is not consistently rejected for the French manufacturing sector. In contrast, Nerlove and Schuermann (1995), applying different latent variable models, firmly reject rational expectations for firms in Switzerland and the UK. However, their analysis also challenges the validity of adaptive and naive expectations models. Further evidence from British business survey data by Low, McIntosh and Schiantarelli (1990) reveals systematic biases in firms' forecasts. Their study indicates a tendency to overpredict changes in prices, costs, and new orders, while underestimating actual production levels.

Labor demand. What drives firms' labor demand? The Ifo business survey data do not include direct information on wages or labor costs, and technical change is often poorly measured. To address this, König and Zimmermann (1984) integrated industry-level wage and nonwage labor costs from macroeconomic sources. Their analysis, based on log-linear probability models, finds that while these costs have a statistically significant effect on employment plans, their influence is surprisingly weak. Instead, labor demand is primarily shaped by capacity utilization and production expectations. To explore this further, Ross and Zimmermann (1993) use a categorical indicator model, leveraging a specific Ifo survey question where firms identify up to two key factors influencing their employment plans. The available options include demand uncertainty, insufficient demand, high labor costs, a shortage of skilled workers, and labor-saving technical progress. Their findings strongly indicate that insufficient demand is the dominant factor driving labor demand. This result remains robust across different model specifications, including adjustments for firms' export market integration and disequilibrium conditions.

International trade. Using Italian business survey data and Probit models, Pupillo and Zimmermann (1991) find evidence that Italian foreign and domestic markets are segmented, as firms can set different prices, with foreign markets displaying greater price elasticity. In a related study, Zimmermann and Pupillo (1992) analyze the factors influencing firms' export activities using OLS and Poisson regressions. Their results show that firm size positively affects relative export levels and the number of export regions, while its impact on export share variability is negative and often insignificant. Market concentration variables yield inconclusive results.

Innovations. Business survey data often provide discrete information on a firm's introduction of product or process innovations, the number of patents, or innovation expenditures. According to industrial organization research, innovative activity is typically linked to firm size, market concentration, and demand pressure. Zimmermann (1985b) was the first to analyze these relationships using business survey data. Employing Ifo data and Probit models, the study integrates industry-level information with firm-level data to capture industry structure more precisely. The results confirm that while firm size and market concentration positively influence innovation, the most decisive factor is firms' expectations of long-term demand. Building on this, König and Zimmermann (1986) merge innovation data from the German business test with information on innovation expenditures from the Ifo innovation test. Using Probit and Tobit models, their analysis further reinforces the conclusion that demand expectations play the dominant role in driving innovative activity.

Seasonality. A technical challenge in analyzing business surveys is accounting for seasonality. Firms are often instructed to exclude seasonal fluctuations from their responses, yet seasonal effects may still persist in the data. Using log-linear probability models and German data,Flaig and Zimmermann (1983) show that production plans and realizations exhibit seasonal patterns, though the extent varies across variables, potentially biasing parameter estimates. Ghysels and Nerlove (1988) examine seasonality in business survey data from Belgium, Germany, and France, also using log-linear probability models. They find substantial seasonal effects but note that responses to seasonally adjusted questions generally reflect a reasonable level of adjustment.

1.6 Conclusions

This chapter examined a significant period in the academic career of Marc Nerlove, documenting his contributions to the economics profession and his broader influence as a researcher and mentor using his long-term project on business test data as a case study. In general, Nerlove's work exemplifies visionary leadership and intellectual breadth, spanning a remarkable array of subdisciplines within economics and econometrics. His research has had a lasting impact on fields such as agricultural and development economics, labor and population studies, time-series and microeconometrics, qualitative data analysis, business cycle theory, and forecasting. His legendary curiosity and openness to new challenges, topics, and collaborations made him an entrepreneurial figure in the academic world. By fostering international networks of scholars, he shaped the careers of numerous PhD students and research partners, many of whom have made it later into influential positions in academia and policy institutions. His extensive publication record in top-tier journals and widely respected books underscores his intellectual rigor but also reflects his commitment to the broad dissemination of ideas. Unlike those who prioritize publishing only in the most prestigious journals, Nerlove seemed to strategically chose diverse outlets, including book chapters and lesser-ranked journals, demonstrating a strong belief in making high-quality research widely accessible across the profession.

One of the major undertakings of the Mannheim group, the Mannheim Business Survey project, co-directed by Heinz König, is a landmark in the early development of microeconometrics for qualitative firm data. The project played a crucial role in advancing qualitative data econometrics at a time when the field was still dominated by time-series analyses of macro-data and a rapidly rising interest in creating individual data-based household-level studies. Introducing log-linear probability models and applying association measures and Pseudo-R^2s provided methodological innovations that expanded the possibilities for empirical research. Moreover, it was the first to apply these techniques to business survey data, thereby integrating micro-level firm data into econometric research in a novel and influential manner. These contributions not only served as methodological milestones, but also influenced subsequent large-scale survey initiatives, such as the German Socio-Economic Panel (GSOEP), which emerged in the same period with support from members of the Mannheim group. While access to Ifo business data was initially restricted and limited to a short time period of the data source, the Ifo Institute has since made these data available for researchers, reflecting a long-term impact on the accessibility and use of business survey data in empirical economics.

The Mannheim project vitalized the CIRET research conferences, fostering an enduring global forum for the exchange of ideas in business cycle analysis and survey-based research. While the impact of the work has been felt across multiple economic subfields, the research contributions of the group have been particularly influential in shaping the microfoundations of key macroeconomic debates. Empirical insights were provided into the evolution of firm-level output and pricing behavior, the nature of disequilibrium adjustments in response to economic shocks, and the role of rational expectations in shaping business decisions.

In the broader context of macroeconomic theory, the Mannheim group offered a data-driven perspective on the Keynesian-neoclassical debate, particularly through the lens of rational expectations. By rigorously analyzing firm-level data on price and production expectations, their research tested the extent to which firms rationally form expectations or whether systematic biases exist. These findings challenged some of the prevailing assumptions in macroeconomic modeling, highlighting the importance of micro-level heterogeneity and the limitations of aggregate models that overlook firm-specific behaviors. These insights have had lasting implications for both theoretical and applied research, influencing how economists conceptualize expectation formation, policy effectiveness, and business cycle dynamics. Through their empirical approach, the Mannheim group not only enriched the discussion on rational expectations but also demonstrated the necessity of grounding macroeconomic debates in robust microeconometric evidence. The legacy of this research lies not only in methodological contributions but also in the persistent advocacy of data-driven economic inquiry, a principle that continues to shape the field today.

Acknowledgements

I wish to thank Meinhard Knoche and Klaus Wohlrabe for very helpful discussions during the preparation of this contribution. I also owe my former colleagues and co-authors at the time, Gebhard Flaig, Seiichi Kawasaki, John McMillan, Lorenzo Pupillo, David Ross, Mike Veall, and in particular my PhD supervisor Heinz König, quite a lot for their support and inspirations.

References

Abberger, K., Graff, M., Müller, O. & Sturm, J.-E. (2022). Composite global indicators from survey data: the global economic barometers. *Review of World Economics*, *158*, 917–945.

Anderson, O. (1952). The business test of the ifo institute for economic research, munich, and its theoretical model. *Review of the International Statistical Institute*, *20*, 1–17.

Becker, S. & Wohlrabe, K. (2008). European data watch: Micro data at the ifo institute for economic research – the "ifo business survey", usage and access. *Schmollers Jahrbuch: Journal of Applied Social Science Studies*, *128*, 307–319.

Bishop, Y. M. M., Fienberg, S. E. & Holland, P. W. (1988). *Discrete multivariate analysis: Theory and practice*. Cambridge: MIT Press.

Flaig, G. & Zimmermann, K. F. (1983). Misspecification and seasonal adjustment of qualitative panel data. *CIRET-Studien*, *33*, 63–95.

Ghysels, E. & Nerlove, M. (1988). Evidence from the belgian business tests on seasonal instability of relationships among responses. In K. H. Oppenländer & G. Poser (Eds.), *Contributions of business cycle surveys to empirical economics*. Aldershot: Gower.

Goodman, L. A. & Kruskal, W. H. (1979). *Measures of association for cross classifications*. New York: Springer.

Ivaldi, M. (1992). Survey evidence on the rationality of expectations. *Journal of Applied Econometrics*, *7*, 225–241.

Kawasaki, S., McMillan, J. & Zimmermann, K. F. (1982). Disequilibrium dynamics: An empirical study. *American Economic Review*, *72*, 992–1004.

Kawasaki, S., McMillan, J. & Zimmermann, K. F. (1983). Inventories and price inflexibility. *Econometrica*, *51*, 599–610.

Kawasaki, S. & Zimmermann, K. F. (1981). Measuring relationships in the log-linear probability model by some compact measures of association. *Statistische Hefte*, *22*, 82–109.

Kawasaki, S. & Zimmermann, K. F. (1986). Testing the rationality of price expectations for manufacturing firms. *Applied Economics*, *18*(12), 1335–1347. doi: 10.1080/00036848600000007

Kirman, A. P. & Sobel, M. J. (1974). Dynamic oligopoly with inventories. *Econometrica*, *42*(2), 279–287.

Knoche, M. (2025, February). *Bestandsaufnahme der geschichte des ifo instituts, teil 2: Aufbauphase des ifo instituts 1949 – 1965.* (mimeo)

Koenig, H. & Oudiz, G. (1979). Modèles log-linéaires pour l'analyse des données qualitatives: Application à l'etude des enquêtes de conjoncture de l'insee et de l'ifo. *Annales de l'INSEE*, *36*, 31–83.

König, H., Nerlove, M. & Oudiz, G. (1981). On the formation of price expectations: An analysis of business test data by log-linear probability models. *European Economic Review*, *16*, 103–138.

Kydland, F. E. & Prescott, E. C. (1982). Time to build and aggregate fluctuations. *Econometrica*, *50*(6), 1345–1370. doi: 10.2307/1913386

König, H. (1979). Zur bildung von preiserwartungen, ein multivariates log-lineares wahrscheinlichkeitsmodell. *Kyklos*, *32*, 380–391.

König, H. & Nerlove, M. (1980). Micro-analysis of realizations, plans and expectations in the ifo business test by multivariate log-linear probability models. In W. H. Strigel (Ed.), *Business cycle analysis, papers presented at the 14th ciret conference proceedings, lisbon, 1979* (pp. 187–226). Westmead, England: Gower Publishing Company.

König, H., Nerlove, M. & Oudiz, G. (1981). On the formation of price expectations: An analysis of business test data by log-linear probability models. *European Economic Review*, *16*, 103–138.

König, H. & Zimmermann, K. F. (1984). Produktionsplanung und arbeitsnachfrage. In H. Siebert (Ed.), *Intertemporale allokation. staatliche allokationspolitik im marktwirtschaftlichen system* (pp. 133–184). Bern: Lang.

König, H. & Zimmermann, K. F. (1986). Innovations, market structure and market dynamics. *Journal of Institutional and Theoretical Economics*, *142*, 184–199.

Low, W., McIntosh, J. & Schiantarelli, F. (1990). What can we learn about firms' output, employment and pricing decisions from business surveys? some evidence for uk companies. In J.-P. Florens, M. Ivaldi, J.-J. Laffont & F. Laisney (Eds.), *Microeconometrics: Surveys and applications* (pp. 145–160). Oxford: Basil Blackwell.

Lucas, R. E. (1972). Expectations and the neutrality of money. *Journal of Economic Theory*, *4*(2), 103-124. Retrieved from https://www.sciencedirect.com/science/article/pii/0022053172901421 doi: https://doi.org/10.1016/0022-0531(72)90142-1

Lucas, R. E. (1976). Econometric policy evaluation: A critique. *Carnegie-Rochester Conference Series on Public Policy*, *1*, 19-46. Retrieved from https://www.sciencedirect.com/science/article/pii/S0167223176800036 doi: https://doi

.org/10.1016/S0167-2231(76)80003-6

Muth, J. F. (1961). Rational expectations and the theory of price movements. *Econometrica*, *29*(3), 315–335.

Nerlove, M. (1958a). Adaptive expectations and cobweb phenomena. *The Quarterly Journal of Economics*, *72*(2), 227–240. doi: 10.2307/1880597

Nerlove, M. (1958b). *The dynamics of supply: Estimation of farmers' response to price*. Baltimore: The Johns Hopkins Press.

Nerlove, M. (1962). A quarterly econometric model for the u.k.: A review article. *American Economic Review*, *52*, 154–176.

Nerlove, M. (1964). Spectral analysis of seasonal adjustment procedures. *Econometrica*, *32*, 241–286.

Nerlove, M. (1966). A tabular survey of macro-econometric models. *International Economic Review*, *7*, 127–175.

Nerlove, M. (1983). Expectations, plans, and realizations in theory and practice. *Econometrica*, *51*, 1251–1279.

Nerlove, M. (1988). Population policy and individual choice. *Journal of Population Economics*, *1*, 17–31.

Nerlove, M., Grether, D. M. & Carvalho, J. L. (1979). *Analysis of economic time series: A synthesis* (Revised Edition 1995 ed.). New York: Academic Press, Inc.

Nerlove, M. & Press, S. J. (1973). *Univariate and multivariate loglinear and logistic models* (Tech. Rep. No. R-1306-EDA/NIH). Rand Corporation.

Nerlove, M. & Press, S. J. (1976). *Multivariate log-linear probability models for the analysis of qualitative data* (Tech. Rep. No. Discussion Paper no. 1). Evanston, IL: Center for Statistics and Probability, Northwestern University.

Nerlove, M., Razin, A. & Sadka, E. (1987). *Household and economy: Welfare economics of endogenous fertility*. New York: Academic Press, Inc.

Nerlove, M., Razin, A. & Sadka, E. (1989). Socially optimal population size and individual choice. In K. F. Zimmermann (Ed.), *Economic theory of optimal population*. Berlin, Heidelberg: Springer. doi: 10.1007/978-3-642-50043-5_2

Nerlove, M. & Schuermann, T. (1995). Expectations: are they rational, adaptive, or naive? an essay in simulation-based inference. *Advances in Econometrics and Quantitive Economics', Basil Blackwell, Oxford*, 354–381.

Nerlove, M. & Zepeda Payeras, M. (1986). Sales, production and prices: The consistency of plans, expectations and realizations of mexican firms. In *Proceedings of the 17th ciret conference* (pp. 499–530). Aldershot, England: Gower Publishing.

Oppenländer, K. H. & Poser, G. (1989). *Handbuch der ifo-umfragen* (1st ed.). Berlin: Springer.

Pupillo, L. & Zimmermann, K. F. (1991). Relative export prices and firm size in imperfect markets. *Open Economies Review*, *2*, 295–304.

Ross, D. R. & Zimmermann, K. F. (1993). Evaluating reported determinants of labor demand. *Labour Economics*, *11*, 71–84.

Sargent, T. J. & Wallace, N. (1975). "rational" expectations, the optimal monetary instrument, and the optimal money supply rule. *Journal of Political Economy*,

83(2), 241–254. Retrieved from http://www.jstor.org/stable/1830921

Sauer, S., Schasching, M. & Wohlrabe, K. (2023). *Handbook of ifo surveys* (Vol. 100). München: ifo Institut.

Theil, H. (1955). Recent experiences with the munich business test: An expository article. *Econometrica*, *23*, 184–192.

Thonstad, T. & Jochems, D. B. (1961). The influence of entrepreneurial expectations and appraisals on production planning: An econometric study of the german leather and shoe industries. *International Economic Review*, *2*, 135–153.

Veall, M. R. & Zimmermann, K. F. (1996). Pseudo-r2 measures for some common limited dependent variable models. *Journal of Economic Surveys*, *10*(3), 241–259.

Zimmermann, K. F. (1985a). *Familienökonomie. theoretische und empirische untersuchungen zur frauenerwerbstätigkeit und geburtenentwicklung*. Berlin, Heidelberg, New York, Tokyo: Springer-Verlag.

Zimmermann, K. F. (1985b). Innovationsaktivität, preisflexibilität, nachfragedruck und marktstruktur. In G. Bombach, B. Gahlen & A. E. Ott (Eds.), *Industrieökonomik: Theorie und empirie* (pp. 67–84). Tübingen: J.C.B. Mohr.

Zimmermann, K. F. (1986). On rationality of business expectations: A micro analysis of qualitative responses. *Empirical Economics*, *11*, 23–40. doi: 10.1007/BF01978143

Zimmermann, K. F. (1997). Analysis of business surveys. In M. H. Pesaran & P. Schmidt (Eds.), *Handbook of applied econometrics, volume ii – microeconometrics* (pp. 407–441). Oxford: Blackwell Publishers.

Zimmermann, K. F. & Pupillo, L. (1992). Determinants of export activity in italian manufacturing industries: results from panel data. In H. Kräger & K. F. Zimmermann (Eds.), *Export activity and strategic trade policy* (pp. 105–119). Berlin: Springer.

Chapter 2
Still the 'Dismal Science' Two Centuries after Malthus? Marc Nerlove's Research on Population and the Environment

John Rust

Abstract I discuss prescient theoretical work by Marc Nerlove and coauthors on population and environmental dynamics, including whether world population will eventually reach a steady state and if so, whether such a steady state will be an dystopic one where the planet is overpopulated and environmentally degraded with low wages and welfare per capita, or more of a utopic one, with a smaller population where higher environmental quality and per capita wages and welfare can be sustained. I consider his theoretical predictions in light of four decades of subsequent research and experience on population, economic growth, and climate change. Though the future remains highly uncertain, my reading of the evidence agrees with the pessimistic conclusion of Nerlove and Meyer (1997) that "the unpriced nature of environmental resources leads parents to fertility decisions which, while optimal from their own selfish point of view, ultimately lead to environmental disaster." I discuss the worldwide slowdown in fertility but express doubt that the deceleration in population growth is sufficient by itself to reduce the likelihood of environmental disaster. The biggest threat to the biosphere is uncontrolled growth in per capita output, absent a 'silver bullet' technological solution to the climate crisis.

2.1 Introduction

Marc Nerlove richly deserves the title of 'Renaissance scholar' for his wide-ranging contributions to economics, spanning fields from econometrics to agricultural and resource economics, using both applied and theoretical modes of analysis. His areas of expertise included development, demography and environmental/resource economics. In this chapter I consider how his work influenced current research and relates to critical problems facing the world today. Though most of Marc's work in this area

John Rust ✉
Georgetown University, Washington DC, USA, e-mail: jrust@editorialexpress.com

B. H. Baltagi and L. Mátyás (eds.), *Seven Decades of Econometrics and Beyond*, Advanced Studies in Theoretical and Applied Econometrics 57,
https://doi.org/10.1007/978-3-031-92699-0_2

was theoretical rather than empirical, his concerns and conclusions were prescient and of central relevance four decades later.

Marc's earliest published work in this area includes Nerlove (1974) where he noted that few of the then-existing models of economic growth accounted for endogenous population growth, and among those that did, "few theories of population growth and family decision-making have gone much beyond the Malthusian model" which posited that "the development of mankind was severely limited by the pressure that population growth exerted on the availability of food" (Abramitzky & Braggion, 2003, p. 423). Of course, the Malthusian prediction of stagnant real wages was belied by the exponential growth in population, total output, and per capita output and wages after the Industrial Revolution. Marc recognized that this phenomenal growth could not be explained solely as a result of investment in physical capital and improved agricultural productivity. Instead, "crucial to the understanding of long-term growth, that much investment which occurs in the economy is made in human beings rather than in physical capital and that fertility itself is shaped in important ways by economic considerations has led to renewed interest in the economics of household decisions." He offered suggestions as to how a developing theory of "new home economics" may be "integrated in a theory of economic growth and development through an understanding of the way in which investment in human capital increases the value of human time and thus changes over time the resource constraints and the relative costs and prices which 'households' face in their decisions on the number and quality of children they attempt to produce." (p. S201).

Marc also realized that the exponential economic and population growth following the Industrial Revolution may not be sustainable, and would ultimately exact a huge environmental toll that could potentially slow or even end the two centuries of phenomenal population and economic growth. In 1991 Marc delivered the inaugural Frederick V. Waugh Memorial Lecture to the American Economics Association with a paper titled "Population and the Environment: A Parable of Firewood and Other Tales" Nerlove (1991). Marc noted that "At bottom, many long-term environmental problems, whether they derive proximately from use of modern agricultural technology to augment food production or too rapid exploitation of exhaustible energy and other natural resources, stem ultimately from the pressure of human population and human desires for subsistence, if not greater, levels of creature comforts." (p. 1334). He developed a 'planar' (i.e., two variables) overlapping generations model with endogenous fertility and environmental degradation, with a focus on less developed countries where children are often treated as unpaid workers helping in household production, and thus valued at least in part for their ability to increase household consumption (for example, through animal husbandry or gathering firewood). In Marc's model, population growth degrades the environment, but the environment also has a feedback effect on fertility. Normally we might expect a degraded environment to reduce fertility and thus constitute an 'automatic stabilizer' against excessive population growth, consistent with Malthus's theories. However, Marc considered the implications of the empirically justified assumption that "in much of the Third World fertility is likely to react positively to environmental degradation because parents perceive the benefits of having more children to be higher under environmentally more adverse circumstances than under more favorable ones." Marc's analysis lead to the dismal conclusion that "the possibilities for a stable equilibrium between human population and its environment are quite limited."

> "Even given a relatively favorable relationship between population pressure and the evolution of environmental degradation over time, a stable equilibrium can be achieved only if fertility responds negatively to environmental degradation and then only if the response is sufficiently large in absolute magnitude in relation to the dynamic response of the environment to population pressure. Under exceptionally adverse environmental circumstances, rising death rates can ultimately bring a halt to further environmental deterioration and/or lead to human extinction."

While Marc's 'parable' was focused on developing countries rather than developed ones, I believe his arguments apply more generally to the entire world due to our failure to achieve a sustainable 'ecological footprint' and particularly our inability to collectively limit carbon dioxide (CO_2) emissions and the global warming it causes. Marc noted "Hardin's justly famous 'The Tragedy of the Commons' of how the unpriced or underpriced character of environmental and other natural resources leads to overexploitation and ultimate degradation" and the "population pressure which lies behind such degradation and depletion especially in developing countries" but recognized that there is no good or easy solution to this problem. Economists can pretend that a Pigouvian carbon tax can solve the greenhouse gas externality, but Marc was wise enough to know that the state of politics is a huge barrier to the mass cooperation (e.g., via the Paris Climate Accords) necessary to impose and enforce such a tax and transfer proceeds to compensate harmed individuals around the world. This is just the Tragedy of the Commons in another form, a political tragedy.

I consider Marc's predictions in light of more than three decades of new empirical evidence and theoretical work on these topics. In Section 2.2 I discuss the dynamics of world population, looking back historically at the causes of the rapid acceleration in the populations of Europe and North America after the Industrial Revolution, and the delayed takeoff in the populations of China and India until after the 'Green Revolution'. The paradox is that the rapid increase in overall population has occurred despite *falling* total fertility rates.

The secular decline in total fertility has fallen to below the 2.1 child replacement rate in most parts of the world outside Africa and other less developed countries in the Middle East. Many forecasts predict that world population will top out at about 10 billion by 2100, but it is not clear whether this will be a stable steady state, or will start to decline — a prospect that many observers find quite alarming. At least in developing countries, increasingly pessimistic expectations about the future state of the world may be one of the causes of the general decline in fertility in virtually all developed countries around the world. Unfortunately, even if population growth stopped immediately, it is insufficient to avoid a 'climate crash' and the suffering that will cause for millions or even billions of people around the globe.

Malthusianism has been widely discredited in failing to predict the astounding epoch of exponentially increasing population and income per capita following the Industrial Revolution (or Green Revolutions in China and India). Malthus made the mistake of extrapolating that past experience would continue for the foreseeable future, and he failed to predict the major 'structural break' that occurred with the Industrial Revolution. It may be just as foolhardy to predict that exponential growth in population and per capita income will also continue for the indefinite future as do, for example, Tupy and Pooley (2022). We know world population growth is slowing dramatically and is likely to level off, so the relevant question now is: can per capita income can continue to grow exponentially without bound? Whether this is possible depends largely on technology and whether continued growth can occur without the huge damage to the environment it has inflicted so far. So a modern-day version of Malthusianism could be stated as: will environmental degradation, particularly global warming, put binding constraints on how much per capita income can grow in the future, or can we count on technology to save us?

In Section 2.3, I argue that slowing population growth is insufficient by itself to solve the climate problems we are facing. From the available evidence I have seen, even if population and per capita income were to remain fixed at their current levels, humanity will still face a growing maelstrom of increasingly severe environmental disasters including hurricanes, floods, forest fires, droughts, and rising sea levels that make major population zones near sea level uninhabitable. Section 2.4 considers technological change and whether it provides a 'silver bullet' that can slow global warming and environmental degradation without requiring current and future generations to endure a substantial cut in their living standards. All of these changes could potentially be managed if humans as a whole were able to cooperate and agree on policies (including taxes and transfers) that could mitigate the effects of climate change, slow the rate of environmental degradation and find ways for humans to coexist with artificial intelligence (AI) and other advanced technologies. But international rivalries, political tribalism, influence and lobbying by powerful business interests including the fossil fuel industry, climate denialism, outright stupidity, and of course the Tragedy of the Commons leave me with little hope that a silver bullet technological fix or the necessary mass cooperation could be achieved before huge irreversible damage is done, something Wagner and Weitzman (2015) call *Climate Shock*. Thus, I agree with the prognosis that Marc reached over 3 decades ago, but not just for a handful of developing countries, but the entire planet.

I warn that I am not a demographer and don't claim to be an expert on these issues. My views are based on my own reading of literatures including demography, economics, environmental and climate science. I am trying, in my own small way, to emulate Marc, whose work inspired me to think both broadly and deeply on a topic of critical importance, even though unlike him, I do not use a mathematical model to discipline my intuitive speculations. Though I agree with Marc that there are no good or easy solutions, in Section 2.5 I discuss some limited but *feasible* policies that do not require mass cooperation (e.g., subsidizing green energy, hardening infrastructure, and making food supplies more resistant to weather and climate shocks). These policies could help blunt some of the worst outcomes of increasingly frequent environmental disasters if global warming can't be halted in the coming decades.

2.2 End of the Population Explosion?

Marc's warning that unabated population growth could lead to environmental disaster was not new: there were other apocalyptic predictions of this nature going back to the controversial book '*The Population Bomb*' (Ehrlich & Ehrlich, 1968). Outdoing even Malthus, the Ehrlichs predicted that hundreds of millions of people would die of starvation by the mid 1970s. Paul Ehrlich warned in 1970 that "sometime in the next 15 years, the end will come." By "the end" Ehrlich meant "an utter breakdown of the capacity of the planet to support humanity." (quoted from Haberman, 2015).

Marc's warnings focused on developing countries such as sub-Saharan Africa where we have seen ecological disasters such as drought resulting in famines and mass starvation. But he also warned of human extinction, so it is not clear whether he was making predictions applicable to the entire world as the Ehrlichs and others had done. Regardless, Marc's work does have implications for how global population growth affects the environment of the entire planet, and in my comments below I take a global perspective.

Fifty years after the publication of the *The Population Bomb* there is still huge polarization and disagreement about whether the exponential population and economic growth is sustainable, especially given the uncontrovertible evidence of substantial environmental damage due to global warming. On one extreme there is a 'pro-growth lobby' that denies that climate change is a threat to humanity and argues that the genies of technology and human creativity will invariably overcome any such temporary obstacles to limitless growth, and categorizing environmentalists as enemies of progress (e.g., Tupy & Pooley, 2022). On the other is an 'anti-growth lobby' that argues the world must restrain economic and population growth to avoid disaster. Some of the most extreme incarnations of this point of view treat capitalism as the enemy and threat to human survival, and therefore advocate for *degrowth communism* (e.g., Saitō, 2020).

As usual the truth, and rational guidance on policymaking, can be found somewhere between these two extremes. In this section, I briefly review how we got to where we are, and discuss some of the best available forecasts of where the world population will go in the future. While it is always hazardous to forecast far into the future, a wide range of demographic projections predict that world population will top out at around 10 billion before 2100. But it is an incontrovertible fact that total fertility rates are falling all across the planet, and that an increasing number of developed countries are now experiencing *negative* population growth. For example, China's population of 1.4 billion fell by 1.39 million over 2024, its third straight year of population decline. The population of Korea, the nation with the world's lowest total fertility rate (TFR) of 0.8, has declined since 2020. Japan has been shrinking for the past 15 years. The populations of over a dozen other countries such as Italy, Greece, Poland, Portugal, and less surprisingly, Venezuela, are all in decline.

The prospect of a falling world population is viewed as a dire threat by some. Elon Musk tweeted that "population collapse due to low birth rates is a much bigger risk to civilization than global warming." Donald Trump has said that "collapsing fertility is a bigger threat to Western civilization than Russia." The Prime Minister of Japan claimed that its low birthrate leaves them "standing on the verge of whether we can continue to function as a society" Ip and Adamy (2024). However, my guess is this is a development Marc would have welcomed as necessary for humanity to enjoy a sustainable and prosperous future. Speaking for myself, except for a transition to an older world with an increase in the dependency ratio starting with the wave of retirements by Baby Boomers (which does present a huge challenge to many countries around the world) I think the concern about lower populations in the long run if the decline in fertility persists is much ado about nothing. I do not view population decline as a threat. Instead, the best evidence is that it is a consequence of improvements in living standards in modern civilization, combined with a change in traditional social norms. Further, who knows? the decline in TFR might reverse in the future. As I will argue in Sections 2.3 and 2.4, climate change and artificial intelligence are the real threats to the future well-being of humanity.

The term *demographic transition* is used, roughly speaking, to describe different regimes in worldwide population growth, taking the pre-industrial 'Malthusian' epoch of slow population growth as the point of departure. Though demographers have finer distinctions of the stages or phases of the demographic transition, I will just lump them into two: 1) the population explosion, and 2) post-industrial stagnation and decline.

2.2.1 Demographic Transition Phase 1: Population Explosion

Let's start by briefly reviewing the basic facts on how we got to where we currently are, with a world population of slightly over 8 billion according to the US Census Bureau's population clock. As a point of comparison, I take the year 1800, two years after Malthus published his famous essay, Malthus (1798). World population in 1800 was roughly 1 billion people, so world population has increased by a factor of 8 in the two centuries after Malthus. The year 1800 is also significant because it was roughly at the start of the demographic transition (i.e., start of the population explosion) and the midpoint of what we now call the Industrial Revolution, which ignited an exponential economic growth in Britain and later the US and other European countries (though not China or India, which did not take off until after the *Green Revolution* in the 1960s as I discuss further below).

Looking back from Malthus's vantage point in 1800, world population in 1600 was about 500 million, so world population doubled in the two centuries prior to Malthus, but world GDP also roughly doubled over the same time (from 77 billion in current dollars in 1600 to 175 billion in 1800 according to the *Wikipedia* on Gross World Product), leaving GDP per capita essentially unchanged. This of course is consistent with the key Malthusian hypothesis: as income and productivity grow, population grows proportionately to keep real output per capita (and real wages) roughly constant.[1]

However, the sustained growth in total and per capita GDP and real wages after 1800 was phenomenal: by 2020 total world GDP had increased by 400-fold, and with population increasing 8-fold, it follows that real GDP per capita increased 50-fold. Not all of the increase in GDP per capita translated into higher real wages and consumption per capita, though Crafts (2022) concludes while "real consumption earnings growth was slower than the growth of labour productivity the difference is not as large as has been suggested" (p. 11). So not only has there been a huge increase in population, but also a huge increase in consumption per capita, which combine multiplicatively into a unsustainable environmental impact of the human race on the planet. Some geologists refer to the current era as the *Anthropocene* in recognition of the massive impact that economic growth and human activity has had on the climate and biodiversity.

A number of different factors lead to the population explosion, but a key explanation is the fact that around 1800 "the average death rate decreased, from an average 30 deaths per 1000 inhabitants in the beginning of the 19th century to around 15 deaths per 1000 citizens by the beginning of the 20th century. In the meantime, the birth rate however stayed at its previous, high level of 30-35 births per 1000 inhabitants" (Bavel, 2013 p. 284). The improvement in mortality can be attributed to the cumulative impact of a number of key scientific discoveries that improved health, sanitation and agricultural output which reduced the incidence and severity of "of epidemic diseases or failed harvests and famine, or a combination of both". As a consequence, "Later on in the 19th century, child survival began to improve. Vaccination against smallpox for example led to an eradication of the disease, with the last European smallpox pandemic dating from 1871."

The population explosion began in Europe and North America, but in other parts of the world it did not happen until much later, particularly in India and China, because the Industrial Revolution did not occur in these countries. There are a multitude of complex reasons for this lag, partly due to cultural differences as well as colonial exploitation by European imperialists. But suffice it to say that transportation and communication were much more costly and slower than they are now, so scientific knowledge and technological know-how did not diffuse nearly as rapidly as they do today. The populations of China and India did not explode until after the 'Green Revolution' in the 1960s.

Prior to the 1960s the food supply in India was insecure, and imported grains and rice were required for much of the population to attain a subsistence diet. A combination of periodic droughts, poor crop yields, inefficient farming, and interruptions in imports lead to periodic famines, such as the Bengal famine of 1943

[1] Bouscasse, Nakamura and Steinsson (2025) argue that in England "productivity growth was zero prior to 1600" but growth started after 1600. They estimate "productivity growth of 2% per decade between 1600 and 1800, increasing to 5% per decade between 1810 and 1860." Thus, there was only small growth in real wages between 1600 and 1800 and "a large and sustained fall between 1450 and 1600, some recovery over the 17th century, stagnation during the 18th century, and finally a sharp increase after 1800."

that killed 2 million. In the 1960s the Nobel Prize winner Norman Borlaug so-called 'father of the Green Revolution' introduced dwarf wheat into India and Pakistan, causing production to increase enormously. He is credited with saving over 1 billion lives in India, Mexico and elsewhere.[2]

However, it is not clear that increased food supply due to the Green Revolution was the primary cause of the population explosion in India. In fact, Gollin, Hansen and Wingender (2021) claim that the high yield grains introduced by Borlaug "increased income and reduced population growth" (p. 2344). This mechanism behind this is not entirely clear since the authors did not specifically model fertility choices but they conjecture that the higher crop yields and income caused by the Green Revolution may "make parents substitute child quantity for child quality, leading to lower fertility and better education outcomes" (p. 2357). An alternative explanation is similar to Marc's Parable of Firewood: the increase in yields allowed agrarian households to be more productive without having to rely on as many children to do chores on the family farm, such as tending livestock or fetching water. This is particularly true for the poorest, as Traeger (2011) notes: "Poor families are typically larger because they use children as a source of generating income via child labor. Parents also have children for insurance purposes because they envision needing help when they get older" (p. 87).

But the main cause of India's population boom is the same one that caused the population explosion in Europe and North America after the Industrial Revolution: improved healthcare and public health measures that significantly reduced death rates (particularly during childhood), while birth rates remained relatively high in the decades after the Green Revolution. TFR was 5.9 per household in 1960, gradually decreasing to 2.2 by 2020. The slow reduction in TFR combined with the more rapid reduction in mortality lead to a rapid increase in population despite efforts at family planning, which was further compounded by factors such as early marriage, illiteracy, and poverty. In 1800 the population of India was 169 million, and by 1960 it had grown to 450 million, which is only slightly slower growth than the 3-fold increase in world population. Today India has a population of 1.4 billion, a 311% percent increase since 1960 compared to a 266% increase for the world as a whole over this same period. India's fastest population growth occurred after the Green Revolution.

In China, the population explosion was also not primarily caused by the Green Revolution, but rather the impact of new fertility policy in China between 1950 and 1970 when Chairman Mao Zhedong promoted large families. Howden and Zhou (2015) show that this policy change, combined with "General improvements to healthcare did have beneficial results on population growth" especially a large decline in infant mortality after the mid 1960s (p. 237). The authors outline another significant cause of the boom that is consistent with Marc's Parable of the Firewood story: "The *hukou* system nationalized all the country's lands and remunerated workers for their labor hours instead of for their output. At the same time, the scarcity that plagued the

---

[2] This is a good example of how technological progress can enable us to overcome what seem to be hard environmental constraints. As Howden and Zhou (2015) note, "Ehrlich's pessimistic forecast was proved wrong, though due mostly to the increased crop yields from the Green Revolution, not to an imminent reduction in the global population."

country after the Great Leap Forward left parents with few options to provide a better life for their families. Paradoxically, perhaps, one way to increase family earnings was to have additional children. Although children were remunerated less than adults, they still provided an important source of resources for their family. Chinese parents tried to escape poverty by having children as a source of income" (p. 246).

Mao's fertility policies were perhaps too successful, increasing the Chinese population by nearly 50% from 655 million in 1960 to 970 million in 1979, while TFR fell from about 6 in the 1960s to just over 2 by 1979. In 1979 China announced a new one child policy (1CP) that is "widely regarded as an effective piece of government legislation that saved the country from a Malthusian fate" (Howden & Zhou, 2015 p. 227). However, the data show that GDP per capita was basically flat from 1960 to 2000, so the 1CP may have prevented excessive population growth that could have otherwise reduced per capita output and real wages.[3] Economic models have been developed to predict the counterfactual outcome on output, wages, and welfare without 1CP. The analysis of Liao (2013) finds that "The results suggest that imposing the one-child policy promotes the accumulation of human capital. In addition, the economy enjoys higher per capita output. However, output per capita fluctuates after the policy is enforced." Another counterfactual analysis by Gu (2022) finds that "the one-child policy increases the human capital of affected agents by about 47% relative to a counterfactual with no fertility restrictions. However, the effect on aggregate income is negative as the size of the labor force falls." It also increased individual welfare, since "the fertility restriction is a binding policy, it is immediate that it lowers the welfare of generations giving birth during the policy implementation. However, for generations born under the policy, higher human capital and a higher physical capital to labor ratio increase their welfare."

To summarize, the population explosion was associated with a rapid change in many countries from the 'Malthusian era' of pre-industrial agrarian and rural economies to post-industrial urbanized economies due to the Industrial Revolution in Europe and North America around 1800 and improvements in child mortality in India during the Green Revolution and Maoist pro-fertility policies in China in the 1960s. Cumulative scientific advances lead to improvements in sanitation and health and increased productivity in food production that enabled mortality to rapidly fall (particularly child mortality), while fertility remained high due to inertia from social mores and customs. Child mortality is known to have a strong effect on fertility decisions due to the *replacement effect* of families having more children when child mortality is higher (see, e.g., Wolpin, 1984).[4]

---

[3] The 'Chinese Economic Miracle' started after 2000, when GDP grew from roughly $1 trillion in 2000 to nearly $18 trillion by 2023.

[4] Nerlove (1991) noted the "pioneering numerical work of Wolpin" for providing "a significant break-through in the development of a satisfactory analytical characterization of the observed empirical regularity in terms of the structure of the parent's utility function and the existence of *ex ante* costs." In homage to Marc's work on adaptive expectations, I speculate that there must be lagged adjustment in households' beliefs about child mortality, since a model such as Wolpin's with rational expectations predicts that TFR falls as soon as child mortality falls.

2.2.2 Demographic Transition Phase 2: Stagnation and Decline

In the second phase, global TFR has steadily dropped from 4.8 in 1950 to 2.2 in 2021 (Fertility & Collaborators, 2024, p. 2075). According to *Wikipedia* the annual growth rate in world population fell from a high of 2.1% in the baby boom year of the 1960s to below 1% today. Currently, 47 countries in the world have negative population growth rates due to TFR below the steady state replacement rate of 2.1. The comprehensive study by Fertility and Collaborators (2024) finds over 100 out of 204 countries/territories have TFR below 2.1. Only 47 countries currently have TFR above 2.1 and most are in Africa and the Middle East, but TFR has been declining in these countries as well. The country with the world's highest TFR, 7, is Chad and for sub-Saharan Africa, the average is 5.

These are facts that virtually everyone agrees with, but there is less consensus on the *causes* of the secular decline in fertility and effectiveness of policies to increase it. The complexity is that there are multiple causal factors at play and these can differ in different societies, so in many cases the decline in fertility defies simple explanations.[5] But the usual list of reasons includes, beside the reduction in child mortality already noted above, a shift in the population from rural to urban reducing the need for children as household laborers, higher income and wealth lead to a greater demand for education and the *quality* of children than *quantity* of them, higher female education and job opportunities and better contraception leading to delayed marriage and age of first birth, changing social norms about marriage and family size, as well as effects of government policies such as the 1CP in China and family planning programs in India that provided low cost access to contraceptives and family planning education.[6] More recently high cost of living and education and lack of affordable housing could be reducing fertility in many densely populated urban areas (e.g., China, Singapore, etc). Further, as I discuss below, increasing pessimism about future living standards due to climate change, political instability, threat of World War, and concerns about the revolution in AI on the demand for labor may lead fewer couples to have children.

In terms of Marc's Parable of Firewood, it is possible that in advanced societies, parental altruism towards their children operates in the expected direction: couples may choose to have fewer or no children if they expect a degradation in the environment in a generalized sense (i.e., where 'environment' includes economic opportunities). However, this relationship seems very context-dependent, and in developing countries where subsistence agriculture is still prevalent, Marc's hypothesis of a positive relationship between environmental degradation and fertility is likely to

---

[5] Kearney and Levine (2022) analyze the causes of a pronounced decline in US TFR after the Great Recession in 2007 and conclude that "In summary, we have had no success finding evidence in favor of any social, economic, or policy factors being important drivers of the recent decline in the US birth rate, other than the appearance of the Great Recession."

[6] Though TFR is above the replacement rate in Africa, it is declining too, though at a slower rate. Barrett and et al. (2020) note that "Sub-Saharan Africa's slower fertility decline has been traced to many reasons, including inheritance rules, the prevalence of polygamy, lack of access to modern methods of contraception, low education among women, and kinship obligations" p. 6304

hold empirically. The empirical study by Haq, Chowdhury, Ahmed and Chowdhury (2023) finds evidence of this nature, but notes "The heterogeneity observed in the effect of the ecological footprint on TFR underscores that both the magnitude and direction of this relationship are intricately tied to socioeconomic conditions and cultural contexts."Casey et al. (2019) developed an OLG model of the global economy to study the impact of climate change on fertility to provide insight into why environmental degradation has different effects on TFR in developed countries compared to developing ones: "Near the equator, where many poor countries are located, climate change has a larger negative effect on agriculture. The resulting scarcity in agricultural goods acts as a force towards higher agricultural prices and wages, leading to a labor reallocation into this sector. Since agriculture makes less use of skilled labor, climate damage decreases the return to acquiring skills, inducing parents to invest less resources in the education of each child and to increase fertility. These patterns are reversed at higher latitudes, suggesting that climate change may exacerbate inequities by reducing fertility and increasing education in richer northern countries, while increasing fertility and reducing education in poorer tropical countries."

Galor (2005) offered a thought-provoking theory of fertility designed to explain both phases of the demographic transition, i.e., the population boom at the start of the Industrial Revolution followed by the stagnation starting in the middle of the 20th century.

> "In the early stages of the transition from the Malthusian regime, the effect of technological progress on parental income dominated, and the population growth rate as well as the average quality increased. Ultimately, further increases in the rate of technological progress that were stimulated by human capital accumulation induced a reduction in fertility rates, generating a demographic transition in which the rate of population growth declined along with an increase in the average level of education. Thus, consistent with historical evidence, the theory suggests that prior to the demographic transition, population growth increased along with investment in human capital, whereas the demographic transition brought about a decline in population growth along with a further increase in human capital formation."

Galor (2022) cast his theory in almost Darwinian terms. He describes how humanity escaped from the pre-industrial Malthusian equilibrium by considering two clans: the *Quantys* (who prefer quantity of children over quality), and *Qualys* (who prefer to have fewer higher quality children by not using them as productive assets but investing more in their education while young). He asks, "Which of the clans, the Qualy or the Quanty, will have more descendants and thus dominate the population in the long run?" Counterintuitively, he argues that the Qualys will because the parental investment in their human capital pays off in terms of their future success: "This increase in earnings capacity would place the Qualy clan at a distinct evolutionary advantage." That is because the higher earnings of children of the Qualy clan will reduce their mortality and enable them to have more offspring compared to children of the Quanty clan, and thus have fewer children of their own that survive into adulthood.

Galor's explanation suggests that members of the richer Qualy clan should have higher fertility than the poorer Quanty clan, but this is not consistent with the empirical evidence in the later stages of the Industrial Revolution. His theory is less clear on what lead to continued reductions in fertility leading to the second stagnation phase of the demographic transition, but overall I agree with Galor's theory that the desire to invest more in children's human capital to enable them to succeed, coupled with the high cost of these investments ultimately lead to reduced fertility. As Will Hutton's review of Galor's book succinctly summarized, the Industrial Revolution ushered an era of "gradual quickening in the introduction of technologies that required mass education for their successful implementation. This triggered a virtuous circle of more innovation, more investment in education, more need to invest in the quality of children rather than quantity, so that birthrates declined sufficiently to allow living standards and life expectancy to rise. Because it was now rational to invest in children's education rather than get them working, child labour and exploitation fell away" Hutton (2022).

Thus, whether parents are motivated to invest more in their children due to altruism or for selfish reasons (e.g., to have wealthier, educated children to support them in old age), Galor's theory (applicable to developed countries) predicts that economic growth caused the decline in fertility. To the extent that economic growth degrades the environment, Marc's model for undeveloped countries predicts the opposite. However, Nerlove (1991) did not allow parents to invest in their children's human capital, whereas his prior work Nerlove (1974) did emphasize the importance of these investments. In any event, it is now widely recognized that higher human capital investment is a central cause of the secular decline in TFR and is one of the 'new stylized facts' about economic and population growth, see Jones and Romer (2010).

A more ominous cause of declining fertility is the increasing pessimism of younger generations about the future. The OECD report *Society at a Glance* notes that fertility decisions are affected both by real and perceived economic uncertainties, and notes that "Most analyses generally find that birth rates react negatively to economic downturns." Other concerns include "for example, climate change, of energy, food and/or housing costs" and "many people anticipate geo-political instability and socio-economic instability and the outlook is markedly more negative over a 10-year timeframe". Finally, it notes that "many people who believe that today's children will grow up to be worse off than their parents: over 50% in most OECD countries, and in the majority of these countries this negative sentiment strengthened over the past decade" (OECD, 2024, p. 25).

2.2.3 Will World Population Stabilize at 10 Billion?

Marc analyzed a dynamic overlapping generations model of population and the environment but focused on steady state outcomes. Though fertility is declining in most countries around the world, total population is still increasing. Will world population ultimately reach a steady state, and if so, how large will it be? The best estimates of future population growth come from the United Nations World Population Prospects. Their 2024 forecasts go out to 2100 and are shown in figure 2.1 along with 80 and 95% confidence bands. Population is projected to peak at 10.4 billion in 2086 and then slowly decline to 10.3 billion by 2100. For all practical purposes, this constitutes a steady state, *if* the fertility assumptions underlying its projections are accurate.

Fig. 2.1: UN 2024 World Population Forecasts

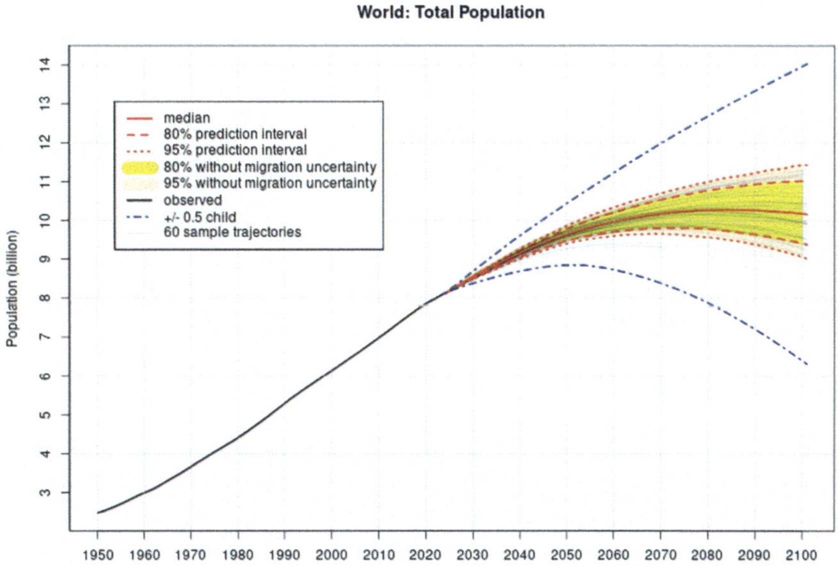

Data source: United Nations, Department of Economic and Social Affairs, Population Division (2024)

However, the size of the 95% confidence bands show there is considerable uncertainty in these forecasts, and these confidence bands do not fully reflect substantial uncertainty about how climate change, pandemics, AI, and other economic uncertainties that could affect future population and fertility. Spears, Vyas, Weston and Geruso (2024) illustrated how sensitive the UN forecasts are just to assumptions about what they call the 'long run TFR' to which the world converges to at the end of the UN forecast interval in 2100: "We show that any stable, long-run size of the world population would persistently depend on when an increase towards replacement fertility begins. Without such an increase, the 400-year span when more than 2 billion people were alive would be a brief spike in history. Indeed, four-fifths of all births—past, present, and future—would have already happened." The assumption that TFR converges to values below replacement rate by 2100 and remains there forever results in the 'population spikes' illustrated in figure 2.2 (taken from Spears et al., 2024).

The scaling of figure 2.2, from 10000BC to 4000AD, seems designed to alarm, though who knows, humanity could be just a temporary blip when you look at things from a longer run though not quite geological timescale. But who can predict what TFR will be by 2100, much less 4000? There are plenty of other things that could

Fig. 2.2: Forecasted population spikes if TFR is below 2.1 in 2100

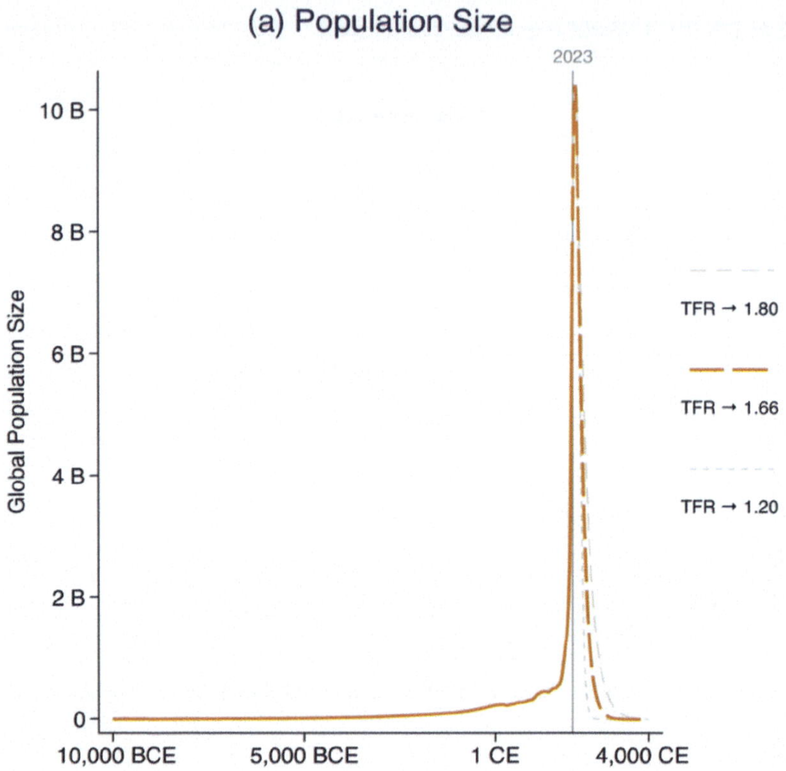

Data source: (Spears et al., 2024)

happen well before 2100 that are far more consequential. Many experts worry that humanity could be wiped out by AI long before 4000, with *homo sapiens* having been superseded by newer generations of super-intelligent beings. For more realistic horizons, the UN's projections are more relevant, but there is huge uncertainty that is not reflected in the confidence bands in figure 2.1. For example, what about the risk of world war, or nuclear war? All of these risks are reflected in the recent decision by the Bulletin of Atomic Scientists to advance their doomsday clock to 89 seconds before midnight, the closest it has ever been. There are other very hard to predict risks to the population such as pandemics, asteroid collisions or other global calamities that are hard to factor into stochastic projections. Another risk, climate change, is factored into the UN population projections but in an informal way, using a more *ad hoc* feedback cycle. The UN world population forecasts are used in longer run climate models to predict climate change, but the UN uses the predictions of climate models to adjust its predictions of future TFR in different countries.

In any event, barring some major disaster, it seems quite unlikely that world population will decline substantially in the next few decades but rather will continue to increase, even given the dramatic declines in TFR around the world. To a first approximation, we can think of the planet as approaching at least a temporary 'steady state' of 10 billion sometime before 2100. How long it will stay there is anyone's guess, but as I discuss below, there is significant risk that such a steady state will be an environmentally degraded dystopia. In the shorter run, slowing TFR creates major transitional problems that need urgent attention. In particular, how well will the world be able to deal with the wave of retirements of the baby boom generation?

2.2.4 Can Migration Mitigate the Baby Boom and Demographic Divide?

The US Census Bureau constructs 'age pyramids' as side by side plots of the age distribution for males and females plotted vertically with the oldest ages at the top. A demographically young society has a triangular age pyramid, where most of the population is young and only a minority old. The aging baby boom generation is apparent in how the age pyramids evolve over time, appearing as a bulge in the age distribution that widens the upper old age part of the age distribution over time. For example, the largest age group in the population was 35 to 39 in 2000 and of course this group are those born toward the end of the baby boom cohort. By 2010 the "wave" in the age distribution corresponding to this cohort appears as a bulge in the share of people aged 45 to 49. By 2020 the widest bulge was for individuals aged 55 to 59, so it is evident how the age distribution is becoming 'top-heavy' from the aging of the baby boom generation. Another way to illustrate the aging of the population is via the *dependency ratio* which is the ratio of the population over 64 to the working age population aged 15 to 64. This ratio has increased from 15% in 1960 to 27% by 2023. The wave of baby boomers is even more pronounced in China and its old age dependency ratio is projected to more than double in the coming decades.

Government social insurance and retirement programs around the world that are funded on a 'pay as you go' basis are increasingly on shaky foundations due to the steadily increasing dependency ratio that is partly due to the post WWII baby boom and the decline in TFR. The US Social Security system is partly funded by a trust fund that is projected to run out in 2035, after which tax increases or benefit cuts will be required. According to the IMF 11 of the largest developed countries in the world have national debt to GDP ratios greater than 100%, with the average for G7 countries being 128%. Combined with 'taxpayer revolt' it is not clear how many nations have the 'fiscal slack' to be able to support their elderly populations in retirement without a significant cut in standard of living. Even countries such as China, with its comparatively lower 77% debt GDP ratio, face these challenges due to its shrinking workforce and underfunded public pension system. As Fertility and Collaborators (2024) observe, "Low levels of fertility have the potential over time to result in inverted population pyramids with growing numbers of older people and declining working-age populations. These changes are likely to place increasing burdens on health care and social systems, transform labour and consumer markets, and alter patterns of resource use" (p. 2058).

A related near term challenge is how to deal with the growing *demographic divide* i.e., the already noted disparity between the aging low fertility developed countries and the younger high fertility developing countries, especially in Africa. A rational solution to these problems would be to allow greater migration of younger workers from low wage developing countries to help support the elderly populations in the high wage developed countries. In principle, this would be a win-win situation that helps the high fertility developing countries as well. As Fertility and Collaborators (2024) observe, the "dramatic shift in the concentration of live births from middle-income and high-income settings to low-income settings will lead to serious challenges related to sustaining and supporting a growing young population in some of the most heat-stressed, politically unstable, economically vulnerable, health system-strained locations" (p. 2091).

Kennan (2013) used an economic model of factor price equalization to predict the welfare gains from unrestricted international labor migration. He finds that "The estimated gains from removing immigration restrictions are huge. Using a simple static model of migration costs, the estimated net gains from open borders are about the same as the gains from a growth miracle that more than doubles the income level in less-developed countries." However, the level of immigration resulting from an 'open borders' policy would also be huge: his model predicts that in the post-immigration steady state, the US would have 354 million new immigrants, relative to its native population of 187 million working people aged 24 to 60.

It seems abundantly clear given the worldwide political backlash to the much more modest immigration flows in recent years that there is little hope that countries will relax immigration restrictions to address our demographic challenges in the foreseeable future. The inability to transcend cultural differences exacerbated by overblown fears of loss of jobs and higher crime by immigrants, combined with the increasingly nationalistic/tribal nature of politics in even the most advanced countries (such as Trump and the strong support he received to "build the Wall" and mass deport undocumented individuals from the US and the increasingly strong support for right wing extremist parties such as AfD in Germany or Orban in Hungary), makes it clear that we will have to look to other policy approaches to address the near-term problems posed by the baby boom transition and demographic shift.

2.2.5 Can 'Pro Fertility Policy' Reverse the Decline in Birth Rates?

There is plenty of evidence that some policies designed to limit or reduce birth rates have been effective, such as the One Child policy in China, which I noted may actually have been 'too successful' in prompting the Chinese government to rescind it and now actively promote larger families. But evidence of the effectiveness of pro-fertility policies that either intentionally or unintentionally promote higher birth rates is much less clear. This is an illustration of where it is possible to 'pull on a string' but not to 'push on a string'.

For example, due to concern about low birthrates China's 1CP started to be relaxed for certain groups in 2011 and extended to an increasing share of the population until it was completely abolished in 2016. In 2021 China adopted a '3-child policy' to try to change cultural norms towards bigger families. However, as I noted, China's population has been decreasing for the last several years. An empirical study by Lin et al. (2024) concludes that "The results suggest that lifting birth restrictions had a short-term effect on the increase in birth rates and rates of natural population increase. However, birth policy with lifting birth restrictions alone may not have sustained impact on population growth in the long run" (p. 364). So far, the Chinese government's moral suasion has not had a measurable impact on societal attitudes toward family size, or overcome obstacles such as the high cost of raising children in urban areas.

Other policies try to promote fertility through financial incentives, childcare subsidies and job protection to families having children. As Hiriscau (2024) notes, "The research literature has identified two primary sets of policies that can influence fertility rates. One strand of the literature examines the impact of leave policies on fertility, considering variations in benefits, duration, job protection, and availability for either parent. Studies have indicated that maternity and parental leave policies have a positive effect on fertility rates" and the other strand of literature "focuses on the effect of financial incentives on fertility, particularly on child cash transfers and child-related taxes." Though many studies find statistically significant positive effects, the cost of these incentives is large but their overall impact on fertility is not big, and not enough to reverse the general decline in TFR around the world.[7]

There is a third class of policies that are not directly intended to increase birth rates but rather restrict womens' control over their fertility such as limits on or bans on abortion (including the use of safe drugs for medicated abortions such as Mifipristone), and reductions in funding and access to agencies that provide family planning and access to birth control. An example is the *Dobbs* ruling by the US Supreme Court in 2022 reversing the right to abortion that women had in all states, allowing individual states to pass restrictions. Dench, Pineda-Torres and Myers (2024) find that "The results indicate that states with abortion bans experienced an average increase in births of 2.3 percent relative to if no bans had been enforced."

Overall, I agree with the conclusions of Fertility and Collaborators (2024) that there is no silver bullet policy that can reverse the secular worldwide downward trend in fertility since 1950 "Social policies to improve birth rates such as enhanced parental leave, free childcare, financial incentives, and extra employment rights, may provide a small boost to fertility rates, but most countries will remain below replacement levels." Even if these policies had more powerful effects, they operate too slowly (i.e., over generations) to reverse the decline in TFR, and hence are not viable policies for

[7] For example, Hiriscau (2024) studies the effect of an extension in paid maternity leave in Romania from 60 days to 1 year and finds that families who are eligible for this benefit experienced only a 2.5 percentage point increase in the probability of having an additional child compared to families who were not eligible.

addressing the more immediate challenges posed by population aging, such as how to deal with the wave of retirements by Baby Boomers the world is experiencing. I also believe individual families should be free to determine how many children they have, and agree with Tupy and Pooley (2022) who are "opposed to government measures that would coerce or otherwise incentivize people to have more children. The human population should reflect the free choices of individual men and women."

However, I disagree with the claim by Fertility and Collaborators (2024) that "once nearly every country's population is shrinking, reliance on open immigration will become necessary to sustain economic growth. Sub-Saharan African countries have a vital resource that aging societies are losing—a youthful population." Slowing population growth can reduce the growth in total GDP, but it does not necessarily reduce the rate of growth in *per capita GDP* as I discuss in the next section.

2.2.6 Will Slowing Population Growth Reduce Economic Growth?

If we are speaking of total GDP, generally yes. For example, in terms of the textbook Solow-Swan growth model, along a 'balanced growth path' the growth in total output or GDP equals the sum of the population growth rate n and the rate of growth in output per worker, g. But along such a path output per worker grows at rate g independent of n. So while it is true mechanically that a reduction in the rate of population growth reduces the rate of growth of GDP, it is not clear why we should worry if output per worker (i.e., GDP per capita) continues to grow at rate g despite a decrease in n. Growth in individual output, wages, and welfare is more important than the growth of the entire economy.

The actual experience of many developed countries with falling population growth rates demonstrates that slowing population growth rate does not reduce wage growth or growth in GDP per capita. For example in China, we noted that the 1CP was a major factor causing population growth rates to fall from 2.8% annual in 1970 to 0.8% in 2000 to slightly negative by 2023, but the China Miracle happened nonetheless, with per capita income growing by a factor of 8.4 from 2000 to 2023, or an average growth rate of 9.4%. Indeed, as I discussed above, the purpose of 1CP was to restrain the growth in population precisely in order to save China from a 'Malthusian fate'.

Despite this success, I have already noted China's abrupt end to 1CP in 2015 followed by further measures to reverse the continued decline in its population, including a 'three child policy' (3CP) in 2021. China's leaders evidently care about the total size of the population and GDP, even though wages and per capita GDP have been growing at a very rapid rate. Russia has adopted a series of policies to increase its birth rate, given that its rate of population decrease ranks 16th among the most rapidly shrinking countries worldwide and its TFR of 1.5 ranks it as the 171st lowest among the 204 countries in the world. The Russian fertility policies started out with using the carrot of incentive payments for having more children (tax breaks and payments to women who have a second or third child), but have grown to include the stick of criminalizing 'child-free propaganda' and limiting access to abortion and contraception. As I already discussed above, these policies have had limited success.

Why does the leadership of the largest countries care about the growth in total population and GDP versus the growth in per capita output, wages and welfare of their citizens? I speculate that it is partly due to fear of losing influence and power if their 'market share' of world population and GDP declines over time. International rivalries drive an interest in higher population growth in part due to a belief that there are significant 'returns to scale' of larger populations leads to a larger consumer base, labor pool, and domestic markets all of which can lead to higher R&D investment to further increase the country's productivity growth and technological edge. For example, there is concern in the US that China is gaining a big lead by graduating many times more engineers with undergraduate degrees as well as PhD degrees in science, technology, engineering and math (STEM). A report by Zwetsloot et al. (2021) warns that "We find that China has consistently produced more STEM doctorates than the United States since the mid-2000s, and that the gap between the two countries will likely grow wider in the next five years. Based on current enrollment patterns, we project that by 2025 Chinese universities will produce more than 77,000 STEM PhD graduates per year compared to approximately 40,000 in the United States. If international students are excluded from the U.S. count, Chinese STEM PhD graduates would outnumber their U.S. counterparts more than three-to-one."

A number of economists also believe that rapid population growth is essential for technological progress. A leading proponent of this point of view is the late economist Julian Simon who believed that "population growth, contrary to Malthusian theory, actually drives technological progress, leading to increased resource abundance and a better standard of living as human ingenuity finds solutions to resource scarcity through innovation and substitution" *Wikipedia*. Simon's 1981 book *The Ultimate Resource* (Simon, 1981) argued that "A larger population influences the production of knowledge by creating more minds to generate new ideas (the supply side) and more consumers to drive up prices and create the financial incentives for the creation of new knowledge (the demand side). This creation of knowledge ultimately makes us wealthier and solves the problems that population growth and rising income may cause" (Ahlburg, 1998, p. 322). Similar views are also echoed in the book *SuperAbundance* Tupy and Pooley (2022). One reason why large populations promote technological progress and therefore economic growth is due to the birth of geniuses such as Edison or Einstein who are 'rare events' that are more likely to occur as tail outcomes in huge populations. Simon concluded that it was valuable to promote large populations to increase the likelihood of new geniuses born in the future whose ideas could radically transform science and technology, which he assumed would benefit all humanity.

While there is some truth in Simon's point of view, it seems less relevant today given an explosion in knowledge brought about by the *Information Revolution* with massive increases in computer power, dramatic reductions in the cost of communication, and the rapid accumulation and dissemination of knowledge via the Internet. In the last few year incredible breakthroughs in artificial intelligence with the emergence of *large language models* (LLMs) trained on the accumulation of data available via the Internet makes the prospect of 'artificial general intelligence' (AGI) more likely and suggests that major scientific breakthroughs and an acceleration in technological progress can occur without the need for steadily growing human populations. We have recently seen a number of dramatic examples of how AI, using deep neural networks and reinforcement learning, are producing game-changing technological breakthroughs. For example, the 2024 Nobel Prize in Chemistry was awarded to a team of computer scientists at Google for their development of 'Alpha Fold' that predicts the 3-dimensional folding of proteins that has huge importance in biology.[8] It is not clear that large populations are the main cause of such breakthroughs, but rather it is an example of 'knowledge building on knowledge' in an accelerating fashion. Instead of ineffectual policies to increase population growth, we can promote technological change more effectively via targeted investments in education and research and development.

Simon's theory that population growth is necessary for technological improvement seems especially questionable given that the fastest population growth is happening in the poorest, least well-developed countries in the world, but this is not where game-changing new technologies are being born. Many developing countries such as in Africa are facing huge challenges just supporting their current populations and averting famine-induced starvation due to climate change, political instability, and wars. Children there are growing up malnourished and poorly educated, and the large waves of emigration from many of the least developed countries suggest that these are not places where we can expect future scientific geniuses to be born and nurtured. In my view, a policy of promoting even higher population growth in these regions given these challenges to slightly increase the chance that the next transformative genius might be born in one of these countries in the future seems both logically and morally dubious.

The view that technological growth will at some point become self-sustaining and disembodied from its human creators was anticipated by many thinkers, including Ray Kurzweil in his 1990 book *The Age of Intelligent Machines* (Kurzweil, 1990). His predictions were remarkably on-target: he argued that humans will eventually be able to build something more intelligent than themselves. He predicted big strides in pattern recognition (a key part of human vision), and knowledge representation (as embodied in language), as two key components of intelligence, and showed how quickly computers were advancing in each of those domains. Now, with the widespread AI advances in language recognition and translation, computer vision and image processing, and early examples of logical reasoning by LLMs, Kurzweil's predictions seem remarkably prescient. He currently predicts that "Artificial intelligence will reach human levels by around 2029. Follow that out further to, say, 2045, we will have multiplied the intelligence, the human biological machine intelligence of our civilization a billion-fold."

[8] A *Nature* article by Callaway (2024) notes that Alpha-fold "has been nothing short of transformative. The tool has made protein structures – often, but not always, highly accurate ones – available to researchers at the touch of a button, and enabled experiments that were unimaginable a decade ago." Other biologists refer to it as a 'major revolution' that will further accelerate progress in biology.

These revolutionary developments suggest that large populations are no longer necessary to support rapid acceleration in technological progress and knowledge. If anything, AI and not low population is what we should be worried about. Many leading experts and thinker (e.g. the late Nobel prize winning physicist Stephen Hawking) worry that the 'genie is out of the bottle' and human livelihoods could be endangered by AI in the future. Indeed, we are already seeing many creative and intellectual professions quite worried that AI will take away their jobs, including graphic artists, musicians, journalists, and lawyers. At the very least, I would agree with the conclusion of Jones and Romer (2010) that "this century will mark a fundamental phase shift in the growth process. Growth in the stock of ideas will likely no longer be supported by growth in the total number of humans."

To summarize the main takeaways from this section: 1) the population boom has ended, 2) economic and technological growth no longer depend on population growth (i.e., human labor is no longer the scarce factor limiting growth), and 3) most pro-growth fertility policies have only small impact. In the next section I argue slower population growth is not to be feared, but potentially welcomed as an 'automatic stabilizer' to mitigate the damage humanity is doing to the ecosystem – but it's not enough.

2.3 Will Slower Population Growth Avert Ecological Disaster?

The key innovation in Nerlove (1991) was to model the dynamics of *environmental capital* E jointly with the total population, N. Higher values of E correspond to a better environment.[9] Marc assumed that the state of the environment evolves as $E_{t+1} = g(E_t, N_t)$ which is monotonic in both arguments, i.e., higher N_t lowers E_{t+1} and higher E_t raises E_{t+1}. He also assumed that the population growth rate depends on E, so total population evolves as $N_{t+1} = h(E_t)N_t$. He argued that plausible models of family decision-making and/or the effects of rising death rates with increasing environmental deterioration implies that the h function is likely to be U-shaped, with high population growth rates in sufficiently low quality environments that initially decrease as the environment E improves but ultimately turn up, so improvement in E leads to faster population growth when E is sufficiently high.

Marc showed that "multiple, at least two, stationary solutions to the dynamic system relating population and environmental quality are likely." One steady state has high N but low E and the other low N but high E so a "lower level of population and better environment." He noted that "The first of these, if there are two, is likely to be characterized by a positive response of fertility and the rate of growth of population to environmental deterioration and the second by a negative response. Only when there is such a negative response is there any possibility of obtaining a stable stationary solution under other plausible assumptions about the parameters of the system." Marc

[9] Marc actually used the variable Z_t to capture environmental degradation at time t, and "Thus, the larger Z_t the lower the level of environmental quality." So we can treat E as roughly the inverse of Marc's variable Z.

showed that the low N and high E steady state is stable but the other high N and low E steady state is locally unstable. Thus, if the world was at the latter steady state, a shock that reduces the state of the environment could lead to a dynamic of higher population growth causing more damage to the environment which in turn generates even higher population growth, potentially leading to an environmental disaster.

Marc showed the stability of steady states depend on the relative sizes of the slopes of the functions h with respect to E and g with respect to N. If the first is large, i.e., the environmental state rapidly deteriorates as population increases, "then the rate of population growth must be rather insensitive to environmental deterioration" to achieve a stable steady state. He noted that the "balance is delicate" – if the environment deteriorates too rapidly as population increases, or population growth rates increase too quickly as the environment deteriorates, then the result is an unstable dynamic between population and the environment.

Marc acknowledged that to keep his analysis tractable he excluded physical and human capital from his model and ignored consumption, saving, and investment decisions. He noted that "There is no doubt in my mind that introduction of physical capital formation to offset the environmentally adverse effects of population pressure and of human capital formation to enhance the quality of individual children would result in far more optimistic conclusions."[10]

Figure 2.3, reproduced from Dasgupta et al. (2023), shows that Marc's optimism may have been misplaced. It shows paths for the global stocks of physical, human and environmental capital on a per capita basis from 1992 to 2014 as estimated by Managi and Kumar (2018). Even though this was a period of declining population growth and rising human capital investment, the rapid rise in physical capital (and the increased consumption associated with it) resulted in a steady decline in environmental (or natural) capital E (blue line in figure 2.3). This suggests that slowing population growth by itself is not enough to arrest ecological decline. This is the main message of the forthcoming book by Spears and Geruso (2025): "It would be easy to think that fewer people would be better-better for the planet, better for the people who remain. This book asks you to think again. Depopulation is not the solution we urgently need for environmental challenges, nor will it raise living standards by dividing what the world can offer across fewer of us." While I agree with their first claim, I disagree with the second. Like Marc and many other leading economists and ecologists including Dasgupta et al. (2023), I believe humanity will live better *sustainably* on a less crowded planet.

---

[10] Nerlove, Razin and Sadka (1989) studied the question of socially optimal population size in a model that includes land as a factor of production but not environmental capital E. They show that even if parents care about the utility of their offspring, equilibrium population size will not generally be optimal with respect to an intergenerational welfare function and are not even Pareto efficient with respect to the present generation. It is known that steady state competitive solutions in overlapping generations models need not be Pareto efficient, and inefficiencies compound if we include E due to the negative externality of population growth and production on the environment. This is the usual Tragedy of the Commons: individuals ignore the environmental impact of their fertility and consumption decisions, both within and across generations.

Fig. 2.3: Trends in Per Capital Physical, Human and Environment Capital

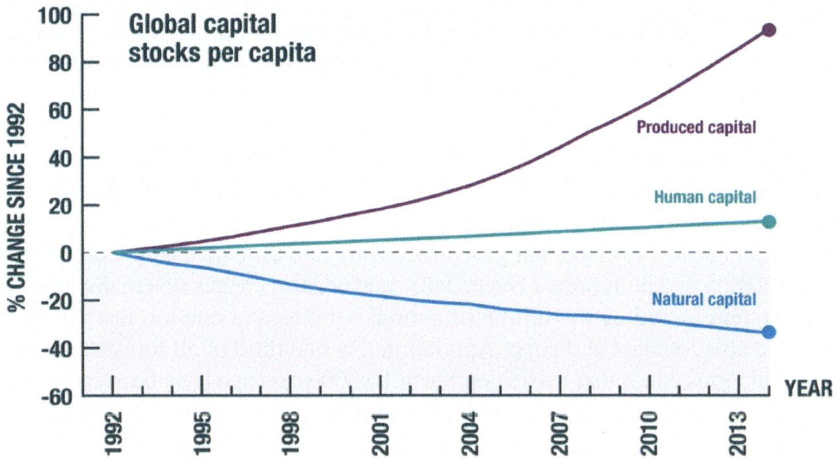

Data source (Dasgupta, Dasgupta & Barrett, 2023)

2.3.1 The Secular Decline in the Environmental Capital Stock

Marc's uni-dimensional idealization of environmental capital E (or its inverse, environmental degradation) is useful for conceptualizing how population and economic growth affect the environment, but it is challenging to define a single numerical summary of the world's environmental state in practice. The environmental state E might be better approximated by a multidimensional vector of latent factors with a set of observable indicators including average atmospheric and ocean temperature, ocean acidification, frequencies of hurricanes, droughts, floods and pandemics, concentrations of greenhouse gases such as CO_2 and methane, measures of biodiversity including fraction of land areas covered by forests and wild areas, and measures of the prevalence of harmful chemical pollutants such as some insecticides as well as plastics (and microplastics) that cannot be decomposed by nature.

Almost all of the observable indicators of the Earth's environment suggest that it is deteriorating at an accelerating rate. The indicators that get the most attention are atmospheric CO_2 concentrations and average global temperature, both of which are steeply rising. Current CO_2 levels of 427 parts per million may seem small but even small concentrations have powerful warming effects and the concentrations are skyrocketing. This seemingly small concentration amounts to over 3300 gigatons (GT) in Earth's atmosphere, an increase of 50% since the start of the Industrial Revolution. Of course, the reason CO_2 is rising so rapidly is the burning of fossil fuels that have powered the amazing growth in population and economic output worldwide over the past two centuries. The oceans are also an important 'carbon sink' that hold 60 times more carbon than the atmosphere and absorb 30% of all CO_2 emissions from human activities. Data from ice core samples show that the current atmospheric CO_2 concentration is 50% higher than the highest previous peak in the past 800,000 years.

Of course the concern about CO_2 (and equivalents) is that it is a greenhouse gas that traps solar radiation, increasing the Earth's temperature. A recent report by Hansen and et al. (2025) concludes that "Global warming has accelerated since 2010 by more than 50% over the 1970-2010 warming rate of 0.18°C per decade." (p. 7). More than 90% of atmospheric heat is absorbed by Earth's oceans, a transfer of energy equivalent to 5 Hiroshima bombs *per second* according to the Bulletin of the Atomic Scientists Nuccitelli (2020).[11]

There is plenty of evidence that global warming and climate change, combined with deforestation and pollution of rivers, lakes and oceans, creates severe disruptions to the ecosystem including disruption of animal habitats at a rate too fast for most species to be able to adapt and cope. Approximately one third of all forests and over half of all wild grasslands that existed on Earth 10,000 years ago have been converted into agricultural land, and two thirds of that is used for grazing of livestock which are an important source of greenhouse gases as well as inefficient in terms of delivery of food calories for human consumption. The loss of wild habitat has, unsurprisingly, lead to a significant reduction in biodiversity. Kolbert (2024) documents what she calls the 'The Sixth Extinction' due to the vast expansion of human impact on the environment from the population explosion and exponential growth in economic activity. She estimates that the extinction of fauna and flora alone constitute between 20 and 50% of all living species on Earth.

Humanity is increasingly feeling the cumulative effects of its exploitation of the environment in the form of reduced food production due to increasingly severe droughts and floods, depletion of fish stocks due to loss of ocean habitat from ocean acidification and heating causing widespread destruction of coral reefs and destruction of spawning grounds, and destruction of physical capital from hurricanes and wildfires.

In an article on one of the last remaining hugely valuable public service websites that the Trump administration has not yet managed to shutter, `climate.gov`, (Smith, 2025) notes that since 1980, the U.S. has sustained 403 weather and climate disasters for which the individual damage costs reached or exceeded $1 billion. The cumulative cost for these 403 events exceeds $2.915 trillion. The frequency and severity of these disasters is increasing. Bhola, Hertelendy, Hart, Adnan and Ciottone (2023) find "strong and increasing correlations between temperature and CO_2 levels, and with the economic cost of disasters in the US. Furthermore, the strength of the correlations seem to be increasing with increases in temperature and CO_2 levels over time. We highlight that the economic impact of natural disasters in the US is staggering, tallying

[11] This level of warming we are experiencing seems inconsistent with the *Gaia hypothesis* "that the Earth's surface is maintained in a habitable state by self-regulating feedback mechanisms involving organisms tightly coupled to their environment" (Lenton, 2002). Though humanity is indeed an organism, the open question is whether humanity collectively is self-regulating, or whether its unchecked activities will heat the Earth's surface to a point where it is no longer habitable for most species, including itself.

over $2.1 trillion over the last 42 years." The website `USAfacts.org` documents how costs of suppressing and fighting wildfires has risen over time, averaging $3.0 billion in the last five years. Worldwide, the amount of land subject to wildfires increased from 2 million hectares in 2001 to over 10 million in 2023, see (USAFacts, 2025). Thawing of permafrost in arctic areas and drying of peat bogs are releasing huge quantities of methane and CO_2 in a dangerous global warming feedback loop. Finally, it is well documented that almost all glaciers around the world are in retreat and many have melted completely. Another hugely valuable government resource that has somehow managed to avoid being shut down by the Trump Administration so far, `NASA.gov`, documents that each year approximately 150 billion tons of Antarctic ice melts and Greenland loses 270 billion tons, see (NASA, 2025).

Much about the complex ecological/climate balance on Earth is still unknown, including how ocean overturning circulations such as the Gulf Stream in the Atlantic are affected by global warming. Glacial melt (such as from the Greenland ice sheet) could disrupt these circulations, affecting the climate in new, unexpected ways. Of special concern is the possibility that at some point the world will reach a climate tipping point. Hansen and et al. (2025) note that "Tipping points are a big concern in popular and scientific discussion of climate change. The most dire belief is that today's accelerated warming is a sign of runaway feedbacks that are pushing climate beyond multiple tipping points, thus causing global warming acceleration that threatens eventual collapse of civilization." This report states that "The greatest climate threat is probably the danger of the West Antarctic ice sheet collapsing catastrophically, raising sea level by several meters and leaving the global coastline in continual retreat for centuries. The West Antarctic ice sheet is vulnerable to collapse because it is a marine ice sheet sitting on bedrock hundreds of meters below sea level" (p. 25).

2.3.2 Reducing the Human Ecological Footprint

The evidence provided above is certainly cause for alarm, or at least for deep concern, but is it proof Earth is headed towards an ecological disaster? Perhaps it depends on how we define 'disaster' – we are certainly seeing an increasing frequency and severity of localized disasters such as hurricanes, floods, droughts, forest fires but it is less clear whether, even if we were to reach a climate tipping point, that humanity will end up completely destroying itself in a 'Seventh Extinction'. It seems more plausible and possible that humanity will respond and take action to avert a widespread ecological disaster, though so far it seems to have dragged its heels and the evidence above suggests we are running out of time.

What are humanity's options? I see three main ways to stop or at least slow the depreciation of the environmental capital stock, E: 1) reduce population (or rate of population growth), 2) reduce per capita output and consumption (or its rate of growth), or 3) develop new technologies that reduce the amount of environmental damage caused by production and consumption activities and/or assist nature's ability to regenerate. Let's focus on CO_2 as the primary cause of environmental degradation, global warming. The following simple equation can be viewed as production function for the net annual global flow of CO_2 equal to the difference between man-made production of CO_2 (principally via burning of fossil fuels) less removal of CO_2 by a combination of natural processes was well as technology:

$$\text{Inflow of } CO_2 = n \times \left(\frac{y}{n}\right) \times \left(\frac{\xi}{y}\right) \times \left(\frac{CO_2}{\xi}\right) - \text{removal of } CO_2. \tag{2.1}$$

In Equation (2.1) n denotes world population, y is world GDP (sum of consumption and investment), and ξ denotes the total energy required to produce world output y. The final term is the sum of natural absorption of CO_2 by plants via photosynthesis (net of respiration of CO_2 by humans and animals and carbon released from organic decay and other sources), inflows into the atmosphere and oceans and other carbon sinks, plus any additional technological reduction in CO_2 through various types of sequestration. Examples of the latter include CO_2 scrubbers or 'vacuums' that store atmospheric CO_2 deep underground, artificial synthesis of carbohydrates via man-made chemical reactions, or promoting natural carbon capture in oceans by adding alkaline compounds such as calcium or magnesium hydroxide that "raise the pH in the surrounding seawater, triggering a chemical reaction that will absorb CO_2 from the atmosphere and convert it to bicarbonate, an ion that can float through the ocean undisturbed for millennia" Cornwall (2023).

As Wagner and Weitzman (2015) pointed out, the primary damage to climate and environment is due to the *stock* of CO_2 in the atmosphere and oceans. Left to natural processes, these accumulated stocks take hundreds or even thousands of years to dissipate, even if human output of CO_2 were to fall to zero. Thus, even if all man-made fossil fuel emissions were to cease immediately, it would take hundreds if not thousands of years for CO_2 to be gradually absorbed by plants and oceanic plankton and other natural carbon sinks before the temperature on Earth would reduce by 1.5°C to its average value prior to the Industrial Revolution. Currently, the gross amount of man-made CO_2 emissions is about 40 Gt annually, but netting out the removal of CO_2 the net addition to the atmosphere and oceans is about 36.8 Gt per year.

The goal of reducing the left-hand side of equation 2.1 to zero is referred to as *net zero*. Under the Paris Climate Accords, just to achieve the goal of keeping global warming limited to 1.5°C above pre-Industrial levels requires "conventional mitigation techniques" that gradually reduce CO_2 emissions from the current value of 36.8 Gigatons/year to 0 by 2050. Actually reducing the huge concentration of CO_2 in the air and oceans would require decades of *negative* CO_2 *emissions* via various carbon removal technologies. Absent some amazing technological innovation, it seems quite unrealistic that the Earth could achieve even the net zero goal by 2050 without major reductions in our standard of living.

Equation (2.1) suggests that reducing world population n should have a powerful impact on emissions. But given current world average CO_2 emissions of 4.8 tons per person, reducing world population by 1 billion (an impossibility in the short run) only reduces net CO_2 by 4.6 Gt, which still puts us far away from the goal of net zero.

Meaningful reductions in total CO_2 emissions seems to require truly Draconian cuts in output per capita *by citizens in the richest countries in the world.* As is well known, there is a strong income gradient in CO_2 emissions: it is a 'luxury good'. An average US citizen contributes 15 tons per year, whereas the average Chinese citizen emits nearly 9 tons, Europeans emit about 7, Sweden about 3.6, South America 2.6, Africa 1, and the lowest-income countries 0.3 tons per capita (Ritchie, Rosado & Roser, 2023). According to the World Bank the 26 lowest income countries contribute only 3.5% of global CO_2 emissions, whereas the US alone produced nearly 14% of total CO_2 emissions. See also Cozzi, Chen and Kim (2023) who show that the world's top 1% greenhouse gas emitters (mostly the rich) emit 1000 times more than the lowest 1% of emitters.

It follows that economic growth, far more than population growth, will be the most important contributor to growth in CO_2 emissions in coming decades. Even though technological improvements are reducing CO_2 emissions per capita, these reductions are happening too slowly to arrest the strong overall growth in atmospheric CO_2

> "So, what does the data tell us? It shows that all is not well in the state of the atmosphere! In order to prevent further warming, the carbon dioxide levels must not grow any further. On the growth curve, this corresponds to the curve having to settle down to 0 ppm/y. There is absolutely no hint in the data that this is happening. On the contrary, the rate of growth is itself growing, having now reached about 2.68 ppm/y the highest growth rate ever seen in modern times. This is not just a 'business as usual' scenario, it is worse than that, we're actually moving backward, becoming more and more unsustainable with every year. This shows unequivocally that the efforts undertaken so-far to limit greenhouse gases such as carbon dioxide are woefully inadequate" (Rasmussen, 2025).

Dasgupta et al. (2023) calculate the size of an ecologically sustainable steady state population under the assumption of equally distributed income at the international price level in 2001. They find that "if humanity were to find ways to husband the biosphere in a sustainable manner and to bring about economic equality, the human population Earth could support at a living standard of 20,000 dollars is approximately 3.3 billion." They also consider what equal standard of living would be to sustain a population of 9 billion people: per capita income could only be 11,480 dollars. They note that "If inequality in the distribution of incomes was judged to be inevitable, the figure would be even smaller."

Further empirical evidence on the massive cuts in standard of living that would be required to put the planet on a path to net zero is provided by Liu and et al. (2022) who analyze the impact of the COVID-19 pandemic on CO_2 emissions. They find that the severe cutback in activity and economic dislocations in the first year of the pandemic, 2020, resulted in a 6.3% reduction in global CO_2 emissions compared to 2019. They conclude that "The extraordinary fall in emissions during 2020 is similar in magnitude to the sustained annual emissions reductions necessary to limit global warming at 1.5°C. This underscores the magnitude and speed at which the energy transition needs to advance".

In the early 1990s Mathis Wackernagel and William Rees introduced the concept of *ecological footprint* to quantify humanity's demand on Earth's ecosystem. They defined the 'demand for biocapacity' as

> "the aggregate area of land and water ecosystems required by specified human populations to produce the ecosystems goods and services they consume and to assimilate their carbon wastes. Footprint accounting is thus based on the premise that the regenerative capacity of the ecosphere is associated with productive ecosystem area. The production of food and fibre; the urbanization of once agricultural or forested lands; and the sequestration of that portion of carbon emissions from fossil fuels that is not already absorbed by oceans or by long-term sequestration strategies in agriculture or forestry, all constitute competing or non-overlapping uses of ecosystems. (Typically, one cannot simultaneously use paved-over land for food production or forest products; today's cropland and commercial forests are usually carbon sources, not sinks). We estimate and sum these separate areas to estimate study populations' total Ecological Footprints" (Rees & Wackernagel, 2013).

Thus, the ecological footprint can be used as a measure of how rapidly humanity is depleting the Earth's environmental capital stock, i.e., the 'excess demand' for Earth's resources. In 2013, they estimated that "that Earth's biocapacity in 2008 was 12 billion hectares (ha) compared to humanity's Footprint of 18.2 billion ha, and that the average Ecological Footprint had reached 2.7 global hectares (gha) per capita compared to only 1.8 gha of available biocapacity per capita". According to the most recent estimates, we would need about 1.7 Earths to satisfy the demands humanity is placing on it in a sustainable manner.

To summarize the main takeaways from this section: 1) the combination of the population explosion and exponential growth in output and consumption per capita has resulted in unsustainable demands on Earth's ecosystem, severely depleting its 'environmental capital' though whether and when this will lead to an 'ecological disaster' is hard to predict, 2) absent a 'silver bullet' technological solution, humanity faces very unpleasant choices if it tries to reach a sustainable level of demand for Earth's resources (e.g., net zero emissions) even by 2050, 3) reducing population is not a realistic policy option, and even if it were, it is far from enough to achieve a sustainable outcome, 4) sustainability requires massive reductions in output per capita, primarily by the richest countries in the world that cause the overwhelming share of environmental damage, and 5) even with technological progress and economic growth, sustainable outcomes involve a tradeoff between population size and living standards, just as Marc had anticipated.

Thus, the answer to the question raised at the start of this section is that slower population growth is not enough to avert ecological disaster. Absent a technological solution that dramatically reduces humanity's ecological footprint, severe reductions in economic growth – indeed negative growth (or 'degrowth') — will be required to reduce living standards and the implied demands on the environment by the richest countries by enough to achieve a sustainable long run outcome.

2.4 Will Technological Progress Avert Ecological Disaster?

It is hard to deny how amazing human ingenuity and technological is, so we can hope that it will lead to breakthroughs that enable production and income and consumption per capita to keep growing rapidly (potentially without bound) as Tupy and Pooley

(2022) and others have argued, while avoiding an ecological disaster, or even more optimistically, without further major environmental damage to the planet. There are three main ways that technology could do this: it could 1) reduce the amount of energy required to produce any given level of output, 2) reduce the amount of CO_2 and other environmental side effects from energy production, and 3) use bioengineering to improve nature's ability to absorb and process wastes. By assisting Mother Nature in her ability to recycle waste and regenerate despite the increasing demands humanity places on her, humans would in effect be tinkering with the ecology and changing it to make it more to their liking.

But whether this is really possible without huge unintended side effects is a really important open question. It is clear humanity has changed the Earth and its ecosystem — that's why this era is called the *Anthropocene* — but so far the evidence in the last section is that these changes have not been for the better, at least as far as all other non-human species facing the Sixth Extinction are concerned. We can hope that growing scientific knowledge can help mankind to 'design' a new 'artificially assisted' ecosystem, just as it has used its ingenuity to create artificial intelligence. But at this point, mere hopes seem more like science fiction. What are the most immediate promising areas where technology can help avert a rapidly approaching climate crisis, and perhaps ecological disaster?

In my opinion, the most immediate crisis facing the planet is global warming caused by uncontrolled greenhouse gas emissions. So the area where technology could have the biggest immediate impact is by reducing the amount of CO_2 and other greenhouse gases per unit of energy generation. According to the World Resources Institute "The energy sector produces the most greenhouse gas emissions by far, accounting for a whopping 75.7% worldwide. The energy sector includes emissions from electricity and heat (29.7% of all emissions), transportation (13.7%), manufacturing and construction (12.7%) and buildings (6.6%)" (Ge, Friedrich & Vigna, 2024). One area where technology has had huge demonstrated success is the rise in solar power from solar photovoltaic (PV) cells. According to the International Energy Agency (IEA), "the cost of electricity generated from solar panels (or solar PV) has fallen dramatically in recent decades. This has contributed to a boom in solar PV deployment, with global capacity now growing at a historic pace. From 2018 to 2023, it tripled. The electricity sector remains the brightest spot for renewables with the strong growth of solar photovoltaics and wind in recent years, building on the already significant contribution of hydropower. But electricity accounts for only a fifth of global energy consumption and finding a greater role for renewable energy sources in transportation and heating remains critical to the energy transition" (International Energy Agency, 2024).

It is clear that we have a long way to go before solar can supplant fossil fuels (coal, natural gas, fuel oil, etc) to power electricity generation: in 2023 it only produced 5.5% of the world's electricity. Though studies show that in principle intermittent renewal sources such as solar and wind power could supply 80% of all electricity, "However, to reliably meet 100% of total annual electricity demand, seasonal cycles and unpredictable weather events require several weeks' worth of energy storage and/or the installation of much more capacity of solar and wind power than is routinely necessary to meet peak demand" (Shaner, Davis, Lewis & Caldeira, 2018). However, as we know, there has been tremendous progress on battery storage technology. According to the IEA battery storage for electricity generation doubled in 2023 alone, and it predicts that total capacity will increase 10-fold to 800 GW by 2030.

Solar is a technological success story and a tremendous reason for optimism. Another technology that could be a game-changer for greenhouse-free electricity generation is fusion power. Fusion tokamak chambers are essentially magnetic bottles that keep the extreme heat of nuclear fusion from melting the generator, but this has required more energy than what the fusion generates. Improved design of fusion reactors have improved over the last decade to the point that many can produce a net positive amount of electricity. However, "Most experts agree that we're unlikely to be able to generate large-scale energy from nuclear fusion before around 2050 (the cautious might add on another decade)" (Ball, 2023) so the consensus is that it will not arrive in time to avert a climate crisis.

Besides electricity generation, I noted that transportation, manufacturing and the heating/cooling of buildings is the next largest source of greenhouse gas emissions. Due to falling costs, improved battery capacity and more charging stations, there has been strong growth in the number of electric vehicles, with over 40 million plug-in electric passenger vehicles on the road, half of which are in China. However, it is less clear whether battery powered jets are feasible for international and other long distance travel, though there has been a proliferation of battery powered drones and electrically powered aircraft for short range travel seems on the horizon. For long distance air travel, there have been technological improvements in the synthesis of sustainable aviation fuels (SAF), including via biofuels and the hydrolysis of water that releases hydrogen that can be combined with CO_2 from the atmosphere to synthesize hydrocarbons, including various types of diesel and jet fuels. Though use of SAF is negligible currently due to its relatively high cost, it could eventually become another technological innovation helping the planet reach net zero.

Hydrogen fuel cells are another promising source of 'green electricity' for powering cars and trucks that avoids the environmental disposal problems of batteries: they combine hydrogen and oxygen and their only emission is water vapor (though this is also a greenhouse gas, it is far less potent than CO_2 or methane). Though they are not yet economically viable for use at large scale, fuel cells are highly efficient, converting between 40% to 60% of the chemical energy in hydrogen into electrical energy, which is significantly more efficient than combustion engines that operate at about 25% thermal efficiency.

There are numerous other areas where technological improvements have significantly reduced the amount of greenhouse gases emitted per unit of energy generation. Light emitting diodes (LED) have been a breakthrough in lighting that have dramatically reduced the energy required per lumen, by 75% compared to incandescent bulbs, and last 25 times longer. Other promising emerging technologies include 'green cement' that can be produced with up to 70% fewer greenhouse gas emissions compared to regular cement.(MIT Department of Materials Science and Engineering, 2025).

There are other technologies that could potentially provide backstop or failsafe options if technologies like the ones discussed were unsuccessful. One such technology is *geoengineering* that includes spreading aerosols high into the atmosphere to reflect solar radiation (much as clouds or volcanic ash already do) to cool the Earth. Wagner and Weitzman (2015) discussed the example of the explosion of Mount Pinatubo, where "20 *million* tons of sulfur dioxide managed to wipe out the global warming effects of 585 *billion* tons of carbon dioxide in the atmosphere." Given this huge leverage, they note that geoengineering would be 'cheap' but only in the "narrow sense of the direct engineering costs of transporting 20 million tons of material into the stratosphere, not necessarily cheap when looking at the full consequences."

I could go on to list dozens of other promising new technologies that can either a) reduce the energy required per unit of output or services produced, or b) reduces the level of greenhouse gases per unit of output/services produced, and per unit of energy produced. This corresponds to potentially large reductions in the ratios ξ/y and CO_2/ξ in the simplified CO_2 production function in Equation (2.1). However, as I noted in the previous section, global warming is accelerating at an alarming pace.

Can these promising new technologies be rolled out and scaled up rapidly enough to avert the worst damage from the climate crisis, and avoid sending the planet beyond a tipping point of no return? I don't think anyone knows the answer to this question, but given the dire consequences we are already experiencing and the danger of passing a tipping point, it would be prudent to follow a combined strategy of 1) subsidizing 'green R&D' and production of green technology such as solar, and 2) taxing or reducing output from fossil fuels even if this implies a temporary reduction in standard of living. If we rely only on option 1), it seems to me that humanity is playing a very risky game, hoping that technological breakthroughs can happen 'just in time' to avert major ecological damage and human suffering.

2.5 What Policies Can Mitigate the Tragedy of the Commons?

Marc Nerlove (1991) warned of the delicate balance between quality of the environment and population growth (and by extension economic growth). His model predicted the possibility of unstable dynamics that could potentially lead to environmental disaster or even human extinction. Marc acknowledged that his simple model ignored technological progress, consumption and savings, and investments in productive and human capital. Extending his model to incorporate these features "would result in far more optimistic conclusions."

My review of the evidence suggests that Marc's optimism may have been misplaced. Despite amazing technological progress and rapid growth in green technologies such as solar power, CO_2 growth continues unabated, global warming is accelerating, and average temperatures have broken through the 1.5°C limit on global warming relative to pre-Industrial times that the Paris Climate Accords determined was necessary to mitigate the worst impacts of climate change. So should we be complacent and hope that a series of miraculous technological breakthroughs will turn things around, or should we try to take additional actions to mitigate these risks?

Marc's answer is that mitigation is possible, provided there is a government or other type of 'social intervention' that can impose taxes and subsidies.

> "Provided the environment can eventually recover from the effects of excessive population growth, such a system of taxes and subsidies could be used to achieve and maintain any specified birth rate. In particular, social intervention could induce parents to determine their fertility in order, first, to reach a stationary solution–or example, the one with the lower level of population and better environment-and then to maintain that equilibrium despite its local instability."

Of course, this is economists' preferred solution: use Pigouvian taxes and subsidies to deal with environmental externalities and then allow the decentralized operation of a competitive economy to take things to a better outcome. This logic applies equally well in models with production and atmospheric pollution and was endorsed as the best solution to the climate crisis by Wagner and Weitzman (2015) in their book *Climate Shock*

> "Far from posing a fundamental problem to capitalism, it's capitalism with all its innovative and entrepreneurial powers that is our only hope of steering clear of the looming climate shock. That's not a call for letting markets run free. *Laissez-faire* may sound good with the right French accent—in theory. But it can't work in a situation where prices don't reflect the true costs of our actions. Unbridled human drive—erroneously bridled drive, really—is what has gotten us into this current predicament. Properly channeled human drive and ingenuity, guided by a high enough price on carbon to reflect its true cost to society, is our best hope for getting us out."

But the presumption of a benevolent, well-functioning and all powerful government that can impose taxes and subsidies seems naive or at least unrealistic. Remember that problems like climate change are *global* and require *global cooperation* by *many if not all governments around the world* to be successful. But this degree of coordination is not happening. How well are the Paris Climate Accords working out now that Donald Trump is President of the United States, forcefully pushing his "Make America Great Again" agenda? Not only has he withdrawn the US from the Paris Accords, he's imposed tariffs on imports of green technology such as solar panels and electric cars from the world's biggest producer of them, China. He is pursuing an unabashed strategy of "drill baby drill" in order to accelerate the burning of fossil fuels and increase America's energy primacy, and removed subsidies for electric vehicles and charging stations and tax breaks for solar panels. Just for good measure, he has shut down USAID and withdrawn from the World Health Organization, putting the poor (especially in vulnerable areas such as Africa, Gaza and other places) at higher risk to malnutrition and starvation and making the world more susceptible to pandemics.

If we cannot expect world governments to cooperate and act in a rational, benevolent manner, it seems even less likely that a 'grassroots' movement to save the environment can be successful. As Wagner and Weitzman (2015) note, "Voluntary coordination is out. Getting seven people to agree on anything is tough; getting seven billion to agree is impossible. That's where governments need to come in, and even there we find global cooperation very difficult."

Global cooperation is another unstable equilibrium, easily toppled. It depends on having strong and wise leadership, but we can't look to Trump to provide it. Instead, he leads the world backwards, erasing hard-won progress already made. He undermines cooperation by asking why the US should tax its CO_2 emissions if China and Russia are not doing the same. This gives license to China and Russia to ask the same question about the US and potentially withdraw from the Paris Accords as well.

We might hope that support for global cooperation might build as the climate crisis becomes more and more severe and more people are directly affected by it and more keenly aware of the ever-growing consequences of continuing to ignore the Tragedy of the Commons. However, humanity collectively so far has behaved in an *extremely myopic fashion* and for the most part seems unwilling to make even minor sacrifices in current consumption in order to make investments to mitigate environmental damage that will reduce consumption of future generations. For example in the US the federal tax on gasoline has remained at just 18 cents per gallon *since 1993.* There would almost certainly be intense voter opposition to any proposal to increase it that reflects anything close to the true social cost of carbon. This sort of behavior is a recipe for 'learning the hard way' i.e., continuing to delay and deny, postponing or refusing to take any costly actions to disincentivize greenhouse gas pollution.

Each year, it is getting more and more obvious to the public that global warming is real, and millions around the globe are being harmed by it. But deprivation, destitution, and destruction does not often bring out the best in human nature. Instead of collective action to remedy the situation, it seems more likely to lead to conflict over the dwindling natural resources and competition for the ability to live in the remaining parts of the world that are less affected by environmental collapse. Thus, while the numbers are hard to predict, we can expect a surge in 'climate migration' in the not too distant future. But there is already a profound lack of cooperation on immigration policy. It seems more likely that countries will turn inward and defend their borders than take costly joint actions such as a world-imposed carbon tax with transfers to compensate the least well off for the higher cost of energy (and all that depends on it). Failure to act sooner rather than later and help poorer countries in the regions most affected by global warming will only exacerbate future levels of climate migration.

Marc recognized the instability of collective action and cited the 'justly famous' Tragedy of the Commons of Hardin (1968). Though in theory it can be solved through government taxation or other binding and strongly enforced social arrangements, there is a regress: the problem of how governments come to agree, or social arrangements come into effect simply opens up the Tragedy of the Commons at a higher level. Despite these challenges, collective action *is* possible. Over 27 countries around the world (including the European Union) have imposed domestic carbon taxes, and over 190 countries have signed on to the (non-binding) Paris Climate Accords. Over 190 countries have signed on to the United Nations Framework Convention on Climate Change, and one of its provisions has been the creation of carbon offset markets.

So there is reason for hope that sufficiently many countries around the world are willing and able to cooperate to adopt policies to limit greenhouse gases and take other actions to arrest the destruction of the environment. But it is crucial to adopt policies that result in meaningful, measurable reductions in CO_2 and for which a high benefit/cost ratio can be documented to the public. Economics can play an important role in this regard, providing analyses that can help the world to avoid adopting ineffective policies or worse, counterproductive ones. For example, Chen, Ryan and Xu (2024) studied the impact of the carbon offset market in China, developed under the Clean Development Mechanism of the Kyoto Protocol. A carbon offset is a payment by a polluting firm (the buyer of the offset) to another firm (the seller) to undertake its own investment to reduce CO_2 pollution on behalf of the buying firm. In theory, these trades should reduce CO_2 emissions by financing investments at firms that have a comparative advantage in mitigating CO_2. However, Chen et al. (2024) find that in China the offset actually *raises* CO_2 emissions: "We find that offset-selling firms increase carbon emissions by 49% in the four years after starting an offset project, relative to a matched sample of non-applicants."

Economic models are also highly influential in policy circles for determining an appropriate 'social cost of carbon' which is a starting point for setting carbon taxes. For example, the Nobel Prize winner Nordhaus (2017) developed a so-called DICE model (Dynamic Integrated Model of Climate and Economy) and calculated a social cost of CO_2 emissions of $31 per ton, and projects that due to continued global warming, this amount will grow by 3% per year until 2050. However, Nordhaus's calculations have come under criticism for grossly underestimating the CO_2 externality. For example, Saitō (2020) notes that "The problem lies with the optimal measures he proposes in his paper. To combat climate change, it is imperative that greenhouse gas emissions decrease. On the other hand, if emissions reduction goals are set too high, it might hinder economic growth. Therefore, he asserts, what we need is 'balance.' But in my view, Nordhaus's proposed 'balance' leans much too far toward the side of economic growth."

Nordhaus acknowledges that his calculations ignore several important factors such as the loss of biodiversity, extreme events (e.g., sea level rise and impact on ocean circulation) and catastrophic events. The DICE model also ignores endogenous responses of population to economic growth and environmental degradation. Lupia and Marsiglio (2021) develop the DICED model, i.e. combining DICE with endogenous demographics. They find that "accounting for endogenous population change substantially increases the estimates of the social costs of environmental policies (measured by both the social cost of carbon and social welfare)" and that fertility policies (such as policies that incentivize female education at the cost of fertility), can reduce the cost of climate change by as much as 16% by limiting population size and the total human ecological footprint.

Neal, Newell and Pitman (2025) showed that the estimated social cost of carbon from models such as DICE are sensitive to assumptions used in underlying econometric models used to predict how weather and climate shocks affect production. "A key assumption inherent in existing econometric models is that a country's economic growth is only related to its own weather shocks, whereas those of their neighbors, trading partners, and the rest of the world are left in the error term." The assumption that economies are unaffected by weather shocks in other countries "causes a mischaracterisation of global weather shocks. Generalising existing models leads them to predict catastrophic damages, where all countries are affected to different degrees." Using an adjusted 'climate damage function' their model predicts much more substantial reductions in GDP per capita due to increases in global mean temperatures compared to the DICE2023 model as shown in figure 2.4.

Fig. 2.4: Climate 'damage functions' – DICE vs Neal et al.

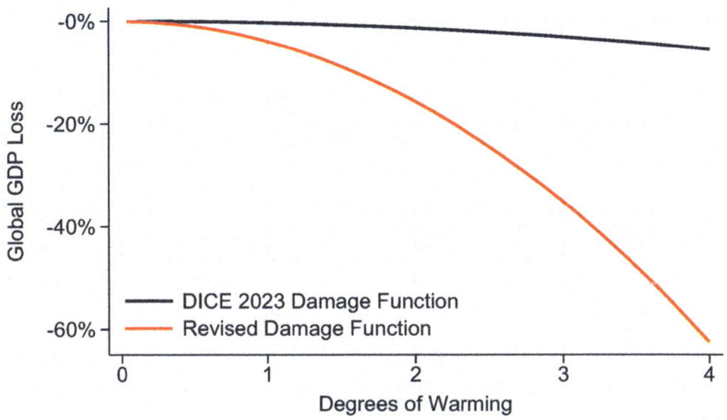

Data source: Neal et al. (2025)

Of course, there are huge uncertainties about predicting the future and how much global warming could reduce GDP, leaving alone the question of valuing the damage to the environment. Perhaps it is not productive to quibble about the precise numbers, but rather to agree that there will be serious damage to the world economy from allowing continued depreciation in the stock of environmental capital. Since the world has already exceeded the target of no more than 1.5°C warming under the Paris Accords, we need to accept that continued global warming is *fait accompli* (or as Wagner and Weitzman (2015) say, it's "baked in") and consider 2nd, 3rd and 4th best damage mitigation strategies that can be undertaken *quickly and unilaterally.*

Examples of such investments include hardening infrastructure (undergrounding power and communications cables against fires and storms), securing food and energy supplies (including storage facilities and avoiding over-dependence on imported food and energy), and diversifying economies to make them less dependent on critical imports and more resilient in the face of interruptions in global supply chains. Since drought is a growing problem in many parts of the world, more investment should be made in desalination plants, and in growing more of the food supply using green houses and low water drip agriculture. Subsidies should be provided to farmers to incentivize them to convert more of their grazing lands to agriculture, which can significantly raise the total caloric output from land allocated to livestock, allowing more land to be reforested. Payments should be increased to countries with large rainforests and other underdeveloped natural resources to incentivize them to preserve these resources and protect biodiversity. Investments should be increased in biotechnology to more rapidly fight future pandemics, and to increase capacity for synthesizing foods in the event of sustained drought. For example, more could be invested to produce food and biofuels from massive floating islands of seaweed such as the Great Atlantic Sargassum Belt.

However, there are technologies we may not *not* want to invest in. One is geoengineering, i.e., injecting sulfur dioxide into the atmosphere as aerosols to reflect solar radiation, similar to the effect of cloud cover or volcanic ash. I share the reservations expressed in Wagner and Weitzman (2015) of the potential unintended consequences of geoengineering including reduced food output and reduced solar radiation to power solar cells. Also, if countries believe this is an effective option they can undertake unilaterally and at relatively low cost, it reduces the incentive to take stronger, less risky but potentially more costly measures today. I also question whether the expansion of battery power creates more environmental problems than it solves. The environmental problems arise from the difficulty of disposing and recycling use batteries and other e-waste, and environmental damage from mining the rare earths needed to produce lithium-ion and other high-performance batteries. Proposals to scrape the ocean floor for poly-metallic nodules needed to meet the rapidly growing demand for batteries seem to pose especially poorly understood risks to the environment. It would be better to invest in improving the cost-efficiency of 'green hydrogen' which is a much more environmentally friendly way to power electric vehicles and store energy than batteries: a closed loop process of splitting water into hydrogen and oxygen, then combusting them to produce electricity and water vapor as the exhaust product.

2.6 Conclusion

Economics is known as the 'dismal science' largely due to the bleak predictions of Thomas Malthus that real wages cannot increase because any overall productivity gain will be offset by a commensurate increase in the population. Malthus's prediction of flat real wages turned out to be flat wrong: total population and real wages both increased exponentially in the two centuries following his death. However, predictions that population and real wages can continue to increase exponentially for the indefinite future could turn out to be equally wrong. Population growth has been decreasing for decades, and total population is predicted to peak before 2100. The real question is whether real wages can continue to grow exponentially for the indefinite future. Those who argue that we can go on with business as usual and count on technology to solve our environmental challenges (as do Tupy & Pooley, 2022, for example), remind me of something Kenneth Boulding said in testimony before Congress: "Anyone who believes that exponential growth can go on forever in a finite world is either a madman or an economist."

Marc realized that humanity can live beyond its means for extended periods of time only by depleting its environmental capital stock. If humanity manages to exhaust its endowment of environmental capital, Mother Nature will impose a harsh budget constraint and real wage growth could suddenly halt or even decline. Billions of people could suffer and millions could die. Malthus's predictions would suddenly be relevant again.

I certainly hope this is not the case. Picking up on Marc's ideas, I have surveyed decades of work in climate change, ecology, and environmental and resource economics that paint a very uncertain picture for the planet and the future of humanity. Due to the Tragedy of the Commons, humanity behaves myopically, as if there were no tomorrow. An alternative rationalization for our inaction is that we are sure that some brilliant new technology will come along just in time to rescue us from any potential climate disaster without any major sacrifice on our part.

Though the future is highly uncertain and one should never rule out the emergence of amazing new 'game-changing' technologies, it seems to me that we are following a very risky path of being unwilling to sacrifice current consumption to increase the precautionary investments that will reduce our ecological footprint. It's as if humanity is too shortsighted to buy fire insurance and unwilling to admit that it might later regret its choice when its house burns down. Of course, it's not as simple as that, and while we each may care deeply about the welfare of the planet and our descendants, the problems of collective action underlying the Tragedy of the Commons prevents meaningful precautionary investments and mitigating actions from being undertaken.

I am not sure whether Marc would agree with my interpretation of the literature and thinking on population and the environment, but I do believe that he would support informed discussion that raises awareness and allows people to express their own views about an uncertain future, even if some are dismal ones that risk bumming people out. Marc would probably agree with Hardin (1968) who noted that "The individual benefits as an individual from his ability to deny the truth even though society as a whole, of which he is a part, suffers. Education can counteract the natural tendency to do the wrong thing, but the inexorable succession of generations requires that the basis for this knowledge be constantly refreshed."

References

Abramitzky, R. & Braggion, F. (2003). Malthusian and neo-malthusian theories. In J. Mokyr (Ed.), *The oxford encyclopedia of economic history* (p. 423-427). Oxford University Press.

Ahlburg, D. A. (1998). Julian simon and the population growth debate. *Population and Development Review*, *24-2*, 317-327.

Ball, P. (2023). What is the future of fusion energy? *Scientific American*, May 2nd issue.

Barrett, S. & et al. (2020). Social dimensions of fertility behavior and consumption patterns in the anthropocene. *Proceedings of the National Academy of Sciences*, *117-12*, 6300-6307.

Bavel, J. V. (2013). The unrealized horrors of population explosion. *Facts, Views and Vision in ObGyn*, *5-4*, 281-291.

Bhola, V., Hertelendy, A., Hart, A., Adnan, S. B. & Ciottone, G. (2023). Escalating costs of billion-dollar disasters in the US: Climate change necessitates disaster risk reduction. *The Journal of Climate Change and Health*, *10*, 1-6.

Bouscasse, P., Nakamura, E. & Steinsson, J. (2025). When did growth begin? new estimates of productivity growth in England from 1250 to 1870. *Quarterly Journal of Economics, forthcoming*.

Callaway, E. (2024). Chemistry Nobel goes to developers of Alphafold AI that predicts protein structures. *Nature, October 9*.

Casey, G., Shayegh, S., Moreno-Cruz, J., Bunzl, M., Galor, O. & Caldeira, K. (2019). The impact of climate change on fertility. *Environmental Resource Letters*, *14*, 1-9.

Chen, Q., Ryan, N. & Xu, D. Y. (2024). Firm selection and growth in carbon offset markets: Evidence from the clean development mechanism in China. *manuscript, Duke University*.

Cornwall, W. (2023). An alkaline solution. *Science*, *382-6674*, 988-992.

Cozzi, L., Chen, O. & Kim, H. (2023). The world's top 1% of emitters produce over 1000 times more co2 than the bottom 1%. *International Energy Agency*. https://www.iea.org/commentaries/the-world-s-top-1-of-emitters-produce-over-1000-times-more-co2-than-the-bottom-1.

Crafts, N. (2022). Slow real wage growth during the industrial revolution: productivity paradox or pro-rich growth? *Oxford Economic Papers*, *74-1*, 1-13.

Dasgupta, P., Dasgupta, A. & Barrett, S. (2023). Population, ecological footprint and the sustainable development goals. *Environmental and Resource Economics*, *84*, 654-675.

Dench, D., Pineda-Torres, M. & Myers, C. (2024). The effects of post-dobbs abortion bans on fertility. *Journal of Public Economics*, *234*, 1-25.

Ehrlich, P. R. & Ehrlich, A. H. (1968). *The population bomb*. Sierra Club/Ballantine Books.

Fertility, G. . & Collaborators, F. (2024). Global fertility in 204 countries and territories, 1950-2021, with forecasts to 2100: a comprehensive demographic analysis for the global burden of disease study 2021. *The Lancet*, *403*, 2057-2099.

Galor, O. (2005). The demographic transition and the emergence of sustained economic growth. *Journal of the European Economic Association*, *3-(2-3)*, 494-504.

Galor, O. (2022). *The journey of humanity: A new history of wealth and inequality and implications for the future*. Penguin Random House.

Ge, M., Friedrich, J. & Vigna, L. (2024). *4 charts explain greenhouse gas emissions by countries and sectors*. Retrieved from https://www.wri.org/insights/4-charts-explain-greenhouse-gas-emissions-countries-and-sectors

Gollin, D., Hansen, C. W. & Wingender, A. M. (2021). Two blades of grass: The impact of the green revolution. *Journal of Political Economy*, *129-8*, 2344-2383.

Gu, J. (2022). Fertility, human capital and income: The effects of China's One Child Policy. *Macroeconomic Dynamics*, *26*, 979-1020.

Haberman, C. (2015). The unrealized horrors of population explosion. *New York Times*, May 31 issue.

Hansen, J. E. & et al. (2025). Global warming has accelerated: Are the United Nations and the public well-informed? *Environment: Science and Policy for Sustainable Development*, *67-1*, 6-44.

Haq, S. M. A., Chowdhury, M. A. F., Ahmed, K. J. & Chowdhury, M. T. A. (2023). Environmental quality and its impact on total fertility rate: an econometric analysis from a new perspective. *BMC Public Health*, *23:2397*, 1-16.

Hardin, G. (1968). The tragedy of the commons. *Science*, *162*, 1243-1248.

Hiriscau, A. (2024). The effect of paid maternity leave on fertility and mothers' labor force participation. *Journal of Labor Research*, *45*, 350-384.

Howden, D. & Zhou, Y. (2015). Why did China's population grow so quickly? *The Independent Review*, *20-2*, 227-248.

Hutton, W. (2022). The journey of humanity review - ambitious bid to explain society's economic development. *The Guardian*, May 2nd issue.

International Energy Agency. (2024). *Solar PV.* Retrieved from https://www.iea.org/energy-system/renewables/solar-pv

Ip, G. & Adamy, J. (2024). Suddenly there aren't enough babies. the whole world is alarmed. *Wall Street Journal*, May 13 issue.

Jones, C. I. & Romer, P. M. (2010). The new Kaldor facts: Ideas, institutions, population, and human capital. *The American Economic Journal: Macroeconomics*, *2-1*, 224-245.

Kearney, M. S. & Levine, P. B. (2022). The causes and consequences of declining US fertility. In *Economic policy in a more uncertain world* (p. 73-101). Aspen Institute.

Kennan, J. (2013). Open borders. *Review of Economic Dynamics*, *16*, L1-L13.

Kolbert, E. (2024). *The sixth extinction: An unnatural history.* Henry Holt and Company.

Kurzweil, R. (1990). *The age of intelligent machines.* MIT Press.

Lenton, T. (2002). Gaia hypothesis. In J. R. Holton (Ed.), *Encyclopedia of atmospheric sciences.* Elsevier.

Liao, P.-J. (2013). The one-child policy: A macroeconomic analysis. *Journal of Development Economics*, *101*, 49-62.

Lin, Y., Zhang, B., Hu, M., Yao, Q., Jiang, M. & Zhu, C. (2024). The effect of gradually lifting the two-child policy on demographic changes in china. *Health Policy and Planning*, *39*, 363-371.

Liu, Z. & et al. (2022). Global patterns of CO_2 emissions reductions in the first year of covid-19. *Nature Geoscience*, *15*, 615-620.

Lupia, V. & Marsiglio, S. (2021). Population growth and climate change: A dynamic integrated climate-economy-demography model. *Ecological Economics*, *184*.

Malthus, T. R. (1798). *An essay on the principle of population.* J. Johnson, London.

Managi, S. & Kumar, P. (2018). *Inclusive wealth report 2018: measuring progress towards sustainability.* Routledge, New York.

MIT Department of Materials Science and Engineering. (2025). *Breaking ground with green cement.* Retrieved from https://dmse.mit.edu/research-impact/

application-impact/breaking-ground-with-green-cement/
NASA. (2025). *Ice sheets.* Retrieved from https://climate.nasa.gov/vital-signs/ice-sheets
Neal, T., Newell, B. R. & Pitman, A. (2025). Reconsidering the macroeconomic damage of severe warming. *Environmental Resource Letters, forthcoming*.
Nerlove, M. (1974). Household and economy: Toward a new theory of population and economic growth. *Journal of Political Economy*, *82-2*, S200-S218.
Nerlove, M. (1991). Population and the environment: A parable of firewood and other tales. *American Journal of Agricultural Economics*, *73-5*, 1334-1337.
Nerlove, M. & Meyer, A. (1997). Endogenous fertility and the environment: A parable of firewood. In P. Dasgupta & K.-G. Mäler (Eds.), *The environment and emerging development issues volume 2.* Oxford: Clarendon Press.
Nerlove, M., Razin, A. & Sadka, E. (1989). Socially optimal population size and individual choice. In K. F. Zimmermann (Ed.), *Economic theory of optimal population.* Springer Verlag, Berlin.
Nordhaus, W. (2017). Revisiting the social cost of carbon. *Proceedings of the National Academy of Sciences*, 1518-1523.
Nuccitelli, D. (2020). Earth is heating at a rate equivalent to five atomic bombs per second. or two hurricane Sandys. *Bulletin of the Atomic Scientists*, February 3 issue.
OECD. (2024). *Society at a glance 2024: OECD social indicators.* https://www.oecd.org/en/publications/society-at-a-glance-2024_918d8db3-en.html. OECD.
Rasmussen, C. E. (2025). Atmospheric carbon dioxide growth rate. *Universal Climate Cooperation.* Retrieved from https://mlg.eng.cam.ac.uk/carl/climate/
Rees, W. E. & Wackernagel, M. (2013). The shoe fits, but the footprint is larger than earth. *PLOS Biology*, *11-11*, 1-2.
Ritchie, H., Rosado, P. & Roser, M. (2023). CO_2 and greenhouse gas emissions. *Our World in Data.* Retrieved from https://ourworldindata.org/co2-and-greenhouse-gas-emissions
Saitō, K. (2020). *Slow down the degrowth manifesto.* Astra House, New York.
Shaner, M. R., Davis, S. J., Lewis, N. S. & Caldeira, K. (2018). Geophysical constraints on the reliability of solar and wind power in the United States. *Environment: Science and Policy for Sustainable Development*, *11*, 914-925.
Simon, J. (1981). *The ultimate resource.* Princeton University Press.
Smith, A. B. (2025). *2024: An active year of U.S. billion-dollar weather and climate disasters.* Retrieved from https://www.climate.gov/news-features/blogs/beyond-data/2024-active-year-us-billion-dollar-weather-and-climate-disasters
Spears, D. & Geruso, M. (2025). *After the spike: Population, progress, and the case for people.* Simon and Schuster.
Spears, D., Vyas, S., Weston, G. & Geruso, M. (2024). Long-term population projections: Scenarios of low or rebounding fertility. *Public Library of Science One*, *19-4*, 1-16.
Traeger, B. (2011). Poverty and fertility in india: Some factors contributing to a positive correlation. *Global Majority E-Journal*, *2-2*, 87-98.

Tupy, M. L. & Pooley, G. L. (2022). *Superabundance: The story of population growth, innovation, and human flourishing on an infinitely bountiful planet.* CATO Institute.

United Nations, Department of Economic and Social Affairs, Population Division. (2024). *UN world population prospects 2024: Summary of results (UN DESA/POP/2024/TR/NO. 9).* https://population.un.org/wpp/graphs?loc=900&type=Probabilistic%20Projections&category=Population&subcategory=1_Total%20Population.

USAFacts. (2025). *How much land do wildfires burn in the US?* Retrieved from https://usafacts.org/articles/how-much-damage-do-wildfires-do-in-the-us/

Wagner, G. & Weitzman, M. L. (2015). *Climate Shock: The economic consequences of a hotter planet.* Princeton University Press.

Wolpin, K. I. (1984). An estimable dynamic stochastic model of fertility and child mortality. *Journal of Political Economy*, *92-5*, 852-874.

Zwetsloot, R., Corrigan, J., Weinstein, E., Peterson, D., Gehlhaus, D. & Fedasiuk, R. (2021). China is Fast Outpacing US STEM Phd Growth. *Center for Security and Emerging Technology.*

Chapter 3
Re-estimating Supply Elasticities of Selected Agricultural Commodities

Felix Chan, Elizabeth L. Jackson, Richard Dwumfour, and László Mátyás

Abstract Marc Nerlove in his seminal work published in 1956 (Nerlove, 1956) explored the relevance of price expectations in agricultural production and how these may affect supply elasticities. This chapter extends the 'Nerlovian model' by taking into account some recent developments in panel data econometrics, volatility modelling and data availability. These new models are then estimated and tested using some FAO data sets. It turns out that although these fresh results shed a slightly different and more nuanced light on Nerlove's original model, his approach is still relevant these days almost seven decades after its original insemination.

3.1 Introduction

In his seminal work, Nerlove (1956), Marc Nerlove highlighted the importance of price expectations in farmers' decision and demonstrated how these expectations may affect the estimate of supply elasticity. Like most seminal works, it inspired many years of future research in both economics and agricultural economics, especially in the development of formulating price expectation. Nerlove was restricted by data and econometric techniques available at the time. Therefore, it seems appropriate to revisit the estimation of supply elasticities, in the spirit of Nerlove (1956), by leveraging the recent developments in econometrics, especially in risk modelling,

Felix Chan ✉
Curtin University, Perth, Australia, e-mail: felix.chan@cbs.curtin.edu.au

Elizabeth Jackson
Curtin University, Perth, Australia, e-mail: Elizabeth.Jackson@curtin.edu.au

Richard Dwumfour
Curtin University, Perth, Australia, e-mail: richard.dwumfour@curtin.edu.au

László Mátyás
Central European University, Budapest, Hungary and Vienna, Austria, e-mail: matyas@ceu.edu

B. H. Baltagi and L. Mátyás (eds.), *Seven Decades of Econometrics and Beyond*, Advanced Studies in Theoretical and Applied Econometrics 57,
https://doi.org/10.1007/978-3-031-92699-0_3

machine learning, and the open data movement, to examine if these advances can provide further insight on the estimation of supply elasticity.

As such, this chapter has three objectives. First, it presents a survey on the supply elasticities of selected agricultural commodities inspired by what is now known as the Nerlovian Model. This part of the chapter can be considered an update of Askari and Cummings (1977), which provided an excellent review of the work inspired by the Nerlovian model until the late 1970s. Second, it extends the existing Nerlovian Model by incorporating the effect of price risk in the formulation of price expectations, as well as the effect on acreage from the risk of different input prices. Perhaps more importantly, the proposed approach also incorporates the interaction between the different agricultural commodities, which could not have been possible in the 1950s due to a lack of data availabilities and appropriate econometric techniques. Given the number of potential risk factors and different methods of generating risk estimates over different time horizons, the third objective of this chapter is to re-estimate the supply elasticities of selected commodities using the more recent feature selection techniques to identify the appropriate risk measures that affect price expectations.

After a concise survey of the literature on supply elasticities of agricultural commodities, the chapter considers an augmentation of price expectation as proposed in Nerlove (1956). The main idea is to incorporate risk into the model of price expectation. In the original formulation, the changes in price expectation are driven only by the difference between the previous price expectation and the actual price. From a decision viewpoint, it has been shown that risk, as reflected by price volatility of the relevant future contracts, may also affect price expectation and farmers' decision on the allocation of land to a particular commodity. In this chapter therefore we also examine the impacts of price volatility on price expectation and whether the inclusion of price risk affects the estimates of supply elasticities of selected agricultural commodities including wheat and barley.

The dependence of prices between different agricultural commodities is important in farmers' decision on land use allocation, investment and product sales. As such, this chapter also extends Nerlove (1956) by incorporating a portfolio approach to risk modelling. Specifically, the interdependence between agricultural commodities is considered in modelling the risk of each commodity. One approach is to utilise the suite of time-varying models for the variance-covariance matrices, including the multivariate Generalized Autoregressive Conditional Heteroskedasticity model as proposed in Diebold and Nerlove (1989).

This chapter is organised as follows. Section 3.2 provides a survey on the literature of estimating supply elasticity based on the Nerlovian model in recent times. Section 3.3 extends the price expectation formulation by incorporating risk variable as well as introducing two risk measures that can be constructed by leveraging information from three dimensional panel data. One of the risk measures introduced in Section 3.3 also generalises the Latent Factor ARCH model as proposed in Diebold and Nerlove (1989), which can be estimated via Kalman Filter. Section 3.4 discusses the data used in this chapter with a focus on the process of linking data from different datasets. This is followed by empirical results in Section 3.5 and Section 3.6 contains some concluding remarks.

3.2 Estimating Supply Elasticities: A Survey

Let us first examine the impact of the Nerlove (1956) paper where he introduced price expectations as the main driver behind farmers' decision. Since price expectations are not directly observed, he proposed a dynamic model that allows them to be expressed as past observed prices. The price expectation model proposed in Nerlove (1956) has the simple form

$$P_t^e - P_{t-1}^e = \beta\left(P_{t-1} - P_{t-1}^e\right), \quad t = 1,\dots,T, \tag{3.1}$$

where P_t^e and P_t are the expected price and the actual price of a particular commodity at time t, respectively. Equation (3.1) can be interpreted as an exponential smoothing model between the expected and realised price if the restriction $0 < \beta < 1$ is imposed. Interestingly, the first official document that discussed exponential smoothing is Brown (1956) and the model was again mentioned more formally in a report to the U.S. Office of Naval Research by Professor Holt (Holt, 1957)[1]. Thus, it can be argued that the econometric novelty in Nerlove (1956) is applying exponential smoothing to model farmers' expectations, which was the cutting edge technique at the time.

The empirical advantage of using exponential smoothing to describe the dynamic of price expectation is that the unobserved price expectation can be expressed as a distributed lag model of the actual prices. Specifically,

$$\begin{aligned} P_t^e - (1-\beta)P_{t-1}^e &= \beta P_{t-1} \\ P_t^e &= \beta \sum_{\tau=1}^{\infty} (1-\beta)^{\tau-1} P_{t-\tau}. \end{aligned} \tag{3.2}$$

The main model of interest, however, is the relation between supply of the commodities and price expectation. That is

$$x_t = \pi_0 + \pi_1 P_t^e + \boldsymbol{\theta}'\mathbf{z}_t + u_t, \tag{3.3}$$

where x_t is the acreage, i.e., the land use of the particular commodity. The main idea here is to develop a relation between the supply of the commodity and the expected price of that commodity along with other explanatory variables. The use of acreage is an interesting choice as it is closely related to the supply of the commodity. Another natural choice is the yield of the commodity as suggested in Askari and Cummings (1977). However, the amount of crop yield depends on the land available for growing that particular commodity. So, acreage needs to be considered as the control variable, i.e., an element in $\mathbf{z}_t$, if crop yields was to be used as x_t to replace acreage.

Given Equation (3.2), Equation (3.3) implies

$$x_t = \pi_0\beta + \pi_1\beta P_{t-1} + (1-\beta)x_{t-1} + (1-\beta)\boldsymbol{\theta}'\Delta\mathbf{z}_t + e_t, \tag{3.4}$$

[1] This article is later reprinted as Holt (2004).

where $e_t = u_t - (1-\beta)u_{t-1}$. Note that Equation (3.4) contains information to estimate both long and short run elasticities. As described in Nerlove and Addison (1958), β describes the short run response, while $1-\beta$ describes the long run response of acreage to the differential between expected and actual prices.

Nerlove (1956) used the model as defined in Equation (3.4) for cotton, wheat and corn in the United States of America. Given the availabilities of panel data, a natural extension to Nerlove (1956) is to consider

$$\begin{aligned} x_{it} =& \pi_0\beta + \pi_1\beta P_{it-1} + (1-\beta)x_{it-1} + (1-\beta)\boldsymbol{\theta}'\Delta\mathbf{z}_{it} + e_{it} \\ =& \beta_0 + \beta_1 P_{it-1} + \beta_2 x_{it-1} + \boldsymbol{\beta}_3'\Delta\mathbf{z}_{it} + e_{it}, \quad i = 1,\ldots,N \quad t = 1,\ldots,T, \end{aligned} \tag{3.5}$$

where P_{it} denotes the price of a particular commodity in country i at time t, and this definition applies naturally to x_{it}, $\mathbf{z}_{it}$ and w_{it}. Note that in this case, u_{it} may be an error components model e.g., $u_{it} = \alpha_i + \lambda_t + \epsilon_t$, which implies $e_{it} = \beta\alpha_i + \lambda_t - (1-\beta)\lambda_{t-1} + \epsilon_t - (1-\beta)\epsilon_{t-1}$. It should be clear that assuming one can obtain consistent estimates for β_1, β_2 and $\boldsymbol{\beta}_3$, denoted $\hat{\beta}_1$, $\hat{\beta}_2$ and $\hat{\boldsymbol{\beta}}_3$, respectively, then β can be estimated by $\hat{\beta} = 1 - \hat{\beta}_2$, the long run response π_1 can be estimated by $\hat{\pi}_1 = \hat{\beta}_1/(1-\hat{\beta}_2)$ and $\boldsymbol{\theta}$ can be estimated by $\hat{\boldsymbol{\theta}} = \hat{\boldsymbol{\beta}}_3/(1-\hat{\beta}_2)$.

Equation (3.5) is a dynamic panel model and classical estimators such as Ordinary Least Squares and Fixed Effects Estimators produce biased and inconsistent estimates for β_2, at least with short T. Nerlove (1967) and Nerlove (1971) are often considered to be the earliest papers to report such bias through Monte Carlo simulations. Inspired by these observations, Nickell (1981) provided one of the first theoretical analysis on the bias of the Fixed Effects Estimator in the content of dynamic panel data model and earned the term *Nickell effect*.

Another econometric challenge associated with estimating Equation (3.5) is the fact that, even without the unobserved heterogeneity, the residuals, $e_{it} = u_{it} - (1-\beta)u_{it-1}$ is a moving average process of order 1. Some of the popular GMM type estimators, such as Arellano-Bond as proposed in Arellano and Bond (1991), do not permit serially correlated errors in general, even though they may be robust against specific forms of serial correlation, such as moving average process, assuming that valid instruments exist.

In general, the econometric challenges in estimating the parameters in Equation (3.5) are (i) robust and consistent estimation of β_2, the coefficient of the autoregression term in a dynamic panel, (ii) robust estimation of a dynamic panel model in the presence of serially correlated errors and (iii) the previous two challenges in the presence of a possibility unbalanced panel. Section 3.5 examines the empirical performance of some of the popular estimators for panel data model in this context.

As previously highlighted, Askari and Cummings (1977) present a thorough review of the literature influenced by the Nerlovian model up to the late 1970s. In the ensuing decades, Nerlove's foundational work has inspired a diverse array of studies across multiple disciplines. In light of this, we extend their review (see Table 3.6 for an overview of the most recent studies) to explore significant advancements in the Nerlovian framework, particularly concerning supply elasticities, with a focused examination of how risk or volatility is integrated within these models.

For instance, Puga and Anderson (2024) examine changes in grape varietal mixes in South Australia's wine regions and their sensitivity to expected revenues. Using the Nerlovian adaptive expectations and partial adjustment framework they analyse changes in varietal mixes, focusing on acreage response to expected revenues. The study found significant sensitivity of acreage decisions to revenue expectations than to what climate conditions in a particular region may be best for the crop. Short-run and long-run supply elasticities turned out to be heterogeneous in the region. The short-run price and revenue elasticities were estimated to range from 0.074 to 0.089, while the long-run elasticities ranged from 0.333 to 0.397. The authors attribute the comparatively low short-run supply elasticities to the fact that grapes, as a perennial crop, are capital-intensive with a longer investment horizon.

Krah (2023) extends Nerlove's model to assess maize price variability and its influence on land use and forest loss in Ghana. By incorporating a measure of price variability as the ratio of the standard deviation to the mean price, he estimates a price elasticity of 0.018. This study underscores the subdued responsiveness of maize producers to price changes, reflective of broader structural constraints.

Diop and Traoré (2023) analyse the asymmetric supply responses in the cotton sector in Mali using a nonlinear ARDL framework based on Nerlove's model. By looking at supply, the area cultivated (production), and responses to both price increases and price decreases, the study found short-run elasticity to be symmetric at 0.63 (0.58), while long-run elasticities differ: 0.87 (0.82) for price decreases vs. 0.43 (0.38) for price increases. These results highlight behavioural asymmetries in response to price volatility.

Nhundu et al. (2022) estimate the supply response for sunflower yields in South Africa using a Nerlovian partial adjustment framework by focusing on both price and non-price incentives over an extended period (1947–2016). From the OLS estimates, the short-run and long-run elasticities were 0.238 and 0.313, respectively. An estimated adjustment speed of 0.272 also shows a slow adjustment to price changes and the importance of non-price factors in influencing the supply of sunflower.

Amine M. Benmehaia (2021) analyses the aggregate supply response of 19 crops in Algeria from 1966 to 2018. Inspired by Nerlove's model, the authors replace the Nerlovian partial adjustment model with an Error Correction Model (ECM). The authors found apple growers to have the highest long-run elasticities (51.1%) among fruit growers, whereas cauliflower had the highest long-run elasticities (99.2%) among vegetable producers. Short-run elasticities were even lower, ranging from 0.161 for Onion (bulb) to 0.393 for cauliflower.

Lemontzoglou and Carmona-Zabala (2024) study supply-side responses in the Greek tobacco sector during 1953–1964, assessing the role of price incentives and state interventions. By incorporating a two-dimensional panel autoregressive distributed lag (ARDL) model, they allow for long-term cointegration between output and market prices while accounting for regional and varietal differences in a non-Nerlove model. This methodological approach enables the estimation of the price elasticity of tobacco supply, which ranges from 1.83 to 4.98, indicating highly elastic responses. The inclusion of regional and varietal fixed effects reveals significant intra-product and regional variations in elasticity, demonstrating the influence of varietal specialization and spatial heterogeneity on supply response.

Similarly, Pates and Hendricks (2021) employ Markov transition regression to estimate rotational supply elasticities for U.S. corn. They find a short-run elasticity of 0.69 and a long-run elasticity of 0.54, demonstrating spatial heterogeneity in rotational response.

Other studies have tried to incorporate climate change and government policy programs in supply responses. These represent some of the non-market factors, *Z* variables, anticipated by Nerlove to be included in the supply response model to handle identification issues (Askari & Cummings, 1977). One such recent study is that of Okou, Keita, N'Dri and Kouakou (2023), who, using perennial crops like cocoa and cashew nuts, forecast the cultivated area using Nerlovian models. The study expands Nerlove's model by integrating rainfall and multi-year lagged acreage adjustments. Even lower than the estimates of Puga and Anderson (2024), the study estimates short-run price elasticity to be around 0.012 for cocoa and cashew nuts.

Yu, Clark, Tian and Yan (2022), while with a non-Nerlove formulation, also examined how climate variables and price policy influence rice yield in high-latitude regions of China. The authors use Kalman filters and a spatial autoregressive combined model to account for heteroscedasticity and spatial correlation. The estimated price elasticity of rice yield using OLS was 0.194, while corn had a cross-price elasticity of -0.097. The study found that climate change has a significant impact on rice yield; higher-rate global warming will decrease the projected rate of increase in rice yield. The results highlight the importance of spatial effects and price policies in yield modelling, emphasizing climate and crop price policy interplay.

This analysis of Nerlovian model applications, although diverse in geography and crop species, demonstrates the wide range of methodologies applied for estimating supply elasticities. These include ordinary least squares (OLS) (Krah, 2023; Puga & Anderson, 2024; Nhundu et al., 2022), autoregressive distributed lag (ARDL) models (Lemontzoglou & Carmona-Zabala, 2024; Diop & Traoré, 2023), error correction models (ECM) (Amine M. Benmehaia, 2021), and generalized method of moments (GMM) (Qian, Ito & Zhao, 2020; Pane & Supriana, 2020; Tenaye, 2020; Zhai, Chen & Wang, 2019; Meyer, 2018; Magrini, Balié & Morales-Opazo, 2018; Haile, Kalkuhl & von Braun, 2015). Additionally, more specialized techniques, such as Markov transition regression (Pates & Hendricks, 2021) and other econometric approaches (Suh & Moss, 2018; Rude & Surry, 2014; Theriault, Serra & Sterns, 2013), have been employed to address specific research contexts.

A notable observation from this body of literature is the predominance of country-specific studies, which, while valuable for capturing localized dynamics, highlight a significant gap in cross-country analyses. Addressing this gap is crucial for understanding broader patterns and differences in supply elasticity determinants across diverse socio-economic and agro-climatic settings.

Despite the substantial methodological advancements and extensions of Nerlove's framework, the explicit incorporation of risk remains relatively limited. Among the few studies addressing this dimension, Krah (2023) model price variability using measures such as the standard deviation of prices, underlining risk as a critical but underexplored factor in supply response modeling.

This review establishes a foundation for extending Nerlove's price expectation model to a cross-country context by incorporating innovative methodologies, such as constructing risk measures using multidimensional panels and Latent Factor ARCH models, discussed in the subsequent section, offer promising avenues for addressing these research gaps and advancing the understanding of agricultural supply responses in a global context.

3.3 Incorporating Risk in Price Expectation

Next, in this section we augmented the price expectation model as defined in Equation (3.1) by considering

$$P^e_{it} - P^e_{it-1} = \beta\left(P_{it-1} - P^e_{it-1}\right) + \lambda\sigma_{it}, \qquad i = 1,\ldots,N, \quad t = 1,\ldots,T, \tag{3.6}$$

where the additional term, σ_{it}, is a measure of risk for the commodity returns at time t, based on the information up to $t-1$, and λ represents the sensitivity of the expected price to market risk for a particularly commodity in country i. The sign of λ therefore provides some indication on the risk attitude and if $\lambda = 0$ then Equation (3.6) reduces to the original price expectation model as in (Nerlove, 1956). This section proposes two approaches to construct σ_{it} based on the techniques from portfolio management theory pioneered by Markowitz (1952). The main idea is to approximate risk using the variance-covariance matrix of commodity returns, weighted by the share of each commodity in the country's agricultural portfolio.

The Nerlovian model as defined in Equation (3.5) can be rewritten as

$$x_{it} = \beta_0 + \beta_1 P_{it-1} + \beta_2 x_{it-1} + \beta_3\sigma_{it} + \boldsymbol{\beta}_4'\Delta\mathbf{z}_{it} + e_{it}, \tag{3.7}$$

where $\beta_0 = \pi_0\beta$, $\beta_1 = \pi_1\beta$, $\beta_2 = 1-\beta$, $\beta_3 = \pi_1\lambda$ and $\beta_4 = \beta\boldsymbol{\theta}$. It should be clear that all parameters are identifiable but consistent and robust estimate of β_2, the coefficient of the lag dependent variable, is crucial in obtaining reliable estimates of other structural parameters.

3.3.1 Constructing Risk Measures by Using Multi-dimensional Panels

Next, we consider a simple approach to construct σ_{it} based on some recent developments in multi-dimensional panel data modelling (see Mátyás, 2024 for further details). The approach is both conceptually and computationally straightforward and it is useful for benchmarking purposes.

Let x_{ijt} and P_{ijt} denote the yield and price of commodity j in country i at time t for $i = 1,\dots,N_1$, $j = 1,\dots,N_2$ and $t = 1,\dots,T$, respectively. Note that $N \equiv N_1$ in the context of the notation used thus far. Define $s_{ijt} = \log(P_{ijt}) - \log(P_{ijt-1})$ as the returns of commodity j in country i at time t and

$$\mathbf{\Omega}_t = (N_1 - N_2)^{-1} \sum_{i=1}^{N_1} (\mathbf{s}_{i\circ t} - \boldsymbol{\mu}_{Jt})(\mathbf{s}_{i\circ t} - \boldsymbol{\mu}_{Jt})', \tag{3.8}$$

where $\mathbf{s}_{i\circ t} = (s_{i1t},\dots,s_{iN_2t})'$, a $N_2 \times 1$ vector containing the returns of commodity for country i at time t, $\boldsymbol{\mu}_{Jt} = (\mu_{1t},\dots,\mu_{N_2t})'$ is a $N_2 \times 1$ vector such that $\mu_{jt} = N_1^{-1} \sum_i s_{ijt}$. That is, μ_{Jt} contains the average cross section returns (over all countries) of each commodity at time t. The weight of the portfolio, $\boldsymbol{\delta}_{it} = (\delta_{i1t},\dots,\delta_{iN_2t})'$, is constructed as $\delta_{ijt} = x_{ijt}/\sum_j x_{ijt}$ $\forall j = 1,\dots,N_2$, for each $i = 1,\dots,N_1$, and each $t = 1,\dots T$. The risk measure σ_{it} is then defined as

$$\sigma_{it} = \boldsymbol{\delta}'_{t-1}\mathbf{\Omega}_{t-1}\boldsymbol{\delta}_{t-1}. \tag{3.9}$$

3.3.2 Constructing Risk Measure by Extending the Latent Factor ARCH Model

The second approach to construct σ_{it} is to utilise the techniques from the conditional variance literature. Since the introduction of the Autogressive Conditional Heteroskedasticity (ARCH) model by Engle (1982) and the Generalised ARCH (GARCH) model in Bollerslev (1986), the study of risk through the modelling of conditional variance have become standard in Financial Econometrics. The multivariate extension of GARCH model has also be a focus in the early 2000s. For comprehensive surveys, see for example, Bauwens, Laurent and Rombouts (2006) and Silvennoinen and Teräsvirta (2008).

Despite the active developments and advances, multivariate Generalised Autoregressive Conditional Heteroskedasticity (M-GARCH) is known to be difficult to estimate in practice when the number of assets is large. Diebold and Nerlove (1989) proposed a latent factor ARCH model aiming to alleviate some of the numerical challenges due to the curse of dimensionality in such cases. Here we extend the model proposed by them and use the new model to create a risk measure for each commodity, which is then used in the augmented Nerlovian Model with risk as a factor that determines crop yield.

Let $\mathbf{s}_t$ be a $N_2 \times 1$ vector containing the return of N_2 commodities at time t and consider the following model for $\mathbf{s}_t$

$$\mathbf{s}_t = \boldsymbol{\mu} + \boldsymbol{\varepsilon}_t, \tag{3.10}$$

$$\boldsymbol{\varepsilon}_t = \mathbf{\Lambda}\mathbf{F}_t + \boldsymbol{\eta}_t, \tag{3.11}$$

where $\mathbf{F}_t$ is a $h \times 1$ vector of latent factors with $h << N_2$ and

$$\begin{aligned}\mathbb{E}(\boldsymbol{\varepsilon}_t) = \mathbb{E}(\boldsymbol{\eta}_t) &= \mathbf{0} \quad \forall t \\ \mathbb{E}(\mathbf{F}_t) &= \mathbf{0} \quad \forall t \\ \mathbb{E}(\mathbf{F}_t'\boldsymbol{\eta}_t) &= \mathbf{0} \quad \forall t \\ \mathbb{E}(\mathbf{F}_t\mathbf{F}_t') &= \mathbf{I} \quad \forall t \\ \mathbb{E}(\boldsymbol{\eta}_t\boldsymbol{\eta}_t') &= \boldsymbol{\Gamma} \quad \forall t.\end{aligned}$$

The conditions above imply that the *unconditional* variance-covariance matrix of $\boldsymbol{\varepsilon}_t$ is $\mathbb{E}(\boldsymbol{\varepsilon}_t\boldsymbol{\varepsilon}_t') = \boldsymbol{\Lambda}\boldsymbol{\Lambda}' + \boldsymbol{\Gamma}$. The *conditional* variance-covariance matrix namely, $\boldsymbol{\Omega}_t$, evolves following the dynamic

$$\mathbb{E}(\mathbf{F}_t\mathbf{F}_t'|\Im_t) = \boldsymbol{\Omega}_t = \boldsymbol{\Omega} + \mathbf{A}\mathbf{F}_{t-1}\mathbf{F}_{t-1}'\mathbf{A}' + \mathbf{B}\boldsymbol{\Omega}_{t-1}\mathbf{B}', \tag{3.12}$$

where $\boldsymbol{\Omega} = \mathbf{I} - \boldsymbol{\Lambda}\boldsymbol{\Lambda}' - \mathbf{B}\mathbf{B}'$, which would ensure that the unconditional variance-covariance matrix of $\mathbf{F}_t$ is the identify matrix.

This setup extends Diebold and Nerlove (1989) in two major ways. First, it generalises the model in Diebold and Nerlove (1989) by including the conditional variance-covariance matrix from pervious period in the dynamic of the conditional variance-covariance. This is similar to how GARCH generalised ARCH by including the conditional variance from previous periods in the dynamic. Second, it allows $\mathbf{F}_t$ to include more than one latent factor, while the number of latent factor in Diebold and Nerlove (1989) was restricted to 1.

Equation (3.12) can be generalised to include high order lags as well. This may not be necessary though in practice as empirical evidence suggested that information from a pervious period seems to be sufficient in describing the dynamic of the conditional variance, and including high order lags does not seem to improve prediction, while including them tend to create numerical difficulties, see for example, Wang, Xiang, Lei and Zhou (2022). Therefore, in this chapter we focus on Equation (3.12) and refrain from including higher order lags.

Given that $\mathbf{F}_t$ is latent, one way to estimate this model is via Kalman Filter. Following the approach proposed in Diebold and Nerlove (1989), the state-space representation of the model above can be written as

$$\begin{aligned}\mathbf{F}_t &= \mathbf{v}_t, \\ \boldsymbol{\varepsilon}_t &= \boldsymbol{\Lambda}\mathbf{F}_t + \mathbf{e}_t.\end{aligned}$$

Since $\mathbf{F}_t$ is not observable, it is replaced by its estimated counterpart denoted $\mathbf{F}_{t|t}$, with $\mathbf{F}_{t|t-1} = 0$ since $\mathbf{F}_t$ is not auto-correlated. The estimated conditional variance-covariance matrix, $\boldsymbol{\Omega}_{t|t-1}$, is defined as

$$\boldsymbol{\Omega}_{t|t-1} = \boldsymbol{\Omega} + \mathbf{A}\mathbf{F}_{t-1|t-1}\mathbf{F}_{t-1|t-1}'\mathbf{A}' + \mathbf{B}\boldsymbol{\Omega}_{t-1|t-1}\mathbf{B}'$$

and $\mathbf{F}_{t|t}$ is defined to be an update from $\mathbf{F}_{t|t-1}$ following

$$\mathbf{F}_{t|t} = \boldsymbol{\Omega}_{t|t-1}\boldsymbol{\Lambda}' \left(\boldsymbol{\Lambda}\boldsymbol{\Omega}_{t|t-1}\boldsymbol{\Lambda}' + \boldsymbol{\Gamma}\right)^{-1} \boldsymbol{\varepsilon}_t.$$

Unlike the case in Diebold and Nerlove (1989), where $\boldsymbol{\Omega}_{t|t-1}$ does not need to be updated, the inclusion of $\boldsymbol{\Omega}_{t-1|t-1}$ in the generation of $\boldsymbol{\Omega}_{t|t-1}$ means $\boldsymbol{\Omega}_{t|t-1}$ needs to be updated as

$$\boldsymbol{\Omega}_{t|t} = \boldsymbol{\Omega}_{t|t-1} - \boldsymbol{\Omega}_{t|t-1}\boldsymbol{\Lambda}' \left(\boldsymbol{\Lambda}\boldsymbol{\Omega}_{t|t-1}\boldsymbol{\Lambda}' + \boldsymbol{\Gamma}\right)^{-1} \boldsymbol{\Lambda}\boldsymbol{\Omega}_{t|t-1}.$$

The derivations of the filter and its updating processes can be found in the Appendix.

The unknown parameters $\boldsymbol{\Theta} = (\boldsymbol{\mu}', \text{vec}\,\mathbf{A}', \text{vec}\,\mathbf{B}', \text{vec}\,\boldsymbol{\Lambda}, \text{vec}\,\boldsymbol{\Gamma}')'$ can be estimated via maximum likelihood, i.e.,

$$\hat{\boldsymbol{\Theta}} = \arg\max_{\boldsymbol{\theta}} l(\boldsymbol{\Theta})$$

where

$$l(\boldsymbol{\Theta}) := -\frac{1}{2}\left[\sum_{t=1}^{T} \log\left|\boldsymbol{\Lambda}\boldsymbol{\Omega}_{t|t-1}\boldsymbol{\Lambda}' + \boldsymbol{\Gamma}\right| + (\mathbf{s}_t - \boldsymbol{\mu})' \left(\boldsymbol{\Lambda}\boldsymbol{\Omega}_{t|t-1}\boldsymbol{\Lambda}' + \boldsymbol{\Gamma}\right)^{-1} (\mathbf{s}_t - \boldsymbol{\mu})\right].$$

Under standard assumptions, see for examples, Harvey, Ruiz and Sentana (1992) and Harvey (2001), $\sqrt{T}\left(\hat{\boldsymbol{\Theta}} - \boldsymbol{\Theta}\right) \overset{d}{\sim} N(\mathbf{0}, \mathcal{I}^{-1})$ where $\mathcal{I} = \frac{\partial^2 l}{\partial \boldsymbol{\Theta} \partial \boldsymbol{\Theta}'}$.

To apply the proposed Latent Factor GARCH model in the present context, let $\mathbf{s}_t = \left(s_{\circ 1t}, \ldots, s_{\circ N_2 t}\right)'$, where $s_{\circ jt} = N_1^{-1}\sum_i s_{ijt}$ is the cross section average of commodity j at time t and the risk measure is therefore

$$\sigma_{it} = \boldsymbol{\delta}_{it-1}' \boldsymbol{\Omega}_{t|t-1} \boldsymbol{\delta}_{it-1} \tag{3.13}$$

with the weight vector, $\boldsymbol{\delta}_{it}$ defined as in Section 3.3.1.

3.4 Data

The main source of data used in this study comes from the Food and Agriculture Organization of the United Nations (FAO, see Food and Agriculture Organisation of the United Nations, 2024). We draw on two different FAO datasets, namely *Producer Prices* (See Food and Agriculture Organization of the United Nations, 2024a) and *Crops and Livestock Products* (See Food and Agriculture Organization of the United Nations, 2024b). The crop yield data covers 108 countries from 1961 to 2022, while the price data covers 181 countries from 1993 to 2022.

In terms of land use data, while the data sources come from FAO, it is not in a usable format. Instead, the data is obtained via *Our World in Data* (see Our World in

Data, 2024), which processed the source data from FAO into a usable format. For further information, see Ritchie and Roser (2019).

The three datasets must be linked (combined) before the data can be used for estimation. The process of linking the three datasets can be found in Chan, Jackson, Duwmfour and Mátyás (2025). Given the estimation of a dynamic panel with serial correlation, we excluded countries with less than 10 time series observations. Since the chapter focus on barley and wheat, the final data consists of 77 countries from 1991 to 2022, which means $N_1 = 77$, $N_2 = 2$ and $T = 31$. Note that the data is an unbalanced panel, so not all countries share the same number of time series observations. The distribution of the number of observations in the T dimension can be found in Figure 3.1.

Fig. 3.1: Distribution of Observations Across Years and Countries

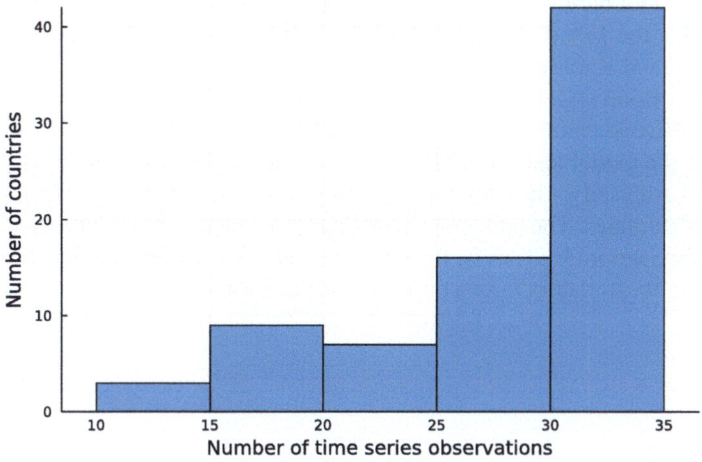

3.5 Empirical Results

Next, we present the estimates of the model as defined in Equation (3.7) for two commodities namely, barley and wheat. The choice of the two crops are based on the fact that these crops are commonly grown together and therefore the choice ensures we have sufficient number of observations across countries and over time. This is particularly important when constructing the risk measures which require observations for both crops from the same countries in any given time period.

The empirical results include the case with the restriction $\beta_3 = 0$, which represents the original Nerlovian Model without the inclusion of any risk measures. This section also presents the results of two cases where β_3 is not restricted to be 0. In these cases, the risk measure, σ_{it} is constructed following Equations (3.9) and (3.13), respectively.

Following the suggestion from Askari and Cummings (1977), the dependent variable, x_{it}, in this case is defined to be the crop yield. In addition to the lag of price, P_{it-1}, and the lag of crop yield, x_{it-1}, we also include the size of land allocated to crop production as a control variable, z_{it}, as argued in Section 3.2. Logarithmic transformation has been applied to x_{it} and P_{it}, before estimation. Thus the coefficients of $\log x_{it-1}$ and $\log P_{it-1}$ can be interpreted as elasticities (or percentage changes).

In terms of computation, the estimation for the Latent GARCH model is conducted via Julia Programming Language (see Bezanson, Edelman, Karpinski & Shah, 2017), while all the panel estimation procedures are conducted via the *plm* Package (see Croissant & Millo, 2018) in Julia via Rcall.jl (see Lai et al., 2024). Related code and Jupyter notebooks can be found in Chan et al. (2025).

Tables 3.1 and 3.2 contain the estimation results for the model as defined in Equation (3.7) for barley and wheat using some of the more conventional estimators. This includes the Fixed Effect estimator (FE) for unbalanced panel as proposed in Wansbeek and Kapteyn (1989),[2] the Generalised Least Square (GLS) estimator, the Arellano-Bond (AB) estimator as proposed in Arellano and Bond (1991) and the Blundell-Bond estimator as proposed in Blundell and Bond (1998). Given the structure of the model leads to a MA(1) process in the residual, only the second lag and beyond are used as instruments in both GMM estimators i.e., AB and BB, to ensure their validities. The tables also contain estimation results with the risk measure as defined in Section 3.3.1, as well as the estimates of the long elasticity as implied by the different estimators for each model specification.

[2] See also Baltagi, 2005, Chapter 9 for details.

Table 3.1: Estimation for Barley using Conventional Estimators

	$\beta_3 = 0$				$\beta_3 \neq 0$			
	FE	AB	BB	GLS	FE	AB	BB	GLS
β_1	0.1535	0.2099	0.0379	0.2022	0.1559	0.2183	0.0383	0.2272
	(10.1415)	(13.3171)	(17.5985)	(81.3512)	(10.326)	(13.7294)	(15.0732)	(128.0209)
β_2	0.3327	0.0796	0.7633	0.2068	0.3332	0.0747	0.7744	0.321
	(15.4408)	(3.3027)	(61.9479)	(90.8944)	(15.5136)	(3.0953)	(66.3997)	(143.383)
β_3	-	-	-	-	-0.0005	-0.0004	4.983e-05	-0.0004
	-	-	-	-	(-3.7279)	(-4.0268)	(0.533)	(-40.9888)
β_4	3.166e-08	2.370e-08	2.531e-08	5.624e-08	3.304e-08	2.443e-08	2.630e-08	7.365e-09
	(1.8082)	(2.8411)	(3.3177)	(25.8701)	(1.8934)	(2.9289)	(3.4505)	(2.9241)
π	0.23	0.228	0.1599	0.2549	0.2337	0.2359	0.1697	0.3346
	(9.1138)	(11.7408)	(9.6686)	(89.6881)	(9.2608)	(12.0385)	(9.2046)	(116.3476)

[1] AB denotes the Arellano and Bond estimator as proposed in Arellano and Bond (1991).
[2] BB denotes the Blundell and Bond estimator as proposed in Blundell and Bond (1998).
[3] t-ratios are in the parenthesis.

Table 3.2: Estimation for Wheat using Conventional Estimators

	$\beta_3 = 0$				$\beta_3 \neq 0$			
	FE	AB	BB	GLS	FE	AB	BB	GLS
β_1	0.1507	0.2134	0.033	0.1354	0.1611	0.2349	0.0326	0.1199
	(11.7963)	(12.0685)	(14.1722)	(26.577)	(12.567)	(12.9608)	(11.2027)	(34.5794)
β_2	0.3509	0.1343	0.8085	0.3497	0.329	0.1067	0.8373	0.328
	(16.5427)	(4.9233)	(64.5181)	(60.7391)	(15.3517)	(3.8456)	(72.4479)	(82.9983)
β_3	-	-	-	-	-0.0005	-0.0006	9.636e-05	-0.0004
	-	-	-	-	(-5.4287)	(-5.4123)	(1.0382)	(-19.3172)
β_4	3.881e-08	3.239e-08	4.707e-08	4.131e-08	4.007e-08	3.286e-08	4.834e-08	3.920e-08
	(2.8166)	(3.9002)	(6.1875)	(17.6245)	(2.9304)	(3.9568)	(6.3549)	(27.8341)
π	0.2321	0.2465	0.1721	0.2082	0.2401	0.2629	0.2006	0.1784
	(10.127)	(10.3967)	(7.6575)	(26.3056)	(10.6425)	(10.9519)	(6.5987)	(33.5831)

[1] AB denotes the Arellano and Bond estimator as proposed in Arellano and Bond (1991).
[2] BB denotes the Blundell and Bond estimator as proposed in Blundell and Bond (1998).
[3] t-ratios are in the parenthesis.

As shown in Tables 3.1 and 3.2, the estimates of β_2 depend heavily on the estimator, but interestingly, the estimates of the long run elasticity, π, is relatively robust across different estimators. In other words, while the estimates of β_2 vary across different estimators, the estimates of the ratio, $\beta_1/(1-\beta_2)$, are relatively stable.

In terms of the importance of risk, it appears that risk is an important factor in price expectation as suggested by the results in Tables 3.1 and 3.2. The coefficient estimates, β_3, are statistically significant generally across different estimators. The negative sign of β_3 estimate provides evidence of risk aversion. This seems to suggest that, while higher than expected price may increase price expectation, the impact is compensated, to some degrees, by the uncertainty of the increase as reflected in the risk measure. The inclusion of risk also inflates the estimates of β_2 slightly, which also leads to slight increase in the estimate of the long run elasticity.

3.5.1 Result from Latent Factor GARCH Model

Given there are two commodities, the specification of the Latent Factor GARCH can be written as

$$\mathbf{s}_t = \begin{bmatrix} s_{1t} \\ s_{2t} \end{bmatrix} = \begin{bmatrix} \mu_1 \\ \mu_2 \end{bmatrix} + \begin{bmatrix} \varepsilon_{1t} \\ \varepsilon_{2t} \end{bmatrix}$$

$$\begin{bmatrix} \varepsilon_{1t} \\ \varepsilon_{2t} \end{bmatrix} = \begin{bmatrix} \lambda_1 \\ \lambda_2 \end{bmatrix} f_t + \begin{bmatrix} \eta_{1t} \\ \eta_{2t} \end{bmatrix}$$

$$\mathbf{\Omega}_t \equiv \omega_t = \omega + a^2 f_t^2 + b^2 \omega_{t-1},$$

where s_{1t} and s_{2t} denote the average cross section returns of barley and wheat at time t, respectively. In order to reduce the number of parameters to be estimated while retaining the ability to capture the persistence of uncertainty, the model further imposed the restriction $b = 1 - a$ following the Integrated GARCH (IGARCH) model as proposed in Engle and Bollerslev (1986). For ease of numerical optimisation, the variance-covariance matrix of $\boldsymbol{\eta} = (\eta_{1t}, \eta_{2t})$, $\mathbf{\Gamma}$, has been re-parameterized, so that $\mathbf{\Gamma} = \tilde{\mathbf{\Gamma}}\tilde{\mathbf{\Gamma}}'$ where $\tilde{\mathbf{\Gamma}}$ is a lower triangular matrix. This approach ensure that $\mathbf{\Gamma}$ is positive definite and hence enhance stability during optimisation routine.

Table 3.3 contains the estimation result of the Latent Factor GARCH Model and Figure 3.2 contains the plot of the estimated common factor, $\hat{f}_t$, over the sample period.

Table 3.3: Parameter Estimates of the Latent Factor Model

μ		Λ		Ω_t		$\tilde{\Gamma}$	
μ_1	1.3632	λ_1	0.3267	ω	462.0702	γ_{11}	8.2362
	(3.1592)		(1.5095)		(0.7705)		(9.6765)
μ_2	0.9025	λ_2	0.3597	a	0.7102	γ_{21}	6.748
	(2.0734)		(1.5146)		(8.1305)		(6.2607)
						γ_{22}	2.1204
							(9.7519)

[1] t-ratios are in the parenthesis.

As shown in Table 3.3, the persistence parameter a is highly significant indicating the latent factor is sensitive to a large shock in the short run but such impact diminishes quickly. The estimates of λ_1 and λ_2 are weakly significant (at 10%). This is perhaps not surprising given the sampling frequency and the number of time series observations in this case are low relative to the typical studies that used this type of models.[3] Therefore, the power of the test is expected to be relatively low.

Fig. 3.2: Estimated Latent Factor

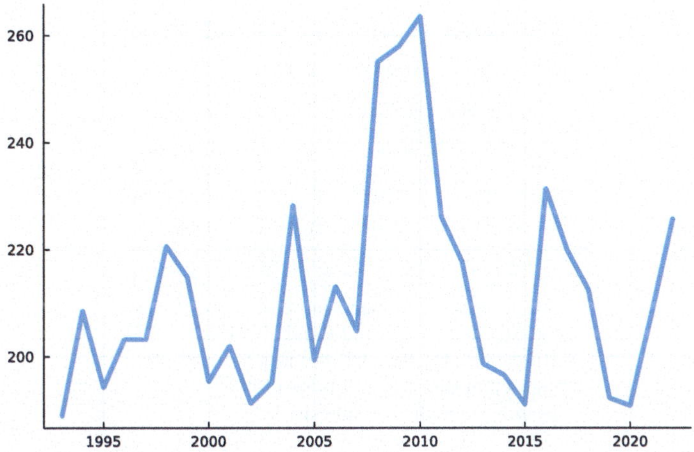

[3] GARCH type models typically utilise data at the daily, or higher, frequency with more than 1000 time series observations. The number of time series observations in this case is only 31. For further discussion see Brooks, Burke and Persand (2003).

The dynamic of the latent factor is insightful as shown in Figure 3.2. It suggests that the latent factor is particularly volatile just before the global financial crisis and while it settles after 2010, it appears to increase again during the COVID-19 pandemic. Although these observations may not be overly surprising, it is encouraging to observe that the proposed model managed to capture such features of the data.

Tables 3.4 and 3.5 contain the estimation results of the Nerlovian model with the risk measure generated by the Latent Factor GARCH model as defined in Equation (3.13). As shown in the Tables, the estimates of β_3 are statistically significant generally across the different estimators. The fact these estimates are also negative re-enforces the earlier result, indicating the importance of risk in determining the expectation of price. It also provides evidence to support risk aversion. Interestingly, the estimated long run elasticities seem to be generally higher with this particular risk measure than the previous two cases across the different estimators. Their variabilities between different estimators also seem higher. Nevertheless, the elasticity estimates from these results are broadly consistent with earlier studies, including those reported in Askari and Cummings (1977).

Table 3.4: Estimation for Barley including Risk from the Latent Factor Model

	FE	AB	BB	GLS
β_1	0.1593	0.2152	0.0679	0.2075
	(10.2665)	(13.8231)	(8.4732)	(57.1193)
β_2	0.3327	0.0768	0.7777	0.3649
	(15.5042)	(3.2591)	(67.634)	(85.7011)
β_3	-0.0005	-0.0005	-0.0006	-0.0035
	(-1.8196)	(-1.9732)	(-3.4093)	(-54.8395)
β_4	3.181e-08	2.408e-08	2.413e-08	7.086e-08
	(1.8183)	(2.9617)	(3.2535)	(26.3821)
π	0.2387	0.2331	0.3053	0.3266
	(9.2242)	(12.1554)	(7.0876)	(53.0468)

[1] AB denotes the Arellano and Bond estimator as proposed in Arellano and Bond (1991).
[2] BB denotes the Blundell and Bond estimator as proposed in Blundell and Bond (1998).
[3] t-ratios are in the parenthesis.

Table 3.5: Estimation for Wheat including Risk from the Latent Factor Model

	FE	AB	BB	GLS
β_1	0.1545	0.2241	0.0609	0.116
	(11.7258)	(12.7647)	(7.3786)	(42.9281)
β_2	0.3516	0.1335	0.8442	0.3919
	(16.6048)	(5.0064)	(76.4978)	(130.0076)
β_3	-0.0003	-0.0008	-0.0007	-0.0004
	(-1.3457)	(-3.4417)	(-3.3751)	(-7.2482)
β_4	3.890e-08	3.297e-08	4.586e-08	4.835e-08
	(2.8048)	(4.0723)	(6.2078)	(50.0946)
π	0.2383	0.2586	0.3907	0.1908
	(10.1169)	(10.9237)	(5.9017)	(42.1395)

[1] AB denotes the Arellano and Bond estimator as proposed in Arellano and Bond (1991).
[2] BB denotes the Blundell and Bond estimator as proposed in Blundell and Bond (1998).
[3] t-ratios are in the parenthesis.

3.5.2 Higher Dimensional Panel

The analysis so far has treated each commodity separately, despite both risk measures leverage the multi-dimensional nature of the data. So another natural extension is to express the Nerlovian model empirically as a three dimensional panel data model. Under the assumption that the parameter vector $\boldsymbol{\beta} = (\beta_0, \beta_1, \beta_2, \beta_3, \boldsymbol{\beta}_4')'$ are the same across all dimensions, Equation (3.6) can be expressed readily as

$$P^e_{ijt} - P^e_{ijt-1} = \beta\left(P_{ijt-1} - P^e_{ijt}\right) + \lambda\sigma_{ijt} \quad i = 1,\ldots,N_1,\ j = 1\ldots,N_2,\ t = 1,\ldots,T$$

and by following the same arguments as those in Section 3.3, Equation (3.7) can be rewritten as

$$x_{ijt} = \beta_0 + \beta_1 P_{ijt} + \beta_2 x_{ijt-1} + \beta_3\sigma_{ijt} + \boldsymbol{\beta}_4'\Delta\mathbf{z}_{ijt} + e_{ijt}.$$

Recall $e_{ijt} = u_{ijt} - (1-\beta)u_{ijt-1}$ and if u_{ijt} represents an error component model, then the structure of e_{ijt} becomes more complicated. The exact form of e_{ijt} would depend on the specification of the error components. For further details and discussions, see Balázsi, Mátyás and Wansbeek (2024) and Chan, Mátyás and Reguly (2024).

While the estimates for β varies across different estimators as shown in previous sections, the estimates are relatively close across the two crops within the same estimators. Here we explore the possibility of estimating the Nerlovian model as a three-dimensional panel data model. For exploratory purposes, it is assumed $u_{ijt} = \alpha_i + \gamma_j + \lambda_t + \epsilon_{ijt}$ which implies $e_{ijt} = \alpha_i^* + \gamma_j^* + \lambda_t^* + \epsilon_{ijt}^*$ such that $\alpha_i^* = \beta\alpha_i$, $\gamma_j^* = \beta\gamma_j$, $\lambda_t^* = \lambda_t - (1-\beta)\lambda_{t-1}$ and $\epsilon_{ijt}^* = \epsilon_{ijt} - (1-\beta)\epsilon_{ijt-1}$.

Let $\hat{\boldsymbol{\beta}}$ be the estimate of $\boldsymbol{\beta}$ from the FE estimator while $\hat{\boldsymbol{\beta}}_1$ and $\hat{\boldsymbol{\beta}}_2$ be the estimates of $\boldsymbol{\beta}$ for barley and wheat, respectively. The hypothesis is therefore

$$
\begin{aligned}
H_0 &: \boldsymbol{\beta}_1 = \boldsymbol{\beta}_2 \\
H_1 &: \boldsymbol{\beta}_1 \neq \boldsymbol{\beta}_2.
\end{aligned}
$$

Let $RSS(\hat{\boldsymbol{\beta}})$ be the residual sum of squares based on $\hat{\boldsymbol{\beta}}$ then one approach to test the hypothesis above is to consider the standard F-test of restrictions

$$
F = \frac{\left[RSS(\hat{\boldsymbol{\beta}}) - RSS(\hat{\boldsymbol{\beta}}_1) - RSS(\hat{\boldsymbol{\beta}}_2)\right]/K}{\left[RSS(\hat{\boldsymbol{\beta}}_1) + RSS(\hat{\boldsymbol{\beta}}_2\right]/(N_1 N_2 T - 2K)} \sim F(K, N_1 N_2 T - 2K),
$$

where K is the number of parameters. The F-test statistics in the case for barley and wheat is 36.365. The 0.05 critical value in this case is 2.607. Thus, there are evidence against the slope coefficients being the same between barley and wheat.

Of course, there are several factors that would affect the reliability of the results above. First, the specification of the error components model may matter. The result above assumed a simple additive model and this may not be valid. For further discussion on the importance of error components model in the context of Fixed Effect Estimator for multi-dimensional panel model see Chan et al. (2024). Second, the Nickell effect may introduce biased estimate, at least for the AR coefficient. While the time dimension in this case is relatively high, such effect cannot be discarded without further investigation. If the effect turns out to be significant, then a generalisation of the GMM estimators may be required. Thus, a more thorough investigation is required if one is serious about estimating the Nerlovian model in a multi-dimensional panel data setting.

3.6 Concluding Remarks

The Nerlovian model remains relevant since its introduction seven decades ago and this chapter has updated it in two major directions. First, it incorporated two different risk measures in the formulation of price expectation, which allowed further investigation on the role of risk in the expectation of price. Second, it extended the empirical version of the original Nerlovian model from a time series model into a three dimensional panel data model. One of the risk measures proposed in this chapter also generalised the Latent Factor ARCH model of Diebold and Nerlove (1989). Although using data at a much lower frequency than the typical use cases, empirical evidence suggests that risk measure based on the proposed Latent Factor

GARCH model provides important insight on the role of risk in price expectation. Specifically, both risk measures reveal that risk reduces the expectation of price in the case of barley and wheat.

This chapter has also examined the performance of four popular estimators in the context of their applications in estimating the Nerlovian model. The panel data version of the Nerlovian model is a dynamic panel data model with serial correlation in the form of an order 1 moving average process. This chapter applied the Fixed Effect (FE) Estimator, two Generalised Method of Moments (GMM) estimators as proposed in Arellano and Bond (1991) and the Blundell and Bond (1998) as well as the Generalised Least Squares (GLS) estimators to estimate the Nerlovian model. The results are consistent with the literature in that the estimate of the Autoregressive coefficient depends heavily on the chosen estimator. However, the implied long run elasticity estimate, which is the focus of the model, remained robust for the two crops considered. The estimated elasticities are also broadly in line with those presented in the literature. The implication is that while it is challenging to produce robust estimate on the short run response of crop yield to the differential between actual and expected price, the elasticity estimate appears to be relatively robust.

As mentioned above, the chapter also briefly discussed the possibility of extending the Nerlovian model in the context of a three-dimensional panel data model. While such extention may lead to more efficient use of information, there exists econometric challenges that yet to be resolved and this could be an interesting direction for future research.

Appendix

Summary of Elasticity Estimates

In Table 3.6 the most recent applications and extensions of the Nerlovian model are summarised. **SR** stands for the short run, while **LR** for the long run elasticities.

Table 3.6: Supply Elasticities by Crop

Authors	Commodity, Location and Data Source	Estimation Method	Elasticity
Puga and Anderson (2024)	Grape varieties. South Australia. Source: Vinehealth Australia and Wine Australia	Ordinary Least Square (OLS)	**SR:** +0.074 to 0.089; **LR:** +0.333 to 0.397
Lemontzoglou and Carmona-Zabala (2024)	Greek tobacco. Greece. National Tobacco Board.	Panel Autoregressive Distributed lag (ARDL)	**SR:** +1.83 to 4.98
Okou et al. (2023)	Cocoa and Cashew nuts. Côte d'Ivoire. Source: Cotton and Cashew Council (CCA), FAO and SODEXAM.	OLS and MLE	**SR:** 0.012
Diop and Traoré (2023)	Cotton. Mali. Source: FAOSTAT.	Non-linear ARDL	**SR:** Area Cultivated (Production): +0.63 (+0.58); **LR:** Area Cultivated (Production): +0.43 (+0.38)
Krah (2023)	Maize. Ghana. Source: Esoko Ghana, FAO, Global Information and Early Warning System (GIEWS).	Pooled OLS and FEs	**SR:** +0.018; **LR:** –
Jongeneel and Gonzalez-Martinez (2022)	Milk and Herd. EU countries. Source: AGMEMOD model database.	OLS and Theil-Goldberger Mixed estimator	**SR:** +0.2 to 0.36 (Milk), +0.1 (Herd); **LR:** –
Yu et al. (2022)	Rice. China. Source: Provincial Statistical Yearbook of China and China Agricultural Products Price Yearbook.	OLS	**SR:** + 0.194; **LR:** –
Nhundu et al. (2022)	Sunflower: South Africa. Source: Centre of Collaboration on Economics of Agriculture Research and Development and DAFF (2017)'s Abstract of Agricultural Statistics	OLS	**SR:** +0.238; **LR:** +0.313
Amine M. Benmehaia (2021)	19 crops: Algeria. Source: FAO	Error Correction Model (ECM)	**SR:** +0.161 to 0.393; **LR:** +to 0.99
Pates and Hendricks (2021)	Corn: U.S. Source: USDA's Cropland Data Layer (CDL).	Markov transition regression	**SR:** +0.69; **LR:** +0.54
Bouraima, Johnson and Atchadé (2020)	Cotton: Benin. Source: Climatology service of Meteo Benin.	Cointegration with structural breaks	**SR:** –; **LR:** +0.97
Naabi and Bose (2020)	Fish: Oman. Source: Various	OLS	**SR:** Export supply: +1.44; **LR:** –
Qian et al. (2020)	Grain (Rice and Wheat). China. Source: China Statistical Yearbook	Generalized method of Moments (GMM)	**SR:** Rice: +0.069 (yield) and +0.083 (planted area); Wheat +0.13 (yield) and +0.12 (planted area); **LR:** –
Pane and Supriana (2020)	Shallots: North Sumatera–Indonesia.	OLS	**SR:** -0.23; **LR:** -0.20
Malaiarasan, Paramasivam, Thomas Felix and Balaji (2020)	Sugar: India. Source:ISMA [Indian Sugar Mills Association], Cooperative Sugar, and Indian Sugar journals	3SLS	**SR:** Production: +0.02; **LR:** –
Tenaye (2020)	Teff, wheat, and barley: Ethiopia. Source: Ethiopia Rural Household Survey (ERHS)	GMM	**SR:** Area: Teff (+5.46), Barley (+0.44), Wheat (0). Yield: Teff (+11.39), Barley (+1.08), Wheat (0); **LR:** Area: Teff (+7.27), Barley (+0.40), Wheat (0). Yield: Teff (+13.89), Barley (+0.70), Wheat (0)
Li, Liu and Song (2020)	Wheat: China. Source: Ministry of Agriculture and Rural Affairs of China; News from the Ministry of finance of China.	Three-stage Least squares (3SLS)	**SR:** Planting area: 0.103; **LR:** –

Table 3.1: Cont'd — Supply Elasticities by Crop

Authors	Commodity, Location and Data Source	Estimation Method	Elasticity
Zhai et al. (2019)	Green fodder: China. Source: Provincial Statistical Yearbooks.	GMM	**SR:** +0.21; **LR:** –
Suh and Moss (2018)	Corn, Cotton, Wheat, and Soybeans.US. Economic Research Service of the US Department of Agriculture.	MLE	**SR:** Corn: +0.53, Cotton: +0.43, Wheat: +0.62, Soybeans: +0.63; **LR:** –
Meyer (2018)	Corn, soybeans, and grassland. US. Source: USDA NASS cropland data layers (CDL's) and Iowa State's Limnology Laboratory website.	Arellano-Bond Difference-GMM	**SR:** Crops (Corn and Soybeans): 0.05; **LR:** –
Iqbal and Babcock (2018)	Corn, soybeans, wheat, and rice. Panel of countries. Source: FASOSTAT,Quandl database, Global Economic Monitor Commodities database, US Bureau of labor Statistics.	Mean Group (MG) estimator	**SR:** Aggregate Crops +0.02, Corn +0.13, Soybeans +0.20, Wheat +0.04, Rice 0; **LR:** Aggregate Crops +0.14, Corn +0.27, Soybeans +0.79, Wheat +0.28, Rice +0.05
Qian, Ito, Mu, Zhao and Wang (2018)	Rice, Wheat, Corn: China. United States Department of Agriculture (USDA), China Agricultural Development Report, National Cost and Return of Agricultural Products in China and China Statistical Yearbook.	OLS and weighted least squares (WLS)	**SR:** Area: Rice +0.07, Wheat 0.13, Corn 0.11. Yield: Rice +0.04, Wheat 0.13, Corn 0.05; **LR:** –
Magrini et al. (2018)	Staple Foods :cereals (maize, wheat, sorghum, rice, millet, and barley); roots and tubers (cassava, yams, and potatoes); and pulses (beans). Ten different SSA countries. Source: FAO (MAFAP database), FAOSTAT, World Bank's Global Economic Monitor (GEM) Commodities Database.and WDI.	Blundell and Bond System-GMM	**SR:** Acreage: farm-gate +0.31, wholesale +0.41, Production: farm-gate +0.60, wholesale +0.63, Yield: farm-gate +0.25, wholesale +0.36; **LR:** –
Le Clech and Fillat-Castejón (2017)	Aggregate grain and oilseed: barley, corn, millet, oats, rape, rice, rye, sorghum, soybean, sunflower, and wheat. FAOSTAT Database. International Fertilizer Industry Association (IFA).	OLS– Driscoll and Kraay (1998) (OLS-DK)	**SR:** +0.10; **LR:** –
Ge and Kinnucan (2018)	Cattle, sheep and goats. Inner Mongolia Autonomous Region (IMAR) - China. Source: Bureau of Statistics in IMAR and mainland China.	Pooled OLS	**SR:** Cattle -0.32, Sheep 0, Goat 0; **LR:** Cattle -0.45, Sheep 0, Goat 0
Kim and Moschini (2018)	Corn and Soybeans. US. Source: National Agricultural Statistics Service (NASS) of the U.S. Department of Agriculture (USDA).	Difference GMM, SUR	**SR:** Yield: Corn +01, Soybeans 0 Acreage: Corn +0.50, Soybeans +0.38; **LR:** Acreage: Corn +0.39, Soybeans +0.26
Haile et al. (2015)	Wheat, rice, corn, soybeans. Cross-country. Source: FAO, World Bank's commodity price database, Bloomberg database.	System–GMM	**SR:** Production: Wheat +0.11, Corn +0.23, Soybean +0.37, Rice +0.06. Acreage: Wheat +0.08, Corn +0.07, Soybean +0.15, Rice +0.02. Yield: Wheat +0.17, Corn +0.09, Soybean +0.15, Rice +0.04; **LR:** Production: Wheat +2.72, Corn +6.28, Soybean +5.07, Rice +0.15. Acreage: Wheat +7.50, Corn +3.14, Soybean +2.15, Rice +0.09. Yield: Wheat +2.08, Corn +2.35, Soybean +1.95

Table 3.1: Cont'd — Supply Elasticities by Crop

Authors	**Commodity, Location and Data Source**	**Estimation Method**	**Elasticity**
Rude and Surry (2014)	Hogs. Canada. Source: Agriculture and AgriFood Canada.	Structural time-series model (STSM)	**SR:** +0.12 to +0.21; **LR:** –
Boussios and Barkley (2014)	Wheat, Corn, Sorghum, Soybean. US. Source: USDA National Agricultural Statistics Service, Kansas State Weather Data Library and the National Climatic Data Center (NCDC)	FE	**SR:** Wheat +0.61, Corn +0.42, Soybean +0.69, Sorghum +0.36; **LR:** Wheat +0.71, Corn +2.48, Soybean +2.18, Sorghum +0.38
Haile, Kalkuhl and von Braun (2014)	Wheat, corn, soybeans, and rice. Cross-Country. Source: FAO and the United States Department of Agriculture (USDA)	OLS	**SR:** Annual: Wheat +0.09, Corn +0.18, Soybean +0.37, Rice +0.02. Intra-annual: Wheat +0.07, Corn +0.11, Soybean +0.14, Rice 0; **LR:** –
Bardal (2013)	Corn, soybeans, rice. Brazil. Source:Centro de Estudos Avançados em Economia Aplicada CEPEA.	GMM	**SR:** +0.11 – 0.14; **LR:** +0.66
Theriault et al. (2013)	Cotton. Mali. Source: Malian Company for Textile Development (CMDT)	Bias-corrected fixed-effect estimator (LSDVC)	**SR:** +0.48 – 0.51; **LR:** +0.64
de Castro and Teixeira (2012)	Cotton, Rice, Bean, Corn, Soybean, Wheat. Brazil.Source: Various	Seemingly unrelated regression (SUR)	**SR:** Cotton (0), Rice (+0.32), Bean (+0.50), Corn (+0.45), Soybean (+0.57), Wheat (+1.31); **LR:** –
Xu, Shengxiong, Zhijian and Wei (2012)	Grape. China. Source: FAO	OLS	**SR:** +0.08; **LR:** +0.80
De Menezes and Piketty (2012)	Soybean. Brazil. Source: Instituto Brasileiro de Geografia e Estatıstica (IBGE) and Municipal Agricultural Production Survey (PAM)	AB GMM	**SR:** +0.18; **LR:** +0.79
Hausman (2012)	Soybeans and Sugarcane. Brazil. Source: IPEA (Instituto de Pesquisa Econômica Aplicada)	Kiviet's bias-corrected estimator	**SR:** Soybean +0.89, Sugarcane 0; **LR:** Soybean +2.23, Sugarcane 0
Ogundeji, Jooste and Oyewumi (2011)	Beef. South Africa. Source:Department of Agriculture.	Error Correction Model (ECM)	**SR:** 0; **LR:** +0.33
Ozkan and Karaman (2011)	Cotton. Turkey. Source: Union of Agricultural Sales Cooperatives	ARDL	**SR:** +0.02 to 0.56; **LR:** +0.44 to 2.01
Asheim, Dahl, Kumbhakar, Oglend and Tveteras (2011)	Farmed salmon. Norway. Source: Norwegian Seafood Export Council, Kontali AS, and Marine Harvest AS.	Three-stage Least squares (3SLS)	**SR:** +0.09; **LR:** +0.14

Derivation of the Kalman Filter

Recall the state-space representation of the model is

$$
\begin{aligned}
\mathbf{s}_t &= \boldsymbol{\mu} + \boldsymbol{\varepsilon}_t \\
\boldsymbol{\varepsilon}_t &= \boldsymbol{\Lambda}\mathbf{F}_t + \mathbf{u}_t \\
\mathbf{F}_t &= \mathbf{v}_t .
\end{aligned}
$$

Define $\boldsymbol{\Omega}_t := \mathbb{E}\left(\mathbf{F}_t\mathbf{F}_t'\right)$, $\boldsymbol{\Gamma} := \mathbb{E}(\mathbf{u}_t\mathbf{u}_t')$, and $\mathbf{H}_{t|t-1} := \mathbb{E}\left(\boldsymbol{\varepsilon}_t - \boldsymbol{\varepsilon}_{t|t-1}\right)\left(\boldsymbol{\varepsilon}_t - \boldsymbol{\varepsilon}_{t|t-1}\right)' = \boldsymbol{\Lambda}\boldsymbol{\Omega}_{t|t-1}\boldsymbol{\Lambda} + \boldsymbol{\Gamma}$. Since $\mathbf{F}_t = \mathbf{v}_t$ with $\mathbb{E}(\mathbf{v}_t) = \mathbf{0}$, this implies $\mathbf{F}_{t|t-1} = \mathbf{0}$ and given $\boldsymbol{\varepsilon}_t$, $F_{t|t-1}$ can be updated as

$$
\begin{aligned}
\mathbf{F}_{t|t} &= \mathbf{F}_{t|t-1} + \mathbf{K}\left(\boldsymbol{\varepsilon}_t - \boldsymbol{\varepsilon}_{t|t-1}\right) \\
&= \mathbf{K}\boldsymbol{\varepsilon}_t
\end{aligned}
$$

for some adjustment matrix $\mathbf{K}$. In this case, $\mathbf{K}$ is selected based on the minimization of $\boldsymbol{\Omega}_{t|t} := \mathbb{E}(\mathbf{F}_t - \mathbf{F}_{t|t})(\mathbf{F}_t - \mathbf{F}_{t|t})'$. That is

$$
\begin{aligned}
\mathbf{K} &= \underset{\mathbf{Q}}{\arg\min}\, \boldsymbol{\Omega}_{t|t}(\mathbf{Q}) \\
&= \mathbf{Q}\mathbf{H}_{t|t-1}\mathbf{Q}' - \boldsymbol{\Omega}_{t|t-1}\boldsymbol{\Lambda}'\mathbf{Q}' - \mathbf{Q}\boldsymbol{\Lambda}\boldsymbol{\Omega}_{t|t-1}.
\end{aligned}
$$

Differentiating with respect to $\mathbf{Q}'$ and set it to zeros gives the First Order Condition that must be satisfied by $\mathbf{K}$, that is

$$
\mathbf{K}\mathbf{H}_t - \boldsymbol{\Omega}_{t|t-1}\boldsymbol{\Lambda}' = 0
$$

which gives the optimal *Kalman Gain* as $\mathbf{K} = \boldsymbol{\Omega}_{t|t-1}\boldsymbol{\Lambda}'\mathbf{H}_t^{-1}$. This provides the update rule for $\mathbf{F}_{t|t-1}$ i.e., $\mathbf{F}_{t|t} = \boldsymbol{\Omega}_{t|t-1}\boldsymbol{\Lambda}'\mathbf{H}_t^{-1}\boldsymbol{\varepsilon}_t$.

Updating $\boldsymbol{\Omega}_{t|t-1}$ follows similar argument. Note that $\boldsymbol{\Omega}_{t|t} = \mathbb{E}\left(\mathbf{F}_t - \mathbf{F}_{t|t})(\mathbf{F}_t - \mathbf{F}_{t|t}\right)$ and

$$
\begin{aligned}
&\mathbb{E}\left(\mathbf{F}_t - \mathbf{F}_{t|t})(\mathbf{F}_t - \mathbf{F}_{t|t})\right) \\
&= \mathbb{E}\left[(\mathbf{F}_t - \mathbf{F}_{t|t-1}) + \mathbf{K}(\boldsymbol{\Lambda}\mathbf{F}_t + \mathbf{u}_t)\right]\left[(\mathbf{F}_t - \mathbf{F}_{t|t-1}) + \mathbf{K}(\boldsymbol{\Lambda}\mathbf{F}_t + \mathbf{u}_t)\right]' \\
&= \mathbb{E}\left[(\mathbf{I} - \mathbf{K}\boldsymbol{\Lambda})\mathbf{F}_t + \mathbf{K}\mathbf{u}_t)\right]\left[(\mathbf{I} - \mathbf{K}\boldsymbol{\Lambda})\mathbf{F}_t + \mathbf{K}\mathbf{u}_t)\right]' \\
&= (\mathbf{I} - \mathbf{K}\boldsymbol{\Lambda})\boldsymbol{\Omega}_{t|t-1}(\mathbf{I} - \mathbf{K}\boldsymbol{\Lambda})' + \mathbf{K}\boldsymbol{\Gamma}\mathbf{K}'
\end{aligned}
$$

and substitute $\mathbf{K} = \boldsymbol{\Omega}_{t|t-1}\boldsymbol{\Lambda}'\mathbf{H}_t^{-1}$ into the last expression above gives the result.

References

Amine M. Benmehaia. (2021, April). Aggregate Supply Response in Algerian Agriculture: The Error Correction Model Applied to Selected Crops. *New Medit*, *20*(1), 85-96. doi: 10.30682/nm2101f

Arellano, M. & Bond, S. (1991). Some tests of specification for panel data: Monte Carlo evidence and an application to employment equations. *The Review of Economic Studies*, *58*(2), 277-297. doi: 10.2307/2297968

Asheim, L. J., Dahl, R. E., Kumbhakar, S. C., Oglend, A. & Tveteras, R. (2011). Are prices or biology driving the short-term supply of farmed salmon? *Marine Resource Economics*, *26*(4), 343–357.

Askari, H. & Cummings, J. T. (1977, June). Estimating agricultural supply response with the Nerlove Model: A survey. *International Economic Review*, *18*(2), 257-292. Retrieved 2024-10-01, from https://www.jstor.org/stable/2525749?origin=crossref doi: 10.2307/2525749

Balázsi, L., Mátyás, L. & Wansbeek, T. (2024). Fixed effects models. In L. Mátyás (Ed.), *The econometrics of multi-dimensional panels: Theory and applications* (pp. 1–37). Cham: Springer International Publishing. Retrieved from https://doi.org/10.1007/978-3-031-49849-7_1 doi: 10.1007/978-3-031-49849-7_1

Baltagi, B. H. (2005). *Econometric analysis of panel data*. John Wiley & Sons.

Bardal, D. (2013). Agricultural production and yield estimation: Two distinctive aspects of brazilian agriculture and a perspective on world food problems. *Future of Food: Journal on Food, Agriculture and Society*, *1*(2), 161–174.

Bauwens, L., Laurent, S. & Rombouts, J. V. K. (2006). Multivariate GARCH models: a survey. *Journal of Applied Econometrics*, *21*(1), 79–109.

Bezanson, J., Edelman, A., Karpinski, S. & Shah, V. B. (2017). Julia: A fresh approach to numerical computing. *SIAM review*, *59*(1), 65–98. Retrieved from https://doi.org/10.1137/141000671

Blundell, R. & Bond, S. (1998). Initial conditions and moment restrictions in dynamic panel data models. *Journal of Econometrics*, *87*(1), 115–143. doi: 10.1016/S0304-4076(98)00009-8

Bollerslev, T. (1986). Generalized autoregressive conditional heteroskedasticity. *Journal of Econometrics*, *31*(3), 307–327.

Bouraima, L. I. M., Johnson, J. W. A. & Atchadé, M. N. (2020). Climate variability and cotton production in the department of Alibori in Benin: An analysis taking of structural breaks. *International Journal of Agricultural and Statistical Sciences, 16(2).* Retrieved from https://www.researchgate.net/publication/346967432_Climate_variability_and_Cotton_production_in_the_Department_of_Alibori_in_Benin_An_analysis_taking_of_structural_breaks_International_Journal_of_Agricultural_and_Statistical_Sciences_Vol_162_783-75_20

Boussios, D. & Barkley, A. (2014). Producer expectations and the extensive margin in grain supply response. *Agricultural and Resource Economics Review*, *43*(3), 335–356.

Brooks, C., Burke, S. P. & Persand, G. (2003). Multivariate GARCH models: software choice and estimation issues. *Journal of Applied Econometrics*, *18*, 725-734.

Brown, R. G. (1956). *Exponential smoothing for predicting demand.* Cambridge Massachusetts: Arthur D. Little, Inc. Retrieved from https://www.industrydocuments.ucsf.edu/docs/jzlc0130

Chan, F., Jackson, J., Duwmfour, R. & Mátyás, L. (2025). Online supplementary materials. In B. H. Baltagi & L. Matyas (Eds.), *Seven decades of econometrics and beyond.* Springer. https://gitlab.com/chansta/nerlovechapter3.

Chan, F., Mátyás, L. & Reguly, Á. (2024). When and how much do fixed effects matter? In L. Mátyás (Ed.), *The econometrics of multi-dimensional panels: Theory and applications* (pp. 39–60). Cham: Springer International Publishing. Retrieved from https://doi.org/10.1007/978-3-031-49849-7_2 doi: 10.1007/978-3-031-49849-7_2

Croissant, Y. & Millo, G. (2018). *Panel data econometrics with R.* Wiley.

de Castro, E. R. & Teixeira, E. C. (2012). Rural credit and agricultural supply in brazil. *Agricultural Economics, 43*(3), 293–302.

De Menezes, T. A. & Piketty, M.-G. (2012). Towards a better estimation of agricultural supply elasticity: the case of soya beans in brazil. *Applied Economics, 44*(31), 4005–4018.

Diebold, F. X. & Nerlove, M. (1989). The dynamics of exchange rate volatility: A multivariate latent factor ARCH model. *Journal of Applied Econometrics, 4*(1), 1–21.

Diop, I. & Traoré, F. (2023). The paradox of the supply elasticity of cotton in Mali. *Applied Economics Letters, 30*(14), 1856–1860. doi: 10.1080/13504851.2022.2083553

Engle, R. F. (1982). Autoregressive conditional heteroscedasticity with estimates of the variance of United Kingdom inflation. *Econometrica: Journal of the Econometric Society, 50*(4), 987–1007.

Engle, R. F. & Bollerslev, T. (1986). Modelling the persistence of conditional variances. *Econometric Reviews, 5*, 1-50.

Food and Agriculture Organisation of the United Nations. (2024). *FAOStat.* https://www.fao.org/faostat/en/. (Accessed: 2024-12-11)

Food and Agriculture Organization of the United Nations. (2024a). *FAOStat.* https://www.fao.org/faostat/en/#data/PP. (Accessed: 2024-12-11)

Food and Agriculture Organization of the United Nations. (2024b). *FAOStat.* https://www.fao.org/faostat/en/#data/QCL. (Accessed: 2024-12-11)

Ge, W. & Kinnucan, H. (2018). Dynamic analysis of the livestock inventory in Inner Mongolia. *China Agricultural Economic Review, 10*(3), 498–515.

Haile, M. G., Kalkuhl, M. & von Braun, J. (2014). Inter-and intra-seasonal crop acreage response to international food prices and implications of volatility. *Agricultural Economics, 45*(6), 693–710.

Haile, M. G., Kalkuhl, M. & von Braun, J. (2015). Worldwide acreage and yield response to international price change and volatility: A dynamic panel data analysis for wheat, rice, corn, and soybeans. *American Journal of Agricultural Economics, 98*(1), 172–190.

Harvey, A. (2001). *Forecasting, structural time series models and the kalman filter.* Cambridge University Press.

Harvey, A., Ruiz, E. & Sentana, E. (1992, April). Unobserved component time series models with Arch disturbances. *Journal of Econometrics*, *52*(1-2), 129–157. Retrieved 2024-10-04, from https://linkinghub.elsevier.com/retrieve/pii/0304407692900683 doi: 10.1016/0304-4076(92)90068-3

Hausman, C. (2012). Biofuels and land use change: sugarcane and soybean acreage response in brazil. *Environmental and Resource Economics*, *51*, 163–187.

Holt, C. C. (1957). *Forecasting seasonals and trends by exponentially weighted moving averages* (Tech. Rep. No. 52). Office of Naval Research Memorandum. Retrieved from https://linkinghub.elsevier.com/retrieve/pii/S0169207003001134

Holt, C. C. (2004). Forecasting seasonals and trends by exponentially weighted moving averages. *International Journal of Forecasting*, *20*(1), 5–10. doi: 10.1016/j.ijforecast.2003.09.015

Iqbal, M. Z. & Babcock, B. A. (2018). Global growing-area elasticities of key agricultural crops estimated using dynamic heterogeneous panel methods. *Agricultural Economics*, *49*(6), 681–690.

Jongeneel, R. & Gonzalez-Martinez, A. R. (2022, March). The role of market drivers in explaining the EU milk supply after the milk quota abolition. *Economic Analysis and Policy*, *73*, 194–209. Retrieved 2024-12-06, from https://linkinghub.elsevier.com/retrieve/pii/S0313592621001703 doi: 10.1016/j.eap.2021.11.020

Kim, H. & Moschini, G. (2018). The dynamics of supply: US corn and soybeans in the biofuel era. *Land Economics*, *94*(4), 593–613.

Krah, K. (2023). Maize price variability, land use change, and forest loss: evidence from Ghana. *Land Use Policy*, *125*, 106472. doi: 10.1016/j.landusepol.2022.106472

Lai, R., Byrne, S., Bates, D., Alday, P., Shah, V. B., Bouchet-Valat, M., . . . Mogensen, P. K. (2024). *Juliainterop/rcall.jl: v0.14.6.* doi: 10.5281/zenodo.13735643

Le Clech, N. & Fillat-Castejón, C. (2017). International aggregate agricultural supply for grain and oilseed: The effects of efficiency and technological change. *Agribusiness*, *33*(4), 569–585.

Lemontzoglou, T. L. & Carmona-Zabala, J. C.-Z. (2024). A poor but efficient crop: Supply-side responses in the Greek tobacco sector, 1953-64. *Historia Agraria Revista de Agricultura e Historia Rural*, 161–190. doi: 10.26882/histagrar.092e02t

Li, J., Liu, W. & Song, Z. (2020, August). Sustainability of the adjustment schemes in China's grain price support policy—an empirical analysis based on the partial equilibrium model of wheat. *Sustainability*, *12*(16), 6447. Retrieved 2024-12-06, from https://www.mdpi.com/2071-1050/12/16/6447 doi: 10.3390/su12166447

Magrini, E., Balié, J. & Morales-Opazo, C. (2018). Price signals and supply responses for staple food crops in sub-saharan africa. *Applied Economic Perspectives and Policy*, *40*(2), 276–296.

Malaiarasan, U., Paramasivam, R., Thomas Felix, K. & Balaji, S. J. (2020, June). Simultaneous equation model for Indian sugar sector. *Journal of Social and Economic Development*, *22*(1), 113–141. Retrieved 2024-12-06, from http://link.springer.com/10.1007/s40847-020-00095-0 doi: 10.1007/s40847-020-00095-0

Markowitz, H. (1952). Portfolio selection. *The Journal of Finance*, *7*, 77-91.

Meyer, K. (2018). The impact of agricultural land use change on lake water quality: evidence from iowa. *Studies in Agricultural Economics*, *120*(2), 105–111.

Mátyás, L. (Ed.). (2024). *The Econometrics of Multi-dimensional Panels*. Springer Nature.

Naabi, A. A. & Bose, S. (2020, July). Do Regulatory Measures Necessarily Affect Oman's Seafood Export-Supply? *Sage Open*, *10*(3), 2158244020950658. Retrieved 2024-12-06, from https://journals.sagepub.com/doi/10.1177/2158244020950658 doi: 10.1177/2158244020950658

Nerlove, M. (1956). Estimates of the elasticities of supply of selected agricultural commodities. *Journal of Farm Economics*, *38*(2), 496-509. doi: 10.2307/1234389

Nerlove, M. (1967). Experimental Evidence on the Estimation of Dynamic Economic Relations from a Time Series of Cross-Section. *The Economic Studies Quarterly (Tokyo. 1950)*, *18*(3), 42–74. doi: 10.11398/economics1950.18.3_42

Nerlove, M. (1971). Further Evidence on the Estimation of Dynamic Economic Relations from a Time Series of Cross Sections. *Econometrica*, *39*(2), 359–382. (Publisher: [Wiley, Econometric Society]) doi: 10.2307/1913350

Nerlove, M. & Addison, W. (1958). Statistical estimation of long-run elasticities of supply and demand. *Journal of Farm Economics*, *40*(4), 861-880. doi: 10.2307/1234772

Nhundu, K., Gandidzanwa, C., Chaminuka, P., Mamabolo, M., Mahlangu, S. & Makhura, M. N. (2022). Agricultural supply response for sunflower in South Africa (1947–2016): The partial Nerlovian framework approach. *African Journal of Science, Technology, Innovation and Development*, *14*(2), 440–450. doi: 10.1080/20421338.2020.1844944

Nickell, S. (1981). Biases in dynamic models with fixed effects. *Econometrica*, *49*(6), 1417–1426.

Ogundeji, A., Jooste, A. & Oyewumi, O. (2011). An error correction approach to modelling beef supply response in south africa. *Agrekon*, *50*(2), 44–58.

Okou, G. C., Keita, K., N'Dri, Y. A. & Kouakou, A. K. (2023). Forecasting agricultural area using Nerlovian Model in Côte d'Ivoire. In *ITISE 2023* (Vols. 39,35, p. 1-9). MDPI. Retrieved from https://www.mdpi.com/2673-4591/39/1/35 doi: 10.3390/engproc2023039035

Our World in Data. (2024). *Food and agriculture - land use*. https://ourworldindata.org/land-use. (Accessed: 2024-12-11)

Ozkan, B. & Karaman, S. (2011). Acreage response for cotton regions in turkey: an application of the bounds testing approach to cointegration. *New Medit: Mediterranean Journal of Economics, Agriculture and Environment*, *10*(2), 42.

Pane, T. C. & Supriana, T. (2020, February). The supply elasticity for North Sumatera shallot in short and long-run. *IOP Conference Series: Earth and Environmental Science*, *454*(1), 012035. Retrieved 2024-12-06, from https://

iopscience.iop.org/article/10.1088/1755-1315/454/1/012035 doi: 10.1088/1755-1315/454/1/012035

Pates, N. J. & Hendricks, N. P. (2021). Fields from Afar: Evidence of Heterogeneity in United States Corn Rotational Response from Remote Sensing Data. *American Journal of Agricultural Economics*, *103*(5), 1759–1782. doi: 10.1111/ajae.12208

Puga, G. & Anderson, K. (2024). What explains changes in grape varietal mixes in Australia's wine regions? *Australian Journal of Agricultural and Resource Economics*, 1467–8489.12594. doi: 10.1111/1467-8489.12594

Qian, J., Ito, S., Mu, Y., Zhao, Z. & Wang, X. (2018). The role of subsidy policies in achieving grain self-sufficiency in china: a partial equilibrium approach. *Agricultural Economics/Zemědělská Ekonomika*, *64*(1).

Qian, J., Ito, S. & Zhao, Z. (2020, October). The effect of price support policies on food security and farmers' income in China. *Australian Journal of Agricultural and Resource Economics*, *64*(4), 1328–1349. Retrieved 2024-12-06, from https://onlinelibrary.wiley.com/doi/10.1111/1467-8489.12398 doi: 10.1111/1467-8489.12398

Ritchie, H. & Roser, M. (2019). Land use. *Our World in Data*. https://ourworldindata.org/land-use.

Rude, J. & Surry, Y. (2014). Canadian hog supply response: A provincial level analysis. *Canadian Journal of Agricultural Economics/Revue Canadienne d'Agroeconomie*, *62*(2), 149–169.

Silvennoinen, A. & Teräsvirta, T. (2008). Multivariate GARCH Models. In T. G. Andersen, R. A. Davis, J.-P. Kreiss & T. Mikosch (Eds.), *Handbook of Financial Time Series* (p. 201-229). Berlin Heidelberg: Springer-Verlag.

Suh, D. H. & Moss, C. B. (2018). Examining crop price effects on production decision and resource allocation: An ex-ante approach. *Applied Economics*, *50*(26), 2909–2919.

Tenaye, A. (2020, Jul). New evidence using a dynamic panel data approach: Cereal supply response in smallholder agriculture in Ethiopia. *Economies*, *8*(3), 61. Retrieved 2024-12-06, from https://www.mdpi.com/2227-7099/8/3/61 doi: 10.3390/economies8030061

Theriault, V., Serra, R. & Sterns, J. A. (2013). Prices, institutions, and determinants of supply in the malian cotton sector. *Agricultural Economics*, *44*(2), 161–174.

Wang, Y., Xiang, Y., Lei, X. & Zhou, Y. (2022). Volatility Analysis based on GARCH-type Models: Evidence from the Chinese stock market. *Economic Research-Ekonomska Istraživanja*, *35*(1), 2530–2554. doi: 10.1080/1331677X.2021.1967771

Wansbeek, T. & Kapteyn, A. (1989, July). Estimation of the error-components model with incomplete panels. *Journal of Econometrics*, *41*(3), 341–361. doi: 10.1016/0304-4076(89)90066-3

Xu, C., Shengxiong, W., Zhijian, Z. & Wei, S. (2012). A model for analysis of supply reaction to price applied to grapes in china. In *2012 international conference on systems and informatics (icsai2012)* (pp. 2559–2562).

Yu, Y., Clark, J. S., Tian, Q. & Yan, F. (2022). Rice yield response to climate and price policy in high-latitude regions of China. *Food Security*, *14*(5), 1143–1157. doi: 10.1007/s12571-021-01253-w

Zhai, S., Chen, Q. & Wang, W. (2019, November). What drives green fodder supply in China?—A Nerlovian analysis with LASSO Variable Selection. *Sustainability*, *11*(23), 6692. Retrieved 2024-12-06, from https://www.mdpi.com/2071-1050/11/23/6692 doi: 10.3390/su11236692

Chapter 4
Discrete Games: A Historical Perspective

Paul A. Bjorn, Isabelle Perrigne and Quang Vuong

Abstract In the seventies, building on the statistical literature, economists have developed interest in the empirical analysis of qualitative variables with the log-linear probability and latent variable models, analyzing individual decisions within a simultaneous equation setting. Starting from the eighties, they began to rely on game theoretic formulations to account for strategic interactions among agents with random utility. This chapter presents the first contributions to the econometrics of discrete games through noncooperative solution concepts, namely Nash and Stackelberg equilibria. This game theoretic approach to the empirical analysis of agents' decisions has led to a rich literature which continues to expand with applications to various domains in economics such as industrial organization, labor, public and development economics as well as beyond the economics field.

4.1 Introduction

Professor Marc Nerlove was Quang Vuong's advisor in the late seventies at Northwestern University and orientated him to work on qualitative variables. This led to Vuong's (1982) dissertation on log-linear probability models. Professor Nerlove had a profound influence on Vuong's research agenda. He recruited him first as a teaching assistant for an undergraduate statistics course in 1977 and then as a research assistant with classmate John Link from 1978 to 1980. Professor Nerlove asked them to develop log-linear probability models building on Nerlove and Press (1973). Vuong was also responsible for the proofreading and indexing of Nerlove, Grether

Paul A. Bjorn
UH Parma, 7007 Powers Blvd, Parma OH 44129, e-mail: pabjorn@hotmail.com

Isabelle Perrigne
Rice University, 6100 Main st, Houston TX 77005, e-mail: iperrigne@gmail.com

Quang Vuong ✉
New York University, 19 W. 4th st, New York NY 10012, e-mail: qvuong@nyu.edu

B. H. Baltagi and L. Mátyás (eds.), *Seven Decades of Econometrics and Beyond*, Advanced Studies in Theoretical and Applied Econometrics 57,
https://doi.org/10.1007/978-3-031-92699-0_4

and Carvalho's (1979) monograph on time series. Professor Nerlove shared his vision of econometrics and empirical research with his students. In particular, he gave Hood and Koopmans (1953) Cowles Foundation monograph to Vuong. The latter still has this book on the shelves in his office at New York University. These two books shaped Vuong's research on structural econometrics, which combines economic theory and statistics. Since his graduation in 1982, he participated in the development of econometric methods ranging from model selection tests to nonparametric estimation procedures that were motivated by empirical questions arising mostly from industrial organization such as auctions.

In this chapter, we present two unpublished papers by Bjorn and Vuong (1984, 1985) following closely their presentation while inserting some updating comments. These papers rely on a game theoretical approach to the analysis of qualitative/discrete variables. They initiated an important line of research as they account for the strategic interactions among multiple decision makers such as spouses in a household or firms in a market. We first put these two papers in the historical context of econometric research in the seventies, which focuses on modeling discrete variables with the log-linear probability and threshold-crossing latent variable models. The difficulty of modeling decisions of two agents in a simultaneous setting called for a different approach that combines econometrics and game theory. Bjorn and Vuong (1984) develop an econometric model for dichotomous/binary variables where the outcome is a Nash equilibrium of a noncooperative game of complete information between two economic agents. Bjorn and Vuong (1985) extend it to a Stackelberg game. These two papers also motivated the development of Vuong's (1989) model selection test for nonnested hypotheses. They are part of Bjorn's (1986) dissertation. The introduction of game theory in econometrics has been a major breakthrough for the analysis of joint discrete decisions such as firms' entry in markets. It further expanded into a vast and blossoming empirical literature with incomplete information games, dynamic games, analysis of networks and bargaining.

The chapter is organized as follows. Section 4.2 provides a historical perspective by reviewing the early literature on qualitative endogenous variables. It also presents a benchmark model of noncooperative game in complete information. Section 4.3 focuses on the Nash solution concept including mixed strategies and multiple equilibria. It provides the likelihood function and discusses identification. Section 4.4 follows a similar pattern with the Stackelberg approach, while Section 4.5 reports the empirical application to labor force participation from Bjorn and Vuong (1984, 1985). Lastly, Section 4.6 briefly reviews four main lines of research on the econometrics of discrete games developed since the early nineties.

4.2 Historical Perspective and Model Set-Up

The first half of this section partly draws from Bjorn's (1986) dissertation and reviews the early literature on qualitative variables up to the introduction of the game theoretic approach. The second half introduces a model with two players that is used in the game theoretic approach in the following sections.

4.2.1 Qualitative Endogenous Variables

Economic agents, firms or individuals, frequently make discrete decisions from a finite set of alternatives. Individuals choose their education such as dropping out of high school, going to college or courses to take. As adults, they decide whether to work full-time or part-time, have children but they also select their residential location, transportation mode, insurance coverage, products they purchase, etc. Firms decide whether to enter a market, the products they launch, their investment, research and development, recruitment, mergers to name a few. Modeling these decisions is crucial to understand the factors that explain or predict them. This can represent a first step to model quantitative variables associated with individuals' discrete decisions. For instance, an individual chooses to work, then he has to decide how many hours he will work; he chooses energy sources for his home and then decides how much energy to consume, etc. This non-exhaustive list of examples relates to a broad range of applied microeconomic fields such as labor, development, industrial organization, health, energy or public economics. Because it is a key component for policy recommendations, modeling agents' decisions goes beyond the economic field with marketing, management, finance, psychology, sociology, political science, urban planning, environment, agronomy, public health, health sciences, etc.

The development of statistical methods to model the probability of an event dates back from the nineteenth century with the logistic function. See Cramer (2004) for historical references. Two models were developed in parallel. The probit model relies on the standard normal distribution to assess the probability of an event, such as'success' or 'failure', as a function of a linear index of some discrete/continuous covariates. The logistic function, which was first used to model population growth, specifies the log-odds of an event as a linear combination of variables. Both models are estimated by maximum likelihood and were further extended to handle polychotomous variables that can be either ordered or unordered leading to the ordered probit and logit models and the multinomial logit and probit models, respectively. Statisticians revive interest for these models in the thirties and research done in the fourties contributed to their development. The logit approach quickly gained popularity because of its simple explicit form.[1]

The introduction of latent variables in those models has been a major leap for the analysis of economic agents' decisions. A latent variable is expressed as a linear combination of exogenous variables with an additive unobserved error term. To generate a dichotomous variable, the sign of the latent variable is associated to each of the two options, for instance an individual chooses to work if his latent variable is

[1] An alternative approach relies on ordinary least squares with the linear probability model by regressing a binary variable on independent variables. Though still popular in empirical studies, this leads to inefficient estimates and imprecise predictions.

positive and will not otherwise. In the socalled conditional logit model, McFadden (1973) combines economic theory and statistical methods to analyze economic agents' choices for their transportation mode. The latent variable is the agent's random utility associated to each transportation option, an individual chooses an option over alternatives because it generates a higher utility. Assuming that the error terms independently follow a Gumbel distribution gives the multinomial logit model because the difference of two Gumbel error terms follows a logistic distribution. This model satisfies the independence of irrelevant alternatives, an axiom from decision theory, which relates to revealed preferences in economics. In simple terms, the choice of an option is not altered if other options are proposed. McFadden's (1973) seminal paper in the analysis of qualitative variables had a profound impact in the empirical analysis of micro data. See surveys by Amemiya (1981), Aigner, Hsiao, Kapteyn and Wansbeek (1984), and McFadden (1984) as well as Manski and McFadden's (1981) monograph and Maddala's (1983) textbook.

Professor Nerlove's research covers a broad range of topics in econometrics. Not surprisingly, he also contributed to the analysis of qualitative variables while adopting a different approach that relies on the analysis of contingency tables and which has a long history in statistics. See Haberman (1974) and Bishop, Fienberg and Holland (1975). To analyze the relationships among variables that are all qualitative, Nerlove and Press (1973) develop the log-linear probability model and discuss its link with logit models. They apply their results to farming practices in a developing economy. As its name indicates, the model decomposes the logarithm of the joint probability of the discrete variables into a linear combination of a main effect and interaction effects. It is especially suitable to understand conditional probabilities and to test joint and conditional independence. See, e.g., Bouissou, Laffont and Vuong (1986) for causality tests with qualitative panel data.

The previous models apply to single agents' choices but individuals do not always take decisions by themselves, they live within a family, a circle of friends. Similarly, firms have competitors. This suggests that their decisions are dependent on other agents' choices. For instance, in a household, a member's decision of working or retiring also depends on the spouse's working or retiring decision. In a market, a firm's entry depends on other firms' entry as well as how their products compare to their competitors in terms of quality and price. Furthermore, because of peer effects, an individual's decision depends on his friends' choices to go to college, to engage in criminal behavior, unhealthy habits, etc. Because it involves multi-agent decisions, the analysis is reminiscent to simultaneous equations which were developed in the mid twentieth century for estimating supply and demand equations. See Hood and Koopmans's (1953) Cowles Commission monograph. In line with this framework, Heckman (1978) develops a model with two equations. The latent variable of each agent is a linear function of the other agent's choice and latent variable. Heckman (1978) points out that simultaneous equations models for dummy endogenous variables appearing as right-hand side variables must satisfy a coefficient restriction that he calls the logical consistency/coherency condition, namely, the product of the two coefficients of the dummy variables must equal to zero. This condition implies that the effect of an individual choice on the other's choice vanishes.

There is a priori no economic reason to impose this condition. As a matter of fact, one expects these two coefficients to be negative when considering firms' entry as a competitor's entry has a negative impact on the firm's profit.

To address this difficulty, Bjorn and Vuong (1984, 1985) introduced a game theoretical approach to capture the interactions among two individuals. Game theory started from the development of two-person zero-sum cooperative games with Von Neumann and Morgenstern's (1944) book, and rapidly found applications in social sciences, and especially economics. The analysis of games was extended to several players and non-zero sum noncooperative games with the Nash equilibrium. See Nash (1950, 1951). The Nash equilibrium applies to noncooperative games with several players under complete information by considering strategies played by individuals who act independently, i.e., there is no binding agreements to enforce their cooperation, and who observe other players' preferences and strategies. Subsequently, Harsanyi (1967) developed the Bayesian Nash equilibrium for games of incomplete information, in which individuals do not possess full information about others. Each player possesses some private information and forms expectations on how others behave. This concept was key for the development of the economics of information with auction theory, contract theory and more generally the theory of incentives with the principal-agent model (see Laffont & Martimort, 2002 and Krishna, 2002).

The Nash and Bayesian Nash equilibria have been a stepping stone in the analysis of relationships among economic agents. Since then, there has been an abundant econometric and empirical literature on qualitative endogenous variables relying on game theoretic solution concepts. We briefly review this literature in Section 4.6. In the next subsection, we present the model set-up for a two-person noncooperative game in complete information following Bjorn and Vuong (1984, 1985).

4.2.2 Model Set-Up: A Game Theoretic Approach

In the game theoretic approach, the endogenous variables are the outcomes of an equilibrium. In view of the previous discussion, the model also includes latent variables and random utility. We restrict our attention to two agents indexed by $i = 1, 2$, each of whom has only two possible and exclusive actions/choices $y_i \in \{0, 1\}$. Let $\tilde{U}_i(y_i, y_j)$ be the utility/payoff that agent i derives from taking action y_i when the other agent $j \neq i$ takes action y_j. Following McFadden (1973, 1981) the utility $\tilde{U}_i(y_i, y_j)$ is treated as random and decomposed into a deterministic component and an unobserved component

$$\tilde{U}_i(y_i, y_j) = U_i(y_i, y_j) + \eta_i(y_i, y_j).$$

Only some utility differences are relevant in the non-cooperative (Nash and Stackelberg) equilibrium concepts invoked below. Without loss of generality, we let

$$U_i(1, y_j) - U_i(0, y_j) = \Delta_i + \alpha_i y_j,$$

where Δ_i represents the systematic utility gain for choosing $y_i = 1$ instead of $y_i = 0$ irrespective of the opponent's action y_j, whereas the structural effect of the opponent's choice reduces to the shift parameter $\alpha_i \in \mathbb{R}$ when $y_j = 1$ and 0 otherwise. Bjorn and Vuong (1984, 1985) assume that the unobserved components satisfy

$$\eta_i(1, y_j) - \eta_i(0, y_j) = \epsilon_i,$$

so that such a difference does not depend on $y_j = 0, 1$. Thus, agent i's utility for choosing $y_i = 1$ instead of $y_i = 0$ when his opponent j chooses action y_j is

$$\tilde{U}_i(1, y_j) = \tilde{U}_i(0, y_j) + \Delta_i + \alpha_i y_j + \epsilon_i, \tag{4.1}$$

for $i = 1, 2$ and $j \neq i$. In general, Δ_1 and Δ_2 are parameterized linear functions of the agents' observed discrete and/or continuous characteristics (X_1, X_2), namely,

$$\Delta_i = X_i' \beta_i \quad \text{for } i = 1, 2, \tag{4.2}$$

where X_1 and X_2 may have some variables in common and β_i are some parameters. Moreover, the pair of unobserved components (ϵ_1, ϵ_2) follows a joint distribution $F(\cdot, \cdot)$ with zero means and density $f(\cdot, \cdot)$ with respect to Lebesgue measure and support $\mathbb{R}^2$ independently from (X_1, X_2). Following Heckman (1978) we refer to (α_1, α_2) as representing the structural effect of the opponent's choice, and the dependence between the unobserved components (ϵ_1, ϵ_2) as the statistical association between the agents' choices. An important task for the analyst is to distinguish these two types of association. The model structure $[\alpha_1, \alpha_2, \beta_1, \beta_2, \epsilon_1, \epsilon_2]$ is common knowledge, namely each agent knows the other's preferences. The setting is a model of complete information.[2]

Let agent i's observed choice be $Y_i \in \{0, 1\}$ for $i = 1, 2$. A standard approach is to assume that Y_i is generated from a latent continuous variable Y_i^* crossing a threshold, namely $Y_i = 1$ if $Y_i^* \geq 0$ and $Y_i = 0$ if $Y_i^* < 0$.[3] When $Y_i^* = \tilde{U}_i(1, Y_j) - \tilde{U}_i(0, Y_j)$, then (4.1) gives the simultaneous equation model in observed and latent variables (Y_1, Y_2, Y_1^*, Y_2^*)

$$\begin{aligned} Y_1^* &= \Delta_1 + \alpha_1 Y_2 + \epsilon_1 \\ Y_2^* &= \Delta_2 + \alpha_2 Y_1 + \epsilon_2. \end{aligned} \tag{4.3}$$

As is well-known from Heckman (1978), Gourieroux, Laffont and Monfort (1980) and Schmidt (1981), the system (4.3) admits a well-defined reduced form, i.e., defines a joint probability distribution for the agents' observed choices (Y_1, Y_2) if and only

[2] The difference with a model of incomplete information is that agent i knows ϵ_i but not ϵ_j, the random term ϵ_j then becomes private information, also commonly called agent's type. However, individual i knows that individual j's private information is drawn from the same distribution $F(\cdot, \cdot)$. The game is then solved relying on a Bayesian Nash equilibrium. See Liu, Vuong and Xu (2017) for an econometric framework of such binary games. See Section 4.6.

[3] When an individual is indifferent between the two alternatives, we assume that he chooses the alternative $y_i = 1$, hence the use of the weak inequality.

if the coherency condition $\alpha_1\alpha_2 = 0$ holds. As argued by Bjorn and Vuong (1984) such a condition renders the model essentially recursive, more precisely, one agent's choice is structurally independent from the other agent's choice. To avoid such a constraint, Bjorn and Vuong (1984, 1985) assume instead that the choices (Y_1, Y_2) are the equilibrium outcomes of the game played by the two agents.

From (4.1), the normal form of the game is given by the following 2×2 table, which displays each agent's payoff for every pair of actions.

Table 4.1: Normal Form

	$y_2 = 0$	$y_2 = 1$
$y_1 = 0$	$\tilde{U}_1(0,0)$; $\tilde{U}_2(0,0)$	$\tilde{U}_1(0,1)$; $\tilde{U}_2(0,0) + \Delta_2 + \epsilon_2$
$y_1 = 1$	$\tilde{U}_1(0,0) + \Delta_1 + \epsilon_1$; $\tilde{U}_2(0,1)$	$\tilde{U}_1(0,1) + \Delta_1 + \alpha_1 + \epsilon_1$; $\tilde{U}_2(0,1) + \Delta_2 + \alpha_2 + \epsilon_2$

Different equilibrium concepts can be invoked to solve this game of complete information. Bjorn and Vuong (1984, 1985) use the noncooperative Nash and Stackelberg equilibrium concepts, respectively. The major difference between the two is that agents move simultaneously in a Nash equilibrium, whereas the Stackelberg game is sequential with one agent, called the leader, moving first, followed by the second agent, called the follower.[4]

4.3 The Nash Approach

Nash (1950, 1951) shows that every finite normal-form game has a Nash equilibrium (NE) in mixed strategies, i.e., a mixed NE. If both mixed strategies are degenerate, the NE is in pure strategies. See e.g., Fudenberg and Tirole (1991). Hence, the preceding finite game (two players with two actions each) has at least one NE. Starting from each player's reaction function (or best response), Bjorn and Vuong (1984) determine all its NEs in pure and mixed strategies. This depends on the signs of the parameters (α_1, α_2) when $\alpha_1\alpha_2 \neq 0$,

[4] For instance, a larger firm with more market power or a spouse with a larger salary/wealth can rationalize the use of a Stackelberg equilibrium. Both equilibria lead to different implications/restrictions on observations that can be tested on data. See Section 4.5 for model selection tests.

• CASE A: $\alpha_1 > 0$ AND $\alpha_2 > 0$.[5]
A1. If $\epsilon_1 < -\Delta_1 - \alpha_1$ and $\epsilon_2 < -\Delta_2$ or if $\epsilon_1 < -\Delta_1$ and $\epsilon_2 < -\Delta_2 - \alpha_2$, there is a unique NE, namely, the pure NE $(0,0)$.
A2. If $\epsilon_1 < -\Delta_1 - \alpha_1$ and $-\Delta_2 < \epsilon_2$, there is a unique NE, namely, the pure NE $(0,1)$.
A3. If $-\Delta_1 - \alpha_1 < \epsilon_1 < -\Delta_1$ and $-\Delta_2 - \alpha_2 < \epsilon_2 < -\Delta_2$, there are three NEs, namely, the pure NEs $(0,0)$ and $(1,1)$ and a nondegenerate mixed NE.
A4. If $-\Delta_1 < \epsilon_1$ and $\epsilon_2 < -\Delta_2 - \alpha_2$, there is a unique NE, namely, the pure NE $(1,0)$.
A5. If $-\Delta_1 - \alpha_1 < \epsilon_1$ and $-\Delta_2 < \epsilon_2$ or if $-\Delta_1 < \epsilon_1$ and $-\Delta_2 - \alpha_2 < \epsilon_2$, there is a unique NE, namely, the pure NE $(1,1)$.

• CASE B: $\alpha_1 > 0$ AND $\alpha_2 < 0$.
B1. If $\epsilon_1 < -\Delta_1$ and $\epsilon_2 < -\Delta_2$, there is a unique NE, namely, the pure NE $(0,0)$.
B2. If $\epsilon_1 < -\Delta_1 - \alpha_1$ and $-\Delta_2 < \epsilon_2$, there is a unique NE, namely, the pure NE $(0,1)$.
B3. If $-\Delta_1 - \alpha_1 < \epsilon_1 < -\Delta_1$ and $-\Delta_2 < \epsilon_2 < -\Delta_2 - \alpha_2$, there is a unique NE, namely, a nondegenerate mixed NE.
B4. If $-\Delta_1 < \epsilon_1$ and $\epsilon_2 < -\Delta_2 - \alpha_2$, there is a unique NE, namely, the pure NE $(1,0)$.
B5. If $-\Delta_1 - \alpha_1 < \epsilon_1$ and $-\Delta_2 - \alpha_2 < \epsilon_2$, there is a unique NE, namely, the pure NE $(1,1)$.

• CASE C: $\alpha_1 < 0$ AND $\alpha_2 > 0$.
C1: $\epsilon_1 < -\Delta_1$ and $\epsilon_2 < -\Delta_2$, there is a unique NE, namely, the pure NE $(0,0)$.
C2: $\epsilon_1 < -\Delta_1 - \alpha_1$ and $-\Delta_2 < \epsilon_2$, there is a unique NE, namely, the pure NE $(0,1)$.
C3: $-\Delta_1 < \epsilon_1 < -\Delta_1 - \alpha_1$ and $-\Delta_2 - \alpha_2 < \epsilon_2 < -\Delta_2$, there is a unique NE, namely, a nondegenerate mixed NE.
C4: $-\Delta_1 < \epsilon_1$ and $\epsilon_2 < -\Delta_2 - \alpha_2$, there is a unique NE, namely, the pure NE $(1,0)$.
C5: $-\Delta_1 - \alpha_1 < \epsilon_1$ and $-\Delta_2 - \alpha_2 < \epsilon_2$, there is a unique NE, namely, the pure NE $(1,1)$.

• CASE D: $\alpha_1 < 0$ AND $\alpha_2 < 0$.
D1. If $\epsilon_1 < -\Delta_1$ and $-\Delta_2 < \epsilon_2$, there is a unique NE, namely, the pure NE $(0,0)$.
D2. If $\epsilon_1 < -\Delta_1$ and $-\Delta_2 < \epsilon_2$ or if $\epsilon_1 < -\Delta_1 - \alpha_1$ and $-\Delta_2 - \alpha_2 < \epsilon_2$, there is a unique NE, namely, the pure NE $(0,1)$.
D3. If $-\Delta_1 < \epsilon_1 < -\Delta_1 - \alpha_1$ and $-\Delta_2 < \epsilon_2 < -\Delta_2 - \alpha_2$, there are three NEs, namely, the pure NEs $(0,1)$ and $(1,0)$ and a nondegenerate mixed NE.
D4. If $-\Delta_1 < \epsilon_1$ and $\epsilon_2 < -\Delta_2$ or if $-\Delta_1 - \alpha_1 < \epsilon_1$ and $\epsilon_2 < -\Delta_2 - \alpha_2$, there is a unique NE, namely, the pure NE $(1,0)$.
D5. If $-\Delta_1 - \alpha_1 < \epsilon_1$ and $-\Delta_2 - \alpha_2 < \epsilon_2$, there is a unique NE, namely, the pure NE $(1,1)$.

These results do not depend on the nuisance quantities $[\tilde{U}_i(0,0), \tilde{U}_i(0,1)]$ for $i = 1,2$. Figures 1a–1d depict the 5 relevant regions in the (ϵ_1, ϵ_2)-space for each case A–D. Let $\mathcal{A}_1, \mathcal{A}_2, \ldots, \mathcal{D}_5$ be the regions in the (ϵ_1, ϵ_2)-space defined by the inequalities appearing in cases A1, A2, ..., D5. For instance, $\mathcal{A}_1 = \{(\epsilon_1, \epsilon_2) : \epsilon_1 < -\Delta_1 - \alpha_1$ and $\epsilon_2 < -\Delta_2$ or $\epsilon_1 < -\Delta_1$ and $\epsilon_2 < -\Delta_2 - \alpha_2\}$. In 16 out of the $4 \times 5 = 20$ cases, there is a unique

[5] We only consider strict inequalities in Cases A–D. Indeed, equalities arise with zero probability because (ϵ_1, ϵ_2) have a joint density with respect to Lebesgue measure. Unlike $\tilde{U}_i(y_i, y_j)$ where the first and second arguments are the agent's and opponent's actions, we use the standard notation (y_1, y_2) for a NE with agent 1's action coming first and agent 2's action coming second.

NE which is in pure strategies. In two cases B3 and C3, there is a unique NE which is in mixed strategies. In the remaining two cases A3 and D3, there are three NEs, namely, one in mixed strategies denoted M and two in pure strategies. Below, we first address mixed strategies and then the multiplicity of NEs.

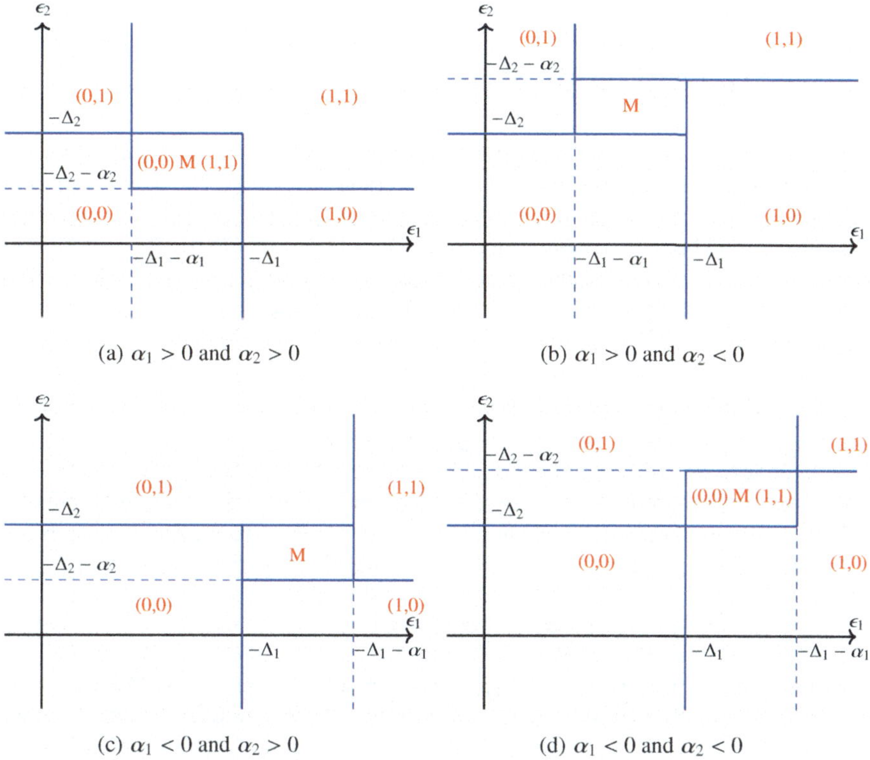

Fig. 4.1: Nash Equilibria

For completeness we consider the case when $\alpha_1\alpha_2 = 0$, i.e., when the coherency condition is satisfied. In this case, there is always a unique NE. This equilibrium coincides with the outcome given by the latent model (4.3) combined with the standard dichotomization $Y_i = 1\!\!1[Y_i^* \geq 0]$, $i = 1,2$. It is in pure strategies and dominant for the player(s) with $\alpha_i = 0$ thereby providing a recursive flavor to the model since player i's best action does not depend on his opponent's choice. For instance, when $\{\alpha_1 = 0, \alpha_2 \neq 0\}$, action $y_1 = 0$ is dominant for agent 1 if $\epsilon_1 < -\Delta_1$ whereas action $y_1 = 1$ is dominant if $\epsilon_1 > -\Delta_1$. Specifically, the unique NE is

- $(0,0)$ if $\epsilon_1 < -\Delta_1$ and $\epsilon_2 < -\Delta_2$,
- $(0,1)$ if $\epsilon_1 < -\Delta_1$ and $\epsilon_2 > -\Delta_2$,
- $(1,0)$ if $\epsilon_1 > -\Delta_1$ and $\epsilon_2 < -\Delta_2 - \alpha_2$

- $(1,1)$ if $\epsilon_1 > -\Delta_1$ and $\epsilon_2 > -\Delta_2 - \alpha_2$.

This agrees with cases A and C when α_1 vanishes. The other case $\{\alpha_1 \neq 0, \alpha_2 = 0\}$ is similarly obtained by switching the players' indices. Lastly, when $\{\alpha_1 = 0, \alpha_2 = 0\}$, i.e., there is no structural effect, there is a unique NE, which is in pure (dominant) strategies. It is obtained as above upon letting $\alpha_2 = 0$.

4.3.1 Nash Equilibria in Mixed Strategies

A NE in mixed strategies only arises in cases A3, B3, C3 and D3. Let p_i be the probability that agent i plays action $y_i = 1$. Appendix 1 provides agent i's mixing probability p_i in a mixed Nash equilibrium. See (4.19). The probabilities (p_1, p_2) depend on (ϵ_1, ϵ_2) but not on the nuisance quantities $[\tilde{U}_i(0,0), \tilde{U}_i(0,1)], i = 1,2$. They satisfy $0 < p_i < 1$ in view of the strict inequalities defining cases A3, B3, C3 and D3. Hence, the probability that agents 1 and 2 play actions y_1 and y_2 in a mixed NE is $\prod_{i=1,2} p_i^{y_i}(1-p_i)^{1-y_i}$ given (ϵ_1, ϵ_2). As a matter of fact, the latter probability is conditional on the mixed NE being played. In cases B3 and C3, the mixed NE is the unique NE and thus is played by assumption. Appendix 1 gives the conditional probabilities $\Pr(y_1, y_2 | \mathcal{B}_3)$ and $\Pr(y_1, y_2 | C_3)$ of observing (y_1, y_2) in cases B3 and C3, respectively. See (4.20) and (4.21).

Because the probabilities $\Pr(y_1, y_2 | \mathcal{B}_3)$ and $\Pr(y_1, y_2 | C_3)$ are complex nonlinear functions of the parameters, Bjorn and Vuong (1984) introduce instead probability weights $b_{y_i y_j}$ and $c_{y_i y_j}$ for $(y_i, y_j) \in \{0,1\}^2$ as additional parameters so that

$$\Pr[y_i, y_j | \mathcal{B}_3] = b_{y_i y_j} \quad \text{and} \quad \Pr[y_i, y_j | C_3] = c_{y_i y_j}, \tag{4.4}$$

where $b_{00} + b_{01} + b_{10} + b_{11} = 1$, $b_{y_i y_j} \geq 0$, $c_{00} + c_{01} + c_{10} + c_{11} = 1$ and $c_{y_i y_j} \geq 0$. This allows the analyst to test the model validity, in particular by testing whether players use the equilibrium mixing probabilities (4.20) and (4.21) through a flexible parameterization of the probability weights $b_{y_i y_j}$ and $c_{y_i y_j}$ that depends on the characteristics (X_1, X_2).

4.3.2 Multiple Nash Equilibria

Multiple NEs arise in cases A3 and D3 with two equilibria in pure strategies and one in mixed strategies in each case. As pointed out by Jovanovic (1989), multiple equilibria raise issues about identification and predictive content of a structural model. To mitigate such a difficulty, Bjorn and Vuong (1984) first eliminate the mixed NE by invoking Pareto dominance.[6] Let $\Delta\tilde{U}_i^0 \equiv \tilde{U}_i(0,0) - \tilde{U}_i(0,1)$. Appendix 2 shows that

[6] A NE Pareto dominates another NE if one player is strictly better off and the other player is not worse off. Bjorn and Vuong (1984) show that the mixed NE is always Pareto dominated by a pure NE or a bargaining solution. Appendix 2 completes their analysis.

the mixed NE is Pareto dominated by at least one pure NE in cases A3 and D3 under some assumptions on $(\Delta\tilde{U}_1^0, \Delta\tilde{U}_2^0)$. More recently, Echenique and Edlin (2004) show that, when they coexist in supermodular games, mixed equilibria are unstable and converging to a pure NE under various adjustment processes. This applies here as the games in cases A3 and D3 are supermodular.[7]

There remain two pure NEs in each case A3 or D3. Appendix 2 shows that a pure NE Pareto dominates the other pure NE (and the mixed NE) under some assumptions on $(\Delta\tilde{U}_1^0, \Delta\tilde{U}_2^0)$ given (ϵ_1, ϵ_2). See (4.22) and (4.24). To illustrate, a frequent assumption in empirical work is $\Delta\tilde{U}_i^0 = 0$ for $i = 1, 2$, i.e., each player's utility from choosing action $y_i = 0$ does not depend on the opponent's choice.[8] When combined with the normalization $\tilde{U}_i(0,0) = 0$ for $i = 1, 2$, the normal formal of the game reduces to that given on Table (4.2).

Table 4.2: Normal Form when $\tilde{U}_i(0,0) = \tilde{U}_i(0,1) = 0$ for $i = 1, 2$

	$y_2 = 0$	$y_2 = 1$
$y_1 = 0$	$0 \;\; ; 0$	$0 \;\; ; \;\; \Delta_2 + \epsilon_2$
$y_1 = 1$	$\Delta_1 + \epsilon_1 \; ; 0$	$\Delta_1 + \alpha_1 + \epsilon_1 \; ; \; \Delta_2 + \alpha_2 + \epsilon_2$

In case A3, where $\alpha_i > 0$ for $i = 1, 2$, (4.22) is satisfied and Figure 4.2a indicates that the pure NE $(1,1)$ Pareto dominates the other pure NE $(0,0)$ and the mixed NE, which are equivalent in expected payoffs for both players. In contrast, in case D3, where $\alpha_i < 0$ for $i = 1, 2$, (4.24) is not satisfied and Figure 4.2d indicates that the mixed NE is Pareto dominated by both pure NEs $(0,1)$ and $(1,0)$ which cannot be Pareto ranked.

Alternatively, the differences $(\Delta\tilde{U}_1^0, \Delta\tilde{U}_2^0)$ can be random with a joint distribution $H(\cdot,\cdot|\epsilon_1, \epsilon_2)$ conditional on (ϵ_1, ϵ_2) that is absolutely continuous with respect to Lebesgue measure. When the support of $H(\cdot,\cdot|\epsilon_1, \epsilon_2)$ given $(\epsilon_1, \epsilon_2) \in \mathcal{A}_3$ is $\{\prod_{i=1,2}[\Delta\tilde{U}_i^0 - (\Delta_i + \alpha_i + \epsilon_i)] \geq 0\}$, then a pure NE, which alternates between $(0,0)$ and $(1,1)$, Pareto dominates the other pure and mixed NEs. See (4.23) which provides the probabilities $\mathsf{Pr}(0,0|\mathcal{A}_3)$ and $\mathsf{Pr}(1,1|\mathcal{A}_3)$ of observing the Pareto dominant NE $(0,0)$ and $(1,1)$ given $\mathcal{A}_3$. A similar result holds in case D3 when the support of $H(\cdot,\cdot|\epsilon_1, \epsilon_2)$ given $(\epsilon_1, \epsilon_2) \in \mathcal{D}_3$ is $\{\prod_{i=1,2}[\Delta\tilde{U}_i^0 + (\Delta_i + \epsilon_i)] \leq 0\}$. See

[7] In case A3, the game is supermodular because $[\tilde{U}_i(1,1) - \tilde{U}_i(0,1)] - [\tilde{U}_i(1,0) - \tilde{U}_i(0,0)] = \alpha_i > 0$, i.e., $\tilde{U}_i(y_i, y_j)$ exhibits increasing differences in (y_i, y_j). See Milgrom and Roberts (1990). In case D3, the game is supermodular by reversing the order of agent 2's choices so that action $1 \prec$ action 0. This gives $[\tilde{U}_1(1,0) - \tilde{U}_1(0,0)] - [\tilde{U}_1(1,1) - \tilde{U}_1(0,1)] = -\alpha_1 > 0$ and $[\tilde{U}_2(0,1) - \tilde{U}_2(1,1)] - [\tilde{U}_2(0,0) - \tilde{U}_2(1,0)] = -\alpha_2 > 0$.

[8] See e.g., Bresnahan and Reiss (1990) and Tamer (2003) in entry models.

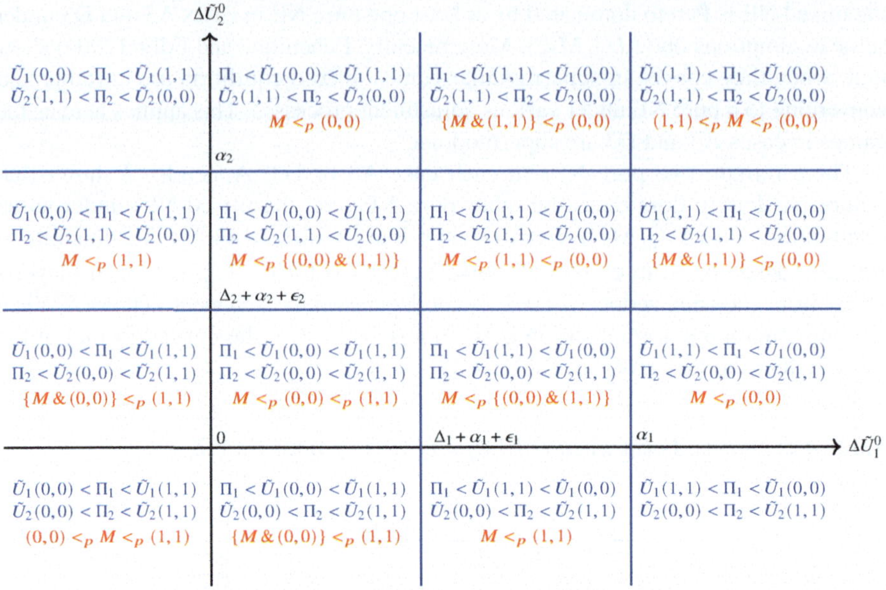

(a) Case A3

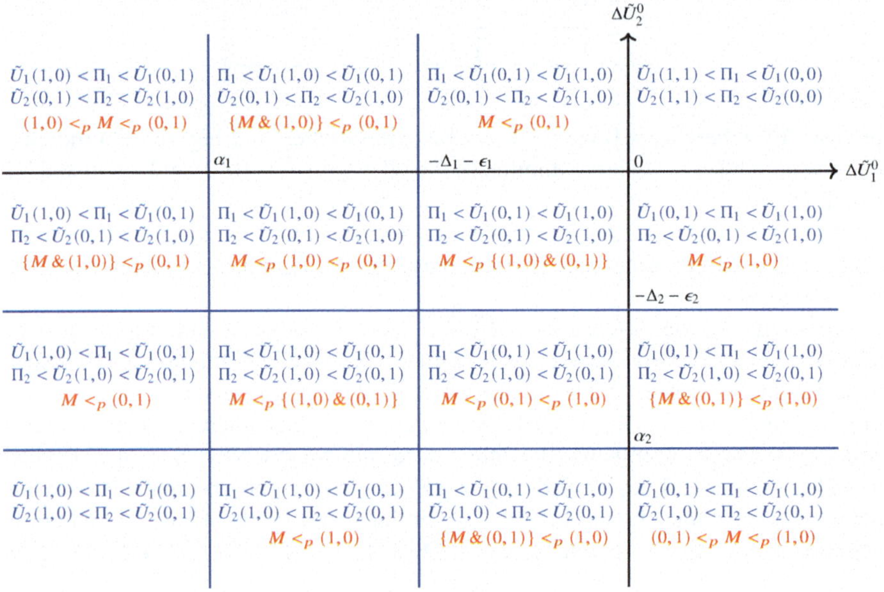

(b) Case D3

Fig. 4.2: Pareto Dominance

(4.25) which provides the probabilities $\text{Pr}(0,1|\mathcal{D}_3)$ and $\text{Pr}(1,0|\mathcal{D}_3)$ of observing the Pareto-dominant NE $(0,1)$ and $(1,0)$ given $\mathcal{D}_3$.

The preceding derivation assumes that the players coordinate through preplay communications on the Pareto dominant equilibrium when it exists. See, however, Fudenberg and Tirole (1991, Section 1.2.4). Moreover, (4.22) and (4.24) may fail so the two pure NEs in case A3 or D3 cannot be Pareto ranked such as when $\Delta\tilde{U}_i^o = 0$ and $\alpha_i < 0$ for $i = 1,2$. In order not to select *a priori* one of the pure NE, Bjorn and Vuong (1984) introduce instead probability weights $a_{y_1 y_2}$ and $d_{y_i y_j}$ over the pure NE (y_i, y_j) in cases A3 and D3, respectively, as additional parameters. Specifically, they let

$$\text{Pr}[0,0|\mathcal{A}_3] = a_{00} \quad \text{and} \quad \text{Pr}[1,1|\mathcal{A}_3] = a_{11} \tag{4.5}$$

$$\text{Pr}[0,1|\mathcal{D}_3] = d_{01} \quad \text{and} \quad \text{Pr}[1,0|\mathcal{D}_3] = d_{10}, \tag{4.6}$$

where $a_{00} + a_{11} = 1$, $a_{y_i y_j} \geq 0$, $d_{01} + d_{10} = 1$ and $d_{y_i y_j} \geq 0$.[9] Similarly to (4.4), the analyst can use these weights to test (4.23) and (4.25) and hence the validity of a selection based on Pareto dominance. More generally, in the absence of a well-accepted theory of NE refinements, (4.5)-(4.6) allow for an empirical understanding of how players choose a pure NE through a logit parameterization of the probability weights (a_{00}, a_{11}) or (d_{01}, d_{10}) that depends on the characteristics (X_1, X_2). Bajari, Hong and Ryan (2010) develop this idea in a setting with $I \geq 2$ players and $K \geq 2$ actions for each player.

4.3.3 Likelihood Functions

The observations are $(Y_{1\ell}, Y_{2\ell}, X_{1\ell}, X_{2\ell})$ for $\ell = 1, \ldots, L$ pairs of players, where $(Y_{1\ell}, Y_{2\ell})$ is the NE outcome of the game theoretic model (4.1)–(4.2) completed with (4.4)–(4.6). The error terms $(\epsilon_{1\ell}, \epsilon_{2\ell})$ are jointly normally distributed $\mathcal{N}_2(0,\rho)$ with mean zeros, unit variances and correlation ρ. Let $\Phi_2(\cdot,\cdot;\rho)$ denote the cdf. The parameter vector is $\theta = (\alpha, \beta, \rho, a, b, c, d)$ where $\alpha \equiv (\alpha_1, \alpha_2)$, $\beta \equiv (\beta_1, \beta_2)$, $a \equiv (a_{00}, a_{11})$, $b \equiv (b_{00}, b_{01}, b_{10}, b_{11})$, $c \equiv (c_{00}, c_{01}, c_{10}, c_{11})$ and $d \equiv (d_{01}, d_{10})$.[10] The random vector $(Y_{1\ell}, Y_{2\ell}, X_{1\ell}, X_{2\ell}, \epsilon_{1\ell}, \epsilon_{2\ell})$ is assumed independent across pairs.

Bjorn and Vuong (1984) show that the likelihood function of the model is

$$\mathcal{L}(\theta) = \prod_{\ell=1}^{L} \text{Pr}(Y_{1\ell}, Y_{2\ell}|X_{1\ell}, X_{2\ell}; \theta), \tag{4.7}$$

[9] When $\alpha_i < 0$ for $i = 1,2$ Bresnahan and Reiss (1990) adopt another approach by using the game theoretic model to explain instead the number of entrants $Y_1 + Y_2$ in entry models thereby collapsing cases D2, D3 and D4 into a single region with one entrant. See Figure 4.1d. Finding an observable that is common to all equilibria is a general strategy for dealing with multiple equilibria. See also Berry (1992) in a setting with more than 2 potential entrants.

[10] Within each pair, one individual/firm is designated as player 1, while the other is player 2. The parameter vector θ, which includes the structural effects (α_1, α_2), is constant across pairs.

where

$$\Pr(0,0|X_{1\ell},X_{2\ell};\theta) = \Phi_2(-X'_{1\ell}\beta_1,-X'_{2\ell}\beta_2;\rho)+I_\ell^{00} \tag{4.8}$$
$$\Pr(0,1|X_{1\ell},X_{2\ell};\theta) = \Phi_2(-X'_{1\ell}\beta_1-\alpha_1,X'_{2\ell}\beta_2;-\rho)+I_\ell^{01} \tag{4.9}$$
$$\Pr(1,0|X_{1\ell},X_{2\ell};\theta) = \Phi_2(X'_{1\ell}\beta_1,-X'_{2\ell}\beta_2-\alpha_2;-\rho)+I_\ell^{10} \tag{4.10}$$
$$\Pr(1,1|X_{1\ell},X_{2\ell};\theta) = \Phi_2(X'_{1\ell}\beta_1+\alpha_1,X'_{2\ell}\beta_2+\alpha_2;\rho)+I_\ell^{11} \tag{4.11}$$

with

$$I_\ell^{00} = \Big[-a_{11}\,1\!\!1(\alpha_1>0,\alpha_2>0)+b_{00}\,1\!\!1(\alpha_1>0,\alpha_2<0)+c_{00}\,1\!\!1(\alpha_1<0,\alpha_2>0)\Big]\,I_\ell,$$

$$I_\ell^{01} = \Big[b_{01}\,1\!\!1(\alpha_1>0,\alpha_2<0)+c_{01}\,1\!\!1(\alpha_1<0,\alpha_2>0)-d_{10}\,1\!\!1(\alpha_1<0,\alpha_2<0)\Big]\,I_\ell,$$

$$I_\ell^{10} = \Big[b_{10}\,1\!\!1(\alpha_1>0,\alpha_2<0)+c_{10}\,1\!\!1(\alpha_1<0,\alpha_2>0)-d_{01}\,1\!\!1(\alpha_1<0,\alpha_2<0)\Big]\,I_\ell,$$

$$I_\ell^{11} = \Big[-a_{00}\,1\!\!1(\alpha_1>0,\alpha_2>0)+b_{11}\,1\!\!1(\alpha_1>0,\alpha_2<0)+c_{11}\,1\!\!1(\alpha_1<0,\alpha_2>0)\Big]\,I_\ell.$$

The integral I_ℓ is the probability that (ϵ_1,ϵ_2) belong to the region $\mathcal{A}_3,\mathcal{B}_3,C_3$ or $\mathcal{D}_3$, i.e.,

$$I_\ell = \int_{-X'_{1\ell}\beta_1-\max\{\alpha_1,0\}}^{-X'_{1\ell}\beta_1-\min\{\alpha_1,0\}}\int_{-X'_{2\ell}\beta_2-\max\{\alpha_2,0\}}^{-X'_{2\ell}\beta_2-\min\{\alpha_2,0\}} d\Phi_2(\epsilon_1,\epsilon_2;\rho).$$

See Figures 4.1a–4.1d. For instance, when $\alpha_1>0$ and $\alpha_2>0$, the outcome $(0,0)$ can be obtained as a single pure NE in case A1 or from one of the two pure NEs with probability a_{00} in case A3. From Figure 4.1a, this gives the probability $\Phi_2(-X'_{1\ell}\beta_1,-X'_{2\ell}\beta_2;\rho)-a_{11}I_\ell$. Similarly, when $\alpha_1>0$ and $\alpha_2<0$, the outcome $(0,0)$ can be obtained as a single pure NE in case B1 or from a realization with probability b_{00} of the mixed NE in case B3. From Figure 4.1b, this gives the probability $\Phi_2(-X'_{1\ell}\beta_1,-X'_{2\ell}\beta_2;\rho)+b_{00}I_\ell$. When the coherency condition $\alpha_1\alpha_2=0$ holds, the additional terms I_ℓ^{00}, I_ℓ^{01}, I_ℓ^{10} and I_ℓ^{11} all vanish because $I_\ell=0$. Hence, the four probabilities (4.8)–(4.11) reduce to those derived by Heckman (1978) for the simultaneous equation model (4.3) with at most one structural shift. The latter is also Maddala and Lee's (1976) recursive model for two dichotomous variables.

An important special case arises when agents' utilities exhibit strategic complementarity, i.e., when $\alpha_i>0$ for $i=1,2$. This is often assumed in models of social interactions and network formation. See e.g., Brock and Durlauf (2001) and Jackson and Wolinsky (1996), respectively, where (α_1,α_2) and ρ represent peer effects and homophily – or endogenous and correlated effects according to Manski (1993a). These effects provide complementary explanations for the common fact that friends tend to behave similarly. From Case A, there is a unique NE, which is in pure strategies, except when $(\epsilon_1,\epsilon_2)\in\mathcal{A}_3$ in which case there are two pure NEs left after eliminating the mixed NE. See Figure 4.1a. The likelihood function is given by (4.7) where (4.8)–(4.11) reduce to

$$\begin{aligned}
\Pr(0,0|X_{1\ell},X_{2\ell};\theta) &= \Phi_2(-X'_{1\ell}\beta_1,-X'_{2\ell}\beta_2;\rho)-a_{11}I_\ell\\
\Pr(0,1|X_{1\ell},X_{2\ell};\theta) &= \Phi_2(-X'_{1\ell}\beta_1-\alpha_1,X'_{2\ell}\beta_2;-\rho)\\
\Pr(1,0|X_{1\ell},X_{2\ell};\theta) &= \Phi_2(X'_{1\ell}\beta_1,-X'_{2\ell}\beta_2-\alpha_2;-\rho)\\
\Pr(1,1|X_{1\ell},X_{2\ell};\theta) &= \Phi_2(X'_{1\ell}\beta_1+\alpha_1,X'_{2\ell}\beta_2+\alpha_2;\rho)-a_{00}I_\ell
\end{aligned}$$

with $I_\ell = \int_{-X'_{1\ell}\beta_1-\alpha_1}^{-X'_{1\ell}\beta_1}\int_{-X'_{2\ell}\beta_2-\alpha_2}^{-X'_{2\ell}\beta_2} d\Phi_2(\epsilon_1,\epsilon_2;\rho)$ and $\theta=(\alpha_1,\alpha_2,\beta_1,\beta_2,\rho,\, a_{00},a_{11})$.

Another important special case arises when agents' utilities exhibit strategic substitutability, i.e., when $\alpha_i < 0$ for $i = 1,2$. This is satisfied in entry models as initiated by Bresnahan and Reiss (1990, 1991) since duopoly profits are lower than monopoly profits. From Case D, there is a unique NE, which is in pure strategies, except when $(\epsilon_1,\epsilon_2) \in \mathcal{D}_3$ in which case there are two pure NEs left after eliminating the mixed NE. See Figure 4.1d. The likelihood function is given by (4.7) where (4.8)–(4.11) reduce to

$$\begin{aligned}
\Pr(0,0|X_{1\ell},X_{2\ell};\theta) &= \Phi_2(-X'_{1\ell}\beta_1,-X'_{2\ell}\beta_2;\rho)\\
\Pr(0,1|X_{1\ell},X_{2\ell};\theta) &= \Phi_2(-X'_{1\ell}\beta_1-\alpha_1,X'_{2\ell}\beta_2;-\rho)-d_{10}I_\ell\\
\Pr(1,0|X_{1\ell},X_{2\ell};\theta) &= \Phi_2(X'_{1\ell}\beta_1,-X'_{2\ell}\beta_2-\alpha_2;-\rho)-d_{01}I_\ell\\
\Pr(1,1|X_{1\ell},X_{2\ell};\theta) &= \Phi_2(X'_{1\ell}\beta_1+\alpha_1,X'_{2\ell}\beta_2+\alpha_2;\rho)
\end{aligned}$$

with $I_\ell = \int_{-X'_{1\ell}\beta_1}^{-X'_{1\ell}\beta_1-\alpha_1}\int_{-X'_{2\ell}\beta_2}^{-X'_{2\ell}\beta_2-\alpha_2} d\Phi_2(\epsilon_1,\epsilon_2;\rho)$ and $\theta=(\alpha_1,\alpha_2,\beta_1,\beta_2,\rho,\, a_{00},a_{11})$.

4.3.4 Identification

The likelihood function (4.7) takes four different functional forms depending on the signs of the structural effects α_1 and α_2. In particular, some parameters (a,b,c,d) disappear within each case A–D, e.g., the probability weights (b,c,d) when $\alpha_1 > 0$ and $\alpha_2 > 0$ (case A). Bjorn and Vuong (1984) and Bjorn (1986) study the (local) parametric identification of the game theoretic model by deriving a rank condition under which the information matrix is non-singular for each case. See Rothenberg (1971). To this end, they focus on the identification of the parameters $\theta = (\alpha_1,\alpha_2,\beta_1,\beta_2,\rho)$ upon setting equal probability weights in cases A3, B3, C3 and D3 so that $a_{00} = a_{11} = 1/2$, $b_{00} = b_{01} = b_{10} = b_{11} = 1/4$, $c_{00} = c_{01} = c_{10} = c_{11} = 1/4$ and $d_{01} = d_{10} = 1/2$. Relying on the chosen functional forms, the authors show that θ is identified except for particular values of the characteristics $(X_{1\ell},X_{2\ell})$. This holds even when the same covariates appear in (4.2), i.e., when $X_1 = X_2$, thereby suggesting that identification in game theoretic models differ from identification in linear simultaneous equation models where exclusion restrictions are required. See Hood and Koopmans (1953). As in the latter, however, some covariates must be present to achieve identification. If not, there are five parameters which are (α_1,α_2,ρ) and two constant terms in (4.2), whereas there are only three independent probabilities among the four observed probabilities $\Pr(y_1,y_2)$, $(y_1,y_2) \in \{0,1\}^2$.

In a general setting with $I \geq 2$ players and $K \geq 2$ actions for each player, Bajari, Hong and Ryan (2010) study identification when the distribution of the unobserved random utility components is known. They consider both pure and mixed NEs which are determined by a software package. See McKelvey and McLennan (1996) for a survey of available algorithms. Using an identification-at-infinity argument, they provide general conditions for identifying the parameters in a linear index specification of the deterministic utility components as well as the probability weights of equilibrium selection. Alternatively, they show how exclusion restrictions leading to some covariates appearing exclusively in agents' deterministic utility components can help identify nonparametrically the deterministic components and the equilibrium selection probabilities. They also propose a computationally convenient Method of Moments based on simulating the conditional probabilities $\Pr(y_1, \ldots, y_K | X_1, \ldots X_K)$. See McFadden (1989) and Pakes and Pollard (1989).

4.4 The Stackelberg Approach

An alternative to the NE concept is the Stackelberg equilibrium (SE) in which one player moves first. See e.g., Fudenberg and Tirole (1991). Without loss of generality, let player 1 be the leader and player 2 the follower. Using the same notation as before $\tilde{U}_i(y_i, y_j)$, where the first and second argument are the agent's and opponent's actions, respectively, the extensive form of the game is displayed in Figure 4.3.

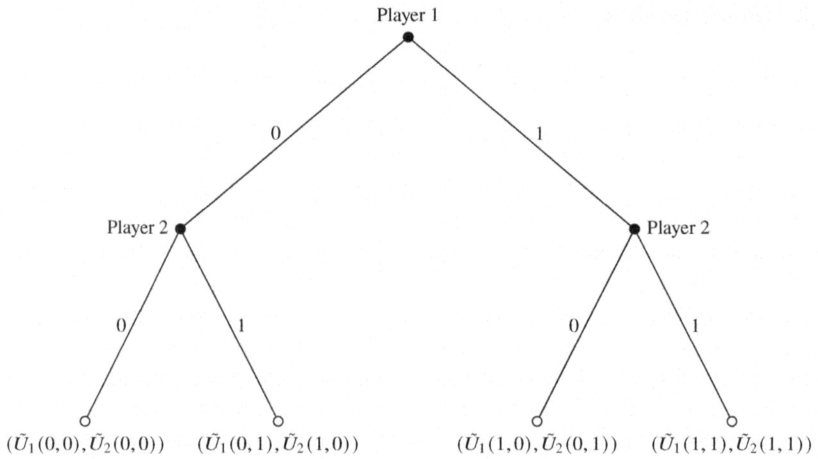

Fig. 4.3: Stackelberg Extensive Form

Because he can commit, player 1 chooses his preferred action taking into account player 2's reaction function (or best response). Bjorn and Vuong (1985) determine the

SE in pure strategies of the above extensive-form game.[11] The outcome (y_1, y_2) arises as a SE depending on $(\Delta_i, \alpha_i, \epsilon_i), i = 1,2$ in view of (4.1). However, because player 1's optimal choice depends on some utility comparisons that are not entertained in a NE, namely $\tilde{U}_1(1,1) - \tilde{U}_1(0,0)$ and $\tilde{U}_1(1,0) - \tilde{U}_1(0,1)$, additional parameters can be identified when the outcome (y_1, y_2) is a SE. Specifically, let

$$\tilde{U}_i(0, y_j) = \Delta_i^0 + \alpha_i^0 y_j + \epsilon_i^0 \text{ and } \tilde{U}_i(1, y_j) = \Delta_i^1 + \alpha_i^1 y_j + \epsilon_i^1 \text{ for } i = 1,2. \quad (4.12)$$

This generalizes (4.1) since taking the difference gives (4.1) with $\Delta_i = \Delta_i^1 - \Delta_i^0$, $\alpha_i = \alpha_i^1 - \alpha_i^0$ and $\epsilon_i = \epsilon_i^1 - \epsilon_i^0$. The new quantity Δ_i^0 and parameter $\alpha_i^0 \in \mathbb{R}$ represent the systematic utility and the structural effect of the opponent's choice when player i chooses action $y_i = 0$, respectively. A similar interpretation applies to Δ_i^1 and $\alpha_i^1 \in \mathbb{R}$ when player i chooses action $y_i = 1$. In general, Δ_i^0 and Δ_i^1 are parameterized linear functions of the observed agents' characteristics (X_1, X_2) as in

$$\Delta_i^0 = X_i'\beta_i^0 \text{ and } \Delta_i^1 = X_i'\beta_i^1 \text{ for } i = 1,2, \quad (4.13)$$

where X_1 and X_2 may have some variables in common and $(\beta_i^0, \beta_i^1), i = 1,2$ are some parameters. This implies (4.2) where $\beta_i = \beta_i^1 - \beta_i^0$.

Four cases are distinguished depending on player 2's reaction function.

• CASE E: $\epsilon_2 < -\Delta_2 - \max\{0, \alpha_2\}$.[12]
E1. If $\epsilon_1 < -\Delta_1$, there is a unique pure SE, namely, $(0,0)$.
E2. If $\epsilon_1 > -\Delta_1$, there is a unique pure SE, namely, $(1,0)$.

• CASE F: $-\Delta_2 - \max\{0, \alpha_2\} < \epsilon_2 < -\Delta_2$.
F1. If $\epsilon_1 < -\Delta_1 - \alpha_1^1$, there is a unique pure SE, namely, $(0,0)$.
F2. If $\epsilon_1 > -\Delta_1 - \alpha_1^1$, there is a unique pure SE, namely, $(1,1)$.

• CASE G: $-\Delta_2 < \epsilon_2 < -\Delta_2 - \min\{0, \alpha_2\}$.
G1. If $\epsilon_1 < -\Delta_1 - \alpha_1^0$, there is a unique pure SE, namely, $(0,1)$.
G2. If $\epsilon_1 > -\Delta_1 - \alpha_1^0$, there is a unique pure SE, namely, $(1,0)$.

• CASE H: $\epsilon_2 > -\Delta_2 - \min\{0, \alpha_2\}$.
H1. If $\epsilon_1 < -\Delta_1 - \alpha_1$, there is a unique pure SE, namely, $(0,1)$.
H2. If $\epsilon_1 > -\Delta_1 - \alpha_1$, there is a unique pure SE, namely, $(1,1)$.

From (4.1) these SEs do not depend on the nuisance quantities $[\tilde{U}_2(0,0), \tilde{U}_2(0,1)]$ as only the utility difference $\tilde{U}_2(1,y_1) - \tilde{U}_2(0,y_1)$ matters to player 2 given player 1's action $y_1 = 0,1$. For instance, case F corresponds to player 2's reaction function of choosing $y_2 = 0$ if $y_1 = 0$ and $y_2 = 1$ if $y_1 = 1$ irrespective of $[\tilde{U}_2(0,0), \tilde{U}_2(0,1)]$. In contrast, accounting for player 2's reaction function, player 1 compares the resulting utilities $\tilde{U}_1(1,1)$ and $\tilde{U}_1(0,0)$ when choosing his optimal action. Such a comparison

[11] Mixed strategies for the follower occur with zero probability. Mixed strategies for the leader are excluded as they raise commitment issues because player 2 sees player 1's realized action before moving. See Conitzer (2016). Recently, SE with mixed strategies for the leader have received attention in security games with applications to airport safety. See Korzhyk, Yin, Kiekintveld, Conitzer and Tambe (2011).

[12] See footnote 5.

then involves α_1^1. A similar situation applies to case G which involves α_1^0 given player 2's reaction function of choosing $y_2 = 1$ if $y_1 = 0$ and $y_2 = 0$ if $y_1 = 1$.

It is worth noting that cases F and G happen with zero probability if $\alpha_2 \leq 0$ or $\alpha_2 \geq 0$, respectively, because ϵ_2 has a density with respect to Lebesgue measure. Hence, when $\alpha_2 \leq 0$, only cases E, G and H can occur with ϵ_2-thresholds $-\Delta_2$ and $-\Delta_2 - \alpha_2$. Figures 4a depicts the regions $\mathcal{E}_{00}, \mathcal{E}_{01}, \mathcal{E}_{10}, \mathcal{E}_{11}$ in the (ϵ_1, ϵ_2)-space corresponding to the SE $(0,0), (0,1), (1,0)$ and $(1,1)$. For instance, the region $\mathcal{E}_{00} = \{(\epsilon_1, \epsilon_2) : \epsilon_1 < -\Delta_1 \text{ and } \epsilon_2 < -\Delta_2\}$ leads to the SE $(0,0)$. Similarly, when $\alpha_2 \geq 0$, only cases E, F and H can occur with ϵ_2-thresholds $-\Delta_2 - \alpha_2$ and $-\Delta_2$. Figures 4b depicts the regions $\mathcal{F}_{00}, \mathcal{F}_{01}, \mathcal{F}_{10}, \mathcal{F}_{11}$ in the (ϵ_1, ϵ_2)-space corresponding to the SE $(0,0), (0,1), (1,0)$ and $(1,1)$. Without loss of generality, these figures are drawn when $\alpha_1^0 < 0 < \alpha_1^1$. In contrast to NE, there always exists a unique SE in pure strategies.

4.4.1 Likelihood Functions

The observations are $(Y_{1\ell}, Y_{2\ell}, X_{1\ell}, X_{2\ell})$ for $\ell = 1, \ldots, L$ pairs of players, where $(Y_{1\ell}, Y_{2\ell})$ is the SE outcome of the extensive-form game of Figure 4.3 with payoffs (4.12)–(4.13). As in the previous section, the error terms $(\epsilon_{1\ell}, \epsilon_{2\ell})$ are jointly normally distributed $\mathcal{N}_2(0, \rho)$ with mean zeros, unit variances and correlation ρ. The parameter vector is $\theta = (\alpha, \beta, \rho)$ where $\alpha \equiv (\alpha_1^0, \alpha_1^1, \alpha_2^0, \alpha_2^1)$ and $\beta \equiv (\beta_1^0, \beta_1^1, \beta_2^0, \beta_2^1)$.[13] The random vector $(Y_{1\ell}, Y_{2\ell}, X_{1\ell}, X_{2\ell}, \epsilon_{1\ell}, \epsilon_{2\ell})$ is assumed independent across pairs.

Bjorn and Vuong (1985) show that the likelihood function of the model is

$$\mathcal{L}(\theta) = \prod_{\ell=1}^{L} \mathsf{Pr}(Y_{1\ell}, Y_{2\ell} | X_{1\ell}, X_{2\ell}; \theta), \tag{4.14}$$

where

$$\mathsf{Pr}(0,0|X_{1\ell}, X_{2\ell}; \theta) = \Phi_2(-X'_{1\ell}\beta_1, -X'_{2\ell}\beta_2; \rho) - I_\ell^{00} \tag{4.15}$$
$$\mathsf{Pr}(0,1|X_{1\ell}, X_{2\ell}; \theta) = \Phi_2(-X'_{1\ell}\beta_1 - \alpha_1, X'_{2\ell}\beta_2; -\rho) + I_\ell^{01} \tag{4.16}$$
$$\mathsf{Pr}(1,0|X_{1\ell}, X_{2\ell}; \theta) = \Phi_2(X'_{1\ell}\beta_1, -X'_{2\ell}\beta_2 - \alpha_2; -\rho) + I_\ell^{10} \tag{4.17}$$
$$\mathsf{Pr}(1,1|X_{1\ell}, X_{2\ell}; \theta) = \Phi_2(X'_{1\ell}\beta_1 + \alpha_1, X'_{2\ell}\beta_2 + \alpha_2; \rho) - I_\ell^{11} \tag{4.18}$$

with $\alpha_i = \alpha_i^1 - \alpha_i^0$, $\beta_i = \beta_i^1 - \beta_i^0$ and

[13] See footnote 10.

$$I_\ell^{00} = \int_{-X'_{1\ell}\beta_1-\alpha_1^1}^{-X'_{1\ell}\beta_1} \int_{-X'_{2\ell}\beta_2-\max\{\alpha_2,0\}}^{-X'_{2\ell}\beta_2} d\Phi_2(\epsilon_1,\epsilon_2;\rho)$$

$$I_\ell^{01} = \int_{-X'_{1\ell}\beta_1-\alpha_1}^{-X'_{1\ell}\beta_1+\alpha_1^0} \int_{-X'_{2\ell}\beta_2}^{-X'_{2\ell}\beta_2-\min\{\alpha_2,0\}} d\Phi_2(\epsilon_1,\epsilon_2;\rho)$$

$$I_\ell^{10} = \int_{-X'_{1\ell}\beta_1+\alpha_1^0}^{-X'_{1\ell}\beta_1} \int_{-X'_{2\ell}\beta_2}^{-X'_{2\ell}\beta_2-\min\{\alpha_2,0\}} d\Phi_2(\epsilon_1,\epsilon_2;\rho)$$

$$I_\ell^{11} = \int_{-X'_{1\ell}\beta_1-\alpha_1}^{-X'_{1\ell}\beta_1-\alpha_1^1} \int_{-X'_{2\ell}\beta_2-\max\{\alpha_2,0\}}^{-X'_{2\ell}\beta_2} d\Phi_2(\epsilon_1,\epsilon_2;\rho).$$

In particular, the integrals I_ℓ^{00} and I_ℓ^{11} vanish when $\alpha_2 \leq 0$ whereas the integrals I_ℓ^{01} and I_ℓ^{10} vanish when $\alpha_2 \geq 0$ because ϵ_2 has a density with respect to Lebesgue measure. The probabilities (4.15)–(4.18) follow from Figures 4a–4b.[14] For instance, the outcome $(0,0)$ is a SE if and only if $(\epsilon_1,\epsilon_2) \in \mathcal{E}_{00}$ when $\alpha_2 > 0$ or $(\epsilon_1,\epsilon_2) \in \mathcal{F}_{00}$ when $\alpha_2 < 0$. This gives (4.15) for the probability of observing $(0,0)$ giving $(X_{1\ell}, X_{2\ell})$.

The first terms in (4.15)–(4.18) are identical to the first terms in (4.8)–(4.11). When the structural effect $\alpha_2 = 0$, the integrals I_ℓ^{00}, I_ℓ^{11}, I_ℓ^{01} and I_ℓ^{10} all vanish.[15] Thus the probabilities (4.15)–(4.18) reduce to those derived by Heckman (1978) for the simultaneous equation model (4.3) with structural shift α_1. In contrast, when the structural effect $\alpha_1 = 0$ so that $\alpha_1^0 = \alpha_1^1$, i.e., when the leader's utility increment $\tilde{U}_1(1,y_2) - \tilde{U}_1(0,y_2) = \Delta_1 + \epsilon_1$ is independent of the follower's action y_2, these four integrals do not necessarily vanish because $\alpha_1^0\alpha_1^1 \neq 0$. Thus, the probabilities (4.15)–(4.18) do not reduce to those obtained by Heckman (1978) for the simultaneous equation model (4.3) with structural shift α_2 or Maddala and Lee's (1976) recursive model where player 2's latent variable Y_2^* depends on player 1's choice Y_1. In other words, relative to the latter, the SE model allows for an effect of player 2's action y_2 on player 1's utility $\tilde{U}_1(1,y_2)$ and $\tilde{U}_1(0,y_2)$ through $\alpha_1^1 = \alpha_1^0$ despite having no effect on player's 1's utility increment $\tilde{U}_1(1,y_2) - \tilde{U}_1(0,y_2)$.

As for NE two important special cases are considered in empirical work. When agents' utilities exhibit strategic complementarity, i.e., when $\alpha_i > 0$ for $i = 1,2$ as in models of peer effects (social interactions) and network formation, the likelihood function is given by (4.14) where (4.15)–(4.18) reduce to

$$\begin{aligned}
\Pr(0,0|X_{1\ell},X_{2\ell};\theta) &= \Phi_2(-X'_{1\ell}\beta_1, -X'_{2\ell}\beta_2;\rho) - I_\ell^{00}\\
\Pr(0,1|X_{1\ell},X_{2\ell};\theta) &= \Phi_2(-X'_{1\ell}\beta_1 - \alpha_1, X'_{2\ell}\beta_2;-\rho)\\
\Pr(1,0|X_{1\ell},X_{2\ell};\theta) &= \Phi_2(X'_{1\ell}\beta_1, -X'_{2\ell}\beta_2 - \alpha_2;-\rho)\\
\Pr(1,1|X_{1\ell},X_{2\ell};\theta) &= \Phi_2(X'_{1\ell}\beta_1 + \alpha_1, X'_{2\ell}\beta_2 + \alpha_2;\rho) - I_\ell^{11}
\end{aligned}$$

[14] When the lower boundary is larger than the upper boundary in the outside integral of (4.15)–(4.18), by convention the integral is the negative of the integrand from the upper boundary to the lower boundary. For instance, if $\alpha_1^1 < 0$ then $I_\ell^{00} = -\int_{-X'_{1\ell}\beta_1}^{-X'_{1\ell}\beta_1-\alpha_1^1}\int_{-X'_{2\ell}\beta_2-\max\{\alpha_2,0\}}^{-X'_{2\ell}\beta_2} d\Phi_2(\epsilon_1,\epsilon_2;\rho)$.

[15] The likelihood (4.14) with probabilities (4.15)–(4.18) also holds when $\alpha_2 = 0$.

with $I_\ell^{00} = \int_{-X'_{1\ell}\beta_1-\alpha_1^1}^{-X'_{1\ell}\beta_1} \int_{-X'_{2\ell}\beta_2-\alpha_2}^{-X'_{2\ell}\beta_2} d\Phi_2(\epsilon_1,\epsilon_2;\rho)$ and $I_\ell^{11} = \int_{-X'_{1\ell}\beta_1-\alpha_1}^{-X'_{1\ell}\beta_1-\alpha_1^1} \int_{-X'_{2\ell}\beta_2-\alpha_2}^{-X'_{2\ell}\beta_2} d\Phi_2(\epsilon_1,\epsilon_2;\rho)$, which can be positive or negative. See Figure 4.4b where these integrals are positive. Similarly, when agents' utilities exhibit strategic substitutability, i.e., when $\alpha_i < 0$ for $i = 1,2$ as in entry models, the likelihood function is given by (4.14) where (4.15)–(4.18) reduce to

$$
\begin{aligned}
\mathsf{Pr}(0,0|X_{1\ell},X_{2\ell};\theta) &= \Phi_2(-X'_{1\ell}\beta_1,-X'_{2\ell}\beta_2;\rho)\\
\mathsf{Pr}(0,1|X_{1\ell},X_{2\ell};\theta) &= \Phi_2(-X'_{1\ell}\beta_1-\alpha_1,X'_{2\ell}\beta_2;-\rho)+I_\ell^{01}\\
\mathsf{Pr}(1,0|X_{1\ell},X_{2\ell};\theta) &= \Phi_2(X'_{1\ell}\beta_1,-X'_{2\ell}\beta_2-\alpha_2;-\rho)+I_\ell^{10}\\
\mathsf{Pr}(1,1|X_{1\ell},X_{2\ell};\theta) &= \Phi_2(X'_{1\ell}\beta_1+\alpha_1,X'_{2\ell}\beta_2+\alpha_2;\rho),
\end{aligned}
$$

with $I_\ell^{01} = \int_{-X'_{1\ell}\beta_1-\alpha_1}^{-X'_{1\ell}\beta_1+\alpha_1^0} \int_{-X'_{2\ell}\beta_2}^{-X'_{2\ell}\beta_2-\alpha_2} d\Phi_2(\epsilon_1,\epsilon_2;\rho)$ and $I_\ell^{10} = \int_{-X'_{1\ell}\beta_1+\alpha_1^0}^{-X'_{1\ell}\beta_1} \int_{-X'_{2\ell}\beta_2}^{-X'_{2\ell}\beta_2-\alpha_2} d\Phi_2(\epsilon_1,\epsilon_2;\rho)$, which can be positive or negative.

4.4.2 Identification

Without additional restrictions, the shift parameters (α_2^0,α_2^1) representing the structural effects in the follower's utility $[\tilde{U}_2(0,y_1),\tilde{U}_2(1,y_1)]$ are not identified because only their difference $\alpha_2 = \alpha_2^1 - \alpha_2^0$ appears in the probabilities (4.15)–(4.18). Similarly, for $i = 1,2$, the coefficients (β_i^0,β_i^1) in the linear specifications (4.13) of Δ_i^0 and Δ_i^1 are not separately identified because only their difference $\beta_i = \beta_i^1 - \beta_i^0$ appears in the probabilities (4.15)–(4.18). However, when there are some variables in X_i, e.g., X_i^{0*} and X_i^{1*}, that are specific to Δ_i^0 and Δ_i^1, then their coefficients $(\beta_i^{0*},\beta_i^{1*})$ in (β_i^0,β_i^1) are identified whenever β_i is identified. Indeed, let $\beta_i^0 = (\beta_i^{0*},\beta_i^{0**})$, $\beta_i^1 = (\beta_i^{1*},\beta_i^{1**})$ and $X_i = (X_i^{0*},X_i^{1*},X_i^{**})$ where X_i^{**} are the variables in X_i that are common to Δ_i^0 and Δ_i^1. Thus, $\beta_i = (-\beta_i^{0*},\beta_i^{1*},\beta_i^{1**}-\beta_i^{0**})$ from the exclusion restrictions.[16]

Bjorn and Vuong (1985) study the (local) parametric identification of the SE model by deriving a rank condition under which the information matrix is non-singular when $\alpha_2 < 0$ and $\alpha_2 > 0$. See Rothenberg (1971). To this end, they focus on the identification of the parameters $\theta = (\alpha_1^0,\alpha_1^1,\alpha_2,\beta_1,\beta_2,\rho)$. Again relying on the chosen functional forms, the authors show that θ is identified except for particular values of the characteristics $(X_{1\ell},X_{2\ell})$. This holds even when the same covariates appear in (4.13), i.e., when $X_1^0 = X_1^1 = X_2^0 = X_2^1$. However, some covariates must be present to achieve identification. If not, there are 6 parameters which are $(\alpha_1^0,\alpha_1^1,\alpha_2,\rho)$ and two constant terms in (Δ_1,Δ_2), whereas there are only three independent probabilities among the four observed probabilities $\mathsf{Pr}(y_1,y_2)$, $(y_1,y_2) \in \{0,1\}^2$.

---

[16] Identification of $(\beta_i^{0*},\beta_i^{1*})$ can be similarly achieved through such exclusion restrictions in the NE model of Section 4.3 when starting from (4.12)-(4.13) instead of (4.1)-(4.2). Only the differences $\alpha_i = \alpha_i^1 - \alpha_i^0$ for $i = 1,2$, however, are identified in the NE model.

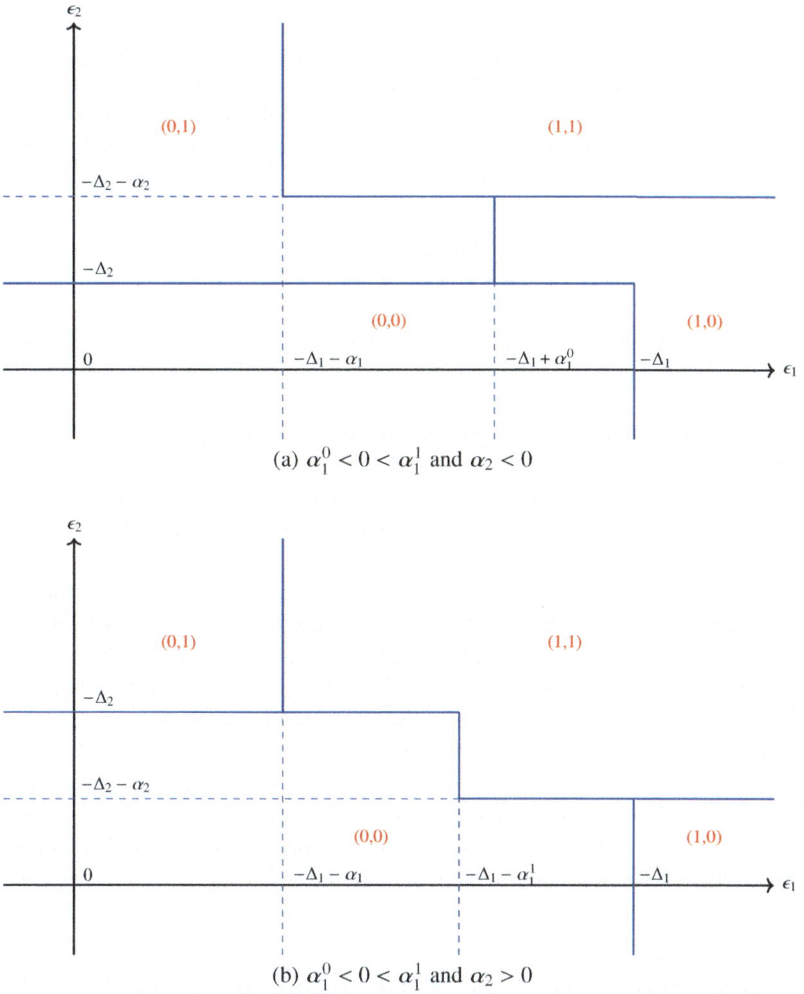

Fig. 4.4: Stackelberg Equilibria

4.5 Application to Labor Force Participation

Bjorn and Vuong (1984, 1985) study the labor force participation of married partners using the 1982 wave of the Panel Study of Income Dynamics.[17] Prior to their papers, Heckman (1974) considers women's decisions while taking the husband's labor force participation as exogenous, or as the outcomes of a joint utility maximization as in Ashenfelter and Heckman (1974). Defining $y_i = 1$ as working and $y_i = 0$ otherwise, the authors define $\tilde{U}_i(1, y_j)$ as partner i's market wage and $\tilde{U}_i(0, y_j)$ as his/her

[17] See Kooreman (1994) for an application to Dutch households.

reservation wage. The set of covariates is standard and include age, education, race, local unemployment rate, nonlabor income and the number of children of given age. The sample contains 2,012 couples with few husbands not working despite nearly two thirds of working wives. Overall, maximum likelihood estimation of the NE model with equal equilibrium selection probabilities and the two SE models with the husband and the wife being alternatively the leader gives sensible results in terms of the coefficients of the covariates. For instance, the presence of young children, which is excluded from the husband's market and reservation wage, has a positive effect (see the coefficient β_i^{0*} in the discussion above) on the wife's reservation wage. Similarly, they reject that the correlation parameter ρ is equal to zero.

Maximum likelihood estimation of the Nash model gives estimates of the structural effects $\hat{\beta}_h < 0$ and $\hat{\beta}_w > 0$ for husband and wife, respectively. The first sign is expected because the husband's reservation wage declines when the wife works. The second sign is unexpected but might be due to the unbalanced contingency table. Estimation of the Stackelberg model with the husband as the leader gives a negative structural effect $\hat{\alpha}_h^0 < 0$ on the husband's reservation wage. This could result from social norms inciting the husband to work when the wife works. It also gives a negative structural effect $\hat{\alpha}_w < 0$, suggesting an increase in the wife's reservation wage when the husband works as expected. Estimation of the Stackelberg model with the wife as the leader gives similar results, i.e. a positive structural effect $\hat{\alpha}_w^0 > 0$ on the wife's reservation wage when the husband works and a negative structural effect $\hat{\alpha}_h < 0$ on the husband's reservation wage which could result from social norms. Using a Wald test, the coherency condition $\alpha_h \alpha_w$ is clearly rejected in the three models.

This work also inspires the development of Vuong's (1989) test for model selection as the NE and the two SE models are nonnested. Bjorn's (1986) dissertation was the first application of this model selection test based on the likelihood ratio. Pairwise comparisons of the three competing models are inconclusive, possibly due to the observed unbalanced contingency table and some model specification features.

4.6 Further Developments and Concluding Remarks

Over the past forty years, the econometrics of discrete games has developed at a quick pace with applications to several domains of economics. We do not intend to provide a complete review here but we outline some major lines of research.

First, in line of Bjorn and Vuong (1984, 1985), static complete information games became popular to analyze strategic interactions among economic agents. The largest number of applications is in industrial organization with the analysis of firms' entry in markets starting with Bresnahan and Reiss (1990, 1991) for monopoly/duopoly markets. Berry (1992) extends this setting to oligopolies in the airline industry where the number of operating airlines is the variable of interest to avoid multiple NE equilibria. He also proposes simulation-based estimators as developed by McFadden (1989) and Pakes and Pollard (1989) to deal with integrals on intricate integration domains. In contrast to early papers, in which entry decisions are independent across markets, Jia (2008) allows for dependence in the discount retailing industry. See Berry and Reiss (2007) for a survey on entry models in industrial organization, in which airline and retailing industries have been extensively studied. Whereas most of these papers rely on parametric models, Bajari, Hong and Ryan (2010) study the semiparametric identification of static complete information games. They show that the model is identified under weaker functional form assumptions using exclusion restrictions and support conditions. Moreover, they develop a simulation-based estimator while allowing for mixed strategies. Because there are multiple NEs, the authors rely on probability weights for equilibrium selection in the spirit of Bjorn and Vuong (1984). They illustrate their results on the analysis of contractors' entry in procurements. The literature on discrete games has proposed alternative strategies to deal with multiple equilibria. The equilibrium selection operates through an external equation in Jia (2008). Bresnahan and Reiss (1990, 1991) and Berry (1992) tackle the problem by considering the number of firms as the dependent variable instead of the firms' entry decisions. Tamer (2003) initiates a different approach based on bounds, which are derived from the NE necessary conditions, leading to set identification. See also Ciliberto and Tamer (2009) and Andrews, Berry and Jia (2004). Estimation relies on moment inequalities as developed by Pakes, Porter, Ho and Ishii (2015) among others. See also Kline, Pakes and Tamer's (2021) survey on moment inequality estimators in industrial organization. A major drawback is the difficulty of performing counterfactuals that typically motivate the structural approach.

A second major line of research is the development of static games with incomplete information in which agents possess private information usually modeled as an error term. The model is solved using the Bayesian NE concept. Here again, industrial organization is at the forefront of this line of research. Assuming independence of private information, Seim (2006) considers firms' entry and location choices in the retailing industry, whereas Sweeting (2009) studies commercial timing among radio stations. Sweeting (2009) exploits the existence of multiple equilibria to identify the payoffs by assuming that at least two equilibria are played by the stations across markets. Aradillas-Lopez (2010) and Bajari, Hong, Krainer and Nekipelov (2010) relax parametric assumptions and propose semiparametric estimators, when analyzing capital investment decisions and stock analysts' recommendations, respectively.[18] Exploiting multiple equilibria, De Paula and Tang (2012) identify the sign of the structural effects, i.e., the impacts of players' actions on each others' payoffs, without any parametric assumption on the players' payoffs, distribution of private information or equilibrium selection mechanism, whereas Lewbel and Tang (2015) show nonparametric identification under exclusion restrictions. However, independence of private information is a restrictive assumption. As noted by Berry and Tamer (2006), entry may occur because of correlated unobserved profitability independently of the effect of competition.[19] In sociology, people who interact tend to be similar with common tastes known as homophily. When private information among players is

[18] See also Aradillas-Lopez and Tamer (2008) for the identification of complete and incomplete information games with alternative equilibrium concepts of limited rationality.

[19] Grieco (2014) considers independence of private information but allows for unobserved heterogeneity to analyze entry in the retailing industry.

dependent, Liu, Vuong and Xu (2017) show that the model structure is nonparametrically identified under exclusion restrictions up to a scale-location normalization. Moreover, they derive the restrictions imposed by the model on observables to assess the model validity. To our knowledge, dependence of private information has not been considered in the empirical literature. One exception is Kim, Perrigne, Vuong and Yan (2025) who combine entry decisions and demand for differentiated products. With peer effect and network models, as discussed below, we expect that such developments will occur in the near future because of homophily.

A third major extension deals with dynamic games. Most of this literature adopts a Markov dynamic setting with infinite horizon. See Rust (1994a, 1994b) in single-agent settings. The complex optimization problem involving multiple periods and uncertainty about future states is reduced into a sequence of deterministic and static optimization problems through the use of a value function and a Bellman equation. This literature also focuses on finding numerical algorithms to alleviate the computational burden of determining the value function. See Ericson and Pakes (1995) and Pakes and McGuire (1994). The introduction of the Markov perfect equilibrium concept by Maskin and Tirole (2001) initiated the estimation of dynamic discrete games.[20] Considering games of incomplete information, a number of papers propose different estimators assuming that private information is independently and identically distributed over time and across agents, and private information does not affect the transition probability of commonly known variables given agents' decisions in each period. Most papers also adopt Hotz and Miller's (1993) two-step procedure which exploits the mapping between the conditional choice probability and the choice-specific value function to avoid computing the equilibrium. Bajari, Benkard and Levin (2007) propose simulation-based estimators, Pesendorfer and Schmidt-Dengler (2008) rely on minimum distance estimators, and Pakes, Ostrovsky and Berry (2007) on method of moments. Aguirregabiria and Mira (2007) propose a nested pseudo maximum likelihood estimator. Empirical applications consider entry and exit of firms, where discrete games are part of a larger model involving continuous endogenous variables to analyze firms' advertising, investment and pricing, joint ventures, and mergers. See Ackerberg, Benkard, Berry and Pakes (2007), Reiss and Wolak (2007) and Aguirregabiria, Collard-Wexler and Ryan (2021) for more empirical references in industrial organization.

A fourth major line of research involves interaction-based models such as peer effects, networks and bargaining. Here the domains of applications are labor, health, education and development and more broadly social sciences. Peer effects and network models rely on the idea that human capital acquisition, i.e., education, labor market outcomes, criminality, or health depends on the behavior and/or characteristics of community members. With possibly a large number of players, as noted by Manski (1993b), the identification problem is challenging as we need to distinguish the exogenous effect due to the propensity of an individual to behave as other individuals

[20] Weintraub, Benkard and Roy (2008) propose an alternative equilibrium concept, i.e., oblivious equilibrium, in which each firm makes decisions based on its own state and knowledge of the long-run average industry state but where firms ignore the current information about their competitors' states simplifying the computation of the Markov perfect equilibrium.

with the same characteristics, the endogenous effect due to the propensity of an individual to behave like his group, and the correlated effect due to the propensity of similar behavior among individuals because they have similar tastes as in homophily. Binary choice models then become useful to analyze social interactions. See Brock and Durlauf (2001, 2007) for the identification and estimation of such models with applications using cross-section and panel data. Important questions are the network formation and the analysis of games within networks such as trade flows, production networks, research collaborations or insurer-provider networks in health markets. A recent literature is rapidly developing based on game theoretic developments by Jackson and Wolinsky (1996), empirical work on peer and neighborhood effects, the econometric analysis of games, and advances in machine learning techniques with large data. As surveyed by Graham (2020), a major contribution lies in the dyadic regression model which rules out interdependencies in link formations. See Graham (2008, 2011) and Menzel (2016) for large networks. As a matter of fact, interdependencies are difficult to model. Jackson and Wolinsky (1996) rely on the pairwise stability equilibrium concept, which is used in matching models. See e.g. Galichon and Salanié (2017). While most papers consider complete information, Leung (2015) considers directed links with private information. This literature is at an early stage and new contributions are expected.

Lastly, Professor Nerlove also pioneered family economics with the analysis of fertility decisions. See Nerlove, Razin and Sadka (1987). In the spirit of strategic interactions, a recent literature focuses on household decisions and intra-household resource distribution using collective bargaining models. Initiated by McElroy and Horney (1981) with Nash bargaining, the collective model developed by Chiappori (1988, 1992) and Browning and Chiappori (1998), in which agents bargain over consumption and other choices, has been a seminal contribution to the literature in family and development economics. Such models have been extended to dynamic and noncooperative solutions. See Almås, Attanasio and Carneiro (2023) for a survey on testing, extensions and applications. Dunbar, Lewbel and Pendakur (2021) derive semiparametric identification results of the distribution of resources across household members thereby allowing to estimate intra-household resource allocations. Using data from developing economies, analysts find that the poverty rates differ across family members because of the large disparity in resources due to different bargaining power. See Dunbar, Lewbel and Pendakur (2013) and Calvi (2020). These models are in complete information but a recent literature provides evidence of private information within the household, thereby calling for new developments.

Needless to say, the empirical analysis of strategic games with the development of econometric methods on identification and estimation remains an active area of research. New tools from machine learning, a larger availability of data including elicitation data and increasing computing facilities will contribute to new advances in this literature and new empirical insights involving important policy questions at the core of our economies and societies.

Acknowledgements We are grateful to Cheng Hsiao and Laszlo Matyas for encouraging us to contribute to this volume and to Ian Kim for outstanding research assistance. For the original working papers by Bjorn and Vuong (1984, 1985), we are indebted to David Grether, Jean-Jacques Laffont, Richard McKelvey and Douglas Rivers for insightful comments as well as to Bjorn's classmates Donald Lien and Ken Mcue for proofs and computational help, respectively. Bjorn and Vuong's initial work was supported by the National Science Foundation through grant SES-8410593.

Appendix 1

Cases A3, B3, C3 and D3 exhibit a NE in mixed strategies. Let p_i be the probability that agent i plays action $y_i = 1$. From Fudenberg and Tirole (1991), in a mixed NE player j must be indifferent between choosing $y_j = 1$ and $y_j = 0$ when his opponent randomizes with probability p_i. From the normal form of the game, it follows that a mixed NE is characterized by the conditions $[\tilde{U}_j(0,1)+\Delta_j+\alpha_j+\epsilon_j]p_i+[\tilde{U}_j(0,0)+\Delta_j+\epsilon_j](1-p_i)=\tilde{U}_j(0,1)p_i+\tilde{U}_j(0,0)(1-p_i)$, for $i = 1,2$ and $j \neq i$. Each condition reduces to $\Delta_j+\epsilon_j+\alpha_j p_i = 0$, which does not depend on $[\tilde{U}_j(0,1),\tilde{U}_j(0,0)]$, $j = 1,2$. This gives the equilibrium mixing probabilities

$$p_i = -\frac{\Delta_j+\epsilon_j}{\alpha_j} \quad \text{for } i = 1,2, \tag{4.19}$$

with $0 < p_i < 1$ as it can be verified from the strict inequalities defining cases A3, B3, C3 and D3 where there is a mixed NE. Hence, in a mixed strategy NE, the probability that agents 1 and 2 play actions y_1 and y_2 is $\prod_{i=1,2} p_i^{y_i}(1-p_i)^{1-y_i}$ given (ϵ_1,ϵ_2).

As a matter of fact, the latter probability is conditional on the mixed NE being played. In cases B3 and C3, the mixed NE is the unique NE and thus is played by assumption. Hence in case B3, it follows from (4.19) that the probability of observing (y_1,y_2) given $\mathcal{B}_3 \equiv \{(\epsilon_1,\epsilon_2) : -\Delta_1-\alpha_1 < \epsilon_1 < -\Delta_1 \text{ and } -\Delta_2 < \epsilon_2 < -\Delta_2-\alpha_2\}$ is

$$\mathsf{Pr}(y_1,y_2|\mathcal{B}_3)=\frac{1}{\mathsf{Pr}(\mathcal{B}_3)}\int_{\mathcal{B}_3}\prod_{\{i=1,2\&j\neq i\}}\left(-\frac{\Delta_j+\epsilon_j}{\alpha_j}\right)^{y_i}\left(\frac{\Delta_j+\alpha_j+\epsilon_j}{\alpha_j}\right)^{1-y_i} dF(\epsilon_1,\epsilon_2) \tag{4.20}$$

for $(y_1,y_2) \in \{0,1\}^2$ where $\mathsf{Pr}(\mathcal{B}_3) = \int_{\mathcal{B}_3} dF(\epsilon_1,\epsilon_2)$. Similarly for case C3, the probability $\mathsf{Pr}(y_1,y_2|C_3)$ of observing (y_1,y_2) given $C_3 \equiv \{(\epsilon_1,\epsilon_2) : -\Delta_1 < \epsilon_1 < -\Delta_1-\alpha_1 \text{ and } -\Delta_2-\alpha_2 < \epsilon_2 < -\Delta_2\}$ is

$$\mathsf{Pr}(y_1,y_2|C_3)=\frac{1}{\mathsf{Pr}(C_3)}\int_{C_3}\prod_{\{i=1,2\&j\neq i\}}\left(-\frac{\Delta_j+\epsilon_j}{\alpha_j}\right)^{y_i}\left(\frac{\Delta_j+\alpha_j+\epsilon_j}{\alpha_j}\right)^{1-y_i} dF(\epsilon_1,\epsilon_2) \tag{4.21}$$

for $(y_1,y_2) \in \{0,1\}^2$ where $\mathsf{Pr}(C_3) = \int_{C_3} dF(\epsilon_1,\epsilon_2)$.

Appendix 2

From (4.19) it follows that player i's expected profit in the mixed NE of case A3 or D3 is

$$\Pi_i = \frac{\Delta_i + \alpha_i + \epsilon_i}{\alpha_i}\tilde{U}_i(0,0) - \frac{\Delta_i + \epsilon_i}{\alpha_i}\tilde{U}_i(0,1).$$

This is a weighted average of $\tilde{U}_i(0,0)$ and $\tilde{U}_i(0,1)$ with weights strictly between 0 and 1.

• Case A3: The pure NEs are (0,0) and (1,1). We have

$$\begin{aligned}
\tilde{U}_i(0,0) - \Pi_i &= -\frac{\Delta_i + \epsilon_i}{\alpha_i}\left[\tilde{U}_i(0,0) - \tilde{U}_i(0,1)\right] \\
\tilde{U}_i(1,1) - \Pi_i &= \frac{\Delta_i + \alpha_i + \epsilon_i}{\alpha_i}\left\{\alpha_i - \left[\tilde{U}_i(0,0) - \tilde{U}_i(0,1)\right]\right\} \\
\tilde{U}_i(1,1) - \tilde{U}_i(0,0) &= \Delta_i + \alpha_i + \epsilon_i - \left[\tilde{U}_i(0,0) - \tilde{U}_i(0,1)\right].
\end{aligned}$$

Because $\Delta_i + \epsilon_i < 0 < \Delta_i + \alpha_i + \epsilon_i$ with $\alpha_i > 0$ in case A3, it is easy to sign the above differences. Let $\Delta\tilde{U}_i^0 \equiv \tilde{U}_i(0,0) - \tilde{U}_i(0,1)$. Figure 4.2a summarizes the results in the $(\Delta\tilde{U}_1^0, \Delta\tilde{U}_2^0)$-plane given $(\epsilon_1, \epsilon_2) \in \mathcal{A}_3$. There are 16 regions according to whether each coordinate $\Delta\tilde{U}_i^0$ crosses the thresholds 0, $\Delta_i + \alpha_i + \epsilon_i$ and α_i. The first two lines (in blue) of each region strictly rank Π_i, $\tilde{U}_i(0,0)$ and $\tilde{U}_i(1,1)$ for $i = 1,2$. The inequality that switches when $\Delta\tilde{U}_i^0$ crosses a threshold becomes an equality at the threshold. For instance, $\Pi_i = \tilde{U}_i(0,0)$ when $\Delta\tilde{U}_i^0 = 0$. The third line (in red) of each region indicates Pareto-dominance among the three NEs of case A3 whenever possible. For instance, $M <_p (1,1)$ states that the mixed NE is Pareto-dominated by the pure NE (1,1).

In particular, Figure 4.2a shows that the mixed NE is Pareto-dominated by at least one of the pure NE except when either $\{\Delta\tilde{U}_1^0 < 0, \Delta\tilde{U}_2^0 > \alpha_2\}$ or $\{\Delta\tilde{U}_1^0 > \alpha_1, \Delta\tilde{U}_2^0 < 0\}$. Moreover, Figure 4.2a shows that the pure Nash equilibria $(0,0)$ and $(1,1)$ can be Pareto-ranked except when $\prod_{i=1,2}[\Delta\tilde{U}_i^0 - (\Delta_i + \alpha_i + \epsilon_i)] < 0$ or $\{\Delta\tilde{U}_i^0 = \Delta_i + \alpha_i + \epsilon_i, i = 1,2\}$. The latter exception region contains the former exception region. Thus, if

$$\prod_{i=1,2}[\Delta\tilde{U}_i^0 - (\Delta_i + \alpha_i + \epsilon_i)] \geq 0 \text{ with one nonzero term when } (\epsilon_1, \epsilon_2) \in \mathcal{A}_3, \quad (4.22)$$

then one of the pure NE Pareto-dominates the other pure and mixed NEs. For instance, when $\Delta\tilde{U}_i^0 = 0$ for $i = 1,2$, then (4.22) holds since $\Delta_i + \alpha_i + \epsilon_i > 0$. Specifically, we have $M \sim_p (0,0) <_p (1,1)$, i.e., the mixed NE and pure NE $(0,0)$ are equivalent in terms of expected payoff but are both Pareto-dominated by the pure NE (1,1).

Alternatively, the differences $(\Delta\tilde{U}_1^0, \Delta\tilde{U}_2^0)$ can be viewed as random with a joint distribution $H(\cdot,\cdot|\epsilon_1,\epsilon_2)$ conditional on (ϵ_1, ϵ_2) that is absolutely continuous with respect to Lebesgue measure. For $(\epsilon_1, \epsilon_2) \in \mathcal{A}_3$, assume that the support of $H(\cdot,\cdot|\epsilon_1,\epsilon_2)$ is $\{\prod_{i=1,2}[\Delta\tilde{U}_i^0 - (\Delta_i + \alpha_i + \epsilon_i)] \geq 0\}$ so that (4.22) is satisfied almost surely. Thus, a pure NE Pareto-dominates the other pure and mixed NE.

Specifically, we have $\{M \text{ and } (0,0)\} <_p (1,1)$ if $\Delta\tilde{U}_i^0 < \Delta_i + \alpha_i + \epsilon_i, i = 1,2$ whereas $\{M \text{ and } (1,1)\} <_p (0,0)$ if $\Delta\tilde{U}_i^0 > \Delta_i + \alpha_i + \epsilon_i, i = 1,2$. Hence, when the players coordinate on the Pareto-dominant NE, the probabilities of observing $(1,1)$ and $(0,0)$ given $\mathcal{A}_3$ are

$$\mathsf{Pr}(1,1|\mathcal{A}_3) = \frac{1}{\mathsf{Pr}(\mathcal{A}_3)} \int_{\mathcal{A}_3} \int_{\{\Delta\tilde{U}_i^0 \le \Delta_i + \alpha_i + \epsilon_i, i=1,2\}} dH(\Delta\tilde{U}_1^0, \Delta\tilde{U}_2^0 | \epsilon_1, \epsilon_2) dF^o(\epsilon_1, \epsilon_2), \tag{4.23}$$

$$\mathsf{Pr}(0,0|\mathcal{A}_3) = 1 - \mathsf{Pr}(1,1|\mathcal{A}_3),$$

where $\mathsf{Pr}(\mathcal{A}_3) = \int_{\mathcal{A}_3} dF^o(\epsilon_1, \epsilon_2)$.

- Case D3: The pure NEs are (1,0) and (0,1). We have

$$\begin{aligned}
\tilde{U}_i(1,0) - \Pi_i &= -\frac{\Delta_i + \epsilon_i}{\alpha_i} \left\{ \left[\tilde{U}_i(0,0) - \tilde{U}_i(0,1) \right] - \alpha_i \right\} \\
\tilde{U}_i(0,1) - \Pi_i &= -\frac{\Delta_i + \alpha_i + \epsilon_i}{\alpha_i} \left[\tilde{U}_i(0,0) - \tilde{U}_i(0,1) \right] \\
\tilde{U}_i(1,0) - \tilde{U}_i(0,1) &= \left[\tilde{U}_i(0,0) - \tilde{U}_i(0,1) \right] + \Delta_i + \epsilon_i.
\end{aligned}$$

Because $\Delta_i + \alpha_i + \epsilon_i < 0 < \Delta_i + \epsilon_i$ with $\alpha_i < 0$ in case D3, it is easy to sign the above differences. Figure 4.2d summarizes the results in the $(\Delta\tilde{U}_1^0, \Delta\tilde{U}_2^0)$-plane given $(\epsilon_1, \epsilon_2) \in \mathcal{D}_3$. There are 16 regions according to whether each coordinate $\Delta\tilde{U}_i^0$ crosses the thresholds α_1, $\Delta_i + \epsilon_i$ and 0. The reading of Figure 4.2d is analogous to Figure 4.2a.

In particular, Figure 4.2d shows that the mixed NE is Pareto-dominated by at least one of the pure NE except when either $\{\Delta\tilde{U}_1^0 < \alpha_1, \Delta\tilde{U}_2^0 < \alpha_2\}$ or $\{\Delta\tilde{U}_1^0 > 0, \Delta\tilde{U}_2^0 > 0\}$. Moreover, Figure 4.2d shows that the pure Nash equilibria $(0,1)$ and $(1,0)$ can be Pareto-ranked except when $\prod_{i=1,2}[\Delta\tilde{U}_i^0 + (\Delta_i + \epsilon_i)] > 0$ or $\{\Delta\tilde{U}_i^0 = -(\Delta_i + \epsilon_i), i = 1,2\}$. The latter exception region contains the former exception region. Thus, if

$$\prod_{i=1,2} [\Delta\tilde{U}_i^0 + (\Delta_i + \epsilon_i)] \le 0 \text{ with one nonzero term when } (\epsilon_1, \epsilon_2) \in \mathcal{D}_3, \tag{4.24}$$

then one of the pure NE Pareto-dominates the other pure and mixed Nash equilibria. For instance, when $\Delta\tilde{U}_i^0 = 0$ for $i = 1,2$, then (4.24) fails since $\Delta_i + \epsilon_i > 0$. Specifically, the mixed Nash equilibrium is Pareto-dominated by both pure Nash equilibria $(0,0)$ and $(1,1)$ but the latter cannot be Pareto-ranked.

Alternatively, the differences $(\Delta\tilde{U}_1^0, \Delta\tilde{U}_2^0)$ can be viewed as random with a joint distribution $H(\cdot,\cdot|\epsilon_1,\epsilon_2)$ conditional on (ϵ_1,ϵ_2) that is absolutely continuous with respect to Lebesgue measure. For $(\epsilon_1,\epsilon_2) \in \mathcal{D}_3$, assume that the support of $H(\cdot,\cdot|\epsilon_1,\epsilon_2)$ is $\{\prod_{i=1,2}[\Delta\tilde{U}_i^0 + (\Delta_i + \epsilon_i)] \le 0\}$ so that (4.24) is satisfied almost surely. Thus, a pure NE Pareto-dominates the other pure and mixed NE. Specifically, we have $\{M \text{ and } (1,0)\} <_p< (0,1)$ if $\Delta\tilde{U}_1^0 < -(\Delta_1 + \epsilon_1)$ and $\Delta\tilde{U}_2^0 > -(\Delta_2 + \epsilon_2)$ whereas $\{M \text{ and } (0,1)\} <_p (1,0)$ if $\Delta\tilde{U}_1^0 > -(\Delta_1 + \epsilon_1)$ and $\Delta\tilde{U}_2^0 < -(\Delta_2 + \epsilon_2)$. Hence, when the players coordinate on the Pareto-dominant NE, the probabilities of observing

(0, 1) and (1, 0) given $\mathcal{D}_3$ are

$$\Pr(0,1 \mid \mathcal{D}_3) = \frac{1}{\Pr(\mathcal{D}_3)} \int_{\mathcal{D}_3} \int_{\{\Delta\tilde{U}_1^0 \le -(\Delta_1+\epsilon_1),\, \Delta\tilde{U}_2^0 \ge -(\Delta_2+\epsilon_2)\}} dH(\Delta\tilde{U}_1^0, \Delta\tilde{U}_2^0 \mid \epsilon_1, \epsilon_2)\, dF(\epsilon_1, \epsilon_2), \tag{4.25}$$

$$\Pr(1,0 \mid \mathcal{D}_3) = 1 - \Pr(0,1 \mid \mathcal{D}_3),$$

where $\Pr(\mathcal{D}_3) = \int_{\mathcal{D}_3} dF(\epsilon_1, \epsilon_2)$.

References

Ackerberg, D., Benkard, C. L., Berry, S. & Pakes, A. (2007). Econometric tools for analyzing market outcomes. In J. J. Heckman & E. Leamer (Eds.), *Handbook of econometrics* (Vol. 6A, pp. 4171–4276). Amsterdam: Elsevier.

Aguirregabiria, V., Collard-Wexler, A. & Ryan, S. P. (2021). Dynamic games in empirical industrial organization. In K. Ho, A. Hortacsu & A. Lizzeri (Eds.), *Handbook of industrial organization* (Vol. 4, pp. 225–343). Amsterdam: Elsevier.

Aguirregabiria, V. & Mira, P. (2007). Sequential estimation of dynamic discrete games. *Econometrica*, *75*, 1–53.

Aigner, D. J., Hsiao, C., Kapteyn, A. & Wansbeek, T. (1984). Latent variable models in econometrics. In Z. Griliches & M. Intriligator (Eds.), *Handbook of econometrics* (Vol. 2, pp. 1321–1393). Amsterdam: Elsevier.

Almås, I., Attanasio, O. & Carneiro, P. (2023). Household decisions and intra-household distributions. In S. Lundberg & A. Voena (Eds.), *Handbook of the economics of the family* (Vol. 1, pp. 111–149). Amsterdam: Elsevier.

Amemiya, T. (1981). Qualitative response models: A survey. *Journal of Economic Literature*, *19*, 1483–1536.

Andrews, D., Berry, S. & Jia, P. (2004). Confidence regions for parameters in discrete games with multiple equilibria. *Working paper, Yale University*.

Aradillas-Lopez, A. (2010). Semiparametric estimation of a simultaneous game with incomplete information. *Journal of Econometrics*, *157*, 409–431.

Aradillas-Lopez, A. & Tamer, E. (2008). The identification power of equilibrium in simple games. *Journal of Business & Economic Statistics*, *26*, 261–283.

Ashenfelter, O. & Heckman, J. J. (1974). The estimation of income and substitution effects in a model of family labor supply. *Econometrica*, *42*, 73–85.

Bajari, P., Benkard, C. L. & Levin, J. (2007). Estimating dynamic models of imperfect competition. *Econometrica*, *75*, 1331–1370.

Bajari, P., Hong, H., Krainer, J. & Nekipelov, D. (2010). Estimating static models of strategic interactions. *Journal of Business & Economic Statistics*, *28*, 469–482.

Bajari, P., Hong, H. & Ryan, S. P. (2010). Identification and estimation of a discrete game of complete information. *Econometrica*, *78*, 1529–1568.

Berry, S. (1992). Estimation of a model of entry in the airline industry. *Econometrica, 60*, 889–917.

Berry, S. & Reiss, P. (2007). Empirical models of entry and market structure. In M. Armstrong & R. H. Porter (Eds.), *Handbook of industrial organization* (Vol. 3, pp. 1845–1886). Amsterdam: Elsevier.

Berry, S. & Tamer, E. (2006). Identification in models of oligopoly entry. In R. Blundell, W. K. Newey & T. Persson (Eds.), *Advances in economics and econometrics* (Vol. 2, pp. 46–85). New York: Cambridge University Press.

Bishop, Y., Fienberg, S. & Holland, P. (1975). *Discrete multivariate analysis: Theory and practice.* Cambridge, Mass.: MIT Press.

Bjorn, P. A. (1986). *Games in econometrics with applications to labor economics* (Unpublished doctoral dissertation). California Institute of Technology.

Bjorn, P. A. & Vuong, Q. H. (1984). Simultaneous equations models for dummy endogenous variables: a game theoretic formulation with an application to labor force participation. *Working Paper, California Institute of Technology.*

Bjorn, P. A. & Vuong, Q. H. (1985). Econometric modeling of a stackelberg game with an application to labor force participation. *Working Paper, California Institute of Technology.*

Bouissou, M. B., Laffont, J.-J. & Vuong, Q. H. (1986). Tests of noncausality under markov assumptions for qualitative panel data. *Econometrica*, 395–414.

Bresnahan, T. F. & Reiss, P. C. (1990). Entry in monopoly market. *The Review of Economic Studies, 57*, 531–553.

Bresnahan, T. F. & Reiss, P. C. (1991). Entry and competition in concentrated markets. *Journal of Political Economy, 99*, 977–1009.

Brock, W. A. & Durlauf, S. N. (2001). Interactions-based models. In J. J. Heckman & E. Leamer (Eds.), *Handbook of econometrics* (Vol. 5, pp. 3297–3380). Ansterdam: Elsevier.

Brock, W. A. & Durlauf, S. N. (2007). Identification of binary choice models with social interactions. *Journal of Econometrics, 140*, 52–75.

Browning, M. & Chiappori, P.-A. (1998). Efficient intra-household allocations: A general characterization and empirical tests. *Econometrica*, 1241–1278.

Calvi, R. (2020). Why are older women missing in india? the age profile of bargaining power and poverty. *Journal of Political Economy, 128*, 2453–2501.

Chiappori, P.-A. (1988). Rational household labor supply. *Econometrica, 56*, 63–90.

Chiappori, P.-A. (1992). Collective labor supply and welfare. *Journal of Political Economy, 100*, 437–467.

Ciliberto, F. & Tamer, E. (2009). Market structure and multiple equilibria in airline markets. *Econometrica, 77*, 1791–1828.

Conitzer, V. (2016). On Stackelberg mixed strategies. *Synthese, 193*, 689–703.

Cramer, J. S. (2004). The early origins of the logit model. *Studies in History and Philosophy of Science Part C, 35*, 613–626.

De Paula, A. & Tang, X. (2012). Inference of signs of interaction effects in simultaneous games with incomplete information. *Econometrica, 80*, 143–172.

Dunbar, G. R., Lewbel, A. & Pendakur, K. (2013). Children's resources in collective households: identification, estimation, and an application to child poverty in Malawi. *American Economic Review*, *103*, 438–471.

Dunbar, G. R., Lewbel, A. & Pendakur, K. (2021). Identification of random resource shares in collective households without preference similarity restrictions. *Journal of Business & Economic Statistics*, *39*, 402–421.

Echenique, F. & Edlin, A. (2004). Mixed equilibria are unstable in games of strategic complements. *Journal of Economic Theory*, *118*, 61–79.

Ericson, R. & Pakes, A. (1995). Markov-perfect industry dynamics: A framework for empirical work. *The Review of Economic Studies*, *62*, 53–82.

Fudenberg, D. & Tirole, J. (1991). *Game theory*. Cambridge: MIT press.

Galichon, A. & Salanié, B. (2017). The econometrics and some properties of separable matching models. *American Economic Review*, *107*, 251–255.

Gourieroux, C., Laffont, J.-J. & Monfort, A. (1980). Coherency conditions in simultaneous linear equation models with endogenous switching regimes. *Econometrica*, *48*, 675–695.

Graham, B. S. (2008). Identifying social interactions through conditional variance restrictions. *Econometrica*, *76*, 643–660.

Graham, B. S. (2011). Efficiency bounds for missing data models with semiparametric restrictions. *Econometrica*, *79*, 437–452.

Graham, B. S. (2020). Network data. In S. N. Durlauf, L. P. Hansen, J. J. Heckman & R. L. Matzkin (Eds.), *Handbook of econometrics* (Vol. 7A, pp. 111–218). Amsterdam: Elsevier.

Grieco, P. L. (2014). Discrete games with flexible information structures: An application to local grocery markets. *The RAND Journal of Economics*, *45*, 303–340.

Haberman, S. J. (1974). *The analysis of frequency data*. University of Chicago Press.

Harsanyi, J. C. (1967). Games with incomplete information played by Bayesian players. *Management Science*, *14*, 159–182.

Heckman, J. J. (1974). Shadow prices, market wages, and labor supply. *Econometrica*, 679–694.

Heckman, J. J. (1978). Dummy endogenous variables in a simultaneous equation system. *Econometrica*, *46*, 931–959.

Hood, W. C. & Koopmans, T. C. (1953). *Studies in econometric method*. New Heaven: Yale University Press.

Hotz, V. J. & Miller, R. A. (1993). Conditional choice probabilities and the estimation of dynamic models. *The Review of Economic Studies*, *60*, 497–529.

Jackson, M. O. & Wolinsky, A. (1996). A strategic model of social and economic networks. *Journal of Economic Theory*, *71*, 44–74.

Jia, P. (2008). What happens when wal-mart comes to town: An empirical analysis of the discount retailing industry. *Econometrica*, *76*, 1263–1316.

Jovanovic, B. (1989). Observable implications of models with multiple equilibria. *Econometrica*, *57*, 1431–1437.

Kim, I., Perrigne, I., Vuong, Q. & Yan, W. (2025). Differentiated products with endogenous choice sets. *Working paper, New York University*.

Kline, B., Pakes, A. & Tamer, E. (2021). Moment inequalities and partial identification in industrial organization. In K. Ho, A. Hortacsu & A. Lizzeri (Eds.), *Handbook of industrial organization* (Vol. 4, pp. 345–431). Amsterdam: Elsevier.

Kooreman, P. (1994). Estimation of econometric models of some discrete games. *Journal of Applied Econometrics*, *9*, 255–268.

Korzhyk, D., Yin, Z., Kiekintveld, C., Conitzer, V. & Tambe, M. (2011). Stackelberg vs Nash in security games: An extended investigation of interchangeability, equivalence and uniqueness. *Journal of Artificial Intelligence Research*, *41*, 297–327.

Krishna, V. (2002). *Auction theory*. London: Academic Press.

Laffont, J.-J. & Martimort, D. (2002). *The theory of incentives: The principal-agent model*. Princeton: Princeton University Press.

Leung, M. (2015). Two-step estimation of network-formation models with incomplete information. *Journal of Econometrics*, *210*, 182–195.

Lewbel, A. & Tang, X. (2015). Identification and estimation of games with incomplete information using excluded regressors. *Journal of Econometrics*, *189*, 229–244.

Liu, N., Vuong, Q. & Xu, H. (2017). Rationalization and identification of binary games with correlated types. *Journal of Econometrics*, *201*, 249–268.

Maddala, G. S. (1983). *Limited-dependent and qualitative variables in econometrics*. New York: Cambridge University Press.

Maddala, G. S. & Lee, L.-F. (1976). Recursive models with qualitative endogenous variables. *Annals of Economic and Social Measurement*, *5*, 525–545.

Manski, C. F. (1993a). Identification of endogenous social effects: The reflection problem. *The Review of Economic Studies*, *60*, 531–542.

Manski, C. F. (1993b). Identification problems in the social sciences. In P. Marsden (Ed.), *Sociological methodology* (Vol. 23, pp. 1–56). Cambridge: Basil Blackwell.

Manski, C. F. & McFadden, D. (1981). *Structural analysis of discrete data with econometric applications*. Cambridge: MIT Press.

Maskin, E. & Tirole, J. (2001). Markov perfect equilibrium: I. observable actions. *Journal of Economic Theory*, *100*, 191–219.

McElroy, M. B. & Horney, M. J. (1981). Nash-bargained household decisions: Toward a generalization of the theory of demand. *International Economic Review*, *22*, 333–349.

McFadden, D. (1973). Conditional logit analysis of qualitative choice behavior. In P. Zarembka (Ed.), *Frontiers of Econometrics* (pp. 105–142). New York: Academic Press.

McFadden, D. (1981). Econometric models of probabilistic choice. In C. F. Manski & D. McFadden (Eds.), *Structural analysis of discrete data with econometric applications* (pp. 198–272). Cambridge: MIT Press.

McFadden, D. (1984). Econometric analysis of qualitative response models. In Z. Griliches & M. Intriligator (Eds.), *Handbook of econometrics Volume 2* (Vol. 2, p. 1395-1457). Amsterdam: Elsevier.

McFadden, D. (1989). A model of simulated moments for estimation of discrete response models without numerical integration. *Econometrica*, *57*, 995–1026.

McKelvey, R. D. & McLennan, A. (1996). Computation of equilibria in finite games. In H. M. Amman, D. A. Kendrick & J. Rust (Eds.), *Handbook of Computational Economics Volume 1* (Vol. 1, pp. 87–142). Amsterdam: Elsevier.

Menzel, K. (2016). Strategic network formation with many agents. *Working Paper, New York University*.

Milgrom, P. & Roberts, J. (1990). Rationalizability, learning and equilibrium in games with strategic complementarities. *Econometrica*, *58*, 1255–1277.

Nash, J. (1950). Equilibrium points in n-person games. *Proceedings of the National Academy of Sciences*, *36*, 48–49.

Nash, J. (1951). Non-cooperative games. *Annals of Mathematics*, *54*, 286–295.

Nerlove, M., Grether, D. & Carvalho, J. L. (1979). *Analysis of economic time series: A synthesis*. New York: Academic Press.

Nerlove, M. & Press, S. J. (1973). *Univariate and multivariate log-linear and logistic models*. (Working paper, RAND Corporation)

Nerlove, M., Razin, A. & Sadka, E. (1987). *Household and economy: Welfare economics of endogenous fertility*. New York: Academic Press.

Pakes, A. & McGuire, P. (1994). Computing Markov-perfect Nash equilibria: Numerical implications of a dynamic differentiated product model. *The RAND Journal of Economics*, *25*, 555–1370.

Pakes, A., Ostrovsky, M. & Berry, S. T. (2007). Simple estimators for the parameters of discrete dynamic games. *The RAND Journal of Economics*, *38*, 373–399.

Pakes, A. & Pollard, D. (1989). Simulation and the asymptotics of optimization estimators. *Econometrica*, *57*, 1027–1057.

Pakes, A., Porter, J., Ho, K. & Ishii, J. (2015). Moment inequalities and their application. *Econometrica*, *83*, 315–334.

Pesendorfer, M. & Schmidt-Dengler, P. (2008). Asymptotic least squares estimators for dynamic games. *The Review of Economic Studies*, *75*, 910–928.

Reiss, P. C. & Wolak, F. A. (2007). Structural econometric modeling: Rationales and examples from industrial organization. In J. J. Heckman & E. Leamer (Eds.), *Handbook of econometrics* (Vol. 6A, pp. 4277–4415). Amsterdam: Elsevier.

Rothenberg, T. (1971). Identification in parametric models. *Econometrica*, *39*, 577–591.

Rust, J. (1994a). Estimation of dynamic structural models. In C. Sims & J.-J. Laffont (Eds.), *Advances in econometrics* (Vol. 2, p. 119-170). Cambridge: Cambridge University Press.

Rust, J. (1994b). Structural estimation of markov decision processes. In R. F. Engle & D. L. McFadden (Eds.), *Handbook of econometrics* (Vol. 4, pp. 3081–3143). Amsterdam: Elsevier.

Schmidt, P. (1981). Constraints on the parameters in simultaneous tobit and probit models. In C. F. Manski & D. McFadden (Eds.), *Structural analysis of discrete data with econometric applications* (pp. 422–434). Cambridge: MIT Press.

Seim, K. (2006). An empirical model of firm entry with endogenous product-type choices. *The RAND Journal of Economics*, *37*, 619–640.

Sweeting, A. (2009). The strategic timing of radio commercials: An empirical analysis using multiple equilibria. *The RAND Journal of Economics*, *40*, 710–742.

Tamer, E. (2003). Incomplete simultaneous discrete response model with multiple equilibria. *The Review of Economic Studies*, *70*, 147–167.

Von Neumann, J. & Morgenstern, O. (1944). *Theory of games and economic behavior*. Princeton: Princeton University Press.

Vuong, Q. H. (1982). *Conditional log-linear probability models: A theoretical development with an empirical application* (Unpublished doctoral dissertation). Northwestern University.

Vuong, Q. H. (1989). Likelihood ratio tests for model selection and non-nested hypotheses. *Econometrica*, *57*, 307–333.

Weintraub, G. Y., Benkard, C. L. & Roy, B. V. (2008). Markov perfect industry dynamics with many firms. *Econometrica*, *76*, 1375–1411.

Chapter 5
Measuring 'Income' Inequality and Distribution of Outcomes

Esfandiar Maasoumi, Yisroel Cahn

Abstract We provide a suggestive examination of state of knowledge on measurement and analysis of 'inequality' of outcomes, especially of incomes and earnings. This perspective aims to describe the state of art techniques for identifying the distribution of outcomes as the central object and consider interesting functions of it, such as inequality measures, poverty and mobility indices. We distinguish between 'scalar', cardinal functions such as indices, as well as weak uniform rankings, such as by stochastic dominance based on modern rigorous tests. A basic theme permeates the discussion, that of decision theoretic foundations within the potential outcomes paradigm. This permits connectivity and advancement of knowledge that is policy relevant, reveals the essential subjectivity of indices and assessments, and allows important consideration of counterfactual distribution of outcomes. Identification of distributions and their functionals exposes the impact of covariates and different contributing factors to outcomes. A recurrent example is the distribution of earnings of different groups within a population, and decomposition of outcomes by group or other characteristics, and counterfactual states. Modern foundations of inequality measures, dominance rankings, quantile differences/effects, as well as quantile models, instrumental variables and 'distribution regressions' are included and analyzed. The hope is to encourage and facilitate adoption of these important new developments at a time of heightened interest in this central socio-political area of policy analysis. The closely related notions of multivariate well being, mobility and poverty, merit separate treatment and are not addressed.

Esfandiar Maasoumi ✉
Emory University, Atlanta, USA, e-mail: emaasou@emory.edu

Yisroel Cahn
New York University, New York, USA, e-mail: yisroel.cahn@nyu.edu

B. H. Baltagi and L. Mátyás (eds.), *Seven Decades of Econometrics and Beyond*, Advanced Studies in Theoretical and Applied Econometrics 57,
https://doi.org/10.1007/978-3-031-92699-0_5

5.1 Introduction

The question of allocation and 'inequality' has engaged social thinkers for centuries and is back as 'The question of our time'. Within economics, the issue of 'allocation' remains central to policy analysis of market and social outcomes. As faith grew stronger in 'modern' formal market theories of economics, interest declined in this central question, based on a presumption of market ability to determine outcome distributions that would be universally desirable. This faith, not entirely misplaced, was perhaps too uncritical and has undergone healthy questioning and revision at the frontier of economic and social research. Civil society can define its objective function to accommodate any aspect of outcomes it deems desirable. These include ethical and longer term allocation objectives and group outcomes.

The consequences of extreme inequality, certainly of 'inequity', for stability of social and market institutions is now more widely acknowledged. The distinctions between opportunity, mobility, poverty, inequality, and concepts of 'well being' more general than incomes/earnings, are increasingly appreciated and examined. Mechanisms by which a distributed outcome is arrived at are extremely complex and evolving. Models that aspire to identify and separate underlying 'causes', while still challenged, are becoming less restrictive in form and more sophisticated in dealing with sampling issues such as 'selection'. They are also increasingly treated rigorously with the latest econometric techniques that go beyond the modest asymptotic inferences highlighted in earlier surveys, for example, Maasoumi (1998).[1]

This paper provides a selective account of some of these developments. For policy and decision making, objective measurement of outcomes, as they are, and formally subjective assessments, remain a central challenge to evidence-based analysis. This review is primarily focused on transparent subjectivity of seemingly objective measurement of inequality in an outcome distribution, be it incomes, health, education, wealth, etc. Modern inference techniques and algorithms are summarized. It is emphasized that all assessments are fundamentally subjective-comparative in nature, starting with a contrast between a given distribution of income (say), and a perfectly 'equal' outcome as a reference. We emphasize the subjective bases of measurement choices and interpretations even of 'what is', let alone comparisons between groups or over time or policy states. The subjective, welfarist, 'utilitarian', characterization of inequality measures flows from the fundamental works of Dalton in 1920s, to Atkinson and Kolm, in the 1960s. The 1970s and 1980s witnessed a faithful attempt to provide some objectivity in the analysis of fundamental welfare axioms/restrictions that underlie all measures and indices of inequality and poverty, For example see, Shorrocks (1978) and Bourguignon (1979).

In a 2011 issue of AER P&P, A. B. Atkinson asserts "[e]conomists need to be more explicit about the relation between welfare criteria and the objectives

[1] Nerlove's contributions belong to the strand of research concerned with mechanisms and empirical models of taxation, human capital and gender differences. See for example, Nerlove, Razin and Sadka (1987), Nerlove, Razin, Sadka and Weizsacker (1993), Nerlove, Razin, Sadka and Weizsäcker (1993), and Nerlove, Razin and Sadka (1993).

of governments, policymakers and individual citizens". Indeed, such normative discussions are sidestepped when only means are considered.

For example, in evaluating 'right to work' laws which forbid unions from interfering with the employment of nonunion workers, some workers may be hurt while others benefit. Looking at the mean alone might indicate that such laws are neutral or favorable while ignoring ethical questions surrounding such laws.

Another example is forming combined classes for students with high test scores and low test scores, in order to evaluate policy outcomes. Examining the mean test scores of the resulting combined class might not give an accurate picture of who, if anyone, benefited and by how much. Reasonable evaluation criteria for such a policy might entail determining whether the gap in test scores between the two groups of students has been reduced or whether test scores are above a given threshold.

A. B. Atkinson (2011) lists several possible ways applied economists have rationalized neglecting welfare economics. These include assuming away differences in outcomes, assuming agreement on the welfare criteria, or even that welfare discussions are better suited to other disciplines. A. B. Atkinson argues against such stances and urges applied economists to renew their focus on welfare discussions. Indeed, a policy's distributional effects have important, non-trivial implications that are contingent on subjective values. This paper reviews clear approaches to explicitly accommodate different opinions regarding inequality and poverty, and heeds A. B. Atkinson's call to put welfare economics back in the spotlight.

There are important aspects of welfare that are often overlooked by much of the policy evaluation literature — the effect of the policy in the short- medium- and long-term. Multidimensional inequality or poverty comparisons can address such issues by using outcomes of individuals in different periods as dimensions. While multidimensional measures are not discussed in this paper, the appendix includes a way of accounting for the value of time in the income distribution.

Furthermore, if there are multiple outcomes of interest, individuals might be affected positively in one outcome but negatively in another. In many cases, welfare might not only be measured in inequality or poverty in one dimension, but rather by some functional of the joint distribution of outcomes. Cahn (2022) examines whether increases in minimum wage reduced the hours worked of those individuals whose wages were increased. Simply looking at average treatment effects of hours and wages separately would not accurately reflect such an outcome.

Generally, there are three stages in the analysis of outcomes: Choice of the wellbeing object, the distribution of that object, and characterization of that distribution.

Considering 'income', decisions must be made on sources of it, such a labor income, investment, transfers,....Then a choice has to be made for the period of time, monthly, annual, or lifetime/permanent incomes. Inflation adjustment is desirable to consider 'real' incomes, but group specific price indices are generally not available and highly subjective. Choice of the unit as an individual, or household matters a great deal. How are households measured is challenging die to the need for equivalent incomes for different members. A good deal of work is based on expenditures as more easily measured object of utility, compared to 'income'. Price indices are almost universally based on expenditure data, not incomes.

The impact of other causes and covariates may be accommodated, before they are integrated out to obtain the 'marginal' distribution of income. This is highly model dependent, and requires going beyond the regression at the conditional means, as befits the context of concern for 'distribution' of an outcome!

There are complex relations between different means of comparing distributed outcomes. Inequality may be a 'relative' concept or 'horizontal'. It may consider real incomes, or nominal, household or individuals, after or before tax and benefits, and may consider dimensions beyond income, such as health and education. It is instructive to begin with an empirical example of, US household (pre-tax) income distribution at two different points in time. This will serve to highlight many of the subjective and challenging issues, including what is represented by measures of inequality, poverty, mobility,...,. In this example, an otherwise uniformly stochastically dominant outcome represents higher 'relative' inequality!

Table 5.1 represents household income quantiles in the US in 1974 and 2004.

Table 5.1: Household incomes. US 1974-2004

θ	θ-quantile		Growth
	1974	2004	
10%	\$9,741	\$10,927	12.2%
20%	\$16,285	\$18,500	13.6%
50%	\$37,519	\$44,389	18.3%
80%	\$64,781	\$88,029	35.9%
90%	\$83,532	\$120,924	44.8%
95%	\$102,534	\$157,185	53.3%

Note: Columns 2 and 3 give upper limit of the bottom 10%, 20%... of the population.

Source: Cowell (2006)

Every quantile in 2004 is higher than the corresponding one in 1974. As we shall see, to a given degree of statistical confidence, 2004 outcome stochastically dominates 1974. But due to much higher growth rates in upper quantiles, inequality has likely increased. In this situation almost all inequality measures would reverse the ranking in favor of 1974. The extent of this reversal depends on the particular inequality index.

A central object for assessing distributions is the Lorenz curve/function. It is a graph of ordered cummulated income shares against the corresponding cummulated 'population shares':

This curve is depicted below for the same data in Table 5.1.

For two income distributions with the same means, the one that is uniformly closer to the 45 degree line (Lorenz) dominates. It is all about 'relative' incomes and equal

Table 5.2: Average incomes for five quintiles and overall. US 1974-2004

Group	Average Income		Growth
	1974	2004	
1st	$9,324	$10,264	10.1%
2nd	$23,176	$26,241	13.2%
3rd	$37,353	$44,455	19.0%
4th	$53,944	$70,085	29.9%
Top	$95,576	$151,593	58.6%
Overall	$43,875	$60,528	38.0%

Note: Columns 2 and 3 give the average income of the bottom fifth, second fifth... of the population. Source: Cowell (2006)

Fig. 5.1: ECDF Diagram for Table 5.1.

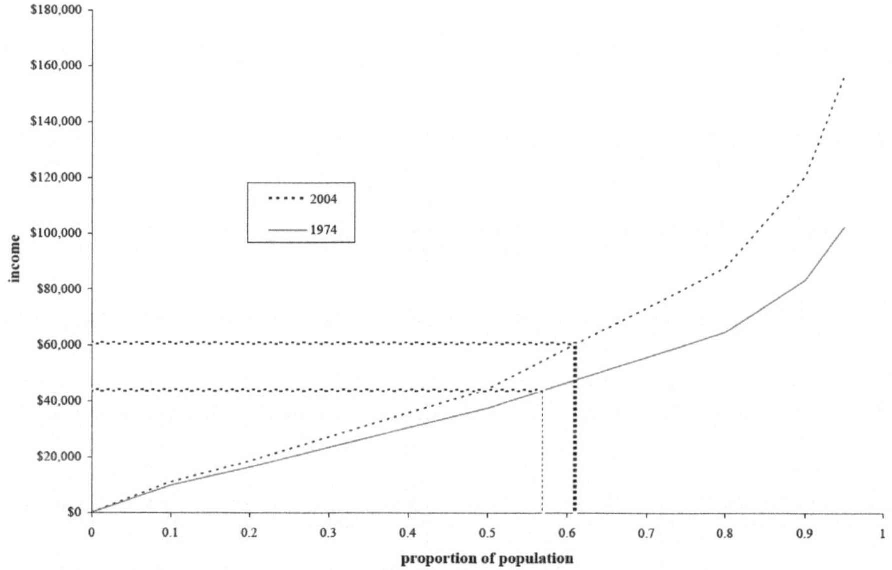

Source: Cowell (2006)

means renders the two relative outcomes comparable. Every 'relative' inequality measure is a function of the area between the Lorenz curve and the 45 degree line.

Fig. 5.2: Lorenz Curve for Table 5.1

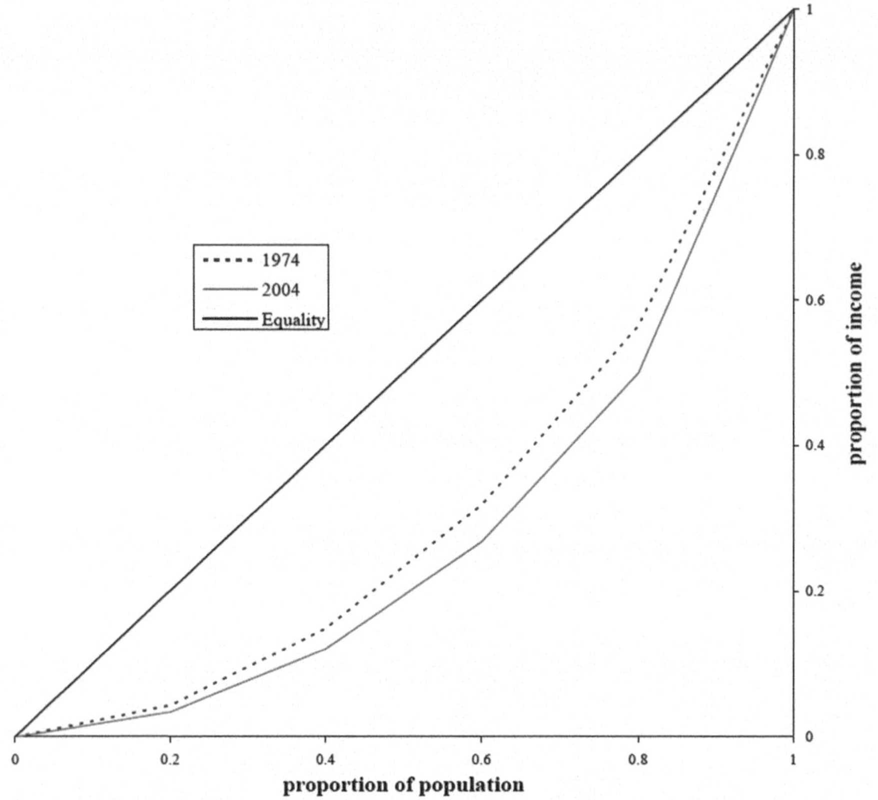

Source: Cowell (2006)

Here, dominance of 1974 in terms of 'equality' is merely a matter of how much, not if.

But the means are seldom equal. And a much 'richer' outcome in 2004, with a higher mean income, and higher quantiles, may be 'dominant' even when degrees of 'aversion to inequality' are introduced. This assessment is accomplished by a simple transformation of Lorenz curve, multiplying it by the corresponding mean income, to obtain the 'Generalized Lorenz' function. This is depicted in the following Figure for the same Table 5.1 data.

This is also known as 'second order stochastic dominance'. The cumulated CDF of one distribution (2004 here) is everywhere to the left of the one for 1974. We expect this, since 'first order dominance' observed in Fig 1 implies higher orders of dominance.

These rankings imply poverty dominance as well, since the CDFs do not cross at lower quantiles, or any that may be viewed as a popular 'poverty line'. There is 'restricted dominance' at every quantile, even though 'degree of poverty' may be

Fig. 5.3: Generalized Lorenz curve for Table 5.1

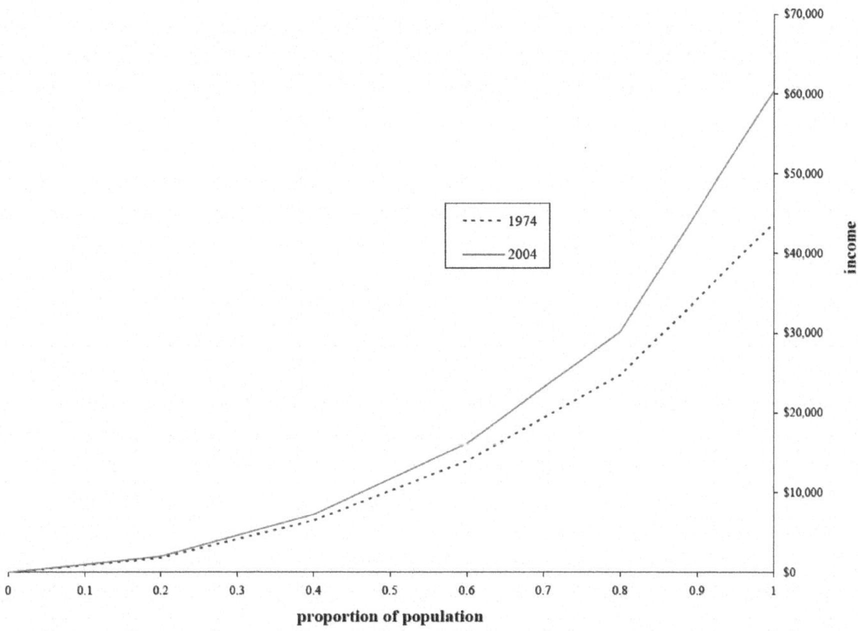

Source: Cowell (2006)

assessed as having increased. By how much, depends on the poverty measure. See Lugo and Maasoumi (2008).

None of the above assessments can guide us clearly to an assessment of mobility. 'Anonymous' views of a distribution of outcomes, as above, are invariant to permutations of the unit outcomes. Mobility measures, be they 'cross sectional' or intergenerational, must shed the property of 'anonymity', which is so central to the axiomatic development of ideal inequality measures.

It is clear that if the CDFs cross, some quantiles would be ranked higher and some lower. If a single crossing occurs at very high incomes, Second Order dominance may still hold. Otherwise, even comparison of the two crossing outcomes will depend on the choice of the inequality index. A Rawlsian will rank based on the lowest incomes, and an extreme trickle-down measure will rank by the highest quantile/income, and all others would fall in between. A conundrum.

Measures of inequality have a long history, and one of the earliest due to Gini continues to be dominant among practitioners and policy makers. The Gini index did not fare well, initially, in the axiomatic search for ideal measures, owing to its well known allowance for intrapersonal comparisons of well being/utility. The axiomatic approach emphasized the apparent objectivity of *welfarist-utilitarian-individualistic* welfare functions (infinitely substitutable utility of individuals), and its emphasis on 'anonymity'. Generalizations of Gini have somewhat rehabilitated Gini, see Donaldson and Weymark (1983). But Gini fails in important ways. One is that it seems to be stuck in a range of about 4.1-4.5 for the US, with recent numbers of 4.2 and 4.1 being used in popular and political press as evidence of movement up or down. These differences are not likely statistically significant, and miss all the movements in the tails. The tails are were most policies are aimed at, and much of the differential evolutions of incomes have occurred! Other measures of inequality, such as Theil's entropy are much more sensitive to tail changes. A second issue with Gini is that it fails to satisfy a very useful 'Aggregation Consistency' property; see Shorrocks (1983), or Maasoumi (1998). If one divides the reference population into R groups, men and women, say, if inequality within one subgroup (women) increases, all else being the same, overall Gini may decline!

Figure 5.4 shows the Gini index of weekly earnings in the US from 1979 to 2019.[2] When the population is split by gender, inequality is lower in either group than when the two are pooled together, implying that much of the inequality displayed in the overall graph results from gaps between the genders. This highlights a shortcoming of the commonly used Gini index because it does not permit a breakdown of the overall inequality in the population into subgroups (additive decomposability).

This additive decomposability property is only satisfied by Theil's two measures, with only one having relatively unambiguous subgroup (population) weights. See Shorrocks (1978) and Bourguignon (1979). In spite of these almost fatal shortcomings of Gini for policy evaluation, it continues to dominate in empirical and government work.

Additionally, many inequality indices, both Gini and Theil's measures, have been criticized since they measure inequality as the relative differences between individuals' allotments (i.e., they are relative inequality measures). For example, if a proposed policy would give all poor individuals a ten percent increase in income while giving wealthy individuals a twenty percent increase, these inequality indexes would assign such a distribution a higher level inequality even though every individual is made better off. Furthermore, many do not find the goal of reducing inequality for its own sake a compelling criterion for improving welfare. For those reasons, some find the goal of reducing poverty (an absolute measure which refers to some threshold defining poverty) more reasonable.

Figure 5.6 shows the Foster, Greer and Thorbecke (1984) poverty index of weekly earnings with different parameter values and a poverty threshold of $400 per week (half the median weekly earnings of individuals in 2019).[3] As opposed to inequality, poverty seems to have remained constant or even declined since the 1970's, although changes are likely not statistically significant. The headcount measure (FGT(0)) measures the percent of individuals falling below the poverty threshold. Extreme

---

[2] Data was collected from US Current Population Survey Merged Outgoing Rotation Group. Following D. H. Autor, Manning and Smith (2016), the sample includes individuals ages 18 through 64 and excludes those who are self-employed. Top-coded values are multiplied by 1.5, and the top two wage percentiles for each state, year, and sex grouping are 'Winsorized' (replaced with the ninety-seventh percentile's value). This sample is commonly used in labor studies. The Gini indexes are lower than the previously discussed 4.1-4.5 range of yearly income due to the Winsorization.

[3] Using the same data as Figure 5.4.

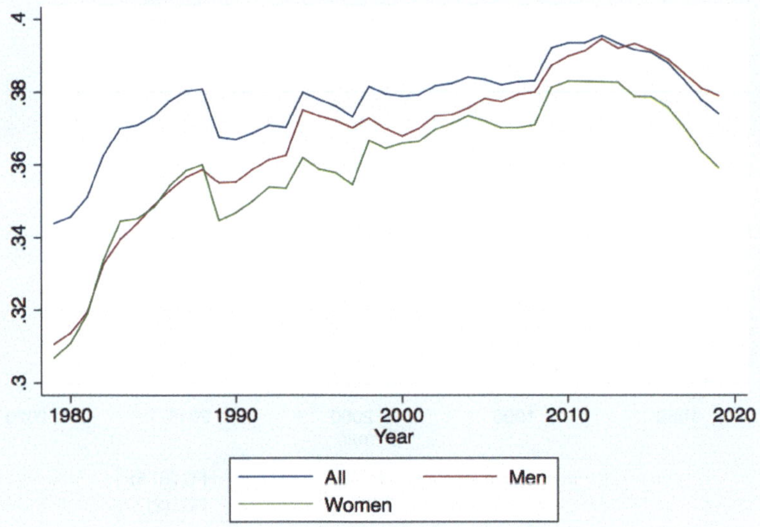

Data Source: National Bureau of Economic Research

Fig. 5.4: Gini Index of Weekly Earning Over time.

poverty (FGT(2) which weighs individuals falling far below the poverty threshold more heavily) was always low with little fluctuation.

The conundrum in the choice of inequality indices may be bypassed, some would argue, with transparent reporting of the data, as in Tables 5.1-5.2 above or Figure 5.5.[4] There are subtle problems with this apparent transparency! It invites problematic and uncritical comparisons of 'dollars' at different quantiles. My favorite example is one of two populations (each with five groups): $\{1,2,3,4,5\}$ vs. $\{1,2,3,4,15\}$. Judged by the median, they are equivalent. Median is seen to be both a robust measure and an 'insensitive' one. The second population may be regarded by some as facing extreme inequality, especially if we associate dollar levels to these groups. The means are 3 and 5. One may argue that the mean of 5 represents no one in the second population, its comparison with the mean of 3 in the first population rendered meaningless and misleading. There are any number of weighting schemes that may value each group differently, and produce different comparatives. In addition, we may not wish to think of a dollar in the highest group as having the same value in the lowest group

[4] Using the same data as Figures 5.4-5.5, Figure 5.5 shows mean weekly earnings, the 90th percentile of weekly earnings minus the 10th, and the 90th percentile of weekly earnings minus the 50th, all in 2019 dollars. Mean weekly earnings increased since the 1980's and seem to be driven by weekly earnings increasing at the top of the distribution; the difference between the 90th percentile and the 10th percentile is proportional to the difference between the 90th percentile and the 50th percentile. Similar to Tables 1-2, this suggests that increases in mean weekly earnings is mainly due to increases at the top of the distribution.

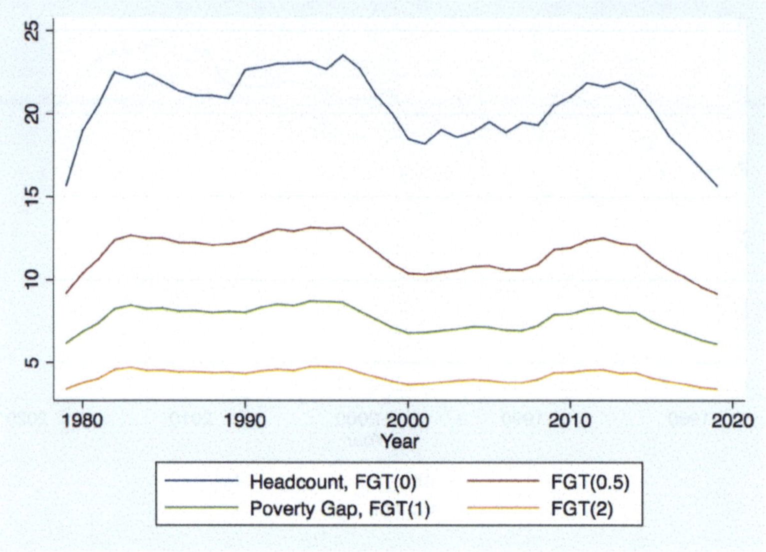

Data Source: National Bureau of Economic Research

Fig. 5.5: Poverty in Weekly Earning Over Time

(1). Intrapersonal comparisons of well being make for even larger set of evaluative functions than the set defined merely on different weight schemes for averaging. Every inequality measure is equivalent (uniquely so up to a monotonic transformation) to an evaluative (welfare) function. Averages, medians and every quantile are highly subjective and exclusionary measures of an outcome distribution!

The non-uniqueness of evaluation functions is a manifestation of the Arrowian Impossibility Theorems. It leads us to consider weaker, uniform rankings which are valued over large classes of 'welfare functions'. We witnessed the pros and cons of this in the US example above. Dominance rankings are powerful when such rankings are present empirically, and equally powerful and informative in policy debates, when such rankings do not exist. Suppose that, in the last example, every entry in the second population was multiplied by a 100 (in real terms). While the *relative* distribution/inequality is unchanged, the second population is now moved so far to the right of population 1 (or 2), as to dominate it uniformly. What is the formal sense in which this uniform ranking obtains?

There are other challenges. Why just income? Forceful criticisms of single dimensional analysis of well being, implicit in *income* inequality analysis emanated from writings of Sen and others, leading to development of multidimensional measures and analysis of well-being, see for example Maasoumi (1986), Tsui (1999). This is a vast area of analysis and much has taken place since my earlier surveys in Maasoumi (1998), Maasoumi (1999). I will not explicitly deal with these otherwise important issues in this paper.

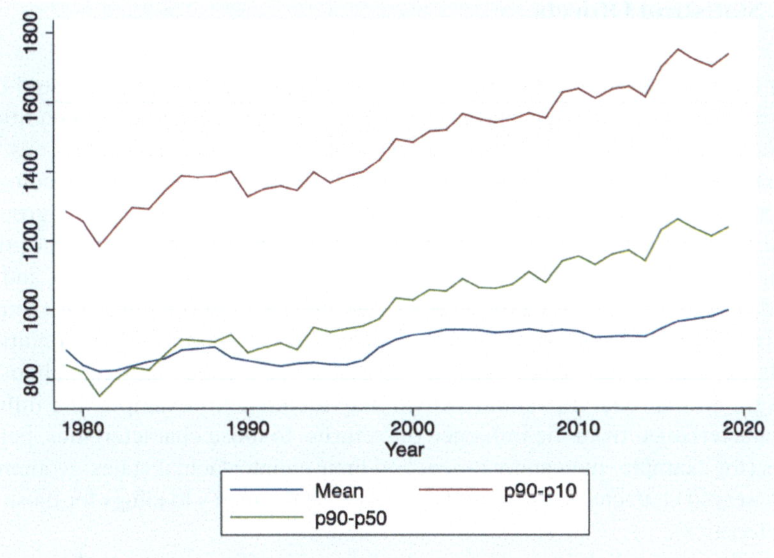

Data Source: National Bureau of Economic Research

Fig. 5.6: Weekly Earning Over Time (2019 Dollars)

Most discussions and conceptions of inequality, especially in public domain, are informal comparisons of Lorenz curves. The share of total incomes going to the top 10 per cent of the population, percentage of assets held by the top 1 per cent. The recent treatise by Piketty (2013) visits historical evidence on the share of capital going to different population groups, and its association, especially at current extremes, with social consequences. Piketty analysis includes different sources of income, including capital gains. The comparative historical analysis, for groups, or factors of production, and over time, is a means of isolating statistical or causal factors. much as a statistical model, or counterfactual statistical analysis is meant to do. The latter methods are formal and capable of identifying and exposing the otherwise implicit assumptions that underlie all such analyses. But both descriptive and rigorous approaches share a common goal: How to measure inequality, as a meaningful functional of a given distribution, and what determines (covariates and causes) those marginal distribution outcomes.

Inequality indices are *expected utilities* for any given utility or weighting function and a given distribution of outcomes. Absent a consensus on utility functions, and any distribution, some distributed outcomes may have higher expected utility for entire classes of utility functions. This is a case for uniform rankings by dominance or other criteria. This was discussed informally above, and will be treated more formally below.

5.1.1 Statistical Objects

Latest developments treat marginal distributions as outcome of integrated conditional distributions, through appropriate and interesting distribution of conditioning covariates (education, experience, neighborhoods, parental characteristics, race, gender). These are the 'causal' and other factors that impact individual outcomes and characterize population subgroups and counterfactual states of those subgroups. A typical Mincer equation is a (conditional mean or quantile regression) description of earnings outcomes, by such factors as education, experience (polynomials), and other individual or location/time characteristics. But the object of interest is the inequality in the (marginal) distribution of earnings, integrating out the effect of the conditioning covariates, or at certain values of those covariates that a policy may care about. This paradigm allows 'decomposition' which may identify, for example, the influence of 'characteristics' from the influence of 'returns' to those characteristics, between groups (for example, men and women, and their counterfactual states: women with their observed characteristics, were they to receive the men's loadings for those same characteristics).

Both inverse probability weighting techniques and regression at the mean and quantiles methods have advanced greatly and rapidly to respond to these questions. We provide a summary of these developments and advocate their adoption for informed, as well as rigorous inference on inequality and related measurement objectives. The advantage of these techniques is that they first identify desired marginal distributions. Once equipped with this marginal distribution, interesting functions of it are computable, including inequality, mobility and poverty measures, quantiles of interest and their distances. Uniform ranking can also be tested for since conditions for it are testable hypotheses about distributions. Comparative measurements and analyses proceed accordingly, as for the 'gap' between distributions and its evolution over time. As we will argue, conception of the 'gap' is essentially the same question as the 'ideal inequality' measures and challenges thereof.

There are generally two approaches currently in vogue for identifying marginal distributons of incomes. The first approach is to compute conditional distributions quantile-by-quantile and conduct comparative analysis with suitable evaluative functions of quantile differences (before or after marginalization by integrating out the desired covariate distributions). A second approach is to first summarize (characterize) each marginal (or conditional) distribution by suitable evaluative functions (like inequality indices), and then compute the difference between these indices. The latter approach has been the mainstay of 'income inequality' literature for better than a century. In more recent times, the first approach has been favored by modern statistically advanced researchers, primarily versed in the potential outcome paradigm. The first approach is also quite helpful in accounting for the perennial problem of 'selection', of many types, and identification arguments that are extremely revealing of the statistical challenges to attribution of 'causes' of outcomes.

The difference between the two approaches is subtle. The first approach requires identification of the distribution of quantile differences. To see the problems of this approach, consider a society with only two men (Males A and B), and two women

(Females A and B). Male wages are ($5,000 , $1,200), respectively; Female wages are ($3,000 , $1,000). Quantile-by-quantile analysis will compare Female A to Male A, and Female B to Male B. However, occupying the same rank in their respective group does not necessarily mean that Female A and Male A are comparable individuals. Implicitly assumed and required in the quantile comparisons is an assumption of rank invariance (or similarity), i.e., one's relative rank is preserved when endowed with each other's skill sets or market returns. Rank invariance requires that male and female ranks refer to the same skills and substitutions thereof, or at least the same intrinsic values of skills.

Rank invariance is unlikely to be satisfied empirically, and has been statistically rejected for several decades of CPS data in the US. Without rank invariance, it is questionable that the first approach ('quantile treatment effect') can deliver meaningful measures of distribution differences. See Maasoumi and Wang (2019). The traditional approach (to summarize the distribution first and then compare the summary measures of inequality, say) is concerned with the distributions for groups, instead of individuals. This is 'anonymous' and does not require identification of individual quantile differences.

Both rank invariance and choice of evaluation functions are problems that have not received sufficient emphasis and scrutiny in the literatures on the gender gap and treatment effects, with a few notable exceptions (Heckman, Smith and Clements (1997); Heckman and Smith (1998); Dehejia (2005)). This paper aims to provide this emphasis. This also helps to connect the inequality literature, and the literature on gender gaps and the literature on treatment effects.

5.1.2 What Does Marginalization Mean?

It is important to distinguish between 'marginalization' to 'derive' the marginal distribution of income *over all values of covariates*, on the one hand, and the distribution of incomes free of some factor influence, like education. Marginalization obtains incomes for *any* value of covariates (characteristics). Residual analysis in which the impact of some factors are removed, is an statistical technique which requires a model (such as for the Mincer equation), in which projections are conducted in order to obtain the 'best' fitted value of income due to any sources other than the ones in the projection space. This residual analysis, while simple in conception and execution, is relatively rare. See Maasoumi and Heshmati (2000). This is in part due to model and covariate dependence in this approach, a perennial challenge in robust econometric model building and inference. Modern 'big data' techniques, such as double machine learning, offer exciting possibilities for robust implementations of this residual based approach.

5.1.3 Addressing Selection and Other issues at the Distribution Level:

A potentially major challenge to empirical evaluative anlaysis of distributed outcomes is non random selection. Regardless of evaluative measures, analysis of inequality can be impacted by selection. Labor force participation (LFP) rates for males have continued to decline for decades, and those for females increased, then peaked, and has decreased slightly in recent years. To the extent that non-working men and women systematically differ from working men and women, measures of inequality would be 'biased' when generalized to the whole populations. For example, if there is positive selection by women over time (high-earning women enter the labor market, and low-earning ones leave). This may lead to a possibly mistaken observation of 'convergence' between distributions of earnings between men and women. This key insight dates back to Heckman (1974), and attention has been paid to it in many studies of women's labor market outcomes. For example, in the gender gap literature, we note a few attempts, mostly on the gap at the mean or median (e.g., Blau & Kahn, 2006; Olivetti & Petrongolo, 2008; Mulligan & Rubinstein, 2008).

It is desirable and informative to address selection at the *entire* distribution beyond the mean and median. We will briefly describe a number of methods to deal with this. Arellano and Bonhomme (2017a) and Maasoumi and Wang (2019) adopt a new quantile-Copula approach to model the joint determination of wages and participation decision for both men and women. This approach allows a recovery of the distributions of wage *offers* for the entire male and female populations. It can also be used to consider 'value of time' in measuring outcomes. Comparing distributions of wage *offers* is informative, but for those who do not work, wage offers do not reveal 'value of time' or the well-being they actually enjoy. Some individuals derive value from not working, and this is captured by their reservation wages. The quantile-copula approach provides a useful structure to recover the reservation wages and its distribution using the potential wage offers and the selection mechanism. This replaces market wages for those non-employed with their *reservation wages* instead, see Maasoumi and Wang (2019).

A more recent 'distribution regressions' approach is due to Chernozhukov, Fernández-Val and Luo (2018). It is based on local Gaussian characterization of distributions, and leads to familiar distribution (MLE, probit) regressions from which both inequality measures and other distribution functionals can be derived. It can handle selection as well, but it depends on different identification assumptions and strategies.

The recent technology for analyzing inequality and gap between subgroups can provide nuanced findings. These findings, while consistent with some findings based on traditional summary measures and regressions, modify and mediate others. For instance, based on the Current Population Survey data from 1976 to 2016, Maasoumi and Wang (2019) find that, firstly, without correcting for selection, while women generally perform worse than men in the labor market, they are catching up with men (Blau & Kahn, 1997, Blau & Kahn, 2006, and Goldin, 2014). The gender differences have decreased over time, especially in the 1980s and early 1990s, although at a much slower rate since the mid-1990s. The perception of the actual 'gap' and its dynamics varies according to measures employed. The quantile evolve differently over the past several decades. Entropic measures of inequality and the gap provide a more nuanced picture of the evolution of the gap between wage distributions. Specifically, Generalized Entropy measures indicate a generally larger convergence until early 1990s, and a more pronounced flattening since then, for full-time workers. Moreover, the gap increases monotonically with the level of inequality aversion for entropy measures.

Selection indeed impacts all measures of inequality and the comparative evolutions. Once selection is accounted for, MW find that convergence of earnings is slower, with a recent reversal in the trend in parts of the wage distribution between mid-1990s and the most recent recession, followed by a further marked decline in the gap, especially among low-skilled workers. Weak uniform ranking of wage distributions between men and women is less likely, and one does not find'uniform' narrowing of the gap at all quantiles.

Maasoumi and Wang (2019) find labor force participation varies by education and race. The relative economic position of less educated women has lacked progress, or even deteriorated, in more recent years, and the existing studies may have understated this because many low-wage earners among less educated women exit the labor force. Similar results hold for black women. Specifically, the wage gap for black women has narrowed less compared to both Hispanics and whites, although the gender gap within minority groups (blacks and Hispanics) is generally smaller than amongst whites. We pay special attention to impact of 'types', such as education and race, within gender groups.

Taking 'value of time' into account has been found to moderate degree of convergence between men and women and other groups. Women's relative well-being, especially among those in the upper tail, may have even worsened over time.

The rest of the paper unfolds to elaborate on the topics discussed in the Introduction, as follows: In Section 5.2 a brief formal description of decision theoretic foundations of assessing outcomes is presented. It includes both the equal equivalent income and the equity-effciency reduced form representation of relations between welfare functions and inequality measures. General axioms are discussed as potentially desirable properties for inequality measures. Analysis of the 'gap' between distributed outcomes derives from this presentation. For this purpose, the role of metrics for measuring distance between entire distributions, for example by entropies, is highlighted. Some well known inequality measures are given and the value (or limitations) additive decomposability (or aggregation consistency) and Theil's measures are explained. The section ends with a discussion of weak uniform ranking of entire outcomes over classes of welfare functions, as a means of avoiding cardinal assessments based on particular inequality indices.

Section 5.3 describes some of the latest developments and techniques for recovering unconditional distributions from conditional ones. This section emphasizes conditional quantile methods which account for explanatory covariates and individual characteristics that interest policy makers. Inequality meaaures assess the marginal distributions, but these are outcomes that are heterogenous. This section also clarifies how counterfactual analysis and decompositions of final outcomes is facilitated by the transition from conditional to marginal distributions, as well as providing a frame for dealing with sample selection. Section 5.3.1 on inverse probability weighting method. It is included in this section to highlight its essential value in obtaining entire counterfactual distributions, and fucntions of potential outcome distributions as natural byproducts. Section 5.3.2 further clarifies the derivation of counterfactual distributions with inverse weighting methods.

Section 5.3.3. describes the latest alternative to inverse probability weighting by means of 'Distribution Regressions'. This is a very promising method which depends on different identification strategies in deriving conditional distributions and counterfactuals. It can handle many data types (discrete or continuous), as well as sample selection and some model misspeficiations.

Section 5.4 provides a modern example of assessing male-female earnings distributions. It demonstrates a new method of correcting for sample selection (labor force participation) based on copulas. It is based on (Mincer) conditional quantile models which is further presented in Section 5.4.1. Section 5.4.2 presents the impact of selection on decomposition and counterfactual findings. An appendix on inference methods is provided, as well as a section on some extensions.

5.2 Ideal Measures of Inequality, Poverty, and Mobility

There is no universally accepted evaluation function of a distribution, and there are many candidates. The field of 'ideal inequality' measures is vast and beautiful! Averages, inequality measures, quantiles and entropies are all well-known functions of distributions that summarize its quantiles *anonymously*. This is typically without regard to identity of those who occupy a given quantile. Each function attributes its own weights to different wage levels and individuals. For example, the average (function) assumes equal weights to all percentiles, treating a dollar of high-wage earners and a dollar of low-wage earners equally. All evaluative functions that underly various measures and indices can be endowed with a decision theoretic basis. I will expand on this and provide a more modern decision theoretic motivation.

The decision theoretic framework typically discovers a flexible family of measures of inequality, poverty, and mobility, generally based on entropy functions (the Generalized Entropy family that includes a normalized Kullback-Leibler-Theil measure, and a normalized Hellinger measure, as well as Atkinson's family). Entropy functions share similarities to characteristic functions, such as a one-to-one relation to the corresponding distribution. More importantly, unlike the average function, entropies satisfy many desirable properties such as aversion to inequality (which assigns more weights to a dollar of transfer at lower wages than at higher wages; i.e, the Pigou-Dalton principle of transfers). Each entropy function in this class is characterized by a different level of inequality aversion (for an impartial observer/evaluator of the wage distributions).

5.2.1 Decision-Theoretic Basis of Evaluation Functions

Let y^f and y^m denote (log) wages of groups f and m, with CDF (density) denoted by F_f (f_f) and F_m (f_m), respectively. Let $F_f(y^f_\tau) = \tau$, and $F_m(y^m_\tau) = \tau$ define the τ-th quantile. Note that when one group/distribution is the idealized perfectly equal case, the Gap between is the measure of inequality. A general definition of the gap between the groups is the difference of respective Evaluation Functions (EFs):

$$\text{Gap } = EF_{\gamma,\epsilon}(y^m) - EF_{\gamma,\epsilon}(y^f). \tag{5.1}$$

The gap at a τth quantile is $y^m_\tau - y^f_\tau$, where the median corresponds to $\tau = \frac{1}{2}$. Measures of the gap may be functions of the quantile gaps. The mean gap is $\mathbb{E}[y^m] - \mathbb{E}[y^f] = \int_0^1 \left[y^m_\tau - y^f_\tau\right] d\tau$. Gap at any quantile, or the mean, is a (linear) weighted function of quantile gaps. Linear functions of quantiles imply infinite substitutability of a dollar at all wage levels. Alternative functions would reflect different types of weights and/or interpersonal evaluations, reflecting degrees of aversion to inequality/dispersion. There are parallel literatures on ideal inequality (and risk) measures, and ideal entropies. The latter is summarized in Maasoumi (1993) and motivates the inequality literature.

Consider the following Evaluation Function (EF):

$$EF_{\gamma,\epsilon} = \int_0^1 R(\tau,\gamma) U_\epsilon(y_\tau) d\tau, \tag{5.2}$$

where $R(\tau,\gamma) = \gamma(1-\tau)^{\gamma-1}$, and $U(\cdot)$ is a concave function of wages. γ is an aversion to inequality/dispersion parameter. This class of functionals is general and underlies the Atkinson and S-Gini families of inequality measures (which satisfy desirable properties such as the Pigou-Dalton transfer and permutation invariance properties).[5] It allows for flexible weights at different percentiles. Holding $\gamma \neq 1$ fixed, the weight function, $R(\cdot)$, is decreasing with respect to τ, thereby assigning greater weights to lower wages in the evaluation of a wage distribution and hence measurement of the gender gap.

If only relative (scale/mean-independent) measures are to be considered, the function $U(\cdot)$ must be of the following (homothetic) form (see Pratt (1964) or A. Atkinson (1970)):

$$U_\epsilon(y_\tau) = \begin{cases} \frac{y_\tau^{1-\epsilon}}{1-\epsilon} & \text{if } \epsilon \neq 1 \\ \log(y_\tau) & \text{if } \epsilon = 1 \end{cases}. \tag{5.3}$$

---

5 Invariance to permutation of individuals produces anonymity of measures with respect to identity of those who occupy a given quantile. The Pigou-Dalton transfer property (or aversion to inequality) emphasizes that one dollar reduction of gap at lower wages is relatively more valuable than one dollar at higher wages. This principle implies that any redistribution from the rich to the poor can reduce inequality. The definition of inequality-loving would be the opposite of this definition.

Note that the wage quantile y_τ itself is a special case of possible utility functions $U(\cdot)$ at $\epsilon = 0$. This leads to a linear summary function of the quantile or quantile gaps: $\int R(\tau,\gamma)(y_\tau^m - y_\tau^f)d\tau$. In the special case when $\epsilon = 0$ and $\gamma = 1$, EF is $\int y_\tau d\tau = \mathbb{E}[y]$ and the gap is the mean gap. In this case, $\gamma = 1$ (and the mean gap) implies no aversion to inequality (neutrality).

A concave and increasing Evaluation Function of an impartial observer (represented by Equation 5.2) is known to be similarly represented as an important money metric Evaluation Function, called the Equal Distributed Equivalent Income (EDEI) wage, given by

$$EDEI_{\gamma,\epsilon} = U^{-1}(EF_{\gamma,\epsilon}) \tag{5.4}$$

$$= \mu_y \left(1 - I_{\gamma,\epsilon}(y)\right), \tag{5.5}$$

where μ_y is the mean and $I_{\gamma,\epsilon}(\cdot)$ is any relative inequality measure. One can also consider alternative EFs such as those in Aaberge, Havnes and Mogstad (2013). Note that dividing both sides by the mean, we can make scale-invariant evaluations of the wage distribution based on 'relative' inequality measures.

There are many inequality measures, including a monotonic transformation of the Atkinson family of inequality indices known as the Generalized Entropy (GE) family. While there exists no unique (or ideal) inequality measure, influential works by Shorrocks (1980), and Bourguignon (1979) have established the 'ideal' properties of GE.

The famed welfare properties/axioms which support GE are:

1. Anonymity or invariance to permutations,
2. Population replication,
3. Scale invariance, and
4. Pigou-Dalton principle of transfers or aversion to inequality.
   If 3 is replaced with
   3'. 'Invariance' to rank preserving equal increment transformations (adding a fixed amount to everyone), one obtains 'horizontal' or rank inequality measures. I do not cover these measures. If one adds:
5. Additive decomposability (Aggregation consistency). If this strong decomposability requirement is added, Theil's population share weighted inequality measure is identified as the 'ideal' inequality measure; see below for further details.

5.2.2 An Alternative Equity-Efficiency Representation

An alternative representation of the essential relation between inequality measures and welfare (evaluative) functions is instructive and exposes the 'equity-Efficiency' trade off which is central to debates on inequality and policy. A good example of the following account is given by Amiel and Cowell (1997):

Let $x = (x_1, x_2, ...x_n) \in$ X be an income vector from the set X of all ordered non negative vectors. $n(x)$ is the number of individuals in x, and the mean is denoted by

$\mu(x)$. Consider a class of 'additive shares inequality indices' as follows:

$$I(x) = \sum_{i=1}^{n} w_i T(s_i(x)), \tag{5.6}$$

where T(·) defines various inequality indices, and the i-th share is defined by

$$s_i(x) = x_i / n. \quad \mu(x) \tag{5.7}$$

This class of inequality measures is quite large, including Generalized Entropy , Gini, relative mean deviation and logarithmic variance. For instance Gini is a special case with,

$$w_i = -(1/n)(n - 2i + 1). \tag{5.8}$$

The Extended Gini indices merely replace the above conditon on w_i with increasingness in i; See Weymark (1981). A Reduced Form expression of EF is available in terms of mean income and inequality (upto monotonic transformations), as follows:

$$EF(x) = H^{EF}(\mu(x), I(x)). \tag{5.9}$$

For differentiable functions H, denoting its derivatives by H_μ and H_I, a very useful monotonicity condition can be expressed for well known inequality indices. For GE it is given as follows:

$$-(H_\mu / H_I) > max([ns_i]^{\alpha-1} - 1)/(\alpha - 1) - \alpha . I(x)/n. \quad \mu(x)) \tag{5.10}$$

Based on the Equity-Effciency frontiers defined from the above representation, one may rank income distribution profiles by EF or by the correspoding mean-inequality profiles. See Amiel and Cowell (1997). This provides a different perspective than the more common expression of EFs in terms of the 'Equal Distributed Equivalent Income' given earlier.

5.2.2.1 Measuring the 'Gap' between Entire Outcome Distributions

The contrast between the entropies of two distributions is an example of how we may assess the divergence between the densities of wages for two groups. Since GE and Theil measures may be seen to be divergences of a given income distribution from the 'uniform' distribuition, the difference between two such inequality indices provides a measure of the entropic gap between them (since the uniform distribution entropy cancels out). Because there are competing normalizations for entropy functions, we write the symmetric GE measure of divergence between the densities of wages for two groups with a single parameter k of inequality aversion.

$$\frac{1}{2} \cdot \left[I_k(f_1, f_2) + I_k(f_2, f_1) \right] \quad \forall k \in [0, 1], \tag{5.11}$$

where $I_k(\cdot,\cdot)$ is a GE measure of divergence given by

$$I_k(f_1, f_2) = \frac{1}{k-1}\Big[\int \Big(\frac{f_1}{f_2}\Big)^{k-1} f_1 dy - 1\Big],$$

$$I_k(f_2, f_1) = \frac{1}{k-1}\Big[\int \Big(\frac{f_2}{f_1}\Big)^{k-1} f_2 dy - 1\Big].$$

Two popular members are:

1. The normalized and symmetrised Kullback-Leibler-Theil measure:

$$\text{KL} = \frac{1}{2}\cdot\Big[\int [\log(\frac{f_f}{f_m})\cdot f_f + \log(\frac{f_m}{f_f})\cdot f_m]dy\Big]. \tag{5.12}$$

2. And, at $k = \frac{1}{2}$, one obtains an entropy *distance* metric that is a normalization of the Bhattacharya-Matusita-Hellinger measure, given by:

$$S_\rho = \frac{1}{2}\int_{-\infty}^{\infty}\Big(f_m^{1/2} - f_f^{1/2}\Big)^2 dx \tag{5.13}$$

$$= \frac{1}{2}\int \left[1 - \frac{f_m^{1/2}}{f_f^{1/2}}\right]^2 dF_f.$$

Varying k corresponds to different levels of inequality aversion. Shannon's entropy is the basis of both the KL measure and Theil's inequality measures, and is more 'inequality averse' than the Gini and the Hellinger (or S_ρ).

The property of aggregation consistency, and related useful additive decomposability of inequality measures deserves further analysis. With a slightly different normalization, we may express the GE family of inequality measures as the sum of two terms, a within group and a between group component, as follows:

$$I_\gamma = \sum_{r=1}^{R} Y_{r.}^{\gamma+1}(n_r/n)^\gamma I_\gamma^r + I_\gamma^b, \tag{5.14}$$

$Y_r/\sum_j^{n_r} y_j = Y_{r.}$ share of group r income Y_r In total income, r=1,2,.....R; n$_r$ is the number of units in the r-th group, I_γ^r, is the 'within group' inequality, and I_γ^b is the 'between group' inequality, measured between the *group* income shares. γ reflects degree of inequality aversion.

The decomposition given above is ambiguous since the group weights are function so of incomes, and lack invariance to income changes. This ambiguity is resolved only at $\gamma = -1$, which defines Theil's second inequality index, given as follows:

$$I_{-1} = \log n + \frac{1}{n}\sum_{i}^{n} \log(y_i/\sum_{j}^{n} y_j), \text{ for any income vector } (y_1, y_2, ...y_n). \quad (5.15)$$

Theil's additive decomposition which is unique to it, is given as follows:

$$I_{-1} = \sum_{r=1}^{R} (n_r/n) I_{-1}^{r} + I_{-1}^{b}. \quad (5.16)$$

Theil's second index is aggregation consistent, a rise in inequality in any group r, all else being equal, will raise total inequality. which will be attributable to events within that group.

Table 5.3: Inequality Indices example in Table 5.1

	1974	2004
$I_A^{.25}$	0.067	0.097
$I_A^{.5}$	0.134	0.190
$I_A^{.75}$	0.207	0.286
$I_A^{1.0}$	0.297	0.418
I_{Gini}	0.395	0.466
I_{GE}^{0}	0.352	0.542
I_{GE}^{1}	0.267	0.406

Source: Cowell (2006)

5.2.3 Uniform Ordering: Stochastic Dominance Tests

Measures of inequality provide 'complete' (cardinal) rankings. When distributions cross (especially at lower tails),[6] different inequality measures will differ in their rankings, depending on the underlying evaluation functions. It is useful to test whether distributions can be uniformly ranked over large classes of (evaluation) functions to a statistical degree of confidence. Absent any uniform dominance relations, all measures of inequality and gap need to be examined relative to the underlying evaluation functions. Examples of such uniform rankings are Lorenz, Generalized Lorenz and other dominance orderings.

[6] As they do for US wages in some years.

Let U_1 denote the class of all *increasing* von Neumann-Morgenstern type utility functions u that are increasing in wages (i.e. $u' \geq 0$), and U_2 the class of utility functions in U_1 such that $u'' \leq 0$ (i.e. concave). Concavity implies an aversion to inequality:

First Order Dominance:

Male wages y^m First Order Stochastically Dominate (FSD) Female wages y^f *if and only if*

1. $Eu(y^m) \geq Eu(y^f)$ for all $u \in U_1$ with strict inequality for some u.
2. Or, $F_m(y) \leq F_f(y)$ for all y with strict inequality for some y.
3. Or, $y^m_\tau \geq y^f_\tau$ for all points on the support.

The last condition is very intuitive. If income is higher at every quantile for one group, then it is first order dominant, whatever inequality measure is employed. Not even a Rawlsian comparison (of the poorest) would reverse the ranking. y^m FSD y^f implies that the mean m-wage is greater than the mean f-wage. Average or median income for one group may be higher without uniform dominance, because other members may be worse off.

Cumulative distributions often cross, making FSD unlikely. If class of utility functions is further restricted, one may still find correspondingly higher order dominance. Inequality measures correspond to preferences that are increasing and inequality averse, to various degrees. Inequality aversion is represented by concavity of the utility function (much like risk aversion): *Second Order Dominance:*

(y^m) Second Order Stochastically Dominates (y^f) (denoted y^m SSD y^f) *if and only if*

1. $Eu(y^m) \geq Eu(y^f)$ for all $u \in U_2$ with strict inequality for some u.
2. Or, $\int_{-\infty}^{y} F_m(t)dt \leq \int_{-\infty}^{y} F_f(t)dt$ for all x with strict inequality for some x.
3. Or, $\int_0^\tau y^m_u du \geq \int_0^\tau y^f_u du$ for all points on the support.

FSD implies SSD. Again, the last condition is very intuitive. Ordered cumulated quantiles (divided by the mean) are the basis of the Lorenz curve. If the lowest q percentile of the population receives less than q percentile of incomes, at every level, there is Lorenz dominance. When multiplied by means, it is Generalized Lorenz dominance, also known as Second Order Stochastic Dominance. Higher order SD rankings are based on narrower classes of preferences, with increasing 'degrees' of aversion to inequality. As noted in the Introduction for the US example, Relative income inequality rankings may not agree with SD rankings.

Advances in SD tests based on a generalized Kolmogorov-Smirnov test discussed in Linton, Maasoumi and Whang (2005), and the recent text by Whang (2019). The tests for FSD and SSD may be based on the following functionals:

$$d = \sqrt{\frac{N_1 N_2}{N_1 + N_2}} \min\sup[F_m(y) - F_f(y)], \tag{5.17}$$

$$s = \sqrt{\frac{N_1 N_2}{N_1 + N_2}} \min\sup \int_{-\infty}^{y} [F_m(t) - F_f(t)]dt, \tag{5.18}$$

N_1 and N_2 are respective sample sizes. Test statistics are based on the sample counterparts of d and s, employing empirical CDFs. Bootstrap and subsampling implementations of the tests are common. Maasoumi (2001) surveys the related tests and techniques, including older tests of quantile (inverse distribution) rankings.

The SSD tests are closely related to the decision-theoretic framework outlined above. The money-metric evaluations can be derived from Equation (5.1), and other monotonic transformations. A representation of $EF_{\gamma,\epsilon}$ (using integration by parts) reveals a useful relation to SSD:

$$EF_{\gamma,\epsilon} = \int_0^1 \gamma(\gamma-1)(1-\tau)^{\gamma-2} GL_U(\tau)d\tau, \tag{5.19}$$

where $GL_U(\tau) = \int_0^\tau U(y_u)du$ is the Generalized Lorenz (GL) function of $U(\cdot)$. When $U(\cdot) = y_\tau$, ranking by GL ($GL_U^m - GL_U^f = \int_0^\tau y_u^m du - \int_0^\tau y_u^f du$) is exactly the test of SSD. This helps in interpretation of the SD tests.

5.3 Recovering Unconditional Distributions

Let $F_{y|x_i} \equiv Pr[y_i \leq y|x_i]$ be the conditional CDF of the wages given $x = x_i$, and $Q_\tau(y|x_i)$ the corresponding τ^{th} conditional quantile. Note that $Q_\tau(y|x_i) = F_{y|x_i}^{-1}(\tau)$, the inverse of the conditional CDF. The marginal distribution is related to conditional distribution as follows [7]

$$F_y(y) = \mathbb{E}[I[y_i \leq y]] = \mathbb{E}[\mathbb{E}[I[y_i \leq y]|x_i]] = \mathbb{E}[F_{y|x_i}], \tag{5.20}$$

where $I[\cdot]$ is an indicator function. The conditional CDF is related to its inverse (the conditional quantile function) as follows (see, e.g., Angrist and Pischke (2009, p.282))

$$F_{y|x_i} = \int_0^1 I[F_{y|x_i}^{-1}(\tau) \leq y]d\tau = \int_0^1 I[Q_\tau(y|x_i) \leq y]d\tau. \tag{5.21}$$

The unconditional CDF may be estimated by

[7] The first equality follows from the definition of unconditional CDF. And the second equality follows directly from the law of iterated expectations. The last is the definition.

$$\widehat{F}_y(y) = \frac{1}{N}\sum_{i=1}^{N}\int_0^1 I[Q_\tau(y|x_i) \le y]d\tau]. \tag{5.22}$$

The corresponding unconditional quantiles can be obtained by inverting the marginal CDF:

$$Q_\tau(y) = \inf\{y : \widehat{F}_y(y) \ge \tau\}.$$

Inequality and other functions , as well as stochastic dominance tests are performed on these estimates.

5.3.1 Inverse Probability Weighting Methods

We are often interested in answering two types of counterfactual situations: First, what if the wage structure of group A (women) is replaced with the wage structure of group B (men), *but holding the distribution of group A characteristics constant*? Conversely, what if we replace the distribution of group A's characteristics to that of group B's, holding the wage structure unchanged? This is the basis of Oaxaca-Blinder decompositions which are applied in context of the conditional mean of linear additive regressions. A similar decomposition is possible for the entire distributions (and at quantiles, with rank invariance type assumptions).

There are a number of available methods to identify the counterfactual distributions of interest. Below I describe the inverse probability weighting method. Suppose we wish to identify the distributions of the following counterfactual outcomes:

$$\ln(w_i^{c1}) = g_0(X_{i1}, \epsilon_{i1}), \quad \text{(Counterfactual Outcome \#1)} \tag{5.23}$$

$$\ln(w_i^{c2}) = g_1(X_{i0}, \epsilon_{i0}). \quad \text{(Counterfactual Outcome \#2)} \tag{5.24}$$

F_{c1} (f_{c1}) denote the corresponding CDF (PDF) of the counterfactual outcome $\ln(w_i^{c1})$. F_{c2} (f_{c2}) represents the corresponding CDF (PDF) of the counterfactual outcome $\ln(w_i^{c2})$. Notice that the differences in the distributions of F_{c1} and F_1 ($\ln(w_i^{c1})$ v.s. $\ln(w_i^{f})$) come solely from differences in wage structures; the comparisons of these two distributions thus provide insight into potential *discrimination*. On the other hand, the differences in the distributions of F_{c1} and F_1 ($\ln(w_i^{c2})$ v.s. $\ln(w_i^{f})$) come solely from differences in the distribution of human capital characteristics; the comparisons thus provide some insight into the 'gap' due to *productivity* differences.

As shown Firpo (2007, Lemma 1), under the following assumptions:

[A1.] Unconfoundedness/Ignorability: Let (D, X, ϵ) have a joint distribution. For all x, ϵ is independent of D conditional on $X = x$.
[A2.] Common Support: For all x, $0 < p(x) = \Pr[D = 1|X = x] < 1$.

The counterfactual outcome CDF of $\ln(w_i^{c1})$ is identified aas follows:

$$F_{c1} = \mathbb{E}[\omega_{c1}(D, X) \cdot I[(\ln(w_i) \le y)],$$

where

$$\omega_{c1}(D,X) = \left(\frac{p(x)}{1-p(x)}\right)\cdot\left(\frac{1-D}{p}\right).$$

The counterfactual outcome CDF of $\ln(w_i^{c2})$, F_{c2}, is similarly identified. The propensity score, $p(x)$, is estimated, typically by probit or logit methods. p is the population propotion of group A.

Both assumptions (A1) and (A2) are commonly used in the literature. Assumption (A1) implies here that given the values of observable human capital characteristics X, the distribution of unobservable human capital characteristics such as ability is independent of gender. Assumption (A2) rules out the possibilities that a particular value x belongs to either group A or B, and that the set of wage determinants, (X, ϵ) differ across groups. Once we identify the counterfactual distributions of interest, counterfactual decomposition analysis can be conducted directly, or based on suitable 'metrics' to measure 'distance' between distributions. The only metric member of the entropy divergence measures discussed earlier is the Hellinger metric. Analysis of these divergences is often reported in terms of inequality measures as well. It should be clear from the earlier decision theoretic discussion in this paper, all such meaures are subjective in terms of their weighting of distances at different quantiles, in addition to the requirement of some degree of 'rank invariance' for quantile counterfactual comparisons.

5.3.2 Distribution Regressions Model

The distributions referenced in prior sections, including the counterfactual ones, may be estimated based on a new method of 'distribution regressions'. The following account is from Chernozhukov and Fernandez-Val (2018):

The first key observation is that a partitioning of the sample space allows a binary regression framework to model and estimate the conditional distribution of an outcome given covariates. The outcome is a real-valued variable that may be continuous (e.g., log wages as above), count (number of patents), nonnegative (duration), discrete or binary as in propensity score models.

The second key observation is that the conditional distribution of Y given the covariates X can be expressed as

$$F_{Y|X}(y \mid x) = \mathbb{E}[1\{Y \leq y\} \mid X = x].$$

Accordingly, all outcomes (binary or not) allow a construction of a collection of binary response variables (partition), of the events that the outcome falls bellow a set of thresholds:

$$1\{Y \leq y\}, \quad y \in T \subset \mathbb{R},$$

where T is a countable subset of $\mathbb{R}$. For estimation and other practical purposes, T is taken to be a finite collection of grid points. Then binary regressions for a collection of binary response variables are used to model the conditional distribution of Y given X,

$$F_{Y|X}(y \mid x) = \Pr(Y \leq y \mid X = x) = F_y(B(x)'\beta(y)),$$

where F_y is a link function, which may vary with the threshold y, $B(X)$ is a dictionary of transformations of X (including a 1 as the first entry), and $\beta(y)$ is the parameter vector that may vary with the threshold y. This is the *distribution regression model.*

The distribution regression model is quite flexible and nests a variety of classical models for conditional distribution functions, including the ones described in this paper. Other examples from Chernozhukov and Fernandez-Val are as follows:

Example 1. Classical Normal Regression Model In the classical normal regression model, $Y \mid X \sim N(B(X)'\gamma, \sigma^2)$, the conditional distribution of Y given X is

$$F_{Y|X}(y \mid x) = \Phi((y - B(x)'\gamma)/\sigma),$$

where Φ is the standard normal distribution. This conditional distribution is a special case of the distribution regression model with F_y equal to the probit link Φ, and $\beta(y) = (y - \gamma_1, -\gamma'_{-1})'/\sigma$, for $\gamma = (\gamma_1, \gamma'_{-1})'$. The slopes here don't vary with y.

Example 2. Cox Proportional Hazard Model The Cox duration regression is popular to model conditional distributions in duration and survival analysis, as well as to model non-negative outcomes, such as capital (in (S,s) models) and wages. The conditional distribution is:

$$F_{Y|X}(y \mid x) = 1 - \exp(-\exp(t(y) - B(x)'\gamma)),$$

where $t(\cdot)$ is an unknown monotonic transformation. It corresponds to the following location-shift representation:

$$t(Y) = B(X)'\gamma + V,$$

where V has an extreme value distribution (Gumbel) and is independent of X:

$$V \mid X \sim \log(-\log(U(0,1))).$$

The hazard rate is given as,

$$h(y \mid x) = -\frac{\partial}{\partial y}\ln(1 - F_{Y|X}(y \mid x)) = \frac{\partial t(y)}{\partial y}\exp(t(y))\exp(-B(x)'\gamma),$$

depends proportionally on $\exp(-B(x)'\gamma)$. The conditional distribution is a special case of the distribution regression model with F_y equal to the complementary log-log link, $F_y(u) = 1 - \exp(-\exp(u))$, and $\beta(y) = (t(y) - \gamma_1, -\gamma'_{-1})'$, for $\gamma = (\gamma_1, \gamma'_{-1})'$. The slopes here don't vary with y, while distributional regression allows for slopes to be varying with y.

Example 3. Poisson Regression Model The Poisson distribution is frequently used to model count variables taking values in $0, 1, 2, \ldots$. The conditional distribution of the count variable Y given X takes the form:

$$F_{Y|X}(y \mid x) = \sum_{k=0}^{y} \frac{\exp(B(x)'\gamma)^k \exp(-\exp(B(x)'\gamma))}{k!} = Q(y, \exp(B(x)'\gamma)),$$

where Q is the incomplete Gamma function. This model can be seen as an special case of the distribution regression model with link function $F_y(u) = Q(y, \exp(u))$ and $\beta(y) = \gamma$. The Poisson regression model is based on a widely criticized assumption that the same index governs the whole distribution. The zero-inflated Poisson regression model is a little more flexible by allowing the coefficients to be different at 0. The distribution regression model does not restrict the heterogeneity of the coefficients at any level.

Given the conditional distribution one can recover the conditional quantiles:

$$F_{Y|X}^{\leftarrow}(u \mid x), \quad u \in [0, 1],$$

where $\leftarrow$ denotes the left-inverse of the map $y \mapsto F_{Y|X}(y \mid x)$ on T. The left-inverse of a function $G : T \rightarrow [0, 1]$ on T is defined as

$$G^{\leftarrow}(u) := \inf\{t \in T : G(t) \geq u\} \wedge \sup\{t \in T\}, \tag{5.25}$$

with the convention $\inf\{\emptyset\} = +\infty$. Given a graph of a distribution function $t \mapsto G(t)$ one may obtain the graph of the quantile function $u \mapsto G^{\leftarrow}(u)$ by simply *flipping* the axes and *mirroring* the resulting image.

There are many ways to estimate the distribution regression model. Chernozukhov and Fernandez-Val (2018) describe one method as follows. We can estimate the conditional distribution by:

$$\hat{F}_{Y|X}(y|x) = F_y(B(x)'\hat{\beta}(y)), \quad y \in T,$$

where for each $y \in T$, $\hat{\beta}(y)$ is the maximum likelihood estimator,

$$\hat{\beta}(y) \in \arg\max_{b(y) \in \mathcal{B}} \mathbb{E}[1(Y \leq y) \ln F_y(B(X)'b(y)) + 1(Y > y) \ln(1 - F_y(B(X)'b(y)))],$$

where $\mathcal{B}$ is the parameter space for $\beta(y)$. For example, $\hat{\beta}(y)$ is the probit estimator with the normal link $F_y = \Phi$, or the logit estimator with the logistic link $F_y = \Lambda$.

Inference on $y \mapsto F_y(B(x)'\hat{\beta}(y))$ for $y \in T$ is standard since one can use the delta method in conjunction with the GMM formulation of the problem. One can view estimation of

$$\theta_0 = \text{vec}(\beta(y) : y \in T)$$

as GMM estimation with the score:

$$g(Z, \theta) = \text{vec}\left(g_y(Z, b(y)) : y \in T\right),$$

$$g_y(Z,b(y)) = \frac{\partial}{\partial b(y)}\{1(Y \le y)\ln F_y(B(X)'b(y)) \\ + 1(Y > y)\ln(1 - F_y(B(X)'b(y)))\},$$

which simply stacks the scores of many binary regressions; the joint parameter vector is

$$\theta = \mathrm{vec}(b(y) : y \in T),$$

and stacks the parameters of many binary regressions. The map $y \mapsto F_y(B(x)'\hat{\beta}(y))$, $y \in T$, is a smooth transformation of the estimators $\hat{\beta}(y)$, $y \in T$, so the delta method delivers the large sample properties of the estimators $F_y(B(x)'\hat{\beta}(y))$, $y \in T$. This also very helpful since it means that one may use the bootstrap for inference.

5.4 An Example Based on CPS, Male-Female Distributions

The period 1976-2013, March Current Population Survey (CPS) data is analyzed. We use log of hourly wages, measured by an individual's wage and salary income for the previous year divided by the number of weeks worked and hours worked per week.[8]

The sample includes individuals aged between 18 and 64 who 1) work only for wages and salary, 2) do not live in group quarters, 3) work more than 20 weeks (inclusive), and more than 35 hours per week in the previous year (e.g., Mulligan and Rubinstein (2008)).

First, we report some baseline results from unconditional distributions (ignoring selection).

5.4.1 Conditional Quantile Selection Models

In the absence of selection, a probability re-weighting approach can be used to recover marginal distributions (see, e.g., Firpo (2007)). Reweighting and quantile approaches are equally valid (Chernozhukov, Fernandez-Val and Melly (2013)). They lead to numerically identical results asymptotically. However, the reweighting approach cannot easily accommodate the selection issue. One cannot identify distributions for groups including unobserved wages for non workers.

[8] Wages are adjusted for inflation based on the 1999 CPI adjustment factors. These are available at https://cps.ipums.org/cps/cpi99.shtml. Following the literature (e.g., Mulligan and Rubinstein (2008); Lemieux (2006)), we exclude extremely low values of wages (less than one unit of the log wages). It has been shown that *inclusion* of imputed wages in wage studies is 'problematic' (Hirsch and Schumacher (2004); Bollinger and Hirsch (2006)). Mulligan and Rubinstein (2008) and Lemieux (2006)) exclude these imputed observations. Such corrections are considered to 'largely eliminate the first-order distortions resulting from imperfect matching' (Bollinger and Hirsch (2013)).

A quantile-copula function approach is proposed in Arellano and Bonhomme (2017b) to correct the entire distribution for selection. Alternative methods will be disucssed in a subsequent section. Selection model add more structure and hence information. The AB approach has greater flexibility in modellng the joint dependence of the marginal variables and leaves marginals unrestricted. In the presence of selection, the AB approach shifts the percentiles as a function of the amount of selection.

Parametric estimation of quantiles is due to Koenker and Bassett (1978), and nonparametric extensions have recently been proposed (e.g., Li and Racine (2008)). In the presence of selection, there are only a few approaches to *point* identify parameters of a quantile function – identification at infinity, the Buchinsky (1998) approach, the Arellano and Bonhomme (2017b), approach, and the more recent distribution regressions of Fernandez-Val et al (2018). Olivetti and Petrongolo (2008) propose another approach but focusing only on median regressions. While they could slightly relax the assumption of selection on unobservables to impute wages for workers who work and have wages for more than a year, they still have to resort to the selection on observable assumption for those who never work.

Identification at infinity is based on the notion that selection bias tends to zero for individuals with certain characteristics who always work and whose probability to work is close to one (Heckman, 1990; Mulligan & Rubinstein, 2008; Chamberlain, 1986). As a result, quantile functions can be identified using the selected sample (even in the absence of exclusion restrictions). However, the definition of 'closeness' to one can be arbitrary in practice and there is a significant trade-off between sample size and the amount of selection bias. Mulligan and Rubinstein (2008) adopt this approach to assess the robustness of their conditional mean results. They define 'closeness' to one as probability of working equal to or greater than .8, and the resulting sample is only about 300 observations per five-year sample, less than 1% of the original sample.[9]

Consider the following quantile wage function (see, e.g., Chernozhukov and Hansen (2008))

$$\ln(w) = g(x,u) \quad u|x \sim Uniform(0,1), \tag{5.26}$$

where $\tau \mapsto g(x_i,\tau)$ is strictly increasing and continuous in τ. This can be a non-separable function of observable characteristics, x, and unobservable disturbances u, normalized and typically interpreted as ability (Doksum, 1974; Chernozhukov & Hansen, 2008).[10] Unobservables, u, are the rank variable or quantile and thus can be

---

[9] Buchinsky (1998) proposed a control function approach to extend Heckman's selection approach to quantiles. He assumed additive separability of observable and unobservables in the wage equation. It also implicitly assumed 'independence between the error term and the regressors conditional on the selection probability.' (Melly & Huber, 2008) Arellano and Bonhomme (2017b) and Arellano and Bonhomme (2017c) note that it is unlikely to specify a data generating process consistent with the Buchinsky assumptions except in the case of either 1) additivity and parallel quantile curves, implying quantile functions are identical and equal to the conditional mean function, or 2) selection is random; see, also, Melly and Huber (2008).

[10] $\Pr[\tau|x] = \tau$. The first equality follows from Equation (5.26). The second follows from the fact that conditional on x, u is uniformly distributed.

fixed in estimations. The participation decision written in a normalized form is given by:

$$S = I(v \leq p(z)) \quad v|x \sim Uniform(0,1), \tag{5.27}$$

where $p(z) = \Pr[S = 1|z]$ is the propensity score, and assuming $p(z) > 0$ with probability one.[11] $I(\cdot)$ is an indicator function (equal to one if the argument is true, zero otherwise). Let $z = (x', \tilde{z}')'$, where $\tilde{z}$ includes a vector of IVs statistically independent of both (u, v) given x. An exclusion restriction is through a variable that affects the selection equation only (see below).
If selection is present,

$$\Pr[\ln(w) \leq g(x,\tau)|s = 1, z] = \Pr[u \leq \tau|v \leq p(z), z] = \frac{C_x(\tau, p(z))}{p(z)}$$
$$\equiv G_x(\tau, p(z)) \neq \tau,$$

where the joint cumulative distribution function (or copula) of (u, v) is defined as $C_x(u, v)$. The observed rank for the τth quantile, $g(x, \tau)$, is no longer the τ in the selected sample. Instead, the observed rank is $G_x(\tau, p(z))$. Knowledge of the mapping between the quantile and its observed rank in the sample allows estimatation of $g(x, \tau)$ using a 'rotated quantile regression'. This is indeed the idea proposed by Arellano and Bonhomme (2017b). [12] Given (a) availability of an exclusion restriction, (b) absolutely continuous bivariate distribution of (U, V) (represented by its copula, $C(u, v)$), (c) continuous outcome, and (d) $p(z) > 0$, $g(\cdot)$ is nonparametrically identified.

5.4.1.1 Practical Implementation

We report results based on a linear index conditional quantile function $g(x, u) = x'\beta(u)$. There is a certain *nonlinearity* allowed by this since it allows x to have differential impact at different quantiles.[13] This index model is a non-separable function of x and u, allowing for interaction between the observable and unobservable characteristics, and is thus preferred to the additive structure that is often assumed in the conditional mean models. Linear quantile regression can provide a weighted least squares approximation to an unknown and potentially nonlinear conditional quantile regression Angrist, Chernozhukov and Fernandez-Val (2006).

Below we provide some graphic evidence of the performance of such linear non-separable models, based on Maasoumi and Wang (2019). The vector, x, is a typical set of wage determinants, including educational attainment dummies, marital status, polynomial terms of age up to third order, racial dummy and regional dummies.

[11] Note that Equation (5.27) is a normalization commonly used in the treatment effects literature. Note that $\mathbb{E}[S|z] = \Pr[S = 1|z] = p(z) = \mathbb{E}[S = 1|p(x)] = \Pr[S = 1|p(z)]$.

[12] The algorithm is provided in detail in 5.5 in the supplemental material. Exclusion restrictions and functional forms regarding $G(\cdot)$ provide identification.

[13] It has been noted, e.g., Melly and Huber (2011), that allowing for arbitrary heterogeneity and non separability does not allow point identification, generally. allowing identification only of bounds of the effects which are 'usually very wide in typical applications'.

This is the common set of covariates in the literature on the gender gap with the CPS data. The corresponding wage equation is similar to what Blau and Kahn (2017) refer to as 'human capital specification'.[14]

MW (2019) estimated the propensity scores probit with a flexible specification that includes polynomial terms of the continuous variables up to third order, as well as interaction terms between them and other discrete variables, in addition to an IV. A linear index model is employed. The main exclusion restriction is the presence/number of young children. The set of variables do not completely overlap with those in the wage equation, providing some additional 'exclusion restrictions' for identification.

Note that identification is further aided by the copula function.Identification analysis in Arellano and Bonhomme (2017b) is general and covers the case where the copula is nonparametric. The Frank copula is a low-dimensional parametric choice. See Maasoumi and Wang (2019). Its single parameter, ρ, captures dependence between $G_x(\tau, p(z)) \equiv G_x(\tau, p(z); \rho)$. Frank copula permits a wide range of potential dependencies, including negative dependence.

The dependence parameter ρ has an additional useful interpretation, indicating the sign of selection. A *negative* ρ indicates *positive* selection into employment, while *positive* ρ implies *negative* selection. This facilitates the comparison to the patterns of selection over time reported in the literature, e.g., Mulligan and Rubinstein (2008). ρ, is further allowed to be gender-specific.

The three-steps of implementation are: Estimate propensity scores, $p(z)$; Estimate the dependence parameter, ρ; and given the estimated ρ and a specified τ, obtain the observed rank, $G_x(\tau, p(z); \rho)$ and estimate β_τ , using the'rotated quantile regression'. To recover the unconditional distribution, we estimate β_τ for $\tau = 0.02, 0.03, \ldots, 0.97, 0.98$.[15]

Maasoumi and Wang (2019) assessed the robustness of these findings relative to Frank copula. They employed the Gaussian copula, another low-dimension Copula that provides dependence parameters that could be compared to ρ by the implied Spearman correlation coefficient. Note that in the special case when both marginal distributions of u and v are normal, the copula is a bivariate normal distribution, as in the Heckman model (Lee, 1983). The Gaussian-copula specification used by Maasoumi and Wang (2019) is based on arbitrary marginals and hence more general.

---

[14] Examples include Blau and Kahn (2006) and Mulligan and Rubinstein (2008). Buchinsky (1998) uses a similar set of variables. Card and DiNardo (2002) and Juhn and Murphy (1997) employ a similar set of wage determinants.

[15] The third step is computationally intensive because, for each year of the data, a large number of quantile regressions must be estimated. The results reported here are based on the 299 replications (as reported in Maasoumi and Wang (2019), which requires estimation of more than a million quantile regressions for the comparison of every two pairs of distributions.).

5.5 Conclusion

Renewed interest in inequality of outcomes elicits three general concerns for the practitioner: (i) what attribute is a good measure of well-being, (ii) how to identify its distribution, and (iii) how to characterize the distribution into a workable statistic. This paper serves as a review of the literature and a selective guide of these stages of the analysis of inequality.

Appendix: Inference and some Algorithms

The parameters, ρ and $\beta(u)$, need to be estimated in the quantile selection models. The propensity scores, $p(z)$, are first estimated using a probit model. Following Arellano and Bonhomme (2017b), consider first how to estimate the quantile selection model given the selection parameter, rho in a copula, and then the estimation of ρ.

1. [Estimation of β_τ] Given a particular $\widehat{\rho}$, $\widehat{\beta_\tau}$ can be estimated by minimizing the following *rotated check function*.

$$\widehat{\beta_\tau} = \operatorname{arg\,min}_{\beta} \sum_{i=1}^{N} S_i[\widehat{G_{\tau,i}}(ln(w_i) - x_i'\beta)^+ + (1 - \widehat{G_{\tau,i}})(ln(w_i) - x_i'\beta)^-]$$

where $(ln(w_i) - x_i'\beta)^+ = \max((ln(w_i) - x_i'\beta), 0)$, and $(ln(w_i) - x_i'\beta)^- = \max(-(ln(w_i) - x_i'\beta), 0)$. $\widehat{G_{\tau,i}} = G(\tau, p(z); \widehat{\rho})$.

2. [Estimation of ρ] To estimate ρ, we follow Arellano and Bonhomme (2017) and exploit the following moment restrictions

$$\mathbb{E}[I(\ln(w) \le x'\widehat{\beta_\tau}) - G(\tau, p(z); \rho) | S = 1, Z = z] = 0.$$

This implies that we can choose ρ that minimize the following objective function

$$\widehat{\rho} = \operatorname{arg\,min}_{\rho} || \sum_{i=1}^{N} \sum_{j=1}^{k} S_i \phi_{\tau_j}(z_i)[I(\ln(w_i) \le x_i'\tilde{\beta}_{\tau_j}(\rho)) - G(\tau_j, p(z_i); \rho)]||,$$

where τ_j takes the finite grid in $\{\frac{3}{10}, \ldots, \frac{7}{10}\}$. $\phi_{\tau_j}(z_i)$ is an instrument function, and

$$\tilde{\beta}_\tau(\rho) = \operatorname{arg\,min}_{\beta} \sum_{i=1}^{N} S_i[G_{\tau,i}(ln(w_i) - x_i'\beta)^+ + (1 - G_{\tau,i})(ln(w_i) - x_i'\beta)^-]$$

where $(ln(w_i) - x_i'\beta)^+ = \max((ln(w_i) - x_i'\beta), 0)$, and $(ln(w_i) - x_i'\beta)^- = \max(-(ln(w_i) - x_i'\beta), 0)$. $G_{\tau,i} = G(\tau, p(z); \rho)$,.

As noted in Arellano and Bonhomme (2017, p.9), if the τ is in $\tau = \{\frac{3}{10}, \ldots, \frac{7}{10}\}$, we actually do not have to repeat this process since we have already obtained these values in the step of estimation of ρ.

Extension: A New Concept of Income Distribution: Value of Time

Wage may not necessarily be a good measure of women's actual well-being for those who do not work, and the comparison of wage offers does not fully serve our purpose.

The presence of young children reduces the probability of a woman being a full-time worker, 'a noteworthy number of these women are married to men who earn relatively high incomes' (Neal, 2004). For individuals, especially women, who do not work full-time, their decisions to stay home do not necessarily reflect low wage offers, but rather 'high shadow prices of time spent at home' (Neal, 2004). In other words, wage offers do not necessarily represent income levels that they may enjoy, or the well-being of those who do not work full-time or work at all. It is then important to take into account the non-market value of time for those who do not work in measuring the gender gap. In economic theory, the actual monetary value of not working (or the best alternative to working full-time) is captured by reservation wages. An interesting yet useful comparison would be based on an alternative wage distribution for men and women, replacing the wage offers with *reservation wages* for those who do not work full-time. Recall that the selection mechanism can be thought of as follows

$$S = I(\ln(w) \geq Y^{\text{reservation wages}}). \tag{5.28}$$

The alternative wage distribution is thus equivalent to the distribution of $\max(\ln(w), Y^{\text{reservation wages}})$. Our quantile approach allows such analysis. With further structure in the selection equation, we can recover the distribution of reservation wages given unemployment. Specifically, we further impose an additive structure of reservation wages given by $R(z) + \eta$, and the labor force participation is based on the comparison of wage offers and reservation wages:

$$S = I(\ln(w) \geq R(z) + \eta). \tag{5.29}$$

As noted in Arellano and Bonhomme (2017), this is equivalent to

$$S = I(v \leq F_{\eta-\ln(w)|Z}(-R(z)|z) \quad v|x \sim Uniform(0,1), \tag{5.30}$$

where $v \equiv F_{\eta-\ln(w)|Z}(-R(z)|z)$ is the standard uniform. Therefore, all the assumptions for qunatile selection models in Section 4.1.2 are met, and the wage function, $g(x,u)$, is identified. Given $g(x,u)$, we can also identify $R(z)$.

In practice, we assume a linear index for $R(z) = z'\gamma$. For a given quantile, τ, (3) becomes

$$\begin{aligned} S &= I(x'\beta_\tau \geq z'\gamma + \eta) \\ &= I(z'\theta \geq \eta), \end{aligned} \tag{5.31}$$

the second equality is due to $x \in z$. Once quantile function is identified, $x'\beta_\tau$, we can estimate reservation wages, $z'\gamma$, via propensity score equation $\Phi(z'\theta)$. This involves a three-step procedure. (1) For every individual with $X = x$, we simulate the complete distribution of potential wages by computing $\ln(w) = x\widehat{\beta}_\tau$ for $\tau = 2, \ldots, 98$ and (2) estimate $z'\widehat{\theta}$, the linear index from the probit model. (3) Reservation wage conditional on non-participation status is identified by $x\widehat{\beta}_\tau - z'\widehat{\theta}$ for $\tau = 2, \ldots, 98$ given $S = 0$. Potential wage conditional on participation status is given by $x\widehat{\beta}_\tau$ for $\tau = 2, \ldots, 98$ given $S = 1$.[16]

Decomposition and Counterfactual Analysis with Selection

Decomposition of observed effects into 'structure' and 'composition' components has a long-standing history in labor economics (see Altonji and Blank (1999) and Fortin, Lemieux and Firpo (2011) for excellent accounts). However, most of this type of analysis usually ignores potential bias due to selection, and is focused on the 'average'.

Structural effects are objects of policies promoting equitable wage-setting; The composition effects concern human capital characteristics such as education. Policy/treatment outcomes may produce 'winners' and 'losers'; structural (or composition) effects could be positive at some parts of the distribution, and negative at others.

Counterfactual distributions may be based on conditional quantile regressions, or on re-weighting by propensity scores, or 'distribution regressions'. Here I describe the quantile approach because we can estimate the (*true*) conditional quantile selection regressions. [17]

Machado and Mata (2005) is among the first to estimate quantiles to recover the counterfactual distribution, and Chernozhukov et al. (2013) discuss the corresponding inferential theory.[18]

[16] In a different context, Bonhomme, Jolivet and Leuven (2014) rely on the selection equation to recover the distribution of agents' underlying preferences in a similar way.

[17] The re-weighting approach cannot be readily extended to address selection for decomposition for the *whole* population. In Maasoumi and Wang (2017), examines the racial gap among women, extending Huber (2014) and propose a re-weighting approach based on *nested* propensity score to recover the counterfactual distributions for the *selected* population.

[18] Albrecht, van Vuuren and Vroman (2009) extends this framework to address the selection issue at the distributional level. However, Albrecht et al. (2009)'s approach is based on Buchinsky (2001)'s quantile selection model, which relies on an unrealistic quantile structure.

The counterfactual distribution can be recovered as follows,

$$F_{Y_c}(y) = F_{Y\langle i|j\rangle}(y) = \int F_{Y_i|X_i}(y|x)dF_{X_j}(x). \tag{5.32}$$

Given Equation (5.21), Equation (5.32) can be re-written as follows

$$F_{Y_c}(y) = F_{Y\langle i|j\rangle}(y) = \int \{\int_0^1 I[Q_\tau(Y_i|X_i) \le y]d\tau\}dF_{X_j}(x) \tag{5.33}$$

$$= \int \{\int_0^1 I[X\beta_i \le y]d\tau\}dF_{X_j}(x). \tag{5.34}$$

The last equality is based on specification of the conditional quantile model. The counterfactual outcome distributions are given by:

$$F_{Y_{c1}}(y) = \int \{\int_0^1 I[X\beta_m \le y]d\tau\}dF_{X_f}(x), \quad \text{(Count. Dist. \#1)} \tag{5.35}$$

$$F_{Y_{c2}}(y) = \int \{\int_0^1 I[X\beta_f \le y]d\tau\}dF_{X_m}(x). \quad \text{(Count. Dist. \#2)} \tag{5.36}$$

F_{c1} represents the counterfactual distribution with male wage structure, with women's human capital characteristics. F_{c2} represents the counterfactual distribution with female wage structure, holding men's human capital characteristics. The differences in the distributions F_{c1} and F_1 provide insight into 'structural effects'. The differences in F_{c2} and F_1 come from differences in the distribution of human capital characteristics; the 'composition effects'.

Decomposition methods, at both the mean and distribution level, are now standard in the graduate economics education (see for example, D. Autor (2015)). Methods for correcting selection bias is a logical additional to the curriculum.

References

Aaberge, R., Havnes, T. & Mogstad, M. (2013, November). *A theory for ranking distribution functions.* https://www.econstor.eu/bitstream/10419/89951/1/dp7738.pdf. IZA Discussion Papers, No. 7738, Bonn.

Albrecht, J., van Vuuren, A. & Vroman, S. (2009). Counterfactual distributions with sample selection adjustments: Econometric theory and an application to the netherlands. *Labour Economics*, *16*(4), 383 - 396.

Altonji, J. & Blank, R. (1999). Race and gender in the labor market. In R. Ashenfelter & D. Card (Eds.), *Handbook of labor economics* (Vol. 3, p. 3143-3259). New York: Elsevier.

Angrist, J., Chernozhukov, V. & Fernandez-Val, I. (2006). Quantile regression under misspecification, with an application to the u.s. wage structure. *Econometrica*, *74*(2), 539-563.

Angrist, J. & Pischke, J. (2009). *Mostly harmless econometrics*. Princeton, NJ: Princeton University Press.

Arellano, M. & Bonhomme, S. (2017a). Quantile selection models with an application to understanding changes in wage inequality. *Econometrica*, *85*(1), 1-28.

Arellano, M. & Bonhomme, S. (2017b). Quantile selection models: with an application to understanding changes in wage inequality. *Econometrica*, *85*(1), 1-28.

Arellano, M. & Bonhomme, S. (2017c). Sample selection in quantile regression: A survey. In R. Koenker, V. Chernozhukov, H. X. & L. Peng (Eds.), *Handbook of quantile regression*. Taylor & Francis Group. https://www.taylorfrancis.com/chapters/edit/10.1201/9781315120256-13/sample-selection-quantile-regression-survey-manuel-arellano-st%C3%A9phane-bonhomme.

Atkinson, A. (1970). On the measurement of inequality. *Journal of Economic Theory*, *II*, 244-63.

Atkinson, A. B. (2011, May). The restoration of welfare economics. *American Economic Review*, *101*(3), 157-61.

Autor, D. (2015). Lecture note 1: Wage density decompositions. In *Graduate labor economics 14.662*. Cambridge MA: MIT OpenCourseWare. https://ocw.mit.edu/courses/14-662-labor-economics-ii-spring-2015/12ff98a88fe3005fec9f81351c40ab73_MIT14_662S15_lecnotes1.pdf.

Autor, D. H., Manning, A. & Smith, C. L. (2016). The contribution of the minimum wage to us wage inequality over three decades: A reassessment. *American Economic Journal: Applied Economics*, *8*(1), 58 - 99.

Blau, F. & Kahn, L. (1997). Swimming upstream: Trends in the gender wage differential in the 1980s. *Journal of Labor Economics*, *15*(1), 1-42.

Blau, F. & Kahn, L. (2006). The u.s. gender pay gap in the 1990s: Slowing convergence. *Industrial and Labor Relations Review*, *60*, 45-66.

Blau, F. & Kahn, L. (2017). The Gender Wage Gap: Extent, Trends, and Explanations. *Journal of Economic Literature*, *55*, 789–865.

Bollinger, C. R. & Hirsch, B. T. (2006). Match bias from earnings imputation in the current population survey: The case of imperfect matching. *Journal of Labor Economics*, *24*(3), 483–519.

Bollinger, C. R. & Hirsch, B. T. (2013). Is earnings nonresponse ignorable? *The Review of Economics and Statistics*, *95*(2), 407–416.

Bonhomme, S., Jolivet, G. & Leuven, E. (2014). School characteristics and teacher turnover: Assessing the role of preferences and opportunities. *Unpublished Manuscript*. http://www.efm.bris.ac.uk/ecgrj/profs_resubmit.pdf.

Bourguignon, F. (1979). Decomposable income inequality measures. *Econometrica*, *47*, 901-20.

Buchinsky, M. (1998). The dynamics of changes in the female wage distribution in the u.s.a.: A quantile regression approach. *Journal of Applied Econometrics*, *13*, 1-30.

Buchinsky, M. (2001). Quantile regression with sample selection: Estimating women's return to education in the u.s. *Empirical Economics*, *26*, 87-113.

Cahn, Y. (2022). *Estimating jointly determined outcomes: How minimum wage affects wages and hours worked* (Tech. Rep.). Emory University. https://github.com/Izzy-Cahn/JointlyDeterminedOutcomes.

Card, D. & DiNardo, J. (2002). Skill-biased technological change and rising wage inequality: Some problems and puzzles. *Journal of Labor Economics*, *20*(4), 733-783.

Chamberlain, G. (1986). Asymptotic efficiency in semi-parametric models with censoring. *Journal of Econometrics*, *32*, 189-218.

Chernozhukov, V., Fernandez-Val, I. & Melly, B. (2013). Inference on counterfactual distributions. *Econometrica*, *81*(6), 2205-2268.

Chernozhukov, V., Fernández-Val, I. & Luo, S. (2018). *Distribution regression with sample selection, with an application to wage decompositions in the UK.* https://arxiv.org/abs/1811.11603. arXiv.

Chernozhukov, V. & Hansen, C. (2008). Instrumental variable quantile regression: A robust inference approach. *Journal of Econometrics*, *142*, 379-298.

Cowell, F. (2006). *Inequality: Measurement* (STICERD - Distributional Analysis Research Programme Papers). Suntory and Toyota International Centres for Economics and Related Disciplines, LSE. Retrieved from https://EconPapers.repec.org/RePEc:cep:stidar:86

Dehejia, R. H. (2005). Program evaluation as a decision problem. *Journal of Econometrics*, *125*, 141-173.

Doksum, K. (1974). Empirical probability plots and statistical inference for nonlinear models in the two-sample case. *Annals of Statistics*, *2*, 267-77.

Donaldson, D. & Weymark, J. (1983). Ethically flexible gini indices for income distributions in the continuum. *Journal of Economic Theory*, *29*(2), 353-358.

Firpo, S. (2007). Efficient semiparametric estimation of quantile treatment effects. *Econometrica*, *75*(1), 259-276.

Fortin, N., Lemieux, T. & Firpo, S. (2011). Decomposition methods in economics. In D. Card & O. Ashenfelter (Eds.), *The handbook of labor economics* (Vol. 4, p. 1-102). Elsevier.

Foster, J., Greer, J. & Thorbecke, E. (1984). A class of decomposable poverty measures. *Econometrica*, *52*(3), 761 - 766.

Goldin, C. (2014). A grand gender convergence: Its last chapter. *American Economic Review*, *104*(4), 1091-1119.

Heckman, J. (1974). Shadow price, market wages, and labor supply. *Econometrica*, *42*(4), 679-694.

Heckman, J. (1990). Varieties of selection bias. *American Economic Review*, *80*(2), 313-318.

Heckman, J. & Smith, J. (1998). Evaluating the welfare state. In S. Strom (Ed.), *Econometrics in the 20th Century: The Ragnar Frisch Centenary.* Cambridge University Press.

Heckman, J., Smith, J. & Clements, N. (1997). Making the most out of programme evaluations and social experiments: Accounting for heterogneity in program impacts. *Review of Economic Studies*, *64*, 487-535.

Hirsch, B. & Schumacher, E. (2004). Match bias in wage gap estimates due to earnings imputation. *Journal of Labor Economics*, *22*(3), 689–722.

Huber, M. (2014). Treatment evaluation in the presence of sample selection. *Econometric Reviews*, *33*(8), 869-905.

Juhn, C. & Murphy, K. (1997). Wage inequality and family labor supply. *Journal of Labor Economics*, *15*(1), 72-97.

Koenker, R. & Bassett, G. (1978). Regression quantiles. *Econometrica*, *46*(1), 33-50.

Lee, L.-F. (1983). Generalized econometric models with selectivity. *Econometrica*, *51*(2), 507-512.

Lemieux, T. (2006). Increasing residual wage inequality: Composition effects, noisy data, or rising demand for skill? *The American Economic Review*, *96*(3), 461–498.

Li, Q. & Racine, J. (2008). Nonparametric estimation of conditional cdf and quantile functions with mixed categorical and continuous data. *Journal of Business & Economic Statistics*, *26*(4), 423-434.

Linton, O., Maasoumi, E. & Whang, Y. (2005). Consistent testing for stochastic dominance: A subsampling approach. *Review of Economic Studies*, *72*, 735-765.

Lugo, M. & Maasoumi, E. (2008). The information basis of multidimensional poverty measurement. In N. Kakwani & J. Silber (Eds.), *Quantitative approaches to multidimensional poverty measurement* (Vol. 2008). Hampshire, UK: Palgrave MacMillan.

Maasoumi, E. (1986). The measurement and decomposition of multi-dimensional inequality. *Econometrica*, *54*(4), 991 - 997.

Maasoumi, E. (1993). A compendium to information theory in economics and econometrics. *Econometric Reviews*, *12*(3), 1-49.

Maasoumi, E. (1998). Empirical analyses of inequality and welfare. In M. H. Pesaran & P. Schmidt (Eds.), *The handbook of applied econometrics, vol ii: Microeconomics.* Hoboken, NJ: Wiley Blackwell Publishers.

Maasoumi, E. (1999). Measuring informativeness by entropy and variance. In D. Slottje (Ed.), *Advances in econometrics, income distribution, and methodology of science: Essays in honor of Camilo Dagum.* Berlin: Springer.

Maasoumi, E. (2001). Parametric and nonparametric tests of limited domain and ordered hypotheses in economics. In B. Baltagi (Ed.), *Companion to econometric theory.* Hoboken, NJ: Wiley Blackwell Publishers.

Maasoumi, E. & Heshmati, A. (2000). Stochastic dominance amongst swedish income distributions. *Econometric Reviews*, *19*, 287-320.

Maasoumi, E. & Wang, L. (2017). What can we learn about the racial gap in the presence of sample selection? *Journal of Econometrics*, *199*(2), 117-130.

Maasoumi, E. & Wang, L. (2019). The gender gap between distributions. *Journal of Political Economy*, *127*, 2438 - 2504.

Machado, J. & Mata, J. (2005). Counterfactual decompositions of changes in wage distributions using quantile regression. *Journal of Applied Econometrics*, *20*, 445-65.

Melly, B. & Huber, M. (2008). *Sample selection, heteroskedasticity, and quantile regression.* Retrieved from https://citeseerx.ist.psu.edu/document?repid=rep1&type=pdf&doi=dffad05ced6db617e7588c9e833539c39fb3c525 (Unpublished Manuscript)

Melly, B. & Huber, M. (2011). *Nonseparable sample selection models.* Retrieved from https://conference.iza.org/conference_files/SPEAC2011/melly_b3683.pdf (Unpublished Manuscript)

Mulligan, C. & Rubinstein, Y. (2008). Selection, investment, and women's relative wages over time. *Quarterly Journal of Economics*, *123*(3), 1061-1110.

National Bureau of Economic Research. (n.d.). *Current population survey (cps) merged outgoing rotation group earnings data.* https://www.nber.org/research/data/current-population-survey-cps-merged-outgoing-rotation-group-earnings-data.

Neal, D. (2004). The measured black-white wage gap among women is too small. *Journal of Political Economy*, *S1*, S1-S28.

Nerlove, M., Razin, A. & Sadka, E. (1987, 12). Intragenerational income distribution policies. In *Household and economy: Welfare economies of endogenous fertility.* Academic Press. https://www.amazon.com/Household-Economy-Endogenous-ECONOMETRICS-MATHEMATICAL/dp/0125157525.

Nerlove, M., Razin, A. & Sadka, E. (1993, 01). Children: A capital good or a base for income redistribution policies. *Public Finance = Finances publiques*, *48*, 78-84.

Nerlove, M., Razin, A., Sadka, E. & Weizsacker, R. (1993, 08). Tax policy, investments in human and physical capital, and productivity. *Journal of Public Economics*, 397-406.

Nerlove, M., Razin, A., Sadka, E. & Weizsäcker, R. (1993, 03). Comprehensive income taxation, investments in human and physical capital, and productivity. *Journal of Public Economics*, *50*, 397-406.

Olivetti, C. & Petrongolo, B. (2008). Unequal pay or unequal employment? a cross-country analysis of gender gaps. *Journal of Labor Economics*, *26*(4), 621-654.

Piketty, T. (2013). *Capital: In the twenty-first century.* Cambridge, MA: Harvard University Press.

Pratt, J. (1964). Risk aversion in the small and in the large. *Econometrica*, *32*, 122-136.

Shorrocks, A. (1978). Income inequality and income mobility. *Journal of Economic Theory*, *19*, 376-93.

Shorrocks, A. (1980). The class of additvely decomposable inequality measures. *Econometrica*, *48*, 613-25.

Shorrocks, A. (1983). Ranking income distributions. *Economica*, *50*, 3-17.

Tsui, K.-y. (1999). Multidimensional inequality and multidimensional generalized entropy measures: An axiomatic derivation. *Social Choice and Welfare*, *16*(1), 145 - 157.

Whang, Y.-J. (2019). *Econometric analysis of stochastic dominance: Concepts, methods, tools, and applications.* Cambridge University Press.

Chapter 6
The Wizard of OZ (Opportunity Zones): Spatial Spillovers in Place-based Programs

Dibya Deepta Mishra, Robin C. Sickles and Yanfei Sun

Abstract The Opportunity Zones (OZ) program, as the largest ongoing place-based development program in the U.S., was intended to stimulate investment and drive economic growth in low-income areas by lowering capital gains tax rates. This paper investigates the spatial spillover effects of the OZ due to their interconnections with high-income neighboring areas. Using two-way fixed effects, synthetic difference-in-differences, and spatial difference-in-differences, we study the impact of OZ on housing prices and nighttime light emissions in the largest state by area in the continental US, Texas. Our empirical results indicate that census tracts located near more developed regions exhibit a stronger response to the OZ program due to the presence of spillover effects. One of the governing factors of these policies is the number of high-income neighbors. However, they play the role of a double-edged sword. A large number of high-income neighbors will make the tract in question not as attractive for investment, even in the presence of tax breaks. This is because the neighbors will provide higher returns. If a census tract is surrounded by some high-income neighbors and there is scope of future return, it may provide incentives for investing. We provide evidence of this trade-off in our paper and also show how these effects should be considered carefully when designing place-based policies, especially when providing location-based tax breaks as in the Opportunity Zone program.

Dibya Deepta Mishra ✉
Department of Economics, Rice University, Houston, USA, e-mail: ddm5@rice.edu

Robin C. Sickles
Department of Economics, Rice University, Houston, USA, e-mail: rsickles@rice.edu

Yanfei Sun
Department of Finance, Toronto Metropolitan University, Toronto, Canada, e-mail: yanfei.sun@torontomu.ca

B. H. Baltagi and L. Mátyás (eds.), *Seven Decades of Econometrics and Beyond*, Advanced Studies in Theoretical and Applied Econometrics 57,
https://doi.org/10.1007/978-3-031-92699-0_6

6.1 Introduction

Much like the Wizard of Oz, whose amazing power and dramatic appearance was the source of awe and inspiration by all of his subjects, Marc Nerlove shaped the intellectual perspectives of countless students and colleague in awe of Marc. The second co-author of this paper was lucky enough to be a Penn colleague of Marc's and also worked with him on projects in Brazil and Washington, DC. His intellectual reach was remarkably broad and intensely deep. His grasp of history, literature, languages, and the world was humbling. I recall Marc commenting on what he learned from a relatively young Mario Vargas Llosa with whom he was bunked in a very distinguished conference for intellectual leaders of different disciplines. Llosa is credited with developing magical realism as a literary genre. Marc's contribution was no less magical. Llosa ultimately won the Nobel Prize in Literature. Marc should have had the same legacy in Economic Sciences.

We have benefited from Marc Nerlove's many contributions in our Chapter, from his care with data, to his use of new applied econometric techniques, to his dedicated interest in the American economy. His classic work on dynamic panels (Balestra & Nerlove, 1966) and our work on spatial panels differ in many ways. However, spatial models are intrinsically dynamic panel models with the subscripts reserved.

Place-based policies have emerged as effective tools for addressing spatial inequality. These policies involve providing incentives to economically disadvantaged areas to stimulate investment. One such place-based policy is opportunity zones. For instance, in 2017, the Tax Cuts and Jobs Act of 2017 (TCJA), introduced the Opportunity Zone (OZ) program. The primary goal of this initiative was to bolster investment in low-income areas by reducing tax rates on capital gains. In so doing it was intended to spur economic growth and job creation within distressed communities. This program mandated state governors to designate approximately 25% of low-income census tracts, or tracts adjacent to such areas, as Opportunity Zones eligible for reduced tax rates.

However, the impact of opportunity zones has been subject to conflicting evidence, partly due to indications of political considerations influencing the selection of eligible regions. For example, Eldar and Garber (2020) find that while state governors often prioritize zones based on distress levels, there is also evidence of favoritism, as manifested by the allocation of tract status to similarly aligned zones. In the same vein, Frank, Hoopes and Lester (2022) find that governors are on average 7.6% more likely to select a census tract as an Opportunity Zone when the tract's state representative is a member of the governor's political party. This complex interplay makes it challenging to evaluate the true impact of opportunity zones. Consequently, numerous studies report limited or negligible effects of the program. Nonetheless, certain studies reveal selective positive impacts of opportunity zones.

For instance, Arefeva, Davis, Ghent and Park (2020) demonstrate that the OZ designation within the TCJA led to amplified employment growth compared to similar tracts. This growth was more pronounced in urban areas than in rural ones. Sage, Langen and Van de Minne (2019) note that while housing prices remained relatively unaffected by opportunity zones, there was a notable increase in redevelopment and vacant land prices. However, it is essential to acknowledge that these changes may be influenced by factors beyond the scope of the OZ status. Kennedy and Wheeler (2021) employ tax filings to uncover spatial concentrations of OZ capital, with evidence of property investments in neighborhoods boasting relatively higher incomes. Moreover, recent journalistic analysis argues that the TCJA prompted real estate investments in gentrifying areas that were already experiencing wealth and demographic shifts (Tankersley, 2021). This highlights the varying sensitivity of different low-income areas to their OZ classification.

The tax benefit of OZ helps to pool capital from potentially broad geographical areas and thus is expected to encourage long-term investment in OZs. However, there are multiple concerns about the OZ program, from the nomination process to its impact on both designated zones and the nation as a whole. States were required to submit nominations for OZs to the Treasury Department. However, there is no detailed and publicly available information from either the federal or state governments describing all the factors used in choosing OZs from all eligible tracts. Barth, Sun and Zhang (2021) question whether the chosen OZs were the most appropriate ones among all the eligible distressed communities, although based on the Government Accountability Office report (2018) OZs have lower incomes and higher poverty than other census tracts, as well as a greater share of the non-white population. Some studies find that political factors play a significant role in OZ designation as the nomination process provides an opportunity for the governor to reward political allies, buy voter support, and help business interests. Alm, Dronyk-Trosper and Larkin (2021) find that census tracts with a higher proportion of representation by Democrats in the state legislative chamber are negatively associated with qualified Opportunity Zone designation, and partisan matching increases the likelihood of OZ designation. The latter finding is also supported by Frank et al. (2022). Eldar and Garber (2020) find that while state governors select zones based on distress levels, favoritism is an important consideration in the nomination process, as tracts in counties that supported the governor in the election are more likely to be chosen as OZs.

Based on the estimation of the Joint Committee on Taxation, OZs would cost $1.6 billion in revenue from 2018 to 2027. GAO (2021) conducted a survey of government officials across all 50 states, Washington D.C., and five U.S. territories to assess the effects of the OZ tax incentive. Out of the 56 respondents, only 20 reported a net positive impact, while 10 claimed a net neutral impact and 5 reported no impact. One respondent reported a net negative impact, and the remaining 20 were uncertain about the impact. Similarly, academic research also shows mixed evidence on the impact of OZs. Interestingly, according to the GAO (2021) report, "[q]ualified Opportunity Funds are making diverse investments. Nearly all of the fund representatives we interviewed were investing in projects principally focused on real estate development." In that case, OZ should observe an increase in real estate prices. Zillow data shows that after the selections were announced, OZs experienced a 20% jump in real estate prices from the previous year. Sage et al. (2019) find a positive price effect of OZ designation for old commercial properties, such as retail and apartment properties, as well as vacant land. Similarly, Frank et al. (2022), Pierzak (2021), and Wiley and Nguyen (2022) also report a higher housing price in OZs. However, while Wiley and Nguyen (2022) and Bekkerman, Cohen, Liu, Maiden and Mitrofanov (2021) find that the OZ program increased real estate prices, it does not have a significant effect on transaction volume. Alm et al. (2021) suggest that while OZs have had a positive impact on non-vacant residential property values, the effect on commercial and vacant property is unclear. In contrast to these findings, Chen, Glaeser and Wessel (2023) argue that OZs appear to have a negligible impact on housing prices.

In terms of economic activities, Frank et al. (2022) document the positive effects of OZ designation on building permits and construction employment. Arefeva et al. (2020) show that the OZ designation increases employment growth relative to comparable tracts, although it does not create jobs in rural areas. Wheeler (2022) finds a substantial impact on promoting new real estate development, and the positive effects extend to surrounding areas. However, Freedman, Khanna and Neumark (2023) criticize the program, stating that its effects are economically small and generally statistically indistinguishable from zero. They report that the employment rates of residents do not change significantly, and while there is a slight increase in average earnings and local poverty rates, but not statistically significant. Feldman and Corinth (2023) find that the impact of OZ is very limited, with no increase in investment for both the number and amount of investment, business activities, or consumer spending in OZs. Snidal and Li (2022) examine small business and residential loan data and do not find OZs have had statistically significant effects on business or residential loan growth.

A potential unintended consequence of OZs that runs counter to their primary goal of bolstering investment in low-income areas and in so doing possibly mitigate income inequality by generating economic growth and job creation within distressed communities, is that OZs could accelerate neighborhood gentrification. As OZs offer investors a reduction in capital gains tax, investors are more likely to seek high-return programs or hot areas to invest, e.g. luxury residential high-rises in Houston and major high-rise residential/luxury retail partnerships in North Miami. This trend leads to the concentration of capital in already gentrifying areas, rather than the most distressed ones. Kennedy and Wheeler (2021) use tax filings to demonstrate that OZ capital is spatially concentrated, and property investments are more likely in neighborhoods with relatively higher incomes. Notably, they find that 84% of OZs receive zero OZ capital. Kurban, Otabor, Cole-Smith and Gautam (2022) also observe that census tracts with positive net migration and lower business vacancy rates are more likely to receive higher financing.

Given these multiple concerns, efficient allocation of opportunity zone status is of utmost importance. In this paper, our objective is to provide evidence that this sensitivity is interconnected with spatial spillovers from neighboring areas. Census tracts in close proximity to more developed regions or other opportunity zones tend to respond more dynamically to place-based policies due to the presence of synergistic effects. To achieve our objective, we begin by estimating the causal treatment impact of OZ status using a two-way fixed effects approach. However, with place-based policies, even untreated units can be exposed to treatment if treated units are nearby. Apart from that the treatment effect is also dependent on nearby infrastructure as it can help facilitate increases in economic activity.

Section 2 provides an overview of the program. Section 3 discusses why spatial spillovers are important, especially within the context of place-based policies, and how to estimate these effects. Section 4 provides an illustrative model of why direct and indirect impacts need to be taken into account when considering allocating subsidies in place-based policies. Section 5 talks about our data and estimation strategy. We provide a discussion about our results in Section 6 and conclude with ideas for future extensions in Section 7.

6.2 The Opportunity Zone Program

The Opportunity Zones (OZs) program, established by the Tax Cuts and Jobs Act of 2017, was designed with the primary objective of catalyzing economic growth and job creation within distressed communities. This program empowered governors in each state to nominate 25% of their eligible low-income tracts for designation as Opportunity Zones. Consequently, numerous low-income communities (LIC) across the United States were designated as Qualified Opportunity Zones, granting investors who allocate eligible capital into Qualified Opportunity Zone assets significant capital gains tax benefits.

The process of selecting these Opportunity Zones initiated with the compilation of a comprehensive list encompassing all low-income or economically distressed communities within each state. To be eligible for consideration, tracts were required to exhibit a poverty rate of at least 20% or possess a median family income not exceeding 80% of the area's median income. Additionally, a limited allowance was made for the selection of up to 5% of the designated Opportunity Zones from contiguous tracts that did not meet the aforementioned criteria but were situated adjacent to a designated LIC with a median family income no greater than 125% of the adjacent LIC.

Governors in each state subsequently nominated up to 25% of these eligible census tracts as OZs. Out of a total of 42,160 eligible tracts, 8,764, representing 21% of the total, were ultimately nominated and subsequently certified. Among these, 8,566 tracts are categorized as LIC tracts, while the remaining 198 are non-LIC contiguous tracts. Notably, approximately 32 million individuals, equivalent to around 10% of the entire U.S. population, reside within these designated Opportunity Zone, when the bill was enacted in 2017.

The OZ tax incentives serve the overarching goal of fostering the development of Opportunity Zones and reducing their poverty rates. These incentives provide taxpayers with the opportunity to defer, and in certain cases permanently exclude, specific gains by investing in a Qualified Opportunity Fund (QOF). Investors can defer taxes on prior capital gains invested in a QOF until either the date of the QOF investment's sale, exchange, or December 31, 2026, provided that all or a portion of the gains are reinvested within 180 days in QOFs.

Additionally, if taxpayers retain investments in QOFs for at least five years, there is a 10% exclusion of the deferred gain, thereby reducing the tax liability to 90% of the rolled-over capital gains. For investments held for at least seven years, an additional

5% exclusion applies. Consequently, investors who want to benefit from the full 15% tax exclusion must invest in QOFs by 2024 to secure the initial 10% exclusion and by 2031 to claim the additional 5% exclusion.

Crucially, investors who maintain investments in QOFs for a minimum of ten years can permanently exclude capital gains taxes on any profits derived from these investments. Those planning to shelter their gains for the entire decade can do so until June 28, 2027, with the proposed regulations suggesting that QOFs can retain these sheltered funds through 2047.

In our analysis, we focus on the state of Texas. Texas is an ideal setting to explore the heterogeneous impact of opportunity zones. It is the largest state in the contiguous United States with an area of approximately a quarter million square miles, making it larger than many countries (Texas is 20% larger than France). Apart from that, there is significant inequality in Texas even though the situation has improved somewhat in recent years (Fisher & Smeeding, 2016). Spatial inequalities are also quite rife in Texas, with 83.7% of the population residing in urban areas and 25% living in the state's five largest cities, according to 2020 Census data. Hence, spatial agglomeration effects should play a large role in the effect of heterogeneous impact of opportunity zones.

In the state of Texas, the second-largest state in the USA by both land area and population, there exist a total of 5,265 census tracts, of which 3,190 are deemed eligible tracts. The office of Texas Governor Greg Abbott designated 628 census tracts as Opportunity Zones, with all of these tracts classified as low-income. In the selection process, Texas emphasized specific criteria, including chronic unemployment, recent natural disasters within the past two years, and low population density, identifying these factors as "significant economic disruptors" likely to benefit from economic stimulus.

Figure 6.1 provides details on all the census tracts in Texas where we identify the opportunity zones, eligible census tracts that were not OZs, and tracts that were not eligible for the OZ program. Figure 6.2 shows the distribution of high income neighbors in Texas for every census tract, where the high income designation comes from the definition of the same from the US Treasury.

Fig. 6.1: Opportunity Zones in Texas

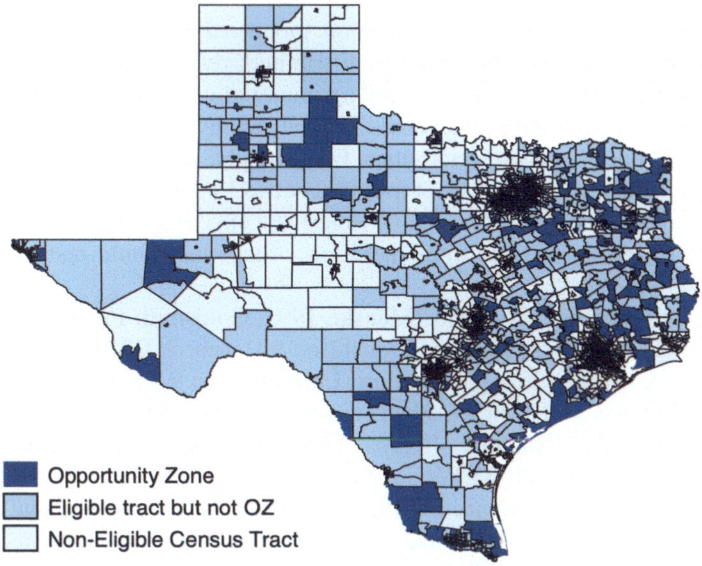

Data source: U.S. Census Bureau (2017); U.S. Department of the Treasury (2018).

Fig. 6.2: High Income Neighbors in Texas

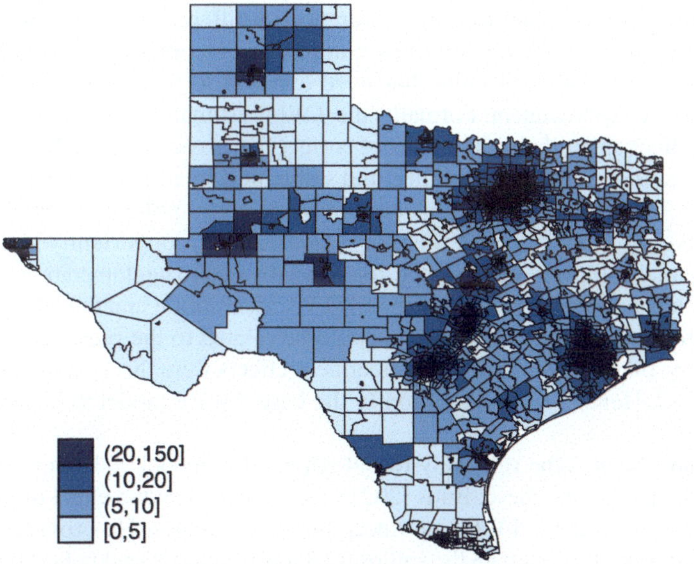

Data source: U.S. Census Bureau (2017).

6.3 Spillover Effects

The effect of location of factors of production, pioneered by Krugman (1991), posits that economic activity will be spatially distributed, resulting in a core-periphery structure. This implies the presence of agglomeration effects. In this case, the Opportunity Zone program's effect could be mediated by how close the opportunity zone is to the core of economic activities. For example, any manufacturing activity would be ineffective in the absence of transport networks that move raw materials and produced goods in and out of the facility. Hence, the effectiveness of any place-based policy would be mediated by agglomeration effects and would result in spatial spillovers.

Spatial spillovers as peer effects (Manski, 1993; Goldsmith-Pinkham & Imbens, 2013) have been well-studied in trade (Donaldson & Hornbeck, 2016), health (Kosfeld, Mitze, Rode & Wälde, 2021), education (Li, Sickles & Williams, 2020; Barrios-Fernández, 2023) as well as in the urban economics literature (Butts, 2021, Figueroa-Armijos & Johnson, 2016,Dubé, Legros, Thériault & Des Rosiers, 2014,Heckert, 2015,Sunak & Madlener, 2014). We use methods from both the two-way fixed effects literature and synthetic difference-in-differences methods to identify spatial spillover effects. We posit that census tracts that are classified as higher income are more likely to be the core of economic activity and have spatial agglomeration effects. Hence, opportunity zones that are closer to a larger number of high-income census tracts will have greater returns to investment as they can take advantage of the positive spillovers. Hence, accounting for these spillovers when deciding which census tracts to classify as opportunity zones could lead to better outcomes.

Given the spatial and temporal variation in the OZ program, the difference-in-differences method is quite apt for evaluating the effect of granting OZ status on economic activity. Note that the difference-in-differences estimator provides the treatment effect on the outcomes by comparing outcomes before and after the treatment date and between units that were exposed to treatment and units that were not exposed to treatment. Formally, the TWFE estimator identifies the Average Treatment Effect under the assumptions of correct linear specification, homogeneous treatment, parallel trends, ignorability, and stable unit treatment value assumption (SUTVA). However, note that in the case of place-based policies, such as the OZ program, even if a census tract was not classified as an opportunity zone, there might be spatial spillovers from treated zones nearby due to agglomeration effects as discussed above. This would violate the SUTVA assumption and thus the potential outcomes for an untreated census tract could be related to the presence of treated status nearby, biasing estimates of the treatment effects were this spatial correlation not addressed. Hence, we need to modify the base TWFE model to identify these effects.

There have been some relatively recent papers that model and estimate spatial spillover effects. Among these, Butts (2021) shows that when the effect of treatment units cross over borders, TWFE produces biased estimates. He provides a semi-parametric approach to estimate the spillover effects of place-based policies. Delgado and Florax (2015) develop an estimator for spatial data that allows for local spatial interaction in potential outcomes, which helps identify direct and indirect treatment effects. Specifically, they modify the standard TWFE model $y_{it} = \alpha_0 + \alpha_1 X_{it} + \alpha_2 D_{it} + \alpha_3 T_{it} + \alpha_4 D_{it} T_{it} + \varepsilon_{it}$ where $D_{it} = 1$ indicates treatment status, and $T_{it} = 1$ indicates pre-post classification. The spatial difference-in-differences model they posit models the effect of treatment spillovers as $y = \alpha_0 + \alpha_1 X + \alpha_2 D + \alpha_3 T + \alpha_4 D \circ T + \alpha_5 W D \circ T + \varepsilon$ where W is the network matrix.

In this paper, we start by estimating a standard TWFE model. Then we augment the model by adding interactions of the treatment status with the number of high-income neighbors near the census tract to model agglomeration effects. We then estimate a spatial difference-in-differences model similar to Delgado and Florax (2015) and then utilize a synthetic difference-in-differences approach to identify spatial spillovers.

6.4 Importance of Indirect Effects for Optimal Allocation

Let N_i represent the set of all census tracts and $i, j \in N_i$ represent two arbitrary census tracts. Let $d_{ij} = 1$ denote the presence of a geographic/business link between i and j and let $d_{ij} = 0$ otherwise. OZ_i represents an indicator for whether census tract i was designated as an opportunity zone. Let θ represent the direct economic effects of a census track being given opportunity zone status, δ_h represent the indirect economic effects of a high income neighbor, and δ_l represent the indirect economic effects of a low income neighbor. Then the net effect on census tract i being designated as an opportunity zone is given by

$$\theta OZ_i + \sum_j \left[\delta_h d_{ij} \mathbf{1}\left(x_j = h\right) OZ_i + \delta_l d_{ij} \mathbf{1}\left(x_j = l\right) OZ_i\right],$$

where x_j represents whether a census tract is high income or not.

The net impact of the program is given by

$$\sum_i \left[\theta OZ_i + \sum_j \left[\delta_h d_{ij} \mathbf{1}\left(x_j = h\right) OZ_j + \delta_\ell d_{ij} \mathbf{1}\left(x_j = \ell\right) OZ_j\right]\right].$$

Hence, an optimal allocation is given by

$$\max_{OZ} \sum_i \left[\theta OZ_i + \sum_j \left[\delta_h d_{ij} \mathbf{1}\left(x_j = h\right) OZ_j + \delta_l d_{ij} \mathbf{1}\left(x_j = \ell\right) OZ_j\right]\right].$$

Based on the previous equation, both direct and indirect impacts are important and need to be considered when we are thinking of the Opportunity Zone allocation and the effectiveness of the Opportunity Zone program.

6.5 Data and Empirical Strategy

We combine data from multiple sources to analyze the effect that designation of Opportunity Zones status has on economic activity. To begin, we derived our initial list of eligible Census Tracts from the US Department of Treasury. Subsequently, we obtained low income and high income designations through data sourced from the Census Bureau. We augmented data on census tracts by incorporating demographic data collected during the span of 2013 to 2022 from the American Community Services surveys. This allows us to control for different demographic trends in census tracts during the analysis period.

We proxy investment levels within each Census Tract using two distinct metrics. The first metric is the Housing Price Index, which offers a comprehensive monthly-level evaluation of single-family housing price fluctuations within census tracts. We use data from Zillow's house price index. Since those indices are available at a zipcode level we required a cross-walk to analyze data at a census tract level.

Complementing our first investment proxy, we use Nightlights data from the National Oceanic and Atmospheric Administration (NOAA) as our second metric. Using an annual-level nightlight raster file, we calculated both the mean and median nightlight intensities within each designated Census Tract. The nightlights data set has been widely used as a measure of economic activity (Gibson, Olivia & Boe-Gibson, 2020). We create a census tract year panel of nightlight intensity fluctuations. The nightlights index is a reliable measure of commercial economic development and provides a highly sensitive measure of economic development. This is because an increase in nightlights is immediately visible since it is measured continuously as compared to other measures of growth, which are not continuously measured.

We use these distinct metrics as outcome variables to measure the economic activity in census tracts. For our estimation, we restrict attention to only the census tracts that were eligible to be classified as opportunity zones. We do this since non-eligible census tracts are demographically different from eligible census tracts as seen in Table 6.1.

Table 6.1: Summary Statistics

	OZ	Eligible	Non-OZ	All
Area in Sq. mi.	24.052	15.431	16.825	17.535
	(50.919)	(102.240)	(97.007)	(93.514)
Estimated Total Population	4,850.634	5,051.156	5,455.968	5396.535
	(2356.951)	(2425.942)	(3092.843)	(3033.649)
Estimated Unemployment	10.932	8.209	5.991	6.476
	(5.515)	(4.235)	(16.332)	(15.674)
Per capita Income	17,771.456	19,651.754	29,885.969	28,696.544
	(6745.084)	(7863.279)	(17699.410)	(17319.350)
Proportion of people above 65	11.633	10.983	11.523	11.534
	(5.295)	(5.255)	(17.118)	(16.340)
Proportion of people under 17	27.131	27.000	25.746	25.882
	(7.175)	(6.850)	(6.691)	(6.752)
SVI Index	9.394	8.730	6.407	6.700
	(1.567)	(1.813)	(22.714)	(21.593)
Poverty Rate	28.884	25.091	15.882	17.159
	(11.422)	(12.072)	(25.995)	(25.241)
Observations	432	2,521	3,968	4,400

Data source: U.S. Census Bureau (2017).

6.5.1 Two-way fixed effects (TWFE)

We start by estimating a TWFE model $y_{it} = \beta_0 + \beta_1 1\{t > \text{Jun } 2017\} + \beta_2 1\{i\ is\ oz\} + \beta_3 1\{t > \text{Jun } 2017\} \times 1\{i\ is\ oz\} + Z_i'\delta + \epsilon_{it}$ where i indexes a tract and t indexes a year. y is the outcome variable and Z_i are census tract level controls. We cluster at the county level.

We first present the effect of being classified as an opportunity zone on the housing price index and the change in the housing price index. We report only β_3 from the previous equation as that is the parameter of interest. The results are shown in Table 6.2. We find a small effect on the growth of the housing price index. However, there is no statistically significant effect of being classified as an opportunity zone on the housing price index. This is consistent with previous findings, for example, Chen et al. (2023), which suggest minimal effects of being classified as an opportunity zone.

Table 6.2: Effect of the Opportunity Zone Program on Housing Price Index

Dep Var	Hindex Growth	Hindex
post × OZ	0.000	-4.720**
	(0.002)	(2.321)
Observations	17,647	17,647
R^2	0.243	0.517

We use data between 2014-2022. post is an indicator variable that takes a value of 1 after the year 2017. OZ is an indicator that takes a value of 1 for any tract that is an opportunity zone. We also restrict our analysis to census tracts that were eligible for being an opportunity zone.

6.5.2 Spatial difference-in-differences

Given our discussion on spatial agglomeration, there are two possible spatial treatment effects. The first is the effect of nearby infrastructure on the outcome variables. This would differ by treatment status. The second spillover effect would be the effect on a census tract due to proximity to treated census tracts. This is regardless of the treatment status of the census tract in question. To disentangle these effects as a function of having access to infrastructure nearby, we first use interactions to motivate our reasoning, which we then examine more formally with a spatial difference-in-differences and synthetic difference-in-differences model.

We use a Geographic Information System (GIS) to create a variable that indicates the number of high income neighbors of a given census tract. The high income classification is based on the same criteria as the Opportunity Zone program. We then interact this variable with $1\{t > t > \text{Jun } 2017\} \times 1\{i\ is\ oz\}$ in the previous equation. We report the parameter on $1\{t > t > \text{Jun } 2017\} \times 1\{i\ is\ oz\}$ as well as the interaction parameter in Table 6.3. We find that there is a statistically significant effect of having a larger number of high income neighbors on the housing index growth. However, there are no significant effects on the housing price index. In Table 6.4 we estimate a similar model but interact $1\{t > t > \text{Jun } 2017\} \times 1\{i\ is\ oz\}$ with the quartiles of the number of high income neighbors as compared to a continuous variable. We find similar results that show that tracts that were classified as opportunity zones had higher housing price effects if they had high income neighbors.

We detect similar patterns with nightlights data as well as Tables 6.5,6.6 and 6.7 make clear.

Table 6.3: Effect of the Opportunity Zone Program on housing price index interacted with high income neighbors

Dep Var	Hindex growth	Hindex
post × OZ	-0.008***	-5.099
	(0.003)	(4.996)
post × OZ × Proportion HighIncome Neighbors	0.021***	-1.049
	(0.008)	(10.004)
R^2	0.255	0.542
Observations	17,647	17,647

We use data between 2014-2022. Post is an indicator variable that takes a value of 1 after the year 2017. OZ is an indicator that takes a value of 1 for any tract that is an opportunity zone. We also restrict our analysis to census tracts that were eligible for being an opportunity zone. We interact post × OZ with a continuous variable, which is equal to the number of high income neighbors of a given census tract. The high income classification is based on the same criteria as the Opportunity Zone program.

Table 6.4: Effect of the Opportunity Zone Program on housing price index interacted with high income neighbors (Quartiles)

Dep Var	Hindex growth	Hindex
post × OZ	-0.004*	-2.648
	(0.002)	(4.075)
post × OZ × high income neighbors(2)	-0.000	-4.577
	(0.005)	(5.453)
post × OZ × high income neighbors(3)	0.016***	-6.671
	(0.005)	(5.589)
post × OZ × high income neighbors(4)	0.011*	3.764
	(0.007)	(6.991)
R^2	0.255	0.541
Observations	17,647	17,647

We use data between 2014-2022. Post is an indicator variable that takes a value of 1 after the year 2017. OZ is an indicator that takes a value of 1 for any tract that is an opportunity zone. We also restrict our analysis to census tracts that were eligible for being an opportunity zone. We interact post × OZ with quartiles of the number of high income neighbors of a given census tract. The high income classification is based on the same criteria as the Opportunity Zone program.

Table 6.5: Effect of the Opportunity Zone Program on nightlights

Dep Var	NL Growth	Mean NL
post × OZ	0.676	-0.053
	(0.448)	(0.159)
R^2	0.198	0.371
Observations	21,917	25,048

We use data between 2014-2022. Post is an indicator variable that takes a value of 1 after the year 2017. OZ is an indicator that takes a value of 1 for any tract that is an opportunity zone. We also restrict our analysis to census tracts that were eligible for being an opportunity zone.

Table 6.6: Effect of the Opportunity Zone Program on nightlights interacted with high income neighbors

Dep Var	NL Growth	Mean NL
post × OZ	4.485***	0.015
	(0.999)	(0.277)
post × OZ × Proportion HighIncome Neighbors	-9.329***	-0.058
	(2.343)	(0.520)
R^2	0.201	0.375
Observations	21,917	25,048

We use data between 2014-2022. Post is an indicator variable that takes a value of 1 after the year 2017. OZ is an indicator that takes a value of 1 for any tract that is an opportunity zone. We also restrict our analysis to census tracts that were eligible for being an opportunity zone. We interact post × OZ with a continuous variable, which is equal to the number of high income neighbors of a given census tract. The high income classification is based on the same criteria as the Opportunity Zone program.

Table 6.7: Effect of the Opportunity Zone Program on nightlights interacted with high income neighbors (Quartiles)

Dep Var	NL Growth	Mean NL
post × OZ	2.595***	-0.171
	(0.688)	(0.219)
post × OZ × high income neighbors(2)	-2.062**	0.058
	(1.016)	(0.445)
post × OZ × high income neighbors(3)	-4.157***	0.854***
	(1.220)	(0.320)
post × OZ × high income neighbors(4)	-4.774**	-0.671
	(2.209)	(0.412)
R sq	0.200	0.382
Observations	21,917	25,048

We use data between 2014-2022. Post is an indicator variable that takes a value of 1 after the year 2017. OZ is an indicator that takes a value of 1 for any tract that is an opportunity zone. We also restrict our analysis to census tracts that were eligible for being an opportunity zone. We interact post × OZ with quartiles of the number of high income neighbors of a given census tract. The high income classification is based on the same criteria as the Opportunity Zone program.

6.5.3 Synthetic difference-in-differences and Results

Although it is not clear how systematically the allocation mechanism is actually implemented, one can certainly question its randomness. As we have pointed out earlier, the designation of a census track as an OZ is influenced by a number of factors, the most important being political ones (Eldar & Garber, 2020) and these might bias the estimated treatment effects. To illustrate this we show the variation in both the housing price index and nightlights as well as their yearly growth with respect to opportunity zone status.

Fig. 6.3: Housing Prices by year

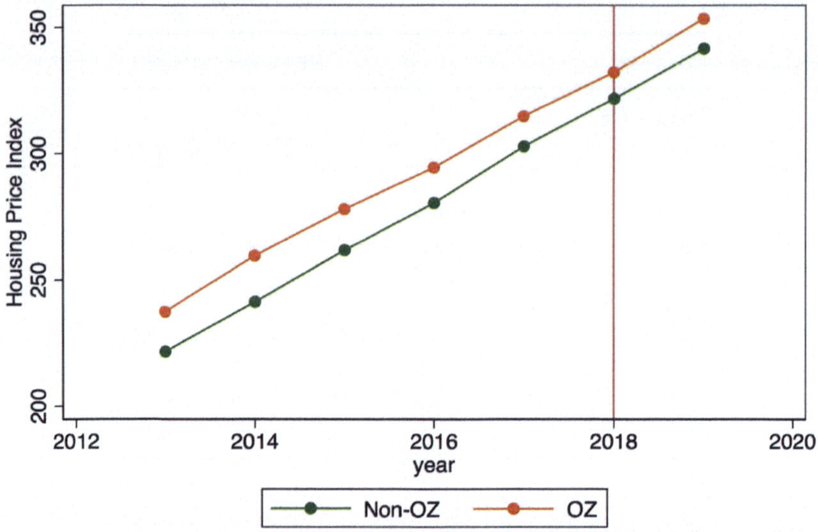

Data source: Contat and Larson (2022)

Fig. 6.4: Housing Prices growth by year

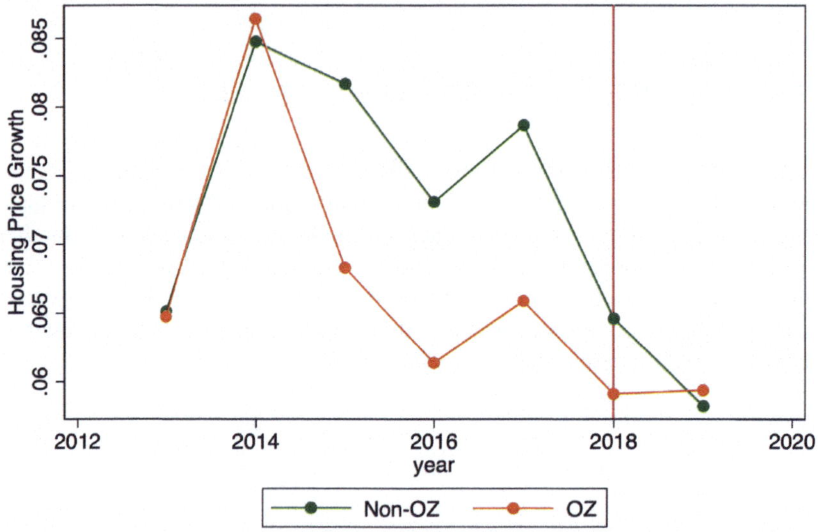

Data source: Contat and Larson (2022)

Fig. 6.5: Mean Nightlights by year

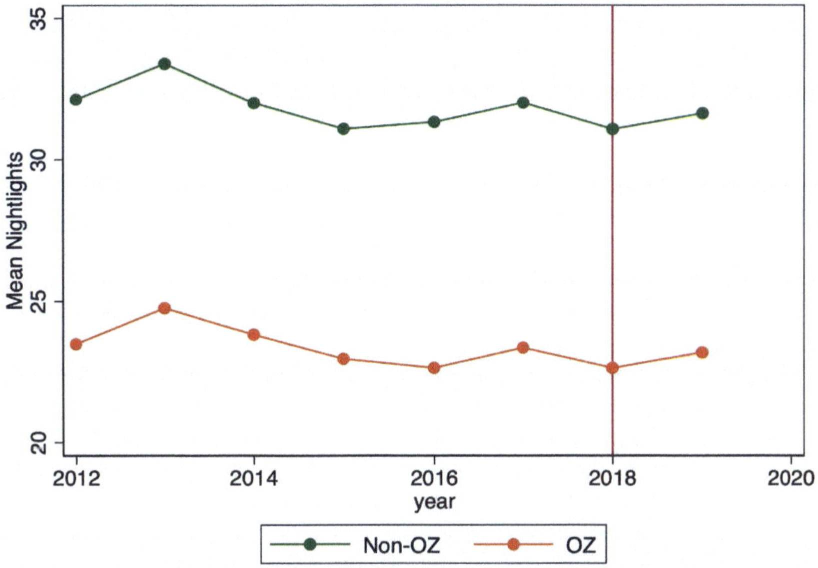

Data source: National Oceanic and Atmospheric Administration (2022)

Fig. 6.6: Nightlights growth by year

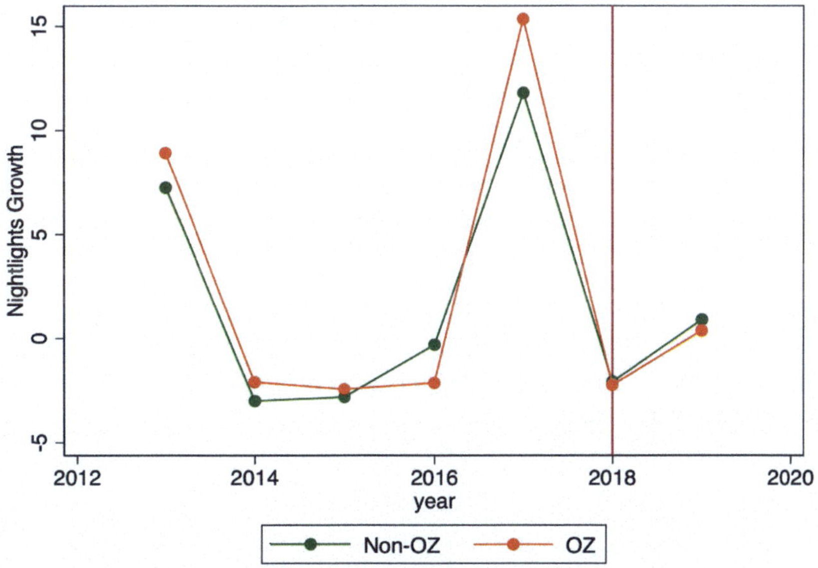

Data source: National Oceanic and Atmospheric Administration (2022)

These apparent differences between eligible non-OZ and OZ census tracts would suggest that in an alternative method that is robust against possible non-random allocation would be helpful in order to provide a check against the standard benchmark TWFE. Hence, we use a synthetic difference-in-differences approach, where for each treatment unit, we compare the outcome variable with a synthetic unit that closely matches the pre trends in outcome and is similar to the treatment unit in other variables (Arkhangelsky, Athey, Hirshberg, Imbens & Wager, 2021). Synthetic difference-in-differences is a methodology that was developed to address these challenges by using statistical techniques to create a synthetic control group that closely matches the characteristics of the treatment group in the pre-treatment period. The synthetic control group is constructed by combining information from multiple control units that have similar pre-treatment outcomes and characteristics to the treatment unit.

The approach involves estimating the counterfactual outcome that would have been observed for the treatment group in the absence of the policy or intervention, by comparing the change in outcomes for the treatment group with the change in outcomes predicted by the synthetic control group. Specifically, consider a balanced panel with N units and T time periods, where the outcome for unit i in period t is denoted by Y_{it}, and exposure to the binary treatment is denoted by $W_{it} \in \{0, 1\}$. Suppose moreover that the first N_{co} (control) units are never exposed to the treatment, while the last $N_{tr} = N - N_{co}$ (treated) units are exposed after time T_{pre}.[1] The synthetic control method begins by finding weights $\hat{\omega}^{\text{sdid}}$ that align pre-exposure trends in the outcome of unexposed units with those for the exposed units, e.g., $\sum_{i=1}^{N_{co}} \hat{\omega}_i^{\text{sdid}} Y_{it} \approx N_{tr}^{-1} \sum_{i=N_{co}+1}^{N} Y_{it}$ for all $t = 1, \ldots, T_{\text{pre}}$. Time weights $\hat{\lambda}_t^{\text{sdid}}$ that balance pre-exposure time periods with postexposure ones (see Section I for details) are then established. These weights are then used in a basic two-way fixed effects regression to estimate the average causal effect of exposure (denoted by τ) :[2]

$$\left(\hat{\tau}^{\text{sdid}}, \hat{\mu}, \hat{\alpha}, \hat{\beta}\right) = \underset{\tau, \mu, \alpha, \beta}{\arg\min} \left\{ \sum_{i=1}^{N} \sum_{t=1}^{T} (Y_{it} - \mu - \alpha_i - \beta_t - W_{it}\tau)^2 \hat{\omega}_i^{\text{sdid}} \hat{\lambda}_t^{\text{sdid}} \right\}.$$

In comparison, DID estimates the effect of treatment exposure by solving the same two-way fixed effects regression problem without either time or unit weights:

$$\left(\hat{\tau}^{\text{did}}, \hat{\mu}, \hat{\alpha}, \hat{\beta}\right) = \underset{\alpha, \beta, \mu, \tau}{\arg\min} \left\{ \sum_{i=1}^{N} \sum_{t=1}^{T} (Y_{it} - \mu - \alpha_i - \beta_t - W_{it}\tau)^2 \right\}.$$

Hence, synthetic difference-in-differences is able to capture ATT even in the absence of parallel trends.

6.5.3.1 Effect on Housing

Table 6.8: Effect of the Opportunity Zone Program — Synthetic Difference-in-Differences

Control Pool	Eligible census tracts		All Census Tracts	
Dep Var	HPI	HPI Growth	HPI	HPI Growth
OZ	1.867	0.007***	5.752***	0.008***
	(1.217)	(0.002)	(1.30)	(0.002)

We use data between 2013-2021. We use synthetic difference-in-differences. In the first column, the synthetic control pool is restricted to eligible census tracts only, and in the second column, we allow for the synthetic controls to be created from all the census tracts. OZ is an indicator for treatment. The standard errors are in brackets.

Table 6.9: Effect of the having more high income neighbors

Control Pool	Opportunity Zones	
Dep Var	HPI	HPI Growth
hinc	-16.9***	-0.043***
	(4.93)	(0.004)

We use data between 2013-2021. We use synthetic difference-in-differences. We restrict the analysis to opportunity zones and compare census tracts, which have more than median high income neighbors to census tracts that have less than median high income neighbors. The ATT estimated is the effect of having more than median high income neighbors. The standard errors are in brackets.

6.5.3.2 Effect on Nightlights

Table 6.10: Effect of the Opportunity Zone Program — Synthetic Difference-in-Differences

Control Pool	Eligible census tracts		All Census Tracts	
Dep Var	Mean_Nighlights	Nightlights Growth	Mean_Nighlights	Nightlights Growth
OZ	0.05	0.15	-0.01	0.026
	(0.118)	(0.270)	(0.137)	(0.291)

We use data between 2013-2021. We use synthetic difference-in-differences. In the first column, the synthetic control pool is restricted to eligible census tracts only, and in the second column, we allow for the synthetic controls to be created from all the census tracts. OZ is an indicator for treatment. The standard errors are in brackets.

Table 6.11: Effect of the having more high income neighbors

Control Pool	Opportunity Zones	
Dep Var	Mean_Nighlights	Nightlights Growth
hinc	1.36***	2.10***
	(0.517)	(0.362)

We use data between 2013-2021. We use synthetic difference-in-differences. We restrict the analysis to opportunity zones and compare census tracts, which have more than median high income neighbors to census tracts that have less than median high income neighbors. The ATT estimated is the effect of having more than median high income neighbors. The standard errors are in brackets.

6.5.4 Spatial difference-in-differences

6.5.4.1 Nightlights Data

To examine the impact of Opportunity Zone (OZ) designations on economic activity, we employed nightlight intensity as a proxy for economic growth. A spatial weights matrix was created based on contiguity, capturing geographical proximity for the year 2019. This matrix allows us to account for spatial dependencies that might influence the outcomes in neighboring census tracts. The data was structured as a panel, enabling us to track changes over time and apply spatial panel regressions. We performed a series of spatial regression analyses using the spxtregress command, where the dependent variable was the logged mean nightlight intensity. We included interaction terms between the OZ designation and a post-2017 indicator to capture any differential effect of the OZ policy after its implementation. These models also controlled for various socioeconomic factors and spatial dependencies, ensuring that the estimated effects of the OZ designation are robust to potential confounding factors related to geographic location and neighboring tract influences.

Table 6.12: Effect on Nightlights using a spatial regression framework

	mean_nl	nlgorwth	mean_nl	nlgrowth
AREA_SQMI	-0.000	0.007	-0.000	0.013
	(0.000)	(0.012)	(0.000)	(0.010)
E_TOTPOP	-0.000***	-0.006***	-0.000***	-0.002***
	(0.000)	(0.000)	(0.000)	(0.000)
EP_UNEMP	-0.000	0.314	0.000**	0.655***
	(0.000)	(0.205)	(0.000)	(0.168)
EP_PCI	-0.000***	-0.000**	0.000***	0.002***
	(0.000)	(0.000)	(0.000)	(0.000)
EP_AGE65	0.000	-0.227	-0.000***	-0.692***
	(0.000)	(0.198)	(0.000)	(0.162)
EP_AGE17	0.000***	-0.185	-0.000	-0.915***
	(0.000)	(0.196)	(0.000)	(0.160)
SPL_THEMES	0.000	-0.502***	-0.000	-0.551***
	(0.000)	(0.156)	(0.000)	(0.128)
EP_POV	-0.000	0.436***	0.000	0.494***
	(0.000)	(0.134)	(0.000)	(0.110)
Designated=1	-0.001	-0.961	0.002**	11.163***
	(0.001)	(3.907)	(0.001)	(3.380)
post=1	-0.003***	16.420***	0.015***	62.346***
	(0.000)	(0.249)	(0.001)	(0.491)
Designated=1 × post=1	0.002**	2.748***	-0.001	0.772
	(0.001)	(0.595)	(0.001)	(0.535)
Constant	0.028***	102.018***	0.037***	120.144***
	(0.002)	(7.583)	(0.001)	(6.244)
Growth	0.852***		0.847***	
	(0.005)		(0.004)	
Index		0.868***		0.981***
		(0.003)		(0.002)

Table 6.12: Cont'd — Effect on Nightlights using a spatial regression framework

	mean_nl	nlgorwth	mean_nl	nlgrowth
Designated=1			-0.022***	-93.472***
			(0.002)	(7.379)
post=1			-0.023***	-61.611***
			(0.001)	(0.579)
Designated=1 × post=1			0.012***	-0.522
			(0.003)	(1.132)
EP_PCI			-0.000***	-0.004***
			(0.000)	(0.000)
σ_u	0.013***	72.101***	0.005***	58.791***
	(0.000)	(0.788)	(0.000)	(0.634)
σ_e	0.029***	15.375***	0.029***	12.520***
	(0.000)	(0.069)	(0.000)	(0.056)
N	30,800	30,800	30,800	30,800

6.5.4.2 Housing Price Analysis

In addition to nightlight intensity, we analyzed the impact of OZ designations on housing prices using data from the Federal Housing Finance Agency (FHFA). The analysis focused on two main outcomes: the Housing Price Index (HPI) and the rate of housing price appreciation. We utilized a similar approach to the nightlight analysis, creating a spatial weights matrix for the year 2019 to account for spatial dependencies between census tracts. Spatial panel regressions were conducted to assess the relationship between OZ status and housing prices, controlling for key economic variables, such as income, unemployment, and population characteristics. Interaction terms were included to test whether the effects of OZ designation on housing prices differed before and after the policy's implementation in 2017. By incorporating these spatial dependencies and interaction effects, our analysis provides a more nuanced understanding of how OZs might influence local housing markets, particularly in terms of price growth and regional disparities.

Table 6.13: Effect of the Opportunity Zone Program on housing price index using Spatial Regression Framework

	Growth	Index	Growth	Index
AREA_SQMI	-0.000	0.007	-0.000	0.013
	(0.000)	(0.012)	(0.000)	(0.010)

Table 6.13: Cont'd — Effect of the Opportunity Zone Program on housing price index using Spatial Regression Framework

	Growth	Index	Growth	Index
E_TOTPOP	-0.000***	-0.006***	-0.000***	-0.002***
	(0.000)	(0.000)	(0.000)	(0.000)
EP_UNEMP	-0.000	0.314	0.000**	0.655***
	(0.000)	(0.205)	(0.000)	(0.168)
EP_PCI	-0.000***	-0.000**	0.000***	0.002***
	(0.000)	(0.000)	(0.000)	(0.000)
EP_AGE65	0.000	-0.227	-0.000***	-0.692***
	(0.000)	(0.198)	(0.000)	(0.162)
EP_AGE17	0.000***	-0.185	-0.000	-0.915***
	(0.000)	(0.196)	(0.000)	(0.160)
SPL_THEMES	0.000	-0.502***	-0.000	-0.551***
	(0.000)	(0.156)	(0.000)	(0.128)
EP_POV	-0.000	0.436***	0.000	0.494***
	(0.000)	(0.134)	(0.000)	(0.110)
Designated=1	-0.001	-0.961	0.002**	11.163***
	(0.001)	(3.907)	(0.001)	(3.380)
post=1	-0.003***	16.420***	0.015***	62.346***
	(0.000)	(0.249)	(0.001)	(0.491)
Designated=1 × post=1	0.002**	2.748***	-0.001	0.772
	(0.001)	(0.595)	(0.001)	(0.535)
Constant	0.028***	102.018***	0.037***	120.144***
	(0.002)	(7.583)	(0.001)	(6.244)
Growth	0.852***		0.847***	
	(0.005)		(0.004)	
Index		0.868***		0.981***
		(0.003)		(0.002)
Designated=1			-0.022***	-93.472***
			(0.002)	(7.379)
post=1			-0.023***	-61.611***
			(0.001)	(0.579)
Designated=1 × post=1			0.012***	-0.522

Table 6.13: Cont'd — Effect of the Opportunity Zone Program on housing price index using Spatial Regression Framework

	Growth	Index	Growth	Index
			(0.003)	(1.132)
EP_PCI			-0.000***	-0.004***
			(0.000)	(0.000)
σ_u	0.013***	72.101***	0.005***	58.791***
	(0.000)	(0.788)	(0.000)	(0.634)
σ_e	0.029***	15.375***	0.029***	12.520***
	(0.000)	(0.069)	(0.000)	(0.056)
N	30,800	30,800	30,800	30,800

6.6 Discussion

Our empirical analysis reveals several interesting patterns in the impact of Opportunity Zones. The most striking finding is the differential effect across our two key outcome measures: while we observe no statistically significant effects of OZ designation on nightlight intensity, we find substantial and statistically significant effects on housing prices. This provides an important insight into how place-based policies manifest in different dimensions of economic activity.

The effects of housing price demonstrate that the OZ designation is effective in attracting capital investment, particularly in the real estate sector. The statistically significant positive effect on housing prices suggests that investors are responding to the tax incentives provided by the program. This aligns with the program's design, which offers capital gains tax benefits that are particularly attractive for real estate investment. However, it is crucial to note that this price appreciation could have ambiguous welfare implications - while it benefits property owners, it might accelerate gentrification and potentially displacement of existing residents.

In contrast, the lack of significant effects on nightlight intensity - our proxy for general economic activity - suggests that the program has not yet generated substantial changes in overall economic vitality. This finding is particularly meaningful because nightlight intensity serves as a high-frequency indicator of economic activity, capturing not only residential development but also commercial and industrial activity. The absence of effects here suggests that while OZ designation may be successful in attracting real estate investment, it has not yet catalyzed broader economic transformation.

A particularly novel contribution of our analysis emerges in the spatial dimension of these effects. Among designated OZs, we find that census tracts with more high-income neighbors exhibit significantly larger positive changes in nightlight levels and growth. This spatial pattern reveals an important mechanism in how place-based policies operate: the effectiveness of these policies appears to be amplified by proximity to existing economic activity. This finding has important implications for policy design, suggesting that the success of place-based interventions may depend critically on the economic geography of the targeted areas.

This spatial heterogeneity is evident in our regression results, where the interaction between the designation of OZ and the presence of high-income neighbors consistently shows positive and significant effects. Specifically, tracts in the third quartile of high-income neighbors show a 0.016 percentage point increase in housing price growth, while those in the fourth quartile show a 0.011 percentage point increase. These effects are economically meaningful and suggest that the benefits of the OZ designation are not uniformly distributed, but rather concentrate in areas with stronger existing economic links.

The spatial heterogeneity of our findings suggests an important policy trade-off. Although designing OZs near high-income areas can maximize the impact of the program on economic activity, this approach might not best serve the program's stated goal of developing economically distressed communities. Our findings indicate that the most disadvantaged areas, those with fewer high-income neighbors, see more limited benefits from the program.

Our use of both synthetic difference-in-differences and spatial regression frameworks provides robust evidence for these patterns. The spatial regression results, in particular, help disentangle direct effects from spillover effects, showing that both the designation itself and the economic characteristics of neighboring areas matter for program outcomes. The synthetic control results further validate these findings while addressing potential selection concerns in OZ designation.

These results contribute to the broader literature on place-based policies by highlighting the importance of spatial context in policy effectiveness. Although previous research has focused primarily on direct effects of such policies, our findings demonstrate that the spatial distribution of economic activity plays a crucial role in mediating policy impacts. This suggests that future place-based policies might benefit from explicitly considering spatial relationships in their design and implementation.

6.7 Conclusion

In this paper, we present evidence of spatial spillovers in outcomes related to place-based policies. One of the governing factors of these policies is the number of high-income neighbors. However, they play the role of a double edged sword. A large number of high-income neighbors will make the tract in question not as attractive for investment, even in the presence of tax breaks. This is because the neighbors will provide higher returns. However, if a census tract is surrounded by some high-income neighbors who are also eligible and there is scope of future return, it might provide incentives for investing.

We provide evidence of this trade-off in our paper and also show how these effects should be considered carefully when designing place-based policies, especially when providing location-based tax breaks as in the Opportunity Zone program. There are several interesting extensions to our paper. Some notable ones imply using a structural model to identify endogenous networks as in De Paula, Rasul and Souza (2019) and extending our analysis to more outcomes.

Acknowledgements We are extremely grateful to the late Dr. James R. Barth for initiating this joint research with us before his untimely death in 2023.

References

Alm, J., Dronyk-Trosper, T. & Larkin, S. (2021). In the land of oz: designating opportunity zones. *Public Choice*, *188*, 503–523.

Arefeva, A., Davis, M. A., Ghent, A. C. & Park, M. (2020). Who benefits from place-based policies? job growth from opportunity zones. *Job Growth from Opportunity Zones (July 7, 2020).*

Arkhangelsky, D., Athey, S., Hirshberg, D. A., Imbens, G. W. & Wager, S. (2021). Synthetic difference-in-differences. *American Economic Review*, *111*(12), 4088–4118.

Balestra, P. & Nerlove, M. (1966). Pooling cross section and time series data in the estimation of a dynamic model: The demand for natural gas. *Econometrica: Journal of the Econometric Society*, 585–612.

Barrios-Fernández, A. (2023). Peer effects in education. In *Oxford Research Encyclopedia of Economics and Finance.* Oxford University Press. https://oxfordre.com/economics/display/10.1093/acrefore/9780190625979.001.0001/acrefore-9780190625979-e-894.

Barth, J. R., Sun, Y. & Zhang, S. (2021). Opportunity zones: do tax benefits go to the most distressed communities? *Journal of Financial Economic Policy*, *13*(3), 301–316.

Bekkerman, R., Cohen, M. C., Liu, X., Maiden, J. & Mitrofanov, D. (2021). *The impact of the opportunity zone program on residential real estate.* https://papers.ssrn.com/sol3/papers.cfm?abstract_id=3780241. SSRN 3780241.

Butts, K. (2021). *Difference-in-differences estimation with spatial spillovers.* https://arxiv.org/pdf/2105.03737. arXiv preprint.

Chen, J., Glaeser, E. & Wessel, D. (2023, January). JUE Insight: The (non-)effect of opportunity zones on housing prices. *Journal of Urban Economics*, *133*, 103451. Retrieved 2023-10-16, from https://linkinghub.elsevier.com/retrieve/pii/S0094119022000286 doi: 10.1016/j.jue.2022.103451

Contat, J. & Larson, W. D. (2022). *A flexible method of house price index construction using repeat-sales aggregates* (Tech. Rep.). SSRN, Working Paper. https://papers.ssrn.com/sol3/papers.cfm?abstract_id=4205810.

Delgado, M. S. & Florax, R. J. (2015). Difference-in-differences techniques for spatial data: Local autocorrelation and spatial interaction. *Economics Letters*, *137*, 123–126.

De Paula, Á., Rasul, I. & Souza, P. (2019). *Identifying network ties from panel data: theory and an application to tax competition.* https://arxiv.org/abs/1910.07452. arXiv preprint.

Donaldson, D. & Hornbeck, R. (2016). Railroads and american economic growth: A "market access" approach. *The Quarterly Journal of Economics*, *131*(2), 799–858.

Dubé, J., Legros, D., Thériault, M. & Des Rosiers, F. (2014). A spatial difference-in-differences estimator to evaluate the effect of change in public mass transit systems on house prices. *Transportation Research Part B: Methodological*, *64*, 24–40.

Eldar, O. & Garber, C. (2020). Does government play favorites? evidence from opportunity zones. *Evidence from Opportunity Zones (September 1, 2020). Duke Law School Public Law & Legal Theory Series*(2020-28).

Feldman, N. & Corinth, K. (2023). *The impact of opportunity zones on commercial investment and economic activity.* https://papers.ssrn.com/sol3/papers.cfm?abstract_id=4086056. SSRN 4086056.

Figueroa-Armijos, M. & Johnson, T. G. (2016). Entrepreneurship policy and economic growth: Solution or delusion? Evidence from a state initiative. *Small Business Economics*, *47*(4), 1033–1047. Retrieved 2023-10-16, from https://www.jstor.org/stable/26154684 (Publisher: Springer)

Fisher, J. & Smeeding, T. M. (2016). Income inequality. *The poverty and inequality report 2016*, 32–38.

Frank, M. M., Hoopes, J. L. & Lester, R. (2022). What determines where opportunity knocks? political affiliation in the selection of opportunity zones. *Journal of Public Economics*, *206*, 104588.

Freedman, M., Khanna, S. & Neumark, D. (2023). Jue insight: The impacts of opportunity zones on zone residents. *Journal of Urban Economics*, *133*, 103407.

Gibson, J., Olivia, S. & Boe-Gibson, G. (2020). Night lights in economics: Sources and uses 1. *Journal of Economic Surveys*, *34*(5), 955–980.

Goldsmith-Pinkham, P. & Imbens, G. W. (2013). Social networks and the identification of peer effects. *Journal of Business & Economic Statistics*, *31*(3), 253–264.

Heckert, M. (2015). A Spatial Difference-in-Differences Approach To Studying the Effect of Greening Vacant Land on Property Values. *Cityscape*, *17*(1), 51–60. Retrieved 2023-10-16, from https://www.jstor.org/stable/26326921 (Publisher: US Department of Housing and Urban Development)

Kennedy, P. & Wheeler, H. (2021). Neighborhood-level investment from the us opportunity zone program: Early evidence. *Available at SSRN 4024514.*

Kosfeld, R., Mitze, T., Rode, J. & Wälde, K. (2021, September). The Covid-19 containment effects of public health measures: A spatial difference-in-differences approach. *Journal of Regional Science*, *61*(4), 799–825. doi: 10.1111/jors.12536

Krugman, P. (1991). Increasing returns and economic geography. *Journal of Political Economy*, *99*(3), 483–499.

Kurban, H., Otabor, C., Cole-Smith, B. & Gautam, G. S. (2022). Gentrification and opportunity zones. *Cityscape*, *24*(1), 149–186.

Li, B., Sickles, R. & Williams, J. (2020). Estimating peer effects on career choice: A spatial multinomial logit approach. In *Essays in Honor of Cheng Hsiao* (pp. 359–381). Emerald Publishing Limited.

Manski, C. F. (1993). Identification of endogenous social effects: The reflection problem. *The Review of Economic Studies*, *60*(3), 531–542.

National Oceanic and Atmospheric Administration. (2022). *Nighttime Lights Dataset.* Retrieved from https://sos.noaa.gov/catalog/datasets/nighttime-lights/

Pierzak, E. F. (2021). Who gains from place-based tax incentives? exploring apartment sales prices in qualified opportunity zones. *The Journal of Portfolio Management*, *47*(10), 145–157.

Sage, A., Langen, M. & Van de Minne, A. (2019). Where is the opportunity in opportunity zones? early indicators of the opportunity zone program's impact on commercial property prices. *Early Indicators of the Opportunity Zone Program's Impact on Commercial Property Prices (May 1, 2019).*

Snidal, M. & Li, G. (2022). The nonimpact of opportunity zones on home and business lending. *Housing Policy Debate*, 1–22.

Sunak, Y. & Madlener, R. (2014, October). *Local Impacts of Wind Farms on Property Values: A Spatial Difference-in-Differences Analysis* [SSRN Scholarly Paper]. Rochester, NY. Retrieved 2023-10-16, from https://papers.ssrn.com/abstract=2500217 doi: 10.2139/ssrn.2500217

Tankersley, J. (2021). Biden administration debating how to overhaul a Trump-era tax break. *The New York Times.*

U.S. Census Bureau. (2017). *American Community Survey (ACS).* Retrieved from https://www.census.gov/programs-surveys/acs

U.S. Department of the Treasury. (2018). *Opportunity Zones.* Retrieved from https://www.cdfifund.gov/opportunity-zones (Accessed: 2025-01-22)

Wheeler, H. (2022). *Locally optimal place-based policies: Evidence from opportunity zones* (Tech. Rep.). Working Paper, November.

Wiley, J. A. & Nguyen, H. (2022). Cherry-picking industrial properties in opportunity zones. *Real Estate Economics*, *50*(5), 1201–1230.

Chapter 7
On the Estimation of Forecaster Loss Functions Using Density Forecasts

Kajal Lahiri, Fushang Liu and Wuwei Wang

Abstract We suggest a novel approach to use density forecasts from surveys to identify asymmetry in forecaster loss functions. We show that we can calculate the loss function parameters for Lin-Lin and Quad-Quad loss functions based on the first order condition of forecast optimality. Since forecasters form their point forecasts based on what they believe to be the data generating processes and their loss functions, we can reverse this process and learn about forecaster loss functions by comparing their point forecasts and density forecasts for the same target. The advantage of this method is that we can relax the two assumptions needed in Elliott, Komunjer and Timmermann's (2008) GMM method: the point forecasts and density forecasts need not to be rational and the loss function parameters need not to be constant over time. Moreover, we do not need to know the actual values of the target variable. This method is applied to density forecasts for annual real output growth and inflation obtained from the Survey of Professional Forecasters (SPF) during 1968-2023. We find that forecasters treat underprediction of real output growth more dearly than overprediction, reverse is true for inflation.

7.1 Introduction

Rationality tests using point forecasts are usually conducted under the assumption of mean squared error (MSE) loss function. Under this assumption, rational point forecasts should be unbiased, and one period forecast errors should be serially

---

Kajal Lahiri ✉
University at Albany, 1400 Washington Avenue, Albany, NY 12222, e-mail: klahiri@albany.edu

Fushang Liu
Massachusetts Department of Revenue, Boston, MA 02114, e-mail: fushangl@gmail.com

Wuwei Wang
Southwestern University of Finance and Economics, Chengdu, China, e-mail: wangwuwei@swufe.edu.cn

B. H. Baltagi and L. Mátyás (eds.), *Seven Decades of Econometrics and Beyond*, Advanced Studies in Theoretical and Applied Econometrics 57,
https://doi.org/10.1007/978-3-031-92699-0_7

uncorrelated. Nerlove (1983), in his pioneering work on the dynamics and optimality of expectations data, has left an indelible mark on the profession. Recently, attention has been devoted to asymmetric loss functions, which are more realistic in many real-life situations. Properties of optimal forecasts under these loss functions have been established and it has been shown that traditional rationality tests based on least square regressions are invalid.[1]

Alternative rationality tests under asymmetric loss have been proposed. For example, Batchelor and Peel (1998) use an ARCH-M model to test rationality of forecasts under a popular loss function – the linear/exponential (Linex) loss. One implicit assumption of this model is that the loss function parameters are constant over time. Granger (1969) points out that loss functions could depend on other variables or the state of the economy, such as the phase of the business cycle or real GDP growth. Elliott, Timmermann and Komunjer (2005); Elliott et al. (2008), hereafter referred to as EKT, point out that an asymmetric loss function may arise from asymmetric stockout and inventory holding costs. Given that the inventory holding costs may change over time due to technological advances and economic conditions, it is reasonable to allow the loss function parameters to vary over time.

Under asymmetric loss, the optimal point forecast may deviate from the central tendency of its underlying density distribution, and the conventional rationality tests based on least squares can be misleading without simultaneously teasing out the effect of asymmetric loss. EKT propose to estimate the parameters of loss function by GMM method and use the GMM over-identification test for forecast rationality. Intuitively, this test asks if there are common loss function parameters that satisfy all moment conditions implied by forecast rationality. Given a family of loss functions indexed by unknown shape parameters, EKT's GMM method provides not only a test of forecast rationality, but also a novel way to estimate loss function parameters directly.[2] However, this method relies on two assumptions. First, it assumes that the loss function parameters are constant over time. Second, and more importantly, it identifies the loss function parameters only under the assumption of forecast rationality. As stated in EKT, they "back out the loss function parameters consistent with the forecast being rational." If the hypothesis of rationality were rejected, the estimated loss function parameters would be biased.[3] Krüger and LeCrone (2019) show that EKT approach leads to precise estimates of the degree of asymmetry that are quite robust to fat tails, serial correlation, and outliers. However, as in other studies, the loss function is assumed to be invariant over time.

In the received forecasting literature, researchers like EKT usually focus on point forecasts in conducting the rationality tests and estimating the loss function

---

[1] See Granger and Pesaran (2000), Christoffersen and Diebold (1996), Diebold, Gunther and Tay (1998), Zellner (1986), and Patton and Timmermann (2007).

[2] The approach has been used in many applications. See for instance, Tsuchiya (2016), Clatworthy, Peel and Pope (2012), Fritsche, Pierdzioch, Rülke and Stadtmann (2015), Wang and Lee (2014) and Döpke, Fritsche and Siliverstovs (2010).

[3] Krüger and LeCrone (2019) using an extensive Monte Carlo study report negative correlation of more than 0.9 between estimated asymmetric loss parameter and absolute average bias across experiments (see their online appendix S3).

parameters. With increasing use of density forecasts in recent years, it is interesting to ask if density forecasts can help improve our knowledge about loss function parameters and can shed light on the validity of the rationality tests using point forecasts alone. We try to answer this question in this chapter. First, we show that for the two common loss functions – Lin-Lin and Quad-Quad loss – the optimal point forecasts will be different from the means or the medians of the underlying density forecasts for any data generating process if the loss is asymmetric. A comparison of these two will suggest if the loss function is asymmetric as well as the direction of asymmetry.[4] This knowledge can then be used to guide the comparison and selection of different estimation methods. Second, by combining the point and the density forecasts for the same target, we can calculate loss function parameters for each period and relax the assumption of constant loss function parameters over time and forecasters. Furthermore, relaxing the assumption of constant loss function allows us to investigate Granger's conjecture that loss function depends on other economic variables or the state of the economy. Finally, using density forecasts, we could modify Batchelor and Peel's ARCH-M method - we no longer need to specify and estimate the process of conditional variance. We could compute the conditional variance from the density forecasts directly and avoid misspecification biases.

With density forecasts and point forecasts both available from U.S. Survey of Professional Forecasters (SPF), we propose to combine point forecasts and density forecasts to estimate the loss function parameters. Unlike previous research, our method does not need to assume unbiasedness of the forecasts because of the availability of subjective probability forecasts produced by the forecasters. We show that professional forecasters view under-prediction of output growth more costly than over-prediction, and the opposite for inflation forecasts. The stylized facts could be mostly explained by the asymmetry in loss functions rather than irrationality. Engelberg, Manski and Williams (2009) also find stylized results that forecasters are overly optimistic in point forecasts than in density forecasts but did not relate their findings to asymmetry in loss functions.

In this chapter, we fit generalized beta distributions and triangular distributions[5] instead of normal distributions to the density forecast histograms. A comparison to results with fitting normal distributions revealed a better performance of the combination of generalized beta distribution and triangular distributions. To compare with Elliott et al. (2005) we apply their GMM method on point forecasts in SPF and compare the estimates of asymmetry of loss function and bias with the mean individual asymmetry derived from our method. Determinants of loss function asymmetry such as the level of the target variable and forecast horizon are discussed.

This chapter is organized as following. In Section 7.2, we review the theoretical background for relationship between optimal point forecast and the central tendency

[4] Lahiri and Liu (2009) have shown that for Linex and Quad-Quad loss functions, a non-zero divergence of the optimal point forecasts from the mean, and for Lin-Lin loss function, a non-zero divergence of the optimal point forecasts from the median of the density forecast is both a necessary and sufficient condition for loss function asymmetry.

[5] Engelberg et al. (2009) and Boero, Smith and Wallis (2008) use generalized beta and triangular distributions in their studies using the same SPF data.

of underlying density forecast, and the model to recover the asymmetry of loss functions by combining density and point forecasts. The GMM method to estimate asymmetry proposed by Elliott et al. (2005) is also reviewed in Section 7.2. In Section 7.3, we describe the features of the data in Survey of Professional Forecasters and the "Real time Data Set for Macroeconomics". In Section 7.4 we present empirical results of the bounds of asymmetric parameters in non-parametric analysis. In Section 7.5 parametric results under Lin-Lin and Quad-Quad loss functions are displayed for the asymmetry parameter. Determinants of loss function asymmetry such as the level of the target variable are included in Section 7.5 too. In Section 7.6 we apply GMM estimation proposed by EKT to SPF point forecasts to estimate time invariant asymmetry for a set of prolific forecasters for each forecast horizon. Section 7.7 summarizes this chapter.

7.2 Review of Estimation of Loss Function Asymmetry

7.2.1 Loss Function and Asymmetry Parameter

A forecaster, while making a point forecast of a target variable, evaluates the unobserved distribution of the variable, and chooses an optimal value given an asymmetric loss function. Thus, the density forecast is the true forecast density which is used in conjunction with a loss function to report a point forecast, cf. Weber (1994).

Suppose a forecaster (forecaster i) with information set I_{ith} believes that the target variable y_t follows a distribution $(p.d.f)$ f_{ith} when he makes the forecast h quarters ahead in year t. The forecaster reports a point forecast y_{ith} that minimizes the expected value of the loss function $L(y_t - y_{ith})$. Loss functions can take different forms, such as linear form (Lin-Lin):

$$L(y_t - y_{ith}) = \alpha_{ith}|y_t - y_{ith}| \qquad if \quad y_t - y_{ith} > 0 \tag{7.1}$$

$$L(y_t - y_{ith}) = (1 - \alpha_{ith})|y_t - y_{ith}| \qquad if \quad y_t - y_{ith} \leq 0$$

or quadratic form (Quad-Quad):

$$L(y_t - y_{ith}) = \alpha_{ith}(y_t - y_{ith})^2 \qquad if \quad y_t - y_{ith} > 0 \tag{7.2}$$

$$L(y_t - y_{ith}) = (1 - \alpha_{ith})(y_t - y_{ith})^2 \qquad if \quad y_t - y_{ith} \leq 0,$$

where $y_{ith} \sim f_{ith}$.

Loss functions can even take distinct forms for different signs of forecast errors, such as the Linex form (loss being linear to forecast error on one side and exponential on the other side).

In the above Equation (7.2), α (alpha) is a parameter of the loss function that measures asymmetry. In conventional research on forecast rationality, the loss function is assumed to be symmetric and $\alpha = 0.5$. In this chapter, we allow α to be any value $[0, 1]$. If $\alpha_{ith} = 0.5$, the forecaster is neutral between over-prediction and under-prediction. However, if $\alpha_{ith} > 0.5$, under-prediction is more costly and the forecaster will likely produce a point forecast that is above the central tendency of f_{ith}. If $\alpha_{ith} < 0.5$, over-prediction is more costly.

7.2.2 Estimation of Asymmetry by Combining Point and Density Forecasts

Estimation of Asymmetry for Lin-Lin Loss Function

Below we show that, under certain loss function forms, the value of α_{ith} could be derived by combining point and density forecasts.

First, consider the case of Lin-Lin (Linear-Linear) loss function. The forecaster would choose an optimal point forecast to minimize the expected value of the loss function.

$$min_{y_{ith}} E[L(y_t - y_{ith})] \tag{7.3}$$

$$\rightarrow min_{y_{ith}} [\alpha_{ith} \int_{y_{ith}}^{\infty} (y_t - y_{ith}) f_{ith}(y_t | I_{ith}) dy_t \tag{7.4}$$

$$-(1 - \alpha_{ith}) \int_{-\infty}^{y_{ith}} (y_t - y_{ith}) f_{ith}(y_t | I_{ith}) dy_t]$$

The first order condition is:

$$\alpha_{ith} \int_{y_{ith}}^{\infty} -f_{ith}(y_t | I_{ith}) dy_t - (y_t - y_{ith}) f_{ith}(y_t | I_{ith}) \tag{7.5}$$

$$-(1 - \alpha_{ith}) \int_{-\infty}^{y_{ith}} -f_{ith}(y_t | I_{ith}) dy_t = 0$$

$$\rightarrow -\alpha_{ith}[1 - F_{ith}(y_{ith} | I_{ith})] + (1 - \alpha_{ith}) F_{ith}(y_{ith} | I_{ith}) = 0 \tag{7.6}$$

$$\rightarrow \alpha_{ith} = F_{ith}(y_{ith}|I_{ith}), \tag{7.7}$$

where F_{ith} is the $c.d.f$ of f_{ith}.

Identity (7.7) indicates that, given the point forecast and underlying subjective distribution, we could recover the loss function parameter. In addition, the relationship between the point forecast and the median of its underlying subjective density distribution could tell the direction of the asymmetry. If the point forecast is above the median, then α_{ith} is above 0.5, and under-prediction is valued more costly. In this framework the asymmetry parameter could vary over time. Thus, we could relax the assumption of constant loss function parameters over time and examine forecaster's asymmetry in every period.

Estimation of Asymmetry for Quad-Quad Loss Function

We first consider the case if the underlying subjective density function is normal. In this case, the asymmetry parameter α_{ith} could also be recovered under the Quad-Quad loss function form as follows:

$$\alpha_{ith} = \frac{D_{ith} - b_{ith}}{2D_{ith} - b_{ith}}, \tag{7.8}$$

where $b_{ith} = \mu_{ith} - y_{ith}$, μ_{ith} is the mean of the underlying subjective density forecast f_{ith}, $D_{ith} = \sigma_{ith}\phi(b_{ith}/\sigma_{ith})) + b_{ith}\Phi((b_{ith}/\sigma_{ith})$, where σ_{ith} is the standard deviation of f_{ith} and ϕ is the density function of standard normal distribution, and Φ is the cumulative density function of the standard normal distribution.

Equation (7.8) is more complicated than the case of Lin-Lin loss function, but the stylized facts still hold that if the point forecast is above the central tendency (in this case, the mean), then α_{ith} is above 0.5, meaning under-prediction is more costly, leading the forecaster to favor an over-the-mean point forecast, and vice versa.

Next consider the case if the underlying subjective density function is generalized beta. EKT prove that, under certain conditions,

$$E[v_{ith}[1(y_t - y_{ith} < 0) - \alpha_{ih}]]|y_t - y_{ith}|^{p-1}] = 0, \tag{7.9}$$

where v_{ith} is a sub-vector of forecaster's information set . p=1 for Lin-Lin loss function and p=2 for Quad-Quad loss function.

Note that, α in Equation (7.9) is time-invariant. This condition could be relaxed if the subjective distribution of y_t is available. Setting v_{ith} to be 1 (vector), Equation (7.9) is reduced to

$$E[1(y_t - y_{ith} < 0) - \alpha_{ih}|y_t - y_{ith}|^{p-1}] = 0. \tag{7.10}$$

We derive a solution to solve for α from Equation (7.10) and then we could estimate the asymmetry parameter alpha of each forecast under Quad-Quad loss.

Under Quad-Quad loss, $p = 2$. When the density distribution of y_t is known as f_{ith} Equation (7.10) becomes

$$\int_{-\infty}^{y_{ith}} (1-\alpha_{ith})(y_{ith}-y)f_{ith}(y|I_{ith})dy] + \int_{y_{ith}}^{\infty} (-\alpha_{ith})(y-y_{ith})f_{ith}(y|I_{ith})dy = 0 \tag{7.11}$$

$$\alpha_{ith} = \frac{\int_{-\infty}^{y_{ith}} (y_{ith}-y)f_{ith}(y|I_{ith})dy}{\int_{-\infty}^{y_{ith}} (y_{ith}-y)f_{ith}(y|I_{ith})dy + \int_{y_{ith}}^{\infty} (y-y_{ith})f_{ith}(y|I_{ith})dy = 0}. \tag{7.12}$$

Equation (7.12) will produce loss function parameter for each forecaster period by period.

7.2.3 Estimation of Loss Function Asymmetry by GMM (EKT)

Equation (7.9) above by EKT provides the basis for estimation of asymmetry and joint testing of rationality and asymmetric loss using time series data. From Equation (7.9) we could derive a GMM estimator of α_{ih} as follows:

$$\hat{\alpha}_{ih} = \frac{[\sum_t v_{ith}|e^*_{ith}|^{p-1}]'\hat{S}_h^{-1}[\sum_t v_{ith}1(e^*_{ith} < 0)|e^*_{ith}|^{p-1}}{[\sum_t v_{ith}|e^*_{ith}|^{p-1}]'\hat{S}_h^{-1}[\sum_t v_{ith}|e^*_{ith}|^{p-1}]]}, \tag{7.13}$$

where $e^*_{ith} = y_t - y_{ith}$ is the ex post forecast error, $\hat{S}_h$ is a consistent estimate of $S_{ih} = E[v_{ith}v'_{ith}(1(e^*_{ith} < 0) - \alpha_{ih})^2|e^*_{ith}|^{2p-2}$.

Standard deviation of α_{ih} is

$$(h'\hat{S}_{ih}^{-1}h)^{-1}/T^{1/2}, \tag{7.14}$$

where $h = E(v_{ith}|y_t - y_{ith}|^{p-1})$.

A joint test of forecast rationality and the flexible loss function asymmetry can be structured as:

$$J = \frac{1}{T}[(\sum_t v_{ith}(1(e^*_{ith} < 0) - \hat{\alpha}_{ih})|e^*_{ith}|^{p-1})' \tag{7.15}$$

$$\hat{S}_h^{-1}(\Sigma_t v_{ith}(1(e^*_{ith} < 0) - \hat{\alpha}_{ih})|e^*_{ith}|^{p-1})] \sim \chi^2_{d-1},$$

where d is the size of v_{ith} and $d > 1$.

Symmetric loss could also be tested by Equation (7.15) if we replace $\hat{\alpha}_{ih}$ by 0.5 and $\hat{S}_h^{-1}$ by $\hat{S}_h^{-1}|(\hat{\alpha}_{ih} = 0.5)$.

7.3 Point and Density Forecasts in Survey of Professional Forecasters

Without information on density forecasts f_{ith}, one needs to assume unbiasedness of the forecast and estimate asymmetry based on forecast error. The asymmetry cannot change across time. The model in Section 7.2 could be implemented when we have information of f_{ith}. The Survey of Professional Forecasters data provide necessary information since 1968: IV. We fit continuous distributions to the histogram density forecasts, following Engelberg et al. (2009) and Lahiri and Wang (2020).

Each forecaster in SPF is asked to provide a point forecast for the level of output and price of this year, not the growth rate. Therefore, to derive the point forecast for the annual growth rate of the variable, we need the value of the variable for the year prior to the forecast year. The value of a variable when a forecaster made his/her forecast may not be the same as it is today for two reasons. First, the value went through several revisions. Second, the base year of the variable (for real variables and price level variables) often has changed. Therefore, we need the vintage observations of such variables as they were available in real time. This is possible by utilizing "The Real-Time Data Set for Macroeconomists" provided by the Philadelphia Fed. It tracks the historical values of macroeconomic variables as they were observed in real time. For example, for real GDP of 1999 Q1, the database records every observation of '1999 Q1 real GDP' from 1999 Q2 until the most recent quarter. These observations are not all the same, due to official data revisions and change of base year. Another helpful fact is that the definition of output and inflation variables in this database is identical to that of SPF throughout our sample period.

The SPF forecasts and Real-Time Data Set for Macroeconomists together provide a longitudinal data set of matched point/density forecasts for output and inflation. They provide sufficient observations for more than four decades at quarterly intervals. We adopt real output forecasts from 1981 Q3 to 2022 Q2, and inflation forecasts from 1968 Q4 to 2022 Q2. We have 5325 observations for real output forecasts where both density forecast and corresponding point forecast are available, and 6751 for inflation. With each matched density and point forecast, we can backout the asymmetry parameter α without using sophisticated estimation method.

7.4 Nonparametric Analysis by Combining Point and Density Forecasts

Under flexible loss function asymmetry, rationality need not be rejected just because the point forecast and the density mean/median/mode are different. Point forecasts that are quite different from the central tendency of the density could be the result of a rational decision-making process with an asymmetric loss function. We now focus on recovering the asymmetry parameter values and a comparison of the point forecast to the density forecast central tendencies.

A challenge in recovering the asymmetry parameter lies in the fact that sometimes the density forecasts contain only a few probability values assigned to a set of predefined bins, forming sparse histograms. In this section, we use non-parametric analysis with regard to the histograms. Thus, we do not aim to recover the value of the asymmetry parameter α but only derive a range or the bounds of α. Here we use a Lin-Lin loss function and Equation (7.7) to derive the bounds of α. Engelberg et al. (2009) compare point forecast to the central tendencies of the densities using SPF in the same way, though they did not point out that their innovative data analysis is equivalent to finding the bounds on the asymmetry parameter.

For each forecast, we compute the upper and lower bounds of α (alpha) from the raw histograms. The upper and lower bounds of alpha are the values of the cumulative distribution function at the right and left bounds of the bin that contains the point forecast. Thus, we make no assumption on the distribution of the density within a bin.

For real output forecasts, 386 point forecasts (7.25%) have a lower bound of alpha above 0.5. 333 point forecasts (6.25%) have an upper bound of alpha below 0.5. For inflation forecasts, 371 point forecasts (5.49%) have a lower bound of alpha above 0.5. 993 point forecasts (14.71%) have an upper bound of alpha below 0.5. Details of these statistics by horizon are reported in table 7.1. We also compare the point forecast and the bounds of central tendency of the density forecast. Results are reported in table 7.2.

We find that, for real output forecasts, there is no systematic asymmetry, and the occasional asymmetry can go on either side. For inflation forecasts, significantly more forecasters view over-predictions more costly and provide lower point forecasts than their density forecasts central tendencies. They are more optimistic when providing the point forecasts than when providing the density forecasts. However, about 90% of the cases suggest bellwether symmetry. Thus, asymmetry is not a dominant feature in these forecast distributions.

7.5 Estimation of Asymmetry with Fitted Forecast Distributions for Each Forecaster

In this section, we fit continuous distributions to each raw histogram in each quarter. We fit generalized beta distributions or triangular distributions (in few unavoidable cases) to forecast histograms following Engelberg et al. (2009). For density forecasts with more than two intervals, we fit generalized beta distributions. When the forecaster attaches probabilities to only one or two bins, we assume that the subjective distribution has the shape of an isosceles triangle. There are many different ways to generalize beta distributions to generate other distributions.[6] The generalized beta distribution we choose has four parameters and its probability density function is defined as follows:

$$f(x,\alpha,\beta,l,r) = \frac{1}{B(\alpha,\beta)(r-l)^{\alpha+\beta-1}}(x-l)^{\alpha-1}(r-x)^{\beta-1}, l \le x \le r, \alpha > 0, \beta > 0, \tag{7.16}$$

where B is the beta function. We further restrict α and β to be greater than one to maintain unimodality of the fitted individual density distribution.

The two parameters α and β define the shape of the distribution, and the other two parameters l and r define the support. Histograms fitted to generalized beta are also divided into four different cases depending on whether the open bin on either end of the support has positive probabilities. Hence as few as two to as many as four parameters may appear in the optimization problem for the fitting process. If all probabilities are attached to closed bins, then for the generalized beta distribution whose density is $f(x,\alpha,\beta,l,r)$, l and r are set to be the lower bound and upper bound of the bins which have positive probabilities. In these cases, only the shape parameters α and β need to be solved in the optimization problem. If the left (right) open bin has positive probabilities, then l (r) needs to be solved in the optimization problem. The generalized beta distribution is preferred the recorded histograms for several reasons. First, when the histograms are treated as discrete distributions with the usual assumption that the probability mass within an interval is assumed to be concentrated at the mid-point of each interval. It does not reflect the expected continuity and uni-modality of the true underlying distributions. Second, compared to the normal distribution, the generalized beta distribution is more flexible to accommodate different shapes in the histograms. The histograms often display excess skewness as well as different degrees of kurtosis. Finally, generalized beta distributions are truncated at both sides, while the normal distribution is defined over an open interval $(-\infty, +\infty)$, which is not true with most of the histograms and counterintuitive to the fact that the target variables have historical bounds.

We have adopted the triangular distributions when only one bin or two bins have positive probability masses (nearly 8% of our histograms). Normal distributions will shrink to a degenerate distribution in these cases. While the use of triangular distribution yields a unique solution for each observation, we should be cognizant of limitations of the assumption. The triangular distribution may exaggerate the spread and uncertainty imbedded in the distribution. In the fitting process we restrict the triangular distribution to be isosceles and allow the support to cover the whole bin which has a probability not less than 50%. There are triangular distributions we

---

[6] See, for instance, Gordy (1998) and Alexander, Cordeiro, Ortega and Sarabia (2012).

could fit with shorter supports and smaller variances if we change the restrictions and assumptions. However, to avoid multiple solutions for densities with only one or two bins and in the absence of additional information, we choose isosceles triangular distributions.

With the fitted density functions, we could estimate the value of alpha for each forecast assuming either Lin-Lin or Quad-Quad loss function according to Equations (7.7), (7.8) and (7.12). The results for Lin-Lin loss scenario are shown below in Section 7.5.1. For Quad-Quad loss the results are shown in Section 7.5.2.

7.5.1 Asymmetry in Lin-Lin Loss Functions

Figure 7.1 shows the distribution of α_{ith} for real output forecasts after fitting the histograms with triangular or generalized beta distributions for each horizon. We can see various levels of asymmetry in the loss functions over individuals, even though symmetric loss function (α close to 0.5) is still the most frequent value. There are slightly more observations with α above 0.5 than below 0.5. It indicates slightly more optimism in point forecasts for real output growth rate than in corresponding density forecasts. Another interesting finding is, in horizon one forecasts (i.e., forecasts in the fourth quarter of each year), the distribution of α is more dispersed. In other horizons, α is more likely to be around 0.5. It is interesting to note that, densities in horizon one are more likely to have fewer bins due to lower uncertainty. For these densities limited information about the distribution is available. It is an issue for histograms with only one or two bins. When only one or two bins contain probability masses, we fit triangular distributions. However, the true distribution may concentrate in one side of the bins or have different bounds than that of the triangular distributions, causing the location of the point forecast to be far from the mean/median of the fitted triangular distribution. When there are more bins, such issues are not relevant. Meanwhile when the density is bounded in a narrow range, a slight change of the point forecast (for instance because of rounding) causes a substantial change in the cumulative probability. This results in increased dispersion of α_{ith}.

Figure 7.2 shows the distribution of α_{ith} for inflation forecasts. Loss functions in inflation forecasts are significantly more asymmetric and the asymmetry is overwhelmingly towards one direction - more alphas are below 0.5. It shows forecasters report significantly lower forecast of inflation rate than the median of the underlying density distribution. Over-prediction is treated as more costly for forecasters with α_{ith} below 0.5. They are significantly more optimistic in their point forecasts than density forecasts when forecasting inflation. These results match with that of Lahiri and Liu (2009) and Krüger and Hoss (2012) using German data also find significant asymmetric loss in inflation forecasts and symmetric loss in output forecasts.

7.5.2 Asymmetry of Quad-Quad Loss Functions

After fitting the histograms with generalized beta / triangular distributions we compute values of alpha assuming Quad-Quad loss function using Equation (7.12). We plot alphas from Quad-Quad loss function against Lin-Lin loss function in Figure 7.3. They are strongly positively correlated, except that the correlation is not linear. Under Quad-Quad loss function, due to higher losses when the forecast is farther from the realization, it is more likely the forecaster reports a point forecast close to the central tendency. Therefore, for the same reported point forecasts that is different from the density mean, Quad-Quad loss function indicates a higher degree of asymmetry than Lin-Lin loss for values of $\alpha > 0.5$ and indicates a lower degree of asymmetry for $\alpha \leq 0.5$. This is true for both growth and inflation. The scatter plot in figure 7.3 also shows that relatively more forecasters report $\alpha > 0.5$ for growth and $\alpha < 0.5$ for inflation forecasts.

7.5.3 Determinants of Loss Function Asymmetry

In the last section we reported asymmetry in forecasters' loss functions and find that the asymmetry in inflation forecasts is more prominent than in real output forecasts. Now we look for determinants of loss function asymmetry as suggested first by Granger and Pesaran (2000).

We are interested in whether forecasters' loss function asymmetry depends on the level of the macroeconomic variable being forecasted and other factors. We run pooled regressions of the asymmetry on macroeconomic variables and horizon dummies. The sample size is around 3000 for each regression. Here we limit the sample to more prolific forecasters who make at least forty valid forecasts. There are more than forty forecasters who qualify by this criterion. For explanatory variables, horizon dummies are included since behavior of forecasters seems different for forecast horizons. Macroeconomic variables are also included. For output growth forecasts, "previous quarter growth" is included. It measures the newly observed growth rate of real output of quarter $q_t - 1$ over $q_t - 5$ when the forecaster makes a forecast in quarter q_t. For inflation forecasts, a similar variable that measures the growth rate of GDP Price Deflator from $q_t - 1$ over $q_t - 5$ is included. Individual fixed effects are captured by individual forecaster dummies.

The results of the regressions for real GDP output growth are shown in Table 7.3. For real output forecasts, the coefficient of previous quarter growth is positive and significant, indicating over-prediction in good times and under-prediction in bad times.

For inflation forecasts, Table 7.4 shows that, the previous quarter's inflation rate significantly affects the asymmetry. It indicates a strong correlation between the asymmetry and inflation rate itself. Its positive sign suggests a trend of over-prediction when it is already high and under-prediction when it is already low. Therefore, it is consistent with the finding that forecasters are more optimistic in good times and more pessimistic in bad times (here 'optimistic' means lower inflation, since a key target is to keep inflation low for policy makers). None of the horizon dummies are significant.

7.6 Time Invariant Alpha, Bias, and Comparison with EKT

Elliott et al. (2005) develop moment conditions implied by forecast rationality to estimate the parameters of loss functions by GMM method, as illustrated in Section 7.2, Equations (7.13) - (7.15). With the GMM method by EKT, one can estimate the value of asymmetry for a forecaster, and the asymmetry is time invariant. The alphas from the combination method we estimate in Section 7.5 are time variant. To compare the results in Section 7.5 to that calculated from the EKT method, we first compute time invariant individual asymmetry from the alpha's we got in the combination method and then compare them to the computed values of asymmetry and test rationality by the EKT method.

7.6.1 Combination Method

In section 7.5 we calculated asymmetry parameter by the combination method for each forecast, after fitting continuous distributions to the raw histograms. Now we extend this practice to derive a time invariant measure of the asymmetry for each forecaster in each horizon. Under Lin-Lin loss function, time invariant alpha can be obtained by regressing $\frac{1}{2}[1 - E^S(\frac{e^*_{ith}}{|e^*_{ith}|})]$ on a constant for each individual, where $e^*_{ith} = y_t - y_{ith}$ is forecast error.

The estimate of asymmetry and its standard deviation in this regression is equivalent to finding the mean and standard deviation of time varying individual alphas that are fixed over time. Therefore, we calculate the first two moments of the asymmetry values from the generalized beta or triangular distributions of each frequent forecaster for each horizon. These estimates can be compared with EKT.

7.6.2 GMM Method from EKT Approach

When we use GMM method to get our estimates, we follow EKT (Equations (7.13) and (7.14)) to estimate asymmetry under flexible loss, and test rationality using Equation (7.15). As a comparison, we also test rationality assuming symmetric loss, still using Equation (7.15), with the value of $\hat{\alpha}_{ih}$ pegged to 0.5.

The null hypothesis of $\beta = 0$ is used for the rationality test. For Lin-Lin loss $p = 1$ and for Quad-Quad loss $p = 2$. When applying EKT approach to SPF data, we adopted the following specifications: 1). v is a 2 by 1 vector containing a constant and

$\hat{y}_{t-h-1}$ (most recent realization of y observed at the time of forecast). 2). Forecast errors are computed from the final values observed at 2022 Q2, not the first or the second release of these macroeconomic variables. The reason is because EKT was implemented over the whole sample.

For frequent forecasters and each forecast horizon (forecasters who provided ten or more forecasts for that horizon), we report how many times the null hypothesis is rejected in Tables 7.5 and 7.6. The results reveal several findings. First, under GMM-EKT approach, without asymmetric loss, rationality is more frequently rejected. However, with asymmetric loss, only a few cases are rejected. This is true regardless of target variable or horizon or the form of loss function. This is convincing evidence to support that conventional tests for rationality without considering asymmetric loss can be misleading, and methods in this section may explain the phenomena better. Second, under GMM-EKT approach, when the null includes symmetric loss, rationality is more frequently rejected in inflation forecasts than in output forecasts. Third, GMM-EKT method produces more asymmetric loss cases than that of the combination method. It is common that the two methods – GMM and the combination method produce different results. EKT estimates of alphas are quite different. When the symmetric loss is rejected under GMM may not be rejected by the "combining density and point forecasts" method and vice versa. There are only a few cases where under both methods symmetric loss is simultaneously rejected. However, there are also a few cases where they are both rejected, while the direction of asymmetry is opposite under the two methods, such as forecaster #535 in horizon three output forecast.[7] Despite this there are many cases the two methods support each other, revealing same direction of asymmetry. Furthermore, under the combination method, most alphas are below 0.5 for inflation forecast. This reconfirms the stylized facts in Section 7.5 that for inflation forecast, the asymmetry towards favoring under-prediction is significant. The GMM-EKT method does not reveal the same information. Lastly, it is quite noteworthy that the standard deviation of alpha under GMM is much smaller than that under "combining density and point forecasts" method. Since in the latter case we do observe time varying alphas and see that for the same forecaster, it varies, it is plausible to assume that the GMM method underestimates the variance of asymmetry parameter by ignoring the cross-sectional variation in the asymmetry parameter.

7.7 Concluding Remarks

In this chapter, we consider how to use information in density forecasts to conduct rationality test under asymmetric loss function and to estimate the loss function parameters based on first order condition of forecast optimality. We estimate the asymmetry of SPF forecasters' loss functions by combining point forecasts and density forecasts. A triangular or generalized beta distributions are fitted to the raw

[7] To maintain the anonymity of the individual forecasters, yet keeping the panel structure of the data set, each forecaster is coded with an identification number. The forecaster #535 stayed on the survey during 2005 Q2 - 2023 Q2.

histograms. We find that forecasters treat underestimation of real output more dearly than over prediction, and the reverse is true for inflation. After computing these measures carefully based on the above distributions, we find that forecasters are more optimistic in their point forecasts than in their density forecasts for the same target variable. This is consistent with other literature. Forecasters tend to predict a higher point forecast of real output than its density central tendency measures and predict a lower point forecast of inflation than its density central tendency. This optimism is more prominent in inflation forecasts. We also find that forecasters are more optimistic in good times and more pessimistic in bad times. Since good quarters are far more frequent than bad quarters in the sample, forecasters show overall optimism.

By restricting asymmetry to be time invariant, we could estimate individual asymmetry from two alternative methods - the combination method and the GMM-EKT approach. For Lin-Lin loss, the two approaches produce quite different results. The GMM estimation reconfirms that allowing for asymmetry in loss function will significantly reduce the number of rejections for rationality test. The estimates of alpha obtained from the two methods often differ. This difference can also be attributed to measurement errors in the histograms, which are after all subjective. The GMM method produces a much smaller variance for alpha than the combination method. We also compute asymmetry parameter for Quad-Quad loss function and find that Quad-Quad loss yields higher degrees of asymmetry than under Lin-Lin loss. Time variation in the asymmetry is partly determined by the level of the target variable and the forecast horizon.

Acknowledgements This chapter was presented at the 2023 IAAE Annual Conference in Oslo, Norway.

Appendix: Tables and Figures

Table 7.1: Nonparametric analysis: counts of asymmetry parameter above or below 0.5

Output, post 1981Q2			
	upper bound<0.5	bound contains 0.5	lower bound>0.5
h=1	62	1230	59
h=2	71	1135	93
h=3	96	1177	105
h=4	104	1064	129
total	333	4606	386
inflation			
	upper bound<0.5	bound contains 0.5	lower bound>0.5
h=1	149	1190	50
h=2	216	1311	127
h=3	308	1482	89
h=4	320	1405	105
total	993	5388	371

Fig. 7.1: Distribution of α_{ith} for real output forecasts (1981 Q3-2022 Q2) when fitting with generalized beta and triangular distributions under Lin-Lin loss

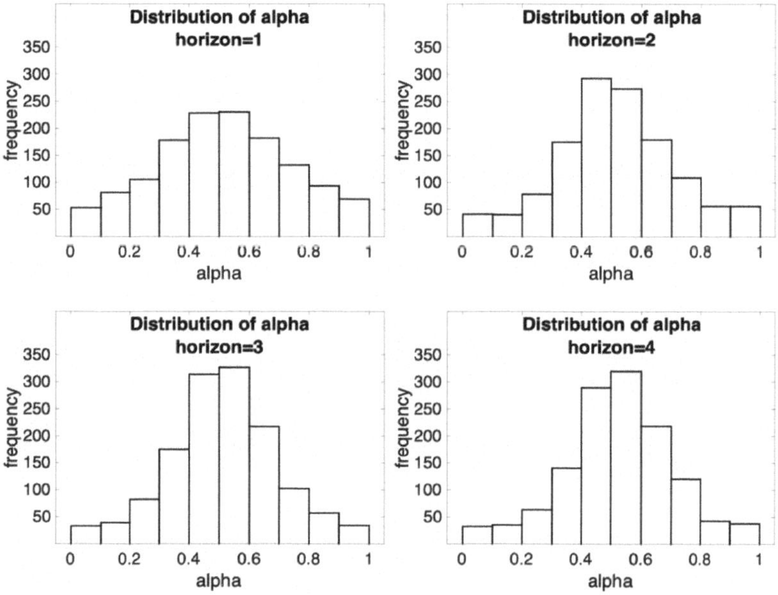

Fig. 7.2: Distribution of α_{ith} inflation forecasts (1969 Q1-2022 Q2) (1981 Q3-2022 Q2) when fitting with generalized beta and triangular distributions under Lin-Lin loss

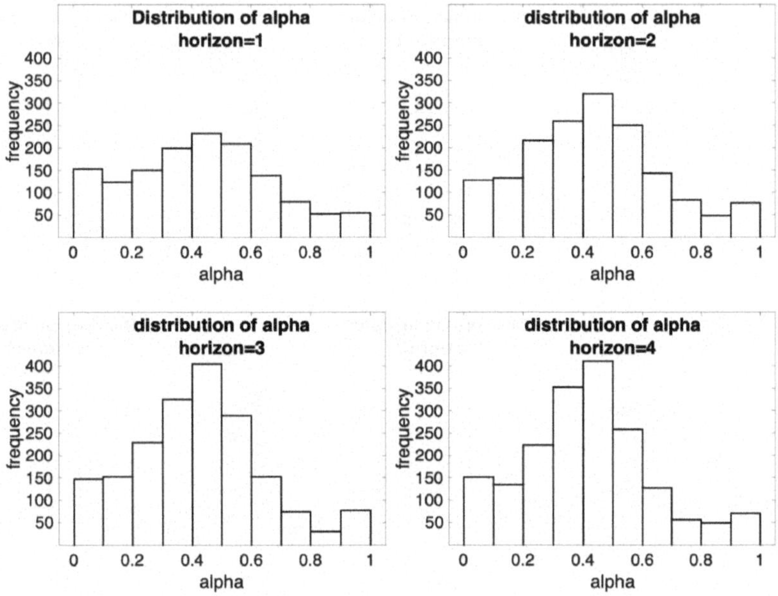

Table 7.2: Comparison of point forecasts and bounds of density mean/median/mode (lower bounds 'LB', upper bounds 'UB', and y^* is the point forecast)

Mean /Output	y* <LB	bounded	y* >UB	median /output	y* <LB	bounded	y* >UB	mode /output	y* <LB	bounded	y* >UB
h=1	38	1266	47	h=1	91	1184	76	h=1	63	1234	54
h=2	46	1157	96	h=2	112	1063	124	h=2	80	1142	77
h=3	57	1214	107	h=3	136	1094	148	h=3	101	1206	71
h=4	55	1114	128	h=4	145	950	202	h=4	106	1106	85
Mean /Inflation	y* <LB	bounded	y* >UB	median /inflation	y* <LB	bounded	y* >UB	mode /inflation	y* <LB	bounded	y* >UB
h=1	122	1222	45	h=1	192	1112	85	h=1	144	1194	51
h=2	199	1334	121	h=2	290	1177	187	h=2	208	1310	136
h=3	284	1508	87	h=3	405	1336	138	h=3	278	1514	87
h=4	295	1445	89	h=4	438	1222	169	h=4	285	1436	108

Table 7.3: Determinants of asymmetry – fixed effect regression, real output forecasts.

Variable	Estimate	Standard Deviation	p-value
(Intercept)	0.6721	(0.0695)	0.0000
Horizon 2 dummy	0.0069	(0.0094)	0.4589
Horizon 3 dummy	0.0001	(0.0092)	0.9931
Horizon 4 dummy	0.0184	(0.0093)	0.0473
Previous quarter growth	0.0030	(0.0026)	0.2581
R^2 0.15. $AdjR^2$: 0.12			
Dependent variable: α_{ith} from Real Output Forecasts			
Independent variables also include individual and year dummies			

Table 7.4: Determinants of asymmetry – fixed effect regression, inflation forecasts.

Variable	Estimate	Standard Deviation	t-statistic	p-value
(Intercept)	0.4535	0.0704	6.4383	0.0000
Horizon 2 dummy	-0.0071	0.0102	-0.6925	0.4887
Horizon 3 dummy	-0.0174	0.0100	-1.7402	0.0819
Horizon 4 dummy	-0.0149	0.0101	-1.4809	0.1387
Previous quarter inflation	0.0479	0.0098	4.8770	0.0000
R^2 0.18. $AdjR^2$ 0.15				
Dependent variable: α_{ith} for Inflation Forecasts				
Independent variables also include individual and year dummies				

Table 7.5: Statistics for Individual time invariant asymmetry – Real output

		GMM-EKT			Combine density and point forecast	
		# (%) of forecasters				
Output /Lin-Lin	# of forecasters	asymmetric loss	rationality rejected	rationality rejected if symmetric loss allowed	asymmetric loss	rationality rejected
h=1	53	27 (51%)	0	11 (21%)	1 (2%)	1 (2%)
h=2	47	26 (55%)	1 (2%)	14 (30%)	10 (21%)	3 (6%)
h=3	49	24 (49%)	0	12 (24%)	5 (10%)	3 (6%)
h=4	48	33 (69%)	3 (6%)	20 (41%)	10 (21%)	2 (4%)
Output /Quad-Quad						
h=1	53	37 (70%)	2 (4%)	10 (19%)	1 (2%)	0
h=2	47	29 (62%)	0	11 (23%)	9 (19%)	14 (30%)
h=3	49	32 (65%)	0	15 (31%)	5 (10%)	6 (12%)
h=4	48	34 (71%)	1 (2%)	20 (42%)	13 (27%)	3 (6%)

Table 7.6: Statistics for Individual time invariant asymmetry - Inflation

		GMM-EKT			Combine density and point forecast	
		# (%) of forecasters				
Inflation /Lin-Lin	# of forecasters	asymmetric loss	rationality rejected	rationality rejected if symmetric loss allowed	asymmetric loss	rationality rejected
h=1	51	29 (57%)	3 (6%)	15 (29%)	11 (22%)	4 (8%)
h=2	55	31 (56%)	3 (5%)	14 (25%)	18 (33%)	2 (4%)
h=3	65	37 (57%)	2 (3%)	20 (31%)	26 (40%)	2 (3%)
h=4	70	39 (56%)	2 (3%)	23 (33%)	32 (46%)	5 (7%)
Inflation /Quad-Quad						
h=1	51	27 (53%)	1 (2%)	12 (24%)	12 (24%)	2 (4%)
h=2	55	30 (55%)	3 (5%)	11 (20%)	18 (33%)	9 (16%)
h=3	65	40 (62%)	3 (5%)	20 (31%)	29 (45%)	4 (6%)
h=4	70	46 (66%)	0	20 (29%)	34 (49%)	6 (9%)

Fig. 7.3: Comparison of Lin-Lin and Quad-Quad loss function asymmetry

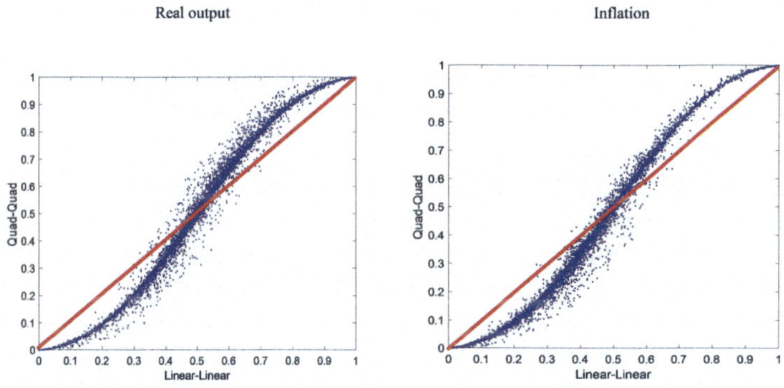

References

Alexander, C., Cordeiro, G. M., Ortega, E. M. & Sarabia, J. M. (2012). Generalized beta-generated distributions. *Computational Statistics & Data Analysis*, *56*(6), 1880–1897.

Batchelor, R. & Peel, D. A. (1998). Rationality testing under asymmetric loss. *Economics Letters*, *61*(1), 49–54.

Boero, G., Smith, J. & Wallis, K. F. (2008). Uncertainty and disagreement in economic prediction: the bank of england survey of external forecasters. *The Economic Journal*, *118*(530), 1107–1127.

Christoffersen, P. F. & Diebold, F. X. (1996). Further results on forecasting and model selection under asymmetric loss. *Journal of Applied Econometrics*, *11*(5), 561–571.

Clatworthy, M. A., Peel, D. A. & Pope, P. F. (2012). Are analysts' loss functions asymmetric? *Journal of Forecasting*, *31*(8), 736–756.

Diebold, F. X., Gunther, T. A. & Tay, A. S. (1998). Evaluating density forecasts with applications to financial risk management. *International Economic Review*, 863–883.

Döpke, J., Fritsche, U. & Siliverstovs, B. (2010). Evaluating german business cycle forecasts under an asymmetric loss function. *OECD Journal: Journal of Business Cycle Measurement and Analysis*, *2010*(1), 1–18.

Elliott, G., Komunjer, I. & Timmermann, A. (2008). Biases in macroeconomic forecasts: irrationality or asymmetric loss? *Journal of the European Economic Association*, *6*(1), 122–157.

Elliott, G., Timmermann, A. & Komunjer, I. (2005). Estimation and testing of forecast rationality under flexible loss. *The Review of Economic Studies*, *72*(4), 1107–1125.

Engelberg, J., Manski, C. F. & Williams, J. (2009). Comparing the point predictions and subjective probability distributions of professional forecasters. *Journal of Business & Economic Statistics*, *27*(1), 30–41.

Fritsche, U., Pierdzioch, C., Rülke, J.-C. & Stadtmann, G. (2015). Forecasting the brazilian real and the mexican peso: Asymmetric loss, forecast rationality, and forecaster herding. *International Journal of Forecasting*, *31*(1), 130–139.

Gordy, M. B. (1998). *A generalization of generalized beta distributions* (Vols. Finance and Economics Discussion Series, Board of Governors of the Federal Reserve System (U.S.)) (No. 1998-18). Retrieved from https://www.federalreserve.gov/Pubs/Feds/1998/199818/199818pap.pdf

Granger, C. (1969). Prediction with a generalized cost of error function. *Journal of the Operational Research Society*, *20*(2), 199–207.

Granger, C. & Pesaran, M. H. (2000). A decision theoretic approach to forecast evaluation. In *Statistics and finance: An interface* (pp. 261–278). World Scientific.

Krüger, J. J. & Hoss, J. (2012). German business cycle forecasts, asymmetric loss and financial variables. *Economics Letters*, *114*(3), 284–287.

Krüger, J. J. & LeCrone, J. (2019). Forecast evaluation under asymmetric loss: A monte carlo analysis of the ekt method. *Oxford Bulletin of Economics and Statistics*, *81*(2), 437–455.

Lahiri, K. & Liu, F. (2009). On the use of density forecasts to identify asymmetry in forecasters' loss function. *Business and Economic Statistics Section-JSM*, 2396–2408.

Lahiri, K. & Wang, W. (2020). Estimating macroeconomic uncertainty and discord using info-metrics. In M. Chen, J. M. Dunn, A. Golan & A. Ullah (Eds.), *Advances in info-metrics: Information and information processing across disciplines* (p. 290-324). Oxford: Oxford University Press.

Nerlove, M. (1983). Expectations, plans, and realizations in theory and practice. *Econometrica: Journal of the Econometric Society*, 1251–1279.

Patton, A. J. & Timmermann, A. (2007). Properties of optimal forecasts under asymmetric loss and nonlinearity. *Journal of Econometrics*, *140*(2), 884–918.

Tsuchiya, Y. (2016). Assessing macroeconomic forecasts for japan under an asymmetric loss function. *International Journal of Forecasting*, *32*(2), 233–242.

Wang, Y. & Lee, T.-H. (2014). Asymmetric loss in the greenbook and the survey of professional forecasters. *International Journal of Forecasting*, *30*(2), 235–245.

Weber, E. U. (1994). From subjective probabilities to decision weights: The effect of asymmetric loss functions on the evaluation of uncertain outcomes and events. *Psychological Bulletin*, *115*(2), 228.

Zellner, A. (1986). Biased predictors, rationality and the evaluation of forecasts. *Economics Letters*, *21*(1), 45–48.

Chapter 8
Estimating Dynamic Probit Models with Higher-order Time- and Network-lag Structure and Correlated Random Effects

Peter H. Egger and Michaela Kesina

Abstract Many strategic choices in the social sciences involve sluggish adjustment with an ex-ante unknown lag structure as well as patterns of interdependency among the cross-sectional units, which call for a flexible parameterization based on multiple networks. This chapter proposes straightforward panel-probit estimation approaches based on control functions for such problems. The paper outlines the estimation approaches and illustrates their suitability by simulation examples.

8.1 Introduction

The importance of sluggish adjustment and inertia in the responses of economic outcomes to shocks is well acknowledged and received tremendous attention in the social sciences at large and economics in particular. The possibility of pooling cross-section and time-series data provides a particularly rich data environment which enables the identification of key structural parameters of dynamic behavior of economic agents (see the seminal articles by Balestra & Nerlove, 1966, Nerlove, 1971a, Nerlove, 1971b, and Nerlove, 1972; and see the overviews in Arellano, 2003, Nerlove, 2005, Baltagi, 2015, 2021, Mátyás & Sevestre, 2008, 2015, Hsiao, 2022).

Social scientists often encounter situations where agents make discrete choices depending on past outcomes (sluggish adjustment) and on the ones of other agents (network interdependence).

For instance, companies may decide upon contracting input suppliers depending on their past experience with them. One reason for inertia in this choice might be that search costs or information costs incentivize them to stay with earlier suppliers. Similarly, companies may consider how their competitors, their suppliers, or their

Peter H. Egger ✉
ETH Zurich, Zurich, Switzerland, e-mail: pegger@ethz.ch

Michaela Kesina
University of Groningen, Groningen, The Netherlands, e-mail: m.kesina@rug.nl

B. H. Baltagi and L. Mátyás (eds.), *Seven Decades of Econometrics and Beyond*, Advanced Studies in Theoretical and Applied Econometrics 57,
https://doi.org/10.1007/978-3-031-92699-0_8

customers behave in that regard. The latter creates network interdependencies. The dynamic (time-lag) pattern and the strength of links to other agents in the game (the order of the network pattern) may not be fully known ex-ante. Then, the researcher might wish to parameterize it by considering higher-order time-lag and network-lag structures to estimate the response function. Higher-order network-lag structures of the envisaged kind are also referred to as ones with multiplex networks. Those are ones where the nodes, such as firms, sectors, countries, etc., are defined to be the same across different networks, but the links between them are generated by different network concepts or different brackets of closeness or distance.

That discrete-choice problems pose specific challenges with dynamic data-generating processes is well understood for decades (see Heckman, 1981). Wooldridge (2005) proposed an elegant solution to dynamic choice problems, where contemporaneous latent outcome is a function of lagged observed binary outcome. He focused on choice problems with a first-order own time-lag structure using an ordinary panel-probit model combined with a control-function approach. However, Wooldridge (2005) did not consider network interdependencies in the agents' choices.

Egger and Kesina (2023) introduce contemporaneous and lagged network interdependencies among agents in a first-order-lag dynamic choice setting. Contemporaneous interdependencies mean that agents play games, reflecting on other agents' payoffs. In such a setting, they cannot condition on the actual choices of other agents, as those will only be simultaneously revealed with theirs'. Such a problem is eventually complex to estimate, and Egger and Kesina (2023) resort to Bayesian Markov Chain Monte Carlo sampling. In any case, their approach focuses on first-order time- and network-lag structures. Hence, agents' contemporaneous choices depend on one type of (ex-ante unobserved) contemporaneous, network-weighted latent outcome. There is one network that is involved in generating that outcome. Moreover, their contemporaneous choices depend on one time lag of own (or network-weighted) outcome. Apart from the relatively computation-intensive routine to estimate such a problem, the considered design is restrictive because it permits only one network and one time lag.

In many problems, the time-lag structure may be richer, and several different networks may connect agents simultaneously. Examples of the latter are rings of neighbors or geographical, cultural, economic, and other networks whose relative importance is not known ex-ante. Then, one might wish to allow for this richer time lag and network context in estimation.

We propose straightforward estimation routines that rely on maximum-likelihood-based panel-probit estimation with correlated random effects. We do so while permitting a rich higher-order time-lag structure and a higher-order network-lag structure (meaning that several networks may be present to link the economic agents). This is done by enriching the control-function approach to account for higher-order initial conditions in the time lag and the network structure.

Specifically, we consider a framework with so-called correlated random effects. Hence, the explanatory variables may all or partly be correlated with the time-invariant residual component. Due to the latter, eventually, none of the model parameters of interest can be estimated consistently and, in finite samples, without bias when not addressed in estimation. Hence, a panel-probit model assuming uncorrelated random effects would be biased and inconsistent.

We demonstrate by way of Monte Carlo simulations that the proposed control-function approach can be used to estimate parameters without significant bias, even in small to medium-sized samples with 250 and 500 cross-sectional units and a time horizon of 5 and 10 periods.

The remainder of the paper is organized as follows. Section 8.2 introduces the process of interest, featuring own-time-lag and network-lag structures, both of a higher order. This is done in a setting where regressors are correlated with the individual time-invariant error component. We propose a control-function approach as a remedy. Section 8.3 describes stylized examples of network interdependence. Section 8.4 outlines a Monte Carlo simulation setting and associated simulation results. We discuss impact estimation in Section 8.5, and we offer a brief conclusion in the last section.

8.2 Variants of a Dynamic Panel-probit Model with Higher-order Lagged Own and Network Effects as well as Correlated Random Effects

We will propose approaches to estimate models that have at least two or all of the three subsequent features, namely the presence of:

- higher order time lags the binary dependent variable,
- higher order time lags of the network lags of the binary dependent variable,
- correlated random effects, whereby the time averages of the explanatory variables are correlated with the time-invariant random effects,

all in the latent process generating a cross-sectional unit's contemporary binary outcome.

Solutions for estimating dynamic binary-choice models where the observables include past realized binary outcomes have been proposed by Wooldridge (2005), Chib and Jeliazkov (2006), Arulampalam and Stewart (2009), Rabe-Hesketh and Skrondal (2013), Arbia, Bille and Leorato (2023), and others. However, that work did not consider the simultaneous presence of network lags of binary outcome in the process.

8.2.1 Notation

We will use the convention to index cross-sectional units by $i = 1, \dots, N$ and time periods by $t = 0, \dots, T$. We will use $n = NT$ to denote the number of observations.

Moreover, we will use $\{y_{i,t}, y^*_{i,t}\}$ to denote the binary and the latent outcome of the choice process with

$$y_{i,t} = 1(y^*_{i,t} > 0). \tag{8.1}$$

The $1 \times K_B$ vector $x_{i,t}$ collects the explanatory variables.

We will furthermore introduce network weights, which parameterize the strength of ties between cross-sectional units i and j. To this end, consider the following two notions of network weights.

- **Various rings of neighbors as multiple network concepts:** Let us consider an example of a lattice, where units $\{i, j\}$ are located and separated by borders. Let us focus on unit i and suppose that $\{o, a\}$ index the coordinates of units on a lattice. Then, with $\{o_i, a_i\}$ being the cell address of unit i on the lattice, all units with cell addresses in the set $\{\{o_i+1, a_i\}, \{o_i, a_i+1\}, \{o_i+1, a_i+1\}\}$ and $\{\{o_i-1, a_i\}, \{o_i, a_i-1\}, \{o_i-1, a_i-1\}\}$ could be called first-order (or direct) neighbors. By the same token, all direct neighbors of those first-order neighbors that do not belong in the first-order-neighbor set and exclude unit i may be called second-order neighbors of i, etc. Figure 8.1 portrays rings of neighbors of the corner unit $i = 1$ on a 5×5 lattice, where darker-gray color indicates higher-order (further-away) rings of neighbors relative to unit 1. We will use $m = 1, ..., M$ to index such rings of neighbors, and, for convenience, we could then think of $m = 1$ as the first ring containing the direct (adjacent) neighbors, and $m = 2, ..., M$ indexes the sets of outer rings of neighbors. In practice, such rings could be generated by literally considering land borders (e.g., with spatial units such as housing blocks, municipalities and cities, prefectures, etc.). Or they could be generated based on some continuous distance metric (e.g., geographical distance, some Mahalanobis distance about continuous variables, etc.) together with some threshold values. E.g., one could dub all firms within a range of 15 kilometers as first-order neighbors, the ones in 15-30 kilometers as second-order neighbors, etc. One could choose the mentioned distance threshold values in terms of fixed distances or make sure that the number of neighbors within a concentric ring is the same across the rings, etc.
- **Various network channels as multiple network concepts:** In contrast to rings of neighbors, one could instead consider a situation with alternative network concepts linking the cross-sectional units. E.g., with units i and j being companies, geographical distance, input-output distance (distance in terms of forward or backward links), or spatial- as well as product-market overlap could be generating distance concepts. Then, each one of those channels could serve to generate respective weights (which decline with the respective distance: in terms of kilometers, in terms of inverse spatial-market overlap, in terms of inverse product-market overlap, etc.).

What will be important in what follows is that we use w to denote pairwise network weights, an index m to denote the network ring or concept, and indices $\{ij\}$ to denote the units the weight pertains to. Accordingly, we will use $w_{m,ij}$ to denote the mth network weight pertaining to the pair of units $\{ij\}$. We will use so-called normalization by degree (also called row-normalization) for those weights by which they will have the properties: (i) $w_{m,ij} = [0, 1]$ and $\sum_{j=1}^{N} w_{m,ij} = 1$.

Finally, we will generally use Greek letters to denote unknown parameters and error components. Specifically, we will use $\nu_{i,t}$ to denote an idiosyncratic normal disturbance term, χ_i to denote an unobservable individual effect which may be correlated with one or all columns of $x_{i,t}$, and $\delta = (\alpha', \lambda', \beta')'$ for the $K \times 1$ parameter vector of interest. Notably, α will be a $K_A \times 1$ vector of parameters on own binary lags, λ is a $K_L \times 1$ vector of parameters on network (as well as time) lags of other units' binary outcomes, and β is a $K_B \times 1$ vector of parameters on $x_{i,t}$.

Using $\ell = 1, ..., L$ to index lags, we can define the ℓth own time lag of binary outcome as $y_{i,t-\ell}$. And we can collect all considered time lags into the $1 \times L$ vector $\underline{y}_{i,t} = (y_{i,t-1}, ..., y_{i,t-L})$. Moreover, we denote the scalar-valued, mth-network-weighted, lagged binary outcome for i at t as $\bar{y}_{m,i,t-1} = \sum_{j=1}^{N} w_{m,ij} y_{m,j,t-1}$. We collect all M types of the latter into the $1 \times M$ vector $\bar{y}_{i,t-1} = (\bar{y}_{i,1,t-1}, ..., \bar{y}_{i,M,t-1})$. Considering L lags with the latter as with $y_{i,t-\ell}$, we can define $\underline{\bar{y}}_{i,t-1} = (\bar{y}_{i,t-1}, ... \bar{y}_{i,t-L})$, noting that $\underline{\bar{y}}_{i,t-1}$ is a $1 \times ML$ vector.

8.2.2 Model Outline

With the above notation at hand, we will consider versions of the stochastic latent process of the form

$$y^*_{i,t} = \zeta + \underline{y}_{i,t-1}\alpha + \underline{\bar{y}}_{i,t-1}\lambda + x_{i,t}\beta + \chi_i + \nu_{i,t}. \tag{8.2}$$

The following considerations are important in the present context. In particular, the parameters of the model in (8.2) cannot be estimated by standard binary-choice cross-sectional or panel-data models.

First, the component structure of the residuals with

$$u_{i,t} = \chi_i + \nu_{i,t} \tag{8.3}$$

including a time-invariant effect χ_i means that cross-sectional models (probit or logit) will not be applicable.

Second, the presence of the time-invariant stochastic term χ_i entails that both the contemporaneous $y_{i,t}$ and $y^*_{i,t}$ as well as all elements of the vector of lagged binary terms $\underline{y}_{i,t-1}$ are functions of χ_i.

Third, in case the random effects χ_i are correlated with some or all of the regressors in $x_{i,t}$, $E(x_{i,t}\chi_i) \neq 0$, not even the parameters β can be estimated consistently.

8.2.3 Control-function (CF) Approach

It turns out that in models of the type in (8.2), consistency of the parameters can be achieved when conditioning on a control function. Two considerations are relevant here.

First, the correlation of the (time-invariant) random effects χ_i with the regressors in $x_{i,t}$ can be addressed by conditioning on the time-invariant component in $x_{i,t}$. This can be done by relying on the so-called Mundlak-Chamberlain-Wooldridge device. Mundlak (1978) introduced the concept in the context of linear panel-data models and demonstrated that when conditioning on the time-average regressors in $x_{i,t}$, say, $\breve{x}_i$, any difference in the parameters between a linear random-effects and a linear fixed-effects model is eliminated.[1] Chamberlain (1984) introduced the concept to the estimation of nonlinear models, including binary-choice models. Specifically, Chamberlain (1984) proposed using a generalized version of the time-invariant terms $\breve{x}_i$, expanding the averaged vector term into a longer vector which includes each year of the data as a separate time-invariant vector. Amemiya and MaCurdy (1986) and Breusch, Mizon and Schmidt (1989) followed Chamberlain in doing so for linear models. The latter approach has been used and popularized by Wooldridge (1995, 2005).

Second, the presence of lagged binary regressors in (8.2) additionally requires conditioning on initial conditions. Wooldridge (2005) proposed a first-order lag model of the form

$$y^*_{i,t} = \zeta + \underline{y}_{i,t-1}\alpha + x_{i,t}\beta + \chi_i + \nu_{i,t}, \tag{8.4}$$

where $\underline{y}_{i,t-1} = y_{i,t-1}$ was a scalar. He suggested including the pre-observation-period binary outcome $y_{i,0}$ into the control function to remove the bias in α. In our case, $\underline{y}_{i,t-1}$ is a vector, and we suggest using the pre-period lag scalar $y_{i,0}$ for it in the control function, as in Wooldridge (2005). For convenience, we will use for the scalar- or vector-valued term of own initial binary lags the term $\underline{y}_{i,0}$ in what follows.

Moreover, the proposed model includes the network-weighted terms $\underline{\bar{y}}_{i,t-1}$ as regressors. Those terms are a source of endogeneity in case χ_i exhibits some network structure. Then, $E(\chi_i\chi_j) \neq 0$, resulting in $E(\breve{x}_i\chi_j) \neq 0$ as well as $E(\underline{\bar{y}}_{i,t-1}\chi_j) \neq 0$. To address the latter form of dynamic network interdependence, we propose including the network-weighted regressor averages $\bar{\breve{x}}_i = \sum_{j=1}^{N} w_{m,ij}\breve{x}_j$ as well as the pre-observation-period term $\bar{y}_{i,0}$ in the control function. For the same convenience as above, we will use for the scalar- or vector-valued term of network-weighted initial binary lags of outcome the notation $\underline{\bar{y}}_{i,0}$ in what follows.

With these arguments at hand, we suggest defining a vector

$$g_i = \left(\breve{x}_i, \underline{y}_{i,0}, \bar{\breve{x}}_i, \underline{\bar{y}}_{i,0}\right). \tag{8.5}$$

[1] As the root of this difference could only be an endogeneity of the regressors in $x_{i,t}$ with respect to χ_i.

The control-function-augmented binary choice model we propose then reads

$$y^{*}_{i,t} = \zeta + \underline{y}_{i,t-1}\alpha + \bar{\underline{y}}_{i,t-1}\lambda + x_{i,t}\beta + g_i\gamma + \vartheta_i + \nu_{i,t}. \tag{8.6}$$

In Section 8.4, we will demonstrate the applicability of this approach in small to medium-sized samples by way of Monte Carlo simulations.

8.3 Stylized Examples of Network Interdependence

It is useful to briefly reflect on higher-order network interdependence in terms of specific stylized examples.

Fig. 8.1: Matrix capturing three rings of neighbors in terms of adjacency on a lattice for unit 1

	Units j					
Units i	1	2	3	4	5	...
1	X	1	2	3		...
2	1	1	2	3		...
3	2	2	2	3		...
4	3	3	3	3		...
5						...
⋮	⋮	⋮	⋮	⋮	⋮	⋱

Figure 8.1 illustrates the notion of rings of neighbors, here on a lattice and from the perspective of cross-sectional unit 1. Unit 1's cell is marked by X. Adjacent to it are first-order neighbors, which are depicted in light-gray color. Outside the ring of first-order neighbors are the second-order neighbors of unit 1 in normal gray color, followed by the third-order neighbors in dark gray color.

In that figure, we assumed that only three rings of neighbors exist. For that reason, units with numbers 5 and higher are not neighbors of unit 1.

When considering a lattice with only nine units (e.g., firms), the first-, second-, and third-order neighborhood matrices corresponding to rings of neighbors, as depicted above, would be banded matrices.

What is specific about these matrices is that (i) every unit has either one or two neighbors (depending on its address) of every type, and (ii) the rings of neighbors are mutually exclusive (whoever is a first-degree neighbor cannot be a second-degree neighbor). While in some contexts, this might make sense, it is too restrictive to think of networks of such kind only.

Fig. 8.2: Matrix capturing first-order neighborliness in terms of adjacency on a lattice with nine units

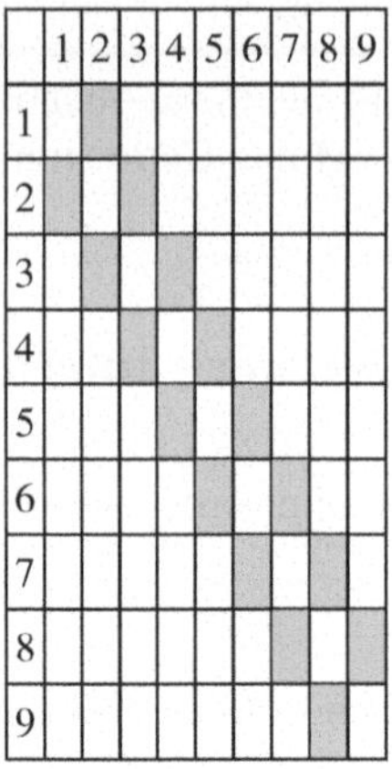

Fig. 8.3: Matrix capturing second-order neighborliness in terms of adjacency on a lattice with nine units

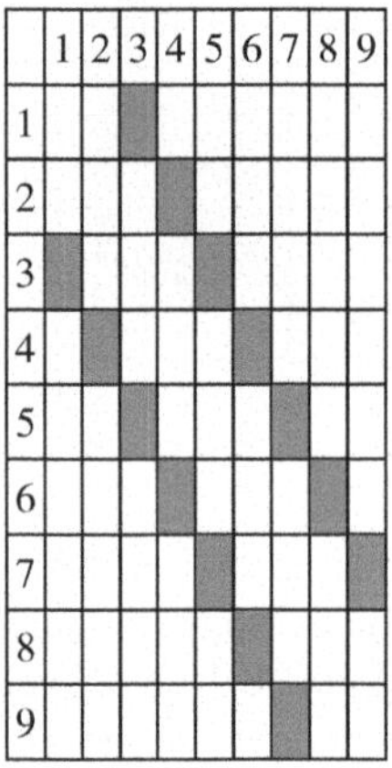

E.g., in input-output networks, one might consider the up-to-ten most important suppliers of a company and its up-to-ten most important customers. Clearly, such relationships would not be ordered as symmetrically as in the above matrices. Moreover, the number of neighbors would be up to 10 for the input network and up to 10 for the output network. However, a company could simultaneously be a key supplier and a key customer to another company. Then, the input-network matrix could have a neighbor entry in the same cell as the output matrix.

The matrices in Figures 8.5 and 8.6 represent examples of a multiplex network with two neighborhood concepts supporting a second-order network process. These

Fig. 8.4: Matrix capturing third-order neighborliness in terms of adjacency on a lattice with nine units

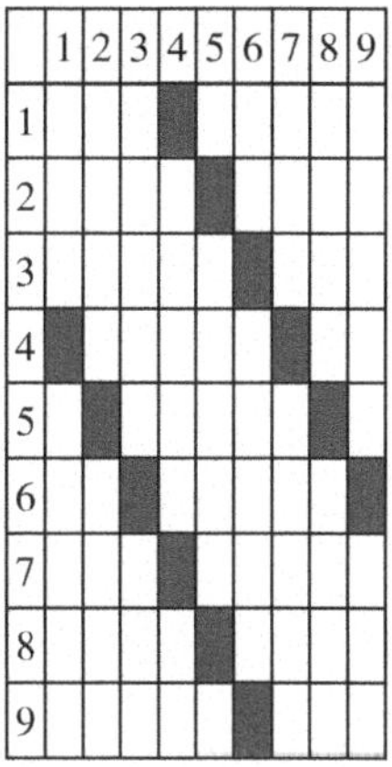

Fig. 8.5: Matrix capturing asymmetric first-order neighborliness in terms of adjacency on a lattice with nine units

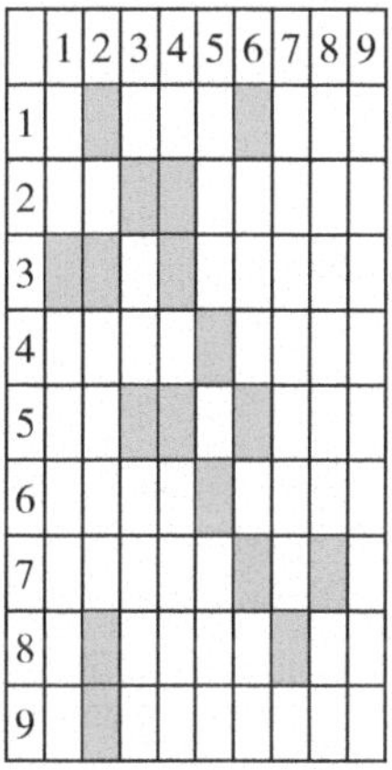

matrices differ from the earlier ones in two to three important regards. First, the number of m-order neighbors is not the same across the units. Second, already the unnormalized matrices are asymmetric. And, third, the two neighborhood concepts are not mutually exclusive. E.g., unit 4 is a first-order as well as a second-order neighbor of unit 2 in these two matrices. Conversely, unit 2 is neither a first-order nor a second-order neighbor of unit 4. As said, an example of such a situation could emerge from first-degree importance in input and in output matrices across sectors or firms.

Fig. 8.6: Matrix capturing asymmetric second-order neighborliness in terms of adjacency on a lattice with nine units

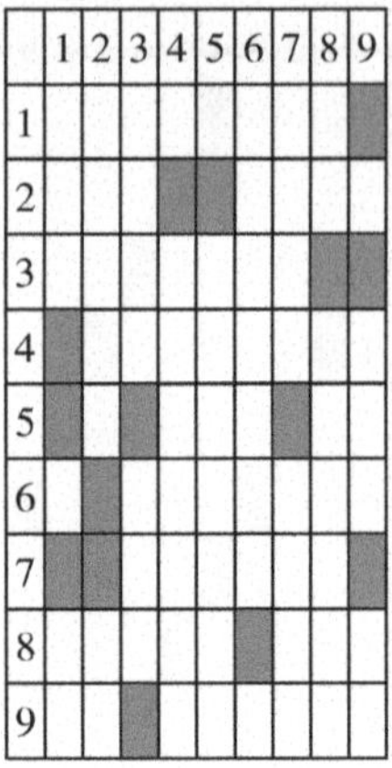

In the simulations, we will consider cases of up to two network matrices with mutually exclusive entries (i.e., they are not overlapping) but are much less structured than the ones above.

8.4 Monte Carlo Simulations

8.4.1 Dimensionality

In this section, we present results from Monte Carlo simulations, with $N = \{250; 500\}$ cross-sectional units and $T = \{5; 10\}$ time periods. Most of the results will be based on $T = 10$.

Consider the following specific remarks regarding the data-generating process.

8.4.2 Concrete Sources of Network Interdependence in the Simulations

It may be illustrative to think of the network processes that emerge from the interaction of companies. Think of firm i as being embedded in a supplier-buyer network. Every company has some input providers and output customers within a neighborhood. We consider three neighborhoods regarding the number of companies they host $\{10; 15; 25\}$. With $N = 250$ companies, we assume that there are 5 regions with three neighborhoods of each type, and with $N = 500$ companies, there are 10 regions with three neighborhoods of each type.

Firms are characterized by three aspects: (i) incorporation (binary; $z_{1,i}$); employment-size class (integer-values $[1;10]$; $z_{2,i}$, ten-pronged); and (iii) standard-normal log productivity ($z_{3,i}$). All of these aspects are assumed to be time-invariant.

We build a non-negative distance metric for all pairs using the form

$$d_{i,j} = |0.4 \cdot (z_{1,i} - z_{1,j})/\sqrt{0.25} + 0.4 \cdot (z_{2,i} - z_{2,j})/\sqrt{99/12} + 0.2 \cdot (z_{3,i} - z_{3,j})|,$$

where the square-root scalars are normalizing factors.

We then assume that companies with a distance of $d_{i,j} \leq 0.3$ exhibit a close input-output (supplier-buyer) relationship and ones in the distance interval $d_{i,j} \in (0.3;0.8]$ exhibit a medium-close one.

We will define unnormalized input-output-closeness links $w^0_{1,i,j} = 1(d_{i,j} \leq 0.3)$ and $w^0_{2,i,j} = 1(d_{i,j} \in (0.3;0.8])$. We will normalize the latter so as to obtain $w_{m,i,j} = \frac{w^0_{m,i,j}}{\sum_{j=1}^N w^0_{m,i,j}}$ for $m = 1,2$.

Given the seed we choose, the median number of first-order input-output-linked companies is 4, and that of second-order input-output-linked companies is 7 with both $N = 250$ and $N = 500$. With $N = 250$, the inter-quartile range of the number of first-linked companies is $[2;7]$ and that of second-linked companies is $[4;10]$. With $N = 500$, the inter-quartile ranges of the number of first-linked and second-linked companies are $[3;6]$ and $[4;9]$.

8.4.3 Continuous Regressors, Correlated Random Effects, and Remainder Residual

We consider a single explanatory variable,

$$x_{i,t} = \check{x}_i + \tilde{x}_{i,t}, \tag{8.7}$$

where we ensure that $\sum_{i=1}^N \check{x}_i = 0$ and $\sum_{t=1}^T \tilde{x}_{i,t} = 0$. Hence, $\check{x}_i$ is cross-sectionally demeaned and $\tilde{x}_{i,t}$ is time-demeaned for each company. Otherwise, $\tilde{x}_{i,t}$ is drawn from a standard normal distribution.

Also, the random effect χ_i exhibits zero mean, whereby $\sum_{i=1}^N \chi_i = 0$ in each Monte Carlo draw. $(\check{x}_i, \chi_i)$ are otherwise drawn as bivariate normal with unitary variance and covariance of 0.25. Hence, the random effects are correlated with the time-invariant component of $x_{i,t}$.

We do not consider the case where χ_i is cross-sectionally correlated. However, we always consider a model that includes a control function that would allow for it as in (8.5) and (8.6).

The remainder residual is drawn independently from a standard normal distribution. By this token, we have $E(x_{i,t}\nu_{i,t}) = 0$.

8.4.4 Initialization of the Process

We consider the case where the lag order is (up to) $L = 3$. Consequently, we specify three initial values for the binary outcome, namely, $\underline{y}_{i,0} = (y_{i,0}, y_{i,-1}, y_{i,-2})$. Each one of the latter is determined by $y_{i,-\ell} = 1(\eta_{i,\ell} \geq 0)$, where $\eta_{i,\ell}$ is standard normally distributed with $E(\eta_{i,\ell}\eta_{j,k}) = 0$ for all (i, j), $(k, \ell) = 1, 2, 3$ and $k \neq \ell$.

We start the process in year $t = 1$, using $\underline{y}_{i,t-1} = \underline{y}_{i,0}$ and $\underline{\bar{y}}_{i,t-1} = \underline{\bar{y}}_{i,0}$. Using the respective model parameters α, λ, and β together with $\{x_{i,1}, \chi_i, \nu_{i,1}\}$ obtains the value of the latent outcome variable $y^*_{i,1}$. The latter generates the binary counterpart as $y_{i,1} = 1(y^*_{i,1} \geq 0)$. We continue with this procedure iteratively up until the last period in the sample ($T = 5$ or $T = 10$). The outcome in periods $t = 2$ and $t = 3$ depends on two and one pre-sample binary outcome terms, respectively. From period $t = 4$ onwards, all binary-outcome lags on the right-hand side of the model are from within the sample period.

8.4.5 Estimated Models

We estimate three alternative models with every Monte Carlo draw.

Model M0 is a panel probit with random effects based on the assumed latent process of

$$y^*_{i,t} = \zeta + \underline{y}_{i,t-1}\alpha + \underline{\bar{y}}_{i,t-1}\lambda + x_{i,t}\beta + \chi_i + \nu_{i,t}, \tag{8.8}$$

which excludes the control function $g_i\gamma$ on the right-hand side.

Models M1 and M2 are panel probits with random effects based on the assumed latent process of

$$y^*_{i,t} = \zeta + \underline{y}_{i,t-1}\alpha + \underline{\bar{y}}_{i,t-1}\lambda + x_{i,t}\beta + g_i\gamma + \chi_i + \nu_{i,t}. \tag{8.9}$$

They differ to the extent that $\underline{\bar{y}}_{i,t-1}$ and $(\bar{\bar{x}}_i, \underline{\bar{y}}_{i,0})$ only include first-order or closest-neighbor links based on $w_{1,i,j}$ in Model 1, whereas Model M2 includes twice as many terms in each, involving interdependence terms based on $w_{1,i,j}$ as well as $w_{2,i,j}$.

We consider data-generating processes where Model M1 is suitable. As Model M2 nests M1, both should eliminate larger biases in that case, but Model M1 will be more efficient. If the true process involves higher-(second-)order interdependence terms based on $w_{1,i,j}$ as well as $w_{2,i,j}$, Model M1 will display a nontrivially large bias.

8.4.6 Parameter Designs

We consider four alternative parameterizations, which we refer to as Designs D1-D4. Table 8.1 provides a summary. In each and every case, the true model excludes a constant ($\zeta = 0$) and assumes that $\beta = 1$. However, the designs differ in terms of the lag structure and the order of the network interdependence in terms of the parameters α and λ.

Design D1 assumes a third-order time-lag structure ($L = 3$) and a first-order network structure ($M = 1$) with as many time lags. In this case, both Models M1 and M2 are suitable, but Model M1 is more efficient. Design D2 assumes a third-order time-lag structure ($L = 3$) and a second-order network structure ($M = 2$) with as many time lags each. There, only Model M2 is suitable to avoid larger parameter biases. Design D3 is a variant of D2 in that it assumes some cyclical lag-parameter pattern (the second time lag on any lagged variable exhibits a negative sign relative to Design D2, all else being equal). Design D4 assumes a second-order time-lag structure ($L = 2$) and a second-order network structure ($M = 2$). However, only the first two time lags matter with regard to the closest-neighbor network lags, whereas only the first time lag matters with regard to the medium-close-neighbor network lags. In this case, only Model M2 is consistent, but it is inefficient because it includes irrelevant regressors.

8.4.7 Main Results

We summarize the results in Tables 8.2-8.6. In Tables 8.2-8.5, the sample period is $T = 10$, whereas it is only $T = 5$ in Table 8.6.

Tables 8.2 and 8.6 are each based on the parameter Design D1. Table 8.3 uses Design D2, Table 8.4 uses Design D3, and Table 8.5 uses Design D4.

Each table exhibits a 2×2 block structure. In the upper block of rows, we report on the median bias (MBias, i.e., the difference in the median of a parameter across 1,000 Monte Carlo draws and the true value). In the lower block, we report on the robust root mean-squared error. The latter is defined as

$$RRMSE = (\text{MBias}^2 + (\text{IQ}/1.35)^2)^{0.5}, \tag{8.10}$$

where IQ is the inter-quartile range of a parameter across 1,000 Monte Carlo draws. The inter-quartile range is normalized by 1.35 because, asymptotically, the inter-quartile range of a normal distribution amounts to 1.35 times the standard error.

The results in Tables 8.2-8.6 are consistent with our expectations. First, Model M2 exhibits a small bias which fades with sample size throughout the experiments. Model M0 is always biased. The bias on the network-lagged terms is absent in Model M0, because we generated the true process assuming $E(\chi_i \chi_j)$ for companies $i \neq j$.

The results consistently suggest that Model M2 – and, where applicable, Model M1 – is well suited to avoid larger parameter biases by including a control function of the suggested form. Whenever both M1 and M2 eliminate larger biases, Model M2 tends to exhibit a larger RRMSE than Model M1, as expected.

8.4.8 Extension 1: Network Interdependence in the Correlated Random Effects

In the discussion above, we acknowledged that the independent distribution of the (correlated) random component χ_i between individuals implied that the parameters λ on the time-and-network-lagged binary outcome variable were unbiased even in the simple panel-probit model without control function. The reason was that, in expectation, $E(y_{i,t-\ell}\bar{y}_{i,t-k}) = 0$ for all $\{\ell, k\}$ due to the assumption of $E(\bar{y}_{i,t-k}\chi_j) = 0$ and $E(y_{i,t-\ell}\chi_j) = 0$ for all $\{i, j\}$ (as well as $E(v_{i,t}v_{j,s}) = 0$ for all $\{t, s\}$).

In this subsection, we consider, for parameter Design D1 and $N = \{250; 500\}$ as well as $T = 10$ as in Table 8.2, a modified structure, where the model without control function is

$$y^*_{i,t} = \zeta + \underline{y}_{i,t-1}\alpha + \underline{\bar{y}}_{i,t-1}\lambda + x_{i,t}\beta + \xi_i + v_{i,t} \tag{8.11}$$

with the random component ξ_i being not only correlated with the regressors but also interdependent among the residuals. Specifically, we assume

$$\xi_i = \chi_i + \bar{x}_i\rho, \quad \bar{x}_i = (\bar{x}_{1,i}, \bar{x}_{2,i}), \tag{8.12}$$

where $\bar{x}_{m,i} = \sum_{j=1}^{N} w_{m,i,j}x_j$ and $\rho = (\rho_1, \rho_2)'$ with $\rho_m = 1$ for $m = 1, 2$.

Clearly, the latter implies that $E(\underline{\bar{y}}_{i,t-1}\xi_i) \neq 0$, and λ cannot be estimated consistently in that case and requires specific treatment in the control function. However, this is already incorporated in what we have proposed in $g_i\gamma$.

Table 8.7 summarizes the respective results. Indeed, we now find that the parameters λ display a substantial bias in Model 0 as well as in Model 1, neither of which is correctly specified and utilizes an appropriate control function. On the contrary, the approach of Model 2 works well in addressing this case.

8.4.9 Extension 2: an Alternative Network Design

Instead of the block-diagonal network design considered above, we consider a so-called rook design here. For this, we consider a grid of 961 units among which we generate a generalized rook design on a lattice following the approach in Drukker, Egger and Prucha (2023). For this, we choose a scalar $\kappa = 31$, and note that $\kappa^2 = 961$. We then consider all first neighbors who can be reached exactly by rules-conform one-step moves from any unit i. We generate a raw (unnormalized) matrix with unit-pair binary entries $w^0_{1,ij}$. We do the same with all second neighbors who can be reached exactly by rules-conform two-step moves from any unit i, obtaining a raw (unnormalized) matrix with unit-pair binary entries $w^0_{2,ij}$. We note that, due to mutual exclusivity, $w^0_{1,ij} + w^0_{2,ij} = \{0,1\}$. Finally, we normalize the network weights so as to obtain $w_{m,ij} = w^0_{m,ij} / \sum_{j=1}^{N} w^0_{m,ij}$ for each neighbor concept $m \in \{1,2\}$.

Then we run the same simulations as with design D1 and present them in Table 8.8. A comparison of the results in this table with the ones in Table 8.2 suggests that the performance of the proposed models is as well as it had been before. The simple probit model (M0) appears to fare somewhat worse in Table 8.8 than in Table 8.2 with $N = 500$, which is owed to the difference in the degree of neighborliness in the network matrices considered.

8.4.10 Extension 3: an Alternative Initial Condition

In the analysis above, we formulated the initial condition as to contain four terms: averages of the explanatory variables in $\breve{x}_i$, the level of the binary dependent variable measured in the first year prior to the sample period ($\underline{y}_{i,0}$), network-weighted averages of $\breve{x}_i$ ($\bar{\breve{x}}_i$), and network-weighted averages of the binary dependent variable in the first year prior to the sample period ($\underline{\bar{y}}_{i,0}$). It turned out that this was sufficient with the proposed data-generating process.

In this subsection, we modify the initial condition so as to include as many lags of the binary dependent variables measured immediately prior to the sample period as there are lags involved in the model. And, additionally, we include all network-weighted averages of the latter as well. Hence, with a third-order time-lag model, there are four additional terms in what we refer to as Model 1A (as this model considers only a single network), and there are six additional terms in what we call Model 2A. Model 1A is the same as Model 1, except for also including the two pre-period lags $(\underline{y}_{i,0-1}, \underline{y}_{i,0-2})$ as well as $(\underline{\bar{y}}_{i,0-1}, \underline{\bar{y}}_{i,0-2})$, all else equal. Model 2A is the same as Model A1, except that each term in $(\underline{\bar{y}}_{i,0-1}, \underline{\bar{y}}_{i,0-2})$ is a $1 \times M$ vector (with the number of network matrices being $M = 2$ in the simulations) instead of a scalar.

With these arguments at hand, we suggest defining the vector

$$g_i^{\mathrm{A}} = \left(\breve{x}_i, \underline{y}_{i,0}, \bar{\breve{x}}_i, \underline{\bar{y}}_{i,0}\right), \tag{8.13}$$

$$\underline{y}_{i,0} = \left(y_{i,0}, y_{i,0-1}, y_{i,0-2}\right), \tag{8.14}$$

$$\underline{\bar{y}}_{i,0} = \left(\bar{y}_{i,0}, \bar{y}_{i,0-1}, \bar{y}_{i,0-2}\right), \tag{8.15}$$

and using it in estimation.

Tables 8.9 and 8.10 summarize the associated results for the same design as in Table 8.2 for $N = 250$ and $N = 500$ cross-sectional units, respectively, and $T = 10$ time periods. In both tables, we report on the bias and RMSE of Models 1 and 2, which are the same as in Table 8.2. Moreover, we contrast those results with the ones based on the modified initial condition, dubbed Models 1A and 2A.

An inspection of the two tables suggests that using the augmented control function is not necessary here, as expected. The reason is that the bias of Model 0 stems from the presence of a time-invariant error component which is correlated with the regressors in x_{it}. This bias can sufficiently be captured when conditioning on one pre-period lag as well as, eventually, its properly network-weighted counterpart. With the assumed data-generating process, it is confirmed that the biases in Models 1 and 2 are small and not substantially further reduced when considering Models 1A and 2A, respectively, instead. These patterns emerge consistently for the case with $N = 250$ cross-sectional units in Table 8.9 as well as the one with $N = 500$ units in Table 8.10.

8.5 Discussion: Effect Estimates

It should be noted that, due to the dynamic nature of the problem, changes in $x_{i,t}$ do not exert contemporaneous effects on latent outcome (and binary outcome) only but will have lagged effects with sluggish adjustment.

For evaluating contemporaneous (short-run) and cumulative (long-run) marginal effects of one-shot changes in $x_{i,t}$, there is no other way than to compute these effects for each period separately.

The long-run effect will be the state of $y_{i,t+f}$, where no further change in $y^*_{i,t+f}$ materializes relative to the earlier period. However, it should also be noted that in empirically relevant settings, this will probably emerge only a few periods after the shock in $x_{i,t}$ occurs. The reason for the latter lies in the fact that $y^*_{i,t+f}$ will change in periods after the shock only as long as $y_{i,t+f}$ keeps changing. However, the latter, as a discrete variable, changes much more sluggishly than this is the case for linear dynamic processes with continuous dependent variables and continuous regressors.

8.6 Conclusions

We outline control-function strategies for dynamic probit models, which include higher-order own time lags as well as higher-order network-and-time lags of the binary dependent outcome variable.

Such models can be useful when estimating entry choices with sluggish adjustment and network interdependencies when data on binary outcomes are repeatedly observed over time.

We demonstrate in Monte Carlo simulations that the proposed approach can successfully absorb the bias in the parameters of lagged dependent binary outcome indicators already in small to medium-sized samples.

Appendix: Tables

Table 8.1: Parameter designs considered in the Monte Carlo simulations

Parameter on	Parameter	Design-specific true values D1	D2	D3	D4
$y_{i,t-1}$	α_1	1	1	1	1
$y_{i,t-2}$	α_2	0.6	0.6	-0.6	0.6
$y_{i,t-3}$	α_3	0.2	0.2	0.2	0
$\overline{y}_{i,1,t-1}$	$\lambda_{1,1}$	0.5	0.5	0.5	0.5
$\overline{y}_{i,1,t-2}$	$\lambda_{1,2}$	0.3	0.3	-0.3	0.3
$\overline{y}_{i,1,t-3}$	$\lambda_{1,3}$	0.1	0.1	0.1	0
$\overline{y}_{i,2,t-1}$	$\lambda_{2,1}$	0	0.8	0.8	0.8
$\overline{y}_{i,2,t-2}$	$\lambda_{2,2}$	0	0.6	-0.6	0
$\overline{y}_{i,3,t-3}$	$\lambda_{2,3}$	0	0.4	0.4	0
$x_{i,t}$	β	1	1	1	1

Table 8.2: Bias and RMSE for 1,000 Monte Carlo runs, parameter Design D1, and T=10

			Median Bias (MBias)					
			N=250			N=500		
Parameter on	Parameter	True value	Model 0	Model 1	Model 2	Model 0	Model 1	Model 2
$y_{i,t-1}$	α_1	1	0.1500	0.0008	0.0059	0.1467	0.0059	0.0090
$y_{i,t-2}$	α_2	0.6	0.0904	-0.0098	-0.0083	0.0907	-0.0028	-0.0006
$y_{i,t-3}$	α_3	0.2	0.0862	0.0026	0.0027	0.0871	0.0049	0.0052
$\overline{y}_{i,1,t-1}$	$\lambda_{1,1}$	0.5	-0.0660	-0.0160	-0.0080	-0.0411	0.0050	0.0033
$\overline{y}_{i,1,t-2}$	$\lambda_{1,2}$	0.3	-0.0526	0.0083	0.0012	-0.0536	0.0050	0.0031
$\overline{y}_{i,1,t-3}$	$\lambda_{1,3}$	0.1	-0.0515	0.0051	0.0049	-0.0399	-0.0018	0.0008
$\overline{y}_{i,2,t-1}$	$\lambda_{2,1}$	0			-0.0081			-0.0074
$\overline{y}_{i,2,t-2}$	$\lambda_{2,2}$	0			0.0052			-0.0024
$\overline{y}_{i,3,t-3}$	$\lambda_{2,3}$	0			-0.0084			0.0002
$x_{i,t}$	β	1	0.0910	0.0063	0.0104	0.0892	0.0098	0.0117
			Robust RMSE (RRMSE)					
			N=250			N=500		
Parameter on	Parameter	True value	Model 0	Model 1	Model 2	Model 0	Model 1	Model 2
$y_{i,t-1}$	α_1	1	0.1847	0.1108	0.1151	0.1635	0.0785	0.0773
$y_{i,t-2}$	α_2	0.6	0.1360	0.1099	0.1083	0.1222	0.0834	0.0839
$y_{i,t-3}$	α_3	0.2	0.1358	0.1090	0.1119	0.1140	0.0793	0.0786
$\overline{y}_{i,1,t-1}$	$\lambda_{1,1}$	0.5	0.2254	0.2329	0.2438	0.1691	0.1760	0.1741
$\overline{y}_{i,1,t-2}$	$\lambda_{1,2}$	0.3	0.2166	0.2169	0.2312	0.1568	0.1623	0.1599
$\overline{y}_{i,1,t-3}$	$\lambda_{1,3}$	0.1	0.2142	0.2256	0.2306	0.1489	0.1514	0.1544
$\overline{y}_{i,2,t-1}$	$\lambda_{2,1}$	0			0.2943			0.1958
$\overline{y}_{i,2,t-2}$	$\lambda_{2,2}$	0			0.2847			0.1877
$\overline{y}_{i,3,t-3}$	$\lambda_{2,3}$	0			0.2671			0.1713
$x_{i,t}$	β	1	0.1124	0.0734	0.0736	0.1011	0.0527	0.0527

Table 8.3: Bias and RMSE for 1,000 Monte Carlo runs, parameter Design D2, and T=10

			Median Bias (MBias)					
			N=250			N=500		
Parameter on	Parameter	True value	Model 0	Model 1	Model 2	Model 0	Model 1	Model 2
$y_{i,t-1}$	α_1	1	0.2606	0.1291	0.0312	0.2111	0.1148	0.0165
$y_{i,t-2}$	α_2	0.6	0.1394	0.0641	-0.0126	0.1287	0.0680	-0.0013
$y_{i,t-3}$	α_3	0.2	0.1148	0.0524	-0.0053	0.1106	0.0640	0.0063
$\overline{y}_{i,1,t-1}$	$\lambda_{1,1}$	0.5	0.0753	0.2099	-0.0220	0.0604	0.1747	-0.0068
$\overline{y}_{i,1,t-2}$	$\lambda_{1,2}$	0.3	0.1486	0.2476	-0.0001	0.1578	0.2345	0.0045
$\overline{y}_{i,1,t-3}$	$\lambda_{1,3}$	0.1	0.0953	0.2335	-0.0091	0.1057	0.2226	0.0084
$\overline{y}_{i,2,t-1}$	$\lambda_{2,1}$	0.8			0.8338			0.8189
$\overline{y}_{i,2,t-2}$	$\lambda_{2,2}$	0.6			0.6458			0.5919
$\overline{y}_{i,3,t-3}$	$\lambda_{2,3}$	0.4			0.4304			0.4059
$x_{i,t}$	β	1	0.0668	-0.0358	0.0313	0.0378	-0.0382	0.0212
			Robust RMSE (RRMSE)					
			N=250			N=500		
Parameter on	Parameter	True value	Model 0	Model 1	Model 2	Model 0	Model 1	Model 2
$y_{i,t-1}$	α_1	1	0.3043	0.2106	0.1865	0.2376	0.1579	0.1149
$y_{i,t-2}$	α_2	0.6	0.2083	0.1721	0.1635	0.1676	0.1289	0.1177
$y_{i,t-3}$	α_3	0.2	0.1912	0.1758	0.1801	0.1511	0.1252	0.1129
$\overline{y}_{i,1,t-1}$	$\lambda_{1,1}$	0.5	0.3287	0.4094	0.3600	0.2359	0.2989	0.2346
$\overline{y}_{i,1,t-2}$	$\lambda_{1,2}$	0.3	0.3373	0.3995	0.3348	0.2838	0.3373	0.2407
$\overline{y}_{i,1,t-3}$	$\lambda_{1,3}$	0.1	0.3000	0.3902	0.3454	0.2214	0.2998	0.2142
$\overline{y}_{i,2,t-1}$	$\lambda_{2,1}$	0.8			0.9373			0.8620
$\overline{y}_{i,2,t-2}$	$\lambda_{2,2}$	0.6			0.7562			0.6527
$\overline{y}_{i,3,t-3}$	$\lambda_{2,3}$	0.4			0.5822			0.4672
$x_{i,t}$	β	1	0.1164	0.1150	0.1255	0.0756	0.0802	0.0803

Table 8.4: Bias and RMSE for 1,000 Monte Carlo runs, parameter Design D3, and T=10

			Median Bias (MBias)					
			N=250			N=500		
Parameter on	Parameter	True value	Model 0	Model 1	Model 2	Model 0	Model 1	Model 2
$y_{i,t-1}$	α_1	1	0.1338	-0.0139	0.0071	0.1266	-0.0168	0.0047
$y_{i,t-2}$	α_2	-0.6	-1.0594	-1.1774	-1.2079	-1.0649	-1.1760	-1.2061
$y_{i,t-3}$	α_3	0.2	0.0989	-0.0135	-0.0052	0.1019	-0.0090	-0.0005
$\overline{y}_{i,1,t-1}$	$\lambda_{1,1}$	0.5	-0.0369	-0.0094	-0.0111	-0.0215	0.0021	0.0034
$\overline{y}_{i,1,t-2}$	$\lambda_{1,2}$	-0.3	-0.5593	-0.5469	-0.5983	-0.5501	-0.5465	-0.6047
$\overline{y}_{i,1,t-3}$	$\lambda_{1,3}$	0.1	-0.0242	-0.0120	-0.0015	-0.0187	-0.0128	-0.0058
$\overline{y}_{i,2,t-1}$	$\lambda_{2,1}$	0.8			0.7938			0.7992
$\overline{y}_{i,2,t-2}$	$\lambda_{2,2}$	-0.6			-0.6002			-0.6045
$\overline{y}_{i,3,t-3}$	$\lambda_{2,3}$	0.4			0.3996			0.3917
$x_{i,t}$	β	1	0.0654	-0.0131	0.0050	0.0669	-0.0106	0.0073
			Robust RMSE (RRMSE)					
			N=250			N=500		
Parameter on	Parameter	True value	Model 0	Model 1	Model 2	Model 0	Model 1	Model 2
$y_{i,t-1}$	α_1	1	0.1634	0.0962	0.0990	0.1422	0.0679	0.0662
$y_{i,t-2}$	α_2	-0.6	1.0644	1.1822	1.2125	1.0674	1.1784	1.2086
$y_{i,t-3}$	α_3	0.2	0.1325	0.0910	0.0942	0.1187	0.0629	0.0632
$\overline{y}_{i,1,t-1}$	$\lambda_{1,1}$	0.5	0.1777	0.1860	0.1841	0.1320	0.1368	0.1371
$\overline{y}_{i,1,t-2}$	$\lambda_{1,2}$	-0.3	0.5869	0.5763	0.6258	0.5640	0.5617	0.6184
$\overline{y}_{i,1,t-3}$	$\lambda_{1,3}$	0.1	0.1706	0.1744	0.1836	0.1137	0.1213	0.1208
$\overline{y}_{i,2,t-1}$	$\lambda_{2,1}$	0.8			0.8241			0.8155
$\overline{y}_{i,2,t-2}$	$\lambda_{2,2}$	-0.6			0.6390			0.6227
$\overline{y}_{i,3,t-3}$	$\lambda_{2,3}$	0.4			0.4542			0.4148
$x_{i,t}$	β	1	0.0871	0.0634	0.0613	0.0765	0.0413	0.0408

Table 8.5: Bias and RMSE for 1,000 Monte Carlo runs, parameter Design D4, and T=10

			Median Bias (MBias)					
			N=250			N=500		
Parameter on	Parameter	True value	Model 0	Model 1	Model 2	Model 0	Model 1	Model 2
$y_{i,t-1}$	α_1	1	0.1805	0.0388	0.0159	0.1546	0.0317	0.0124
$y_{i,t-2}$	α_2	0.6	0.1069	0.0134	-0.0124	0.1038	0.0223	0.0005
$y_{i,t-3}$	α_3	0	-0.0962	-0.1807	-0.2006	-0.1057	-0.1784	-0.1953
$\overline{y}_{i,1,t-1}$	$\lambda_{1,1}$	0.5	0.0240	0.1193	-0.0029	0.0334	0.1037	-0.0113
$\overline{y}_{i,1,t-2}$	$\lambda_{1,2}$	0.3	0.0148	0.0937	0.0151	0.0165	0.0792	0.0008
$\overline{y}_{i,1,t-3}$	$\lambda_{1,3}$	0	-0.1252	-0.0599	-0.1122	-0.1150	-0.0610	-0.1019
$\overline{y}_{i,2,t-1}$	$\lambda_{2,1}$	0.8			0.7844			0.8031
$\overline{y}_{i,2,t-2}$	$\lambda_{2,2}$	0			0.0307			0.0078
$\overline{y}_{i,3,t-3}$	$\lambda_{2,3}$	0			0.0070			-0.0053
$x_{i,t}$	β	1	0.0836	-0.0028	0.0182	0.0730	-0.0078	0.0105

			Robust RMSE (RRMSE)					
			N=250			N=500		
Parameter on	Parameter	True value	Model 0	Model 1	Model 2	Model 0	Model 1	Model 2
$y_{i,t-1}$	α_1	1	0.2209	0.1363	0.1341	0.1749	0.0921	0.0874
$y_{i,t-2}$	α_2	0.6	0.1602	0.1142	0.1200	0.1311	0.0859	0.0833
$y_{i,t-3}$	α_3	0	0.1523	0.2168	0.2367	0.1307	0.1967	0.2110
$\overline{y}_{i,1,t-1}$	$\lambda_{1,1}$	0.5	0.2536	0.2978	0.2697	0.1742	0.2040	0.1861
$\overline{y}_{i,1,t-2}$	$\lambda_{1,2}$	0.3	0.2394	0.2695	0.2687	0.1621	0.1893	0.1729
$\overline{y}_{i,1,t-3}$	$\lambda_{1,3}$	0	0.2502	0.2409	0.2737	0.1893	0.1724	0.1869
$\overline{y}_{i,2,t-1}$	$\lambda_{2,1}$	0.8			0.8516			0.8292
$\overline{y}_{i,2,t-2}$	$\lambda_{2,2}$	0			0.2985			0.2083
$\overline{y}_{i,3,t-3}$	$\lambda_{2,3}$	0			0.2701			0.1856
$x_{i,t}$	β	1	0.1105	0.0790	0.0839	0.0881	0.0521	0.0543

Table 8.6: Bias and RMSE for 1,000 Monte Carlo runs, parameter Design D1, and T=5

			Median Bias (MBias)					
			N=250			N=500		
Parameter on	Parameter	True value	Model 0	Model 1	Model 2	Model 0	Model 1	Model 2
$y_{i,t-1}$	α_1	1	0.1913	0.0161	0.0254	0.1787	0.0076	0.0132
$y_{i,t-2}$	α_2	0.6	0.0948	0.0101	0.0153	0.0876	0.0061	0.0080
$y_{i,t-3}$	α_3	0.2	0.0525	0.0052	0.0076	0.0530	0.0043	0.0063
$\overline{y}_{i,1,t-1}$	$\lambda_{1,1}$	0.5	-0.1064	-0.0080	-0.0134	-0.0606	-0.0054	-0.0015
$\overline{y}_{i,1,t-2}$	$\lambda_{1,2}$	0.3	-0.0869	-0.0108	-0.0132	-0.0571	0.0027	0.0050
$\overline{y}_{i,1,t-3}$	$\lambda_{1,3}$	0.1	-0.0254	0.0265	0.0306	-0.0289	0.0017	0.0087
$\overline{y}_{i,2,t-1}$	$\lambda_{2,1}$	0			-0.0204			-0.0119
$\overline{y}_{i,2,t-2}$	$\lambda_{2,2}$	0			0.0251			-0.0027
$\overline{y}_{i,3,t-3}$	$\lambda_{2,3}$	0			0.0091			-0.0014
$x_{i,t}$	β	1	0.1432	0.0221	0.0324	0.1267	0.0091	0.0133
			Robust RMSE (RRMSE)					
			N=250			N=500		
Parameter on	Parameter	True value	Model 0	Model 1	Model 2	Model 0	Model 1	Model 2
$y_{i,t-1}$	α_1	1	0.2360	0.1479	0.1518	0.2030	0.1088	0.1105
$y_{i,t-2}$	α_2	0.6	0.1773	0.1586	0.1576	0.1310	0.1076	0.1072
$y_{i,t-3}$	α_3	0.2	0.1527	0.1552	0.1580	0.1060	0.1006	0.1004
$\overline{y}_{i,1,t-1}$	$\lambda_{1,1}$	0.5	0.2959	0.2994	0.3084	0.1926	0.2026	0.2109
$\overline{y}_{i,1,t-2}$	$\lambda_{1,2}$	0.3	0.2800	0.2883	0.2945	0.1774	0.1734	0.1853
$\overline{y}_{i,1,t-3}$	$\lambda_{1,3}$	0.1	0.2480	0.2776	0.2776	0.1651	0.1750	0.1830
$\overline{y}_{i,2,t-1}$	$\lambda_{2,1}$	0			0.3304			0.2548
$\overline{y}_{i,2,t-2}$	$\lambda_{2,2}$	0			0.3178			0.2262
$\overline{y}_{i,3,t-3}$	$\lambda_{2,3}$	0			0.2997			0.1948
$x_{i,t}$	β	1	0.1717	0.1087	0.1127	0.1427	0.0730	0.0747

Table 8.7: Bias and RMSE for 1,000 Monte Carlo runs, parameter Design D1, and T=10 with network-correlated random effects

			Median Bias (MBias)					
			N=250			N=500		
Parameter on	Parameter	True value	Model 0	Model 1	Model 2	Model 0	Model 1	Model 2
$y_{i,t-1}$	α_1	1	0.1007	0.0036	0.0156	0.0937	-0.0015	0.0111
$y_{i,t-2}$	α_2	0.6	0.0380	-0.0185	-0.0093	0.0433	-0.0137	-0.0023
$y_{i,t-3}$	α_3	0.2	0.0499	-0.0041	0.0082	0.0364	-0.0134	-0.0017
$\overline{y}_{i,1,t-1}$	$\lambda_{1,1}$	0.5	0.4205	0.1285	-0.0144	0.3892	0.1947	-0.0087
$\overline{y}_{i,1,t-2}$	$\lambda_{1,2}$	0.3	0.1702	0.0849	0.0062	0.1481	0.0964	-0.0008
$\overline{y}_{i,1,t-3}$	$\lambda_{1,3}$	0.1	0.0865	0.0506	-0.0030	0.0769	0.0659	0.0004
$\overline{y}_{i,2,t-1}$	$\lambda_{2,1}$	0			-0.0180			-0.0062
$\overline{y}_{i,2,t-2}$	$\lambda_{2,2}$	0			0.0074			-0.0006
$\overline{y}_{i,3,t-3}$	$\lambda_{2,3}$	0			-0.0056			0.0003
$x_{i,t}$	β	1	0.0768	0.0138	0.0172	0.0631	0.0055	0.0076
			Robust RMSE (RRMSE)					
			N=250			N=500		
Parameter on	Parameter	True value	Model 0	Model 1	Model 2	Model 0	Model 1	Model 2
$y_{i,t-1}$	α_1	1	0.1509	0.1223	0.1207	0.1220	0.0835	0.0817
$y_{i,t-2}$	α_2	0.6	0.1270	0.1266	0.1240	0.0849	0.0801	0.0770
$y_{i,t-3}$	α_3	0.2	0.1245	0.1232	0.1251	0.0878	0.0836	0.0848
$\overline{y}_{i,1,t-1}$	$\lambda_{1,1}$	0.5	0.4846	0.2862	0.2516	0.4205	0.2608	0.1758
$\overline{y}_{i,1,t-2}$	$\lambda_{1,2}$	0.3	0.2822	0.2573	0.2410	0.2140	0.1948	0.1671
$\overline{y}_{i,1,t-3}$	$\lambda_{1,3}$	0.1	0.2186	0.2264	0.2220	0.1713	0.1724	0.1638
$\overline{y}_{i,2,t-1}$	$\lambda_{2,1}$	0			0.3117			0.1907
$\overline{y}_{i,2,t-2}$	$\lambda_{2,2}$	0			0.2811			0.1840
$\overline{y}_{i,3,t-3}$	$\lambda_{2,3}$	0			0.2633			0.1695
$x_{i,t}$	β	1	0.1048	0.0752	0.0760	0.0806	0.0530	0.0533

Table 8.8: Bias and RMSE for 1,000 Monte Carlo runs, parameter Design D1, and T=10 with second-order rook-design network

			Median Bias (MBias)					
			N=250			N=500		
Parameter on	Parameter	True value	Model 0	Model 1	Model 2	Model 0	Model 1	Model 2
$y_{i,t-1}$	α_1	1	0.1587	0.0008	0.0084	0.1530	0.0067	0.0089
$y_{i,t-2}$	α_2	0.6	0.0915	-0.0134	-0.0113	0.0956	-0.0031	-0.0022
$y_{i,t-3}$	α_3	0.2	0.0868	-0.0004	0.0018	0.0821	0.0023	0.0022
$\overline{y}_{i,1,t-1}$	$\lambda_{1,1}$	0.5	-0.0589	0.0010	0.0141	-0.0534	0.0003	0.0005
$\overline{y}_{i,1,t-2}$	$\lambda_{1,2}$	0.3	-0.0580	0.0070	0.0180	-0.0538	0.0036	0.0062
$\overline{y}_{i,1,t-3}$	$\lambda_{1,3}$	0.1	-0.0589	0.0048	0.0033	-0.0499	0.0030	-0.0017
$\overline{y}_{i,2,t-1}$	$\lambda_{2,1}$	0			-0.0123			-0.0117
$\overline{y}_{i,2,t-2}$	$\lambda_{2,2}$	0			-0.0104			-0.0024
$\overline{y}_{i,3,t-3}$	$\lambda_{2,3}$	0			-0.0103			0.0110
$x_{i,t}$	β	1	0.0891	0.0065	0.0112	0.0870	0.0072	0.0098
			Robust RMSE (RRMSE)					
			N=250			N=500		
Parameter on	Parameter	True value	Model 0	Model 1	Model 2	Model 0	Model 1	Model 2
$y_{i,t-1}$	α_1	1	0.1924	0.1137	0.1163	0.1698	0.0802	0.0799
$y_{i,t-2}$	α_2	0.6	0.1413	0.1096	0.1106	0.1240	0.0790	0.0798
$y_{i,t-3}$	α_3	0.2	0.1404	0.1166	0.1197	0.1110	0.0791	0.0779
$\overline{y}_{i,1,t-1}$	$\lambda_{1,1}$	0.5	0.2448	0.2587	0.2631	0.1848	0.1851	0.1816
$\overline{y}_{i,1,t-2}$	$\lambda_{1,2}$	0.3	0.2218	0.2330	0.2450	0.1647	0.1589	0.1693
$\overline{y}_{i,1,t-3}$	$\lambda_{1,3}$	0.1	0.2128	0.2134	0.2185	0.1590	0.1549	0.1585
$\overline{y}_{i,2,t-1}$	$\lambda_{2,1}$	0			0.2394			0.1792
$\overline{y}_{i,2,t-2}$	$\lambda_{2,2}$	0			0.2247			0.1683
$\overline{y}_{i,3,t-3}$	$\lambda_{2,3}$	0			0.2077			0.1403
$x_{i,t}$	β	1	0.1114	0.0739	0.0749	0.0983	0.0527	0.0530

Table 8.9: Bias and RMSE for 1,000 Monte Carlo runs, parameter Design D1, N=250, and T=10, using alternative initial conditions for Models M1 and M2 each

			Median Bias (MBias), N=250			
Parameter on	Parameter	True value	Model 1	Model 1A	Model 2	Model 2A
$y_{i,t-1}$	α_1	1	0.0008	0.0007	0.0059	0.0030
$y_{i,t-2}$	α_2	0.6	-0.0098	-0.0102	-0.0083	-0.0080
$y_{i,t-3}$	α_3	0.2	0.0026	0.0035	0.0027	0.0018
$\overline{y}_{i,1,t-1}$	$\lambda_{1,1}$	0.5	-0.0160	-0.0120	-0.0080	-0.0226
$\overline{y}_{i,1,t-2}$	$\lambda_{1,2}$	0.3	0.0083	0.0029	0.0012	-0.0016
$\overline{y}_{i,1,t-3}$	$\lambda_{1,3}$	0.1	0.0051	0.0091	0.0049	0.0031
$\overline{y}_{i,2,t-1}$	$\lambda_{2,1}$	0			-0.0081	0.0023
$\overline{y}_{i,2,t-2}$	$\lambda_{2,2}$	0			0.0052	0.0005
$\overline{y}_{i,3,t-3}$	$\lambda_{2,3}$	0			-0.0084	0.0005
$x_{i,t}$	β	1	0.0063	0.0069	0.0104	0.0117
			Robust RMSE (RRMSE), N=250			
Parameter on	Parameter	True value	Model 1	Model 1A	Model 2	Model 2A
$y_{i,t-1}$	α_1	1	0.1847	0.1108	0.1151	0.1175
$y_{i,t-2}$	α_2	0.6	0.1360	0.1099	0.1083	0.1094
$y_{i,t-3}$	α_3	0.2	0.1358	0.1090	0.1119	0.1131
$\overline{y}_{i,1,t-1}$	$\lambda_{1,1}$	0.5	0.2254	0.2329	0.2438	0.2619
$\overline{y}_{i,1,t-2}$	$\lambda_{1,2}$	0.3	0.2166	0.2169	0.2312	0.2323
$\overline{y}_{i,1,t-3}$	$\lambda_{1,3}$	0.1	0.2142	0.2256	0.2306	0.2300
$\overline{y}_{i,2,t-1}$	$\lambda_{2,1}$	0			0.2943	0.3064
$\overline{y}_{i,2,t-2}$	$\lambda_{2,2}$	0			0.2847	0.2891
$\overline{y}_{i,3,t-3}$	$\lambda_{2,3}$	0			0.2671	0.2697
$x_{i,t}$	β	1	0.1124	0.0734	0.0736	0.0742

Table 8.10: Bias and RMSE for 1,000 Monte Carlo runs, parameter Design D1, N=500, and T=10, using alternative initial conditions for Models M1 and M2 each

			Median Bias (MBias), N=500			
Parameter on	Parameter	True value	Model 1	Model 1A	Model 2	Model 2A
$y_{i,t-1}$	α_1	1	0.0008	0.0007	0.0059	0.0030
$y_{i,t-2}$	α_2	0.6	-0.0098	-0.0102	-0.0083	-0.0080
$y_{i,t-3}$	α_3	0.2	0.0026	0.0035	0.0027	0.0018
$\overline{y}_{i,1,t-1}$	$\lambda_{1,1}$	0.5	-0.0160	-0.0120	-0.0080	-0.0226
$\overline{y}_{i,1,t-2}$	$\lambda_{1,2}$	0.3	0.0083	0.0029	0.0012	-0.0016
$\overline{y}_{i,1,t-3}$	$\lambda_{1,3}$	0.1	0.0051	0.0091	0.0049	0.0031
$\overline{y}_{i,2,t-1}$	$\lambda_{2,1}$	0			-0.0081	
$\overline{y}_{i,2,t-2}$	$\lambda_{2,2}$	0			0.0052	
$\overline{y}_{i,3,t-3}$	$\lambda_{2,3}$	0			-0.0084	
$x_{i,t}$	β	1	0.0063	0.0069	0.0104	0.0117
			Robust RMSE (RRMSE), N=500			
Parameter on	Parameter	True value	Model 1	Model 1A	Model 2	Model 2A
$y_{i,t-1}$	α_1	1	0.1847	0.1108	0.1151	0.1635
$y_{i,t-2}$	α_2	0.6	0.1360	0.1099	0.1083	0.1222
$y_{i,t-3}$	α_3	0.2	0.1358	0.1090	0.1119	0.1140
$\overline{y}_{i,1,t-1}$	$\lambda_{1,1}$	0.5	0.2254	0.2329	0.2438	0.1691
$\overline{y}_{i,1,t-2}$	$\lambda_{1,2}$	0.3	0.2166	0.2169	0.2312	0.1568
$\overline{y}_{i,1,t-3}$	$\lambda_{1,3}$	0.1	0.2142	0.2256	0.2306	0.1489
$\overline{y}_{i,2,t-1}$	$\lambda_{2,1}$	0			0.2943	
$\overline{y}_{i,2,t-2}$	$\lambda_{2,2}$	0			0.2847	
$\overline{y}_{i,3,t-3}$	$\lambda_{2,3}$	0			0.2671	
$x_{i,t}$	β	1	0.1124	0.0734	0.0736	0.1011

References

Amemiya, T. & MaCurdy, T. E. (1986). Instrumental-variable estimation of an error-components model. *Econometrica: Journal of the Econometric Society*, 869–880.

Arbia, G., Bille, A. G. & Leorato, S. (2023). Concentrated partial ml estimation for dynamic spatial panel data probit models with fixed effects. *SSRN 4514415*.

https://papers.ssrn.com/sol3/papers.cfm?abstract_id=4514415.

Arellano, M. (2003). *Panel data econometrics*. Oxford University Press.

Arulampalam, W. & Stewart, M. B. (2009). Simplified implementation of the heckman estimator of the dynamic probit model and a comparison with alternative estimators. *Oxford Bulletin of Economics and Statistics*, *71*(5), 659–681.

Balestra, P. & Nerlove, M. (1966). Pooling cross section and time series data in the estimation of a dynamic model: The demand for natural gas. *Econometrica: Journal of the Econometric Society*, 585–612.

Baltagi, B. H. (2015). *The Oxford handbook of panel data*. Oxford University Press.

Baltagi, B. H. (2021). *Econometric analysis of panel data*. Springer.

Breusch, T. S., Mizon, G. E. & Schmidt, P. (1989). Efficient estimation using panel data. *Econometrica: Journal of the Econometric Society*, 695–700.

Chamberlain, G. (1984). Panel data. *Handbook of Econometrics*, *2*.

Chib, S. & Jeliazkov, I. (2006). Inference in semiparametric dynamic models for binary longitudinal data. *Journal of the American Statistical Association*, *101*(474), 685–700.

Drukker, D. M., Egger, P. H. & Prucha, I. R. (2023). Simultaneous equations models with higher-order spatial or social network interactions. *Econometric Theory*, *39*(6), 1154–1201.

Egger, P. H. & Kesina, M. (2023). *Dynamic probit models with network interdependence and unobserved heterogeneity*. Unpublished manuscripts presented at the New York Camp Econometrics 2023.

Heckman, J. E. (1981). The incidental parameters problem and the problem of initial conditions in estimating a discrete time-discrete data stochastic processs. In C. F. Manski & D. McFadden (Eds.), *Structural analysis of discrete data with econometric applications* (p. 114-178).

Hsiao, C. (2022). *Analysis of panel data*. Cambridge University Press.

Mátyás, L. & Sevestre, P. (2008). *The econometrics of panel data: fundamentals and recent developments in theory and practice*. Springer Science & Business Media.

Mátyás, L. & Sevestre, P. (2015). *The econometrics of panel data: handbook of theory and applications*. Kluwer Academic Publishers.

Mundlak, Y. (1978). On the pooling of time series and cross section data. *Econometrica: Journal of the Econometric Society*, 69–85.

Nerlove, M. (1971a). Further evidence on the estimation of dynamic economic relations from a time series of cross sections. *Econometrica: Journal of the Econometric Society*, 359–382.

Nerlove, M. (1971b). A note on error components models. *Econometrica: Journal of the Econometric Society*, 383–396.

Nerlove, M. (1972). Lags in economic behavior. *Econometrica: Journal of the Econometric Society*, 221–251.

Nerlove, M. (2005). *Essays in panel data econometrics*. Cambridge University Press.

Rabe-Hesketh, S. & Skrondal, A. (2013). Avoiding biased versions of wooldridge's simple solution to the initial conditions problem. *Economics Letters*, *120*(2), 346–349.

Wooldridge, J. M. (1995). Selection corrections for panel data models under conditional mean independence assumptions. *Journal of Econometrics*, *68*(1), 115–132.

Wooldridge, J. M. (2005). Simple solutions to the initial conditions problem in dynamic, nonlinear panel data models with unobserved heterogeneity. *Journal of Applied Econometrics*, *20*(1), 39–54.

Chapter 9
Horizontal Regression or Vertical Regression to Generate Counterfactuals?

Cheng Hsiao, Jing Kong, Yimeng Xie and Qiankun Zhou

Abstract Generating counterfactuals through treating a variable as a function of its own past values or treating a variable as a function of other units, typically being referred to as horizontal or vertical regression, respectively, is widely used in the panel measurement of treatment effects. However, their inferences are often based on different assumptions for the data generating process. We consider unifying the underlying assumptions of the two approaches by a factor approach and compare their respective predictive power in terms of the sample configuration of the cross-section dimension N and the time dimension T.

9.1 Introduction

Treatment effects for the ith unit at time t, Δ_{it} are typically defined as the difference between the outcomes receiving the treatment, denoted as y_{it}^1, and the outcome in the absence of treatment, denoted as y_{it}^0, such that $\Delta_{it} = y_{it}^1 - y_{it}^0$. However, it is impossible to simultaneously observe both y_{it}^1 and y_{it}^0. Under the assumption that y_{it}^1 is observed, we need to predict y_{it}^0, say $\hat{y}_{it}^0$, to obtain the estimated treatment effects

$$\hat{\Delta}_{it} = y_{it}^1 - \hat{y}_{it}^0.$$

Cheng Hsiao ✉
University of Southern California, Los Angeles, USA, and Paula and Gregory Chow Center for Studies in Economics, Xiamen University, China, e-mail: chsiao@usc.edu

Jing Kong
University of Southern California, Los Angeles, USA, e-mail: jingkong@usc.edu

Yimeng Xie
Xiamen University, Xiamen, China, e-mail: yimengxie@xmu.edu.cn

Qiankun Zhou
Louisiana State University, Baton Rouge, USA, e-mail: qzhou@lsu.edu

B. H. Baltagi and L. Mátyás (eds.), *Seven Decades of Econometrics and Beyond*, Advanced Studies in Theoretical and Applied Econometrics 57,
https://doi.org/10.1007/978-3-031-92699-0_9

With panel data , a linear projection approach is often used to predict the missing y^0_{it} (for example, Hsiao & Zhou, 2024). There is no loss of generality to assume all the observed values are untreated for $i = 1,\ldots,N$ and $t = 1,\ldots,T$. Suppose at time $T+1$, the first unit received the treatment, but the rest of the units remain untreated, then

$$\mathbf{Y} = \left(\begin{array}{ccc|c} y_{11} & \cdots & y_{1T} & ? \\ \hline y_{21} & \cdots & y_{2T} & y_{2,T+1} \\ \vdots & \ddots & \vdots & \vdots \\ y_{N1} & \cdots & y_{NT} & y_{N,T+1} \end{array} \right) = \left(\begin{array}{c|c} \mathbf{y}_1^{0\prime} & ? \\ \hline \mathbf{Y}_0 & \mathbf{y}^0_{T+1} \end{array} \right), \tag{9.1}$$

where $\mathbf{Y}_0 = (\mathbf{y}^0_2,\ldots,\mathbf{y}^0_N)'$ and $\mathbf{y}^0_i = \left(y^0_{i1},\ldots,y^0_{iT}\right)'$, "?" denotes the data is unavailable. Since in this paper, we consider y^1_{1t} is observed for $T+1,\ldots,T+m$, then the issue of estimating Δ_{it} conditional on $\mathbf{y}^1_{T+1}$ is an issue of predicting y^0_{1t} (e.g., Hsiao & Zhou, 2019, 2024). For notational simplicity, we shall drop the superscript y^0_{it} for $i = 1,\ldots,N$ for the prediction of $y^0_{1,T+1}$ such that the data matrix takes the form

$$\mathbf{Y} = \left(\begin{array}{c|c} \mathbf{y}_1' & ? \\ \hline \mathbf{Y}_0 & \mathbf{y}_{T+1} \end{array} \right). \tag{9.2}$$

Under the nonconfoundedness assumption

$$f\left(y_{it}|d_{1t}\right) = f\left(y_{it}\right) \quad \text{for } i = 1,\ldots,N; t = 1,\ldots,T+m, \tag{9.3}$$

one may either project $y_{1,T+1}$ on its own past value,[1]

$$y_{1,T+1} = \sum_{s=1}^{T} a_s y_{1,s} + v_{1,T+1}. \tag{9.4}$$

or on other units,

$$y_{1,T+1} = \sum_{j=2}^{N} b_j y_{j,T+1} + u_{1,T+1}, \tag{9.5}$$

The former is referred to as the horizontal regression model (HR) and the latter is referred to as the vertical regression model (VR) by Athey, Bayati, Doudchenko, Imbens and Khosravi (2021).

Shen, Ding, Sekhon and Yu (2023) showed that the HR and VR regressions gave identical point estimates of $\hat{y}_{1,T+1}$. However, the results of Shen et al. (2023) only demonstrate the numerical equivalence of the point estimates of $\hat{y}_{1,T+1}$. To obtain the inferential properties of VR and HR estimates, $\hat{y}_{1,T+1}$, we need to postulate

[1] We consider our y_{it} as the observed sample value deviating from the cross-sectional mean at t or the time series mean for each i, i.e., there is no intercept for model (9.4) and (9.5).

explicitly the data generating process of y_{it}. Moreover, Shen et al. (2023) only consider the prediction of $y_{1,T+1}$. In panel measurement literature, often there are multiple measures of the outcomes of the first unit under the treatment, $y_{1,T+1},\ldots,y_{1,T+h}$, and multiple outcomes of control units in the absence of treatment, i.e., the data are in the form

$$\mathbf{Y}=\left(\begin{array}{c|c}\mathbf{y}_1' & ? \\ \hline \mathbf{Y}_0 & \mathbf{Y}_1\end{array}\right), \tag{9.6}$$

where $\mathbf{Y}_1=\left(\mathbf{y}^m_{2,T_0+1},\ldots,\mathbf{y}^m_{N,T_0+1}\right)'$ denotes the $(N-1)\times m$ matrix of observations with $\mathbf{y}^m_{j,T+1}=\left(y_{j,T+1},\ldots,y_{j,T+m}\right)'$ for $j=2,\ldots,N$, and "?" denotes the data $\mathbf{y}^m_{1,T+1}=\left(y_{1,T+1},\ldots,y_{1,T+m}\right)'$ are missing for $m>1$. Since $y_{1,T+j}$ are not available for the HR regression while $y_{i,T+h}$ are available, the identity of the point prediction of $\hat{y}_{1,T+h}$ between HR and VR no longer holds. This raises the questions: (i) What are the statistical properties of HR vs VR? (ii) How best to generate HR prediction for $h>1$?

In this paper, we give conditions for the DGP for the expected values of $\{a_j\}$ and $\{b_s\}$ to be independent of N or T, respectively in Section 2. In Section 3, we suggest to formulate the data generating process of y_{it} in terms of a factor model as a unified framework for considering the statistical properties of HR or VR under (9.4) or (9.5). Section 4 considers the predictive power of VR and HR . Section 5 provides some Monte Carlo results. Concluding remarks are in Section 6.

9.2 Assumptions Underlying HR or VR for Statistical Inference

Shen et al. (2023) showed the identity of HR and VR point prediction for $y_{1,T+1}$,

$$\hat{y}^{HR}_{1,T+1}=\mathbf{y}_{T+1}'\mathbf{Y}_0\left(\mathbf{Y}_0'\mathbf{Y}_0\right)^{-}\mathbf{y}_1, \tag{9.7}$$

and

$$\hat{y}^{VR}_{1,T+1}=\mathbf{y}_1'\mathbf{Y}_0'\left(\mathbf{Y}_0\mathbf{Y}_0'\right)^{-}\mathbf{y}_{T+1}, \tag{9.8}$$

in terms of the l_2-norm, where $\mathbf{y}_1,\mathbf{y}_{T+1}$ and $\mathbf{Y}_0$ are defined in (9.2) and $(\cdot)^{-}$ denotes the generalized inverse.

For statistical inference of HR or VR predictors, the underlying data generating process of y_{it} needs to be postulated. One set of assumptions postulated by Shen et al. (2023) are

Assumption A1:

$$y_{i,T+1}=\sum_{s=1}^{T}\alpha_s y_{i,s}+\varepsilon_{1,T+1}=\alpha'\mathbf{y}_i+\varepsilon_{i,T+1},\quad i=1,\ldots,N, \tag{9.9}$$

where $\alpha = (\alpha_1, \ldots, \alpha_T)'$ is a $T \times 1$ vector of constants and $\mathbf{y}_i = (y_{i1}, \ldots, y_{iT})'$ is the observation for ith unit in the pre-treatment periods. The error $\varepsilon_{i,T+1}$ satisfies $E(\varepsilon_{i,T+1}|\mathbf{y}_i) = 0$ and are independent over i.

Assumption A2:

$$y_{1,t} = \sum_{i=2}^{N} \beta_i y_{i,t} + \epsilon_{1,t} = \beta' \tilde{\mathbf{y}}_t + \epsilon_{1,t}, \quad t = 1, \ldots, T+1, \tag{9.10}$$

where $\beta = (\beta_2, \ldots, \beta_N)'$ is a $(N-1) \times 1$ vector of constants and $\tilde{\mathbf{y}}_t = (y_{2t}, \ldots, y_{Nt})'$ is the observation for control units at time t. The error $\epsilon_{1,t}$ satisfies $E(\epsilon_{1,t}|\tilde{\mathbf{y}}_t) = 0$ and are independent over t.

However, Assumption A1 and A2 are specific assumptions conditional on sample configuration of N and T. Assumption A1 assumes that y_{it} is an autoregressive process of order T. Assumption A2 assumes y_{it} are cross-correlated over N cross-sectional units.

Under Assumption A1, the algebraic identity between HR and VR predictors no longer holds when N is fixed and T increased to T^*, similarly, for N increases to N^* under assumption A2. Furthermore, the weight vector α and β are sample configuration specific. It is not clear what they will approach when N and/or T increase. Moreover, the randomness of y_{it} is due to $\varepsilon_{i,T+1}$ or $\epsilon_{1,t}$. To justify $\hat{\alpha} = (\mathbf{Y}_0'\mathbf{Y}_0)^- \mathbf{Y}_0'\mathbf{y}_{T+1}$ or $\hat{\beta} = (\mathbf{Y}_0\mathbf{Y}_0')^- \mathbf{Y}_0\mathbf{y}_1'$ being the l_2-norm estimates of the weights α and β, we will need additional assumptions, where $()^-$ denotes generalized inverse. To justify (9.7) and (9.8) being the l_2-norm estimates under the panel structure, we propose to replace Assumption A1 and A2 by[2]

Assumption B1. For any N, the $N \times 1$ vector $\mathbf{y}_t = (y_{1t}, \ldots, y_{Nt})'$ is randomly distributed with mean μ_N and constant nonsingular covariance matrix Σ_N.

Assumption B2. For any T, the $T \times 1$ vector $\mathbf{y}_i = (y_{i1}, \ldots, y_{iT})'$ is randomly distributed with mean μ_T and constant nonsingular covariance matrix Σ_T.

Under Assumption B1, the l_2-norm estimate of the weight vector β converges to a constant when $T \to \infty$,

$$\left[E(\tilde{\mathbf{y}}_t\tilde{\mathbf{y}}_t')\right]^- E(\tilde{\mathbf{y}}_t y_{1,t}). \tag{9.11}$$

Under Assumption B2, the l_2-norm estimate of the weight vector α converges to a constant when $N \to \infty$,

$$\left[E(\mathbf{y}_i\mathbf{y}_i')\right]^- E(\mathbf{y}_i y_{i,T+1}). \tag{9.12}$$

9.3 Factor Modeling to Unify the Derivation of Statistical Properties of HR and VR

The HR model is defined as

$$y_{i,T+1} = E(y_{i,T+1}|\mathbf{y}_i) + \eta_{i,T+1} = \alpha'\mathbf{y}_i + \eta_{i,T+1}. \tag{9.13}$$

[2] Assumption B1 and B2 can allow α and β being random (e.g., Swamy, 1971, Hsiao, 1974, 1975).

The VR model is defined as

$$y_{i,T+1} = E\left(y_{i,T+1}|\tilde{\mathbf{y}}_{T+1}^{-i}\right) + \eta_{i,T+1}^{*} = \beta'\tilde{\mathbf{y}}_{T+1}^{-i} + \eta_{i,T+1}^{*}, \tag{9.14}$$

where $\mathbf{y}_i = (y_{i1}, \dots y_{iT})'$, $\tilde{\mathbf{y}}_{T+1}^{-i} = \left(y_{1t}, \dots, y_{i-1,t}, y_{i+1,t}, \dots y_{Nt}\right)'$ denotes $(N-1)\times 1$ vector of y_{jt} excluding y_{it}.the data for control units at time t. $E(\cdot|\cdot)$ denotes linear projection or conditional expectation if y_{it} are jointly normally distributed. We consider the probability distribution of HR or VR predictor $\hat{y}_{1,T+1}$ under the assumption that the data generating process of y_{it} is given by [3]

$$y_{it} = \lambda_i'\mathbf{f}_t + u_{it}, \qquad i = 1, \dots, N; t = 1, \dots, T+m, \tag{9.15}$$

where the time-varying factors , $\mathbf{f}_t$, are common across i with their impact on the i-th individual measured by λ_i, where λ_i are time-invariant but vary across i due to difference in endowment and innate ability. The idiosyncratic component u_{it} varies over i and t with $E(u_{it}|\lambda_i, \mathbf{f}_t) = 0$. The number of factors is unknown, but can be identified following the procedure of Bai and Ng (2002) or Ahn and Horenstein (2013).

Let $\Lambda = (\lambda_1, \lambda_2, \dots, \lambda_N)' = (\lambda_1, \tilde{\Lambda})'$, $\mathbf{F} = (\mathbf{f}_1, \dots, \mathbf{f}_T)'$, $\mathbf{u}_i = (u_{i1}, \dots, u_{iT})'$, and $\mathbf{u}_t = (u_{1t}, u_{2t}, \dots, u_{Nt})' = (u_{1t}, \tilde{\mathbf{u}}_t')'$. Stacking all N cross-sectional units one after another at time t, $\mathbf{y}_t = (y_{1t}, y_{2t}, \dots, y_{Nt})' = (y_{1t}, \tilde{\mathbf{y}}_t')'$, we have

$$\mathbf{y}_t = \Lambda\mathbf{f}_t + \mathbf{u}_t, \qquad t = 1, \dots, T+m. \tag{9.16}$$

Alternatively, we can stack ith individual's T time series observations as $\mathbf{y}_i = (y_{i1}, \dots, y_{iT})'$, then

$$\mathbf{y}_i = \mathbf{F}\lambda_i + \mathbf{u}_i, \qquad i = 1, \dots, N. \tag{9.17}$$

For model (9.15), the common assumption for λ_i and $\mathbf{f}_t$ are (e.g., Bai & Ng, 2006, 2021, Bai, 2009, Li, Shen & Zhou, 2024):

(i) both λ_i and $\mathbf{f}_t$ are fixed constants (e.g., Hsiao, Ching & Wan, 2012);

(ii) λ_i is fixed and $\mathbf{f}_t$ is randomly distributed with $E(\mathbf{f}_t) = \mu_f$;[4]

(iii) $\mathbf{f}_t$ is fixed and λ_i is randomly distributed with $E(\lambda_i) = \mu_\lambda$;[5]

Following Bai (2003, 2009) and Bai and Ng (2002, 2021), we assume:

Assumption C1: The factor process satisfies $E\|\mathbf{f}_t\|^4 \le M < \infty$ and $\frac{1}{T}\sum_{t=1}^{T}\mathbf{f}_t\mathbf{f}_t' \to_p \Sigma_f$, if $\mathbf{f}_t$ is random or $\left\|\frac{1}{T}\sum_{t=1}^{T}\mathbf{f}_t\mathbf{f}_t' - \Sigma_f\right\| \to 0$ if $\mathbf{f}_t$ is fixed constant, where Σ_f is an $r \times r$ positive definite matrix with bounded eigenvalues.[6]

[3] It may be worthwhile to explore treatment effects for sub-layer units within the context of 3D factor models , as discussed by Jin, Lu and Su (2024).

[4] Here, we are considering y_{it} as the deviation from its time series mean, it is equivalent to assuming $\mu_f = 0$.

[5] When we are considering y_{it} as the deviation from its cross-sectional mean, it is equivalent to assuming $\mu_\lambda = 0$.

[6] We let $\|\mathbf{A}\| = \sqrt{tr(\mathbf{AA}')}$.

Assumption C2: The loading λ_i is either fixed constant with $\|\lambda_i\| < \infty$ or is stochastic with $E\|\lambda_i\|^4 \leq M < \infty$. If λ_i is random, then $\frac{1}{N}\sum_{i=1}^{N}\lambda_i\lambda_i' \rightarrow_p \Sigma_\lambda$. If λ_i is fixed, then $\left\|\frac{1}{N}\sum_{i=1}^{N}\lambda_i\lambda_i' - \Sigma_\lambda\right\| \rightarrow 0$, where Σ_λ is an $r \times r$ positive definite matrix with bounded eigenvalues.

and either

Assumption C3: The random error terms $\mathbf{u}_t$ is independently identically distributed over t with nonsingular covariance matrix

$$E(\mathbf{u}_t\mathbf{u}_t') = \tilde{\Omega} = \begin{pmatrix} \sigma_1^2 & \mathbf{c}' \\ \mathbf{c} & \Omega \end{pmatrix}, \tag{9.18}$$

where $\sigma_1^2 = E\left(u_{1t}^2\right)$, $\Omega = E(\tilde{\mathbf{u}}_t\tilde{\mathbf{u}}_t')$, and $\mathbf{c} = E(\tilde{\mathbf{u}}_t u_{1t})$ with $\tilde{\mathbf{u}}_t = (u_{2t},\ldots,u_{Nt})'$. Moreover, all N nonzero eigenvalues of $\tilde{\Omega}$ are $O(1)$.

or

Assumption C4: The random error terms $\mathbf{u}_i$ is independently identically distributed over i with nonsingular covariance matrix

$$E\begin{pmatrix} u_{i,T+1} \\ \mathbf{u}_i \end{pmatrix}(u_{i,T+1}, \mathbf{u}_i') = \begin{pmatrix} \sigma_1^2 & \mathbf{c}^{*\prime} \\ \mathbf{c}^* & \Omega^* \end{pmatrix}.$$

where $\sigma_1^2 = E\left(u_{i1}^2\right)$, $\Omega^* = E(\mathbf{u}_i\mathbf{u}_i')$, and $\mathbf{c}^* = E(\mathbf{u}_i u_{i,T+1})$ with $\mathbf{u}_i = (u_{i1},\ldots,u_{iT})'$. Moreover, all T nonzero eigenvalues of Ω^* are $O(1)$.

Assumptions C1 and C2 are standard assumptions for factor models . They allow Λ and $\mathbf{F}$ to be either fixed constants or random. Assumption C3 allows u_{it} to be heteroskedastic and weakly cross-correlated but independent over t. Assumption C4 allows u_{it} to be serially correlated but independent over i. In principle, one can allow u_{it} to be both weakly cross-correlated and weakly time-dependent as Bai (2003, 2009) or Hsiao, Shi and Zhou (2022).

There is no loss of generality to let $i = 1$ and $t = T+1$ in (9.13) or (9.14), i.e., only the first unit is treated and there is only one post-treatment period.

Lemma 9.1 *For cases (i), (ii) and (iii), conditional on* $(\mathbf{Y}_0, \mathbf{y}_1, \tilde{\mathbf{y}}_{T+1})$,

(a) Under Assumption C1, C2 and C4, the HR model is a constant parameter regression model

$$y_{1,T+1} = E\left(y_{1,T+1}|\mathbf{y}_1\right) + \eta_{T+1} = \alpha'\mathbf{y}_1 + \eta_{T+1}, \tag{9.19}$$

where

$$\alpha = (\mathbf{F}\Sigma_\lambda\mathbf{F}' + \Omega^*)^{-1}(\mathbf{F}\Sigma_\lambda\mathbf{f}_{T+1} + \mathbf{c}^*), \tag{9.20}$$

and the variance of η_{T+1} *equals to*

$$\sigma_\eta^2 = \sigma_1^2 + \mathbf{f}_{T+1}'\Sigma_\lambda\mathbf{f}_{T+1} - (\mathbf{f}_{T+1}'\Sigma_\lambda\mathbf{F}' + \mathbf{c}^{*\prime})(\mathbf{F}\Sigma_\lambda\mathbf{F}' + \Omega^*)^{-1}(\mathbf{F}\Sigma_\lambda\mathbf{f}_{T+1} + \mathbf{c}^*). \tag{9.21}$$

The l_2*-norm weight vector for* α *is estimated by*

$$\hat{\alpha} = \left(\sum_{i=2}^{N} \mathbf{y}_i \mathbf{y}_i'\right)^{-1} \sum_{i=2}^{N} \mathbf{y}_i y_{i,T+1} = \left(\mathbf{Y}_0'\mathbf{Y}_0\right)^{-1} \mathbf{Y}_0' \mathbf{y}_{T+1}. \tag{9.22}$$

When $N \longrightarrow \infty$, $\hat{\alpha} \longrightarrow_p \alpha$.

(b) Under Assumption C1-C3, the VR model is a constant parameter regression model

$$y_{1,T+1} = E\left(y_{1,T+1}|\tilde{\mathbf{y}}_{T+1}\right) + \eta_{T+1}^* = \beta'\tilde{\mathbf{y}}_{T+1} + \eta_{T+1}^*, \tag{9.23}$$

where

$$\beta = (\tilde{\Lambda}\Sigma_f\tilde{\Lambda}' + \Omega)^{-1}(\tilde{\Lambda}\Sigma_f\lambda_1 + \mathbf{c}), \tag{9.24}$$

and

$$\sigma_{\eta*}^2 = \sigma_1^2 + \lambda_1'\Sigma_f\lambda_1 - (\lambda_1'\Sigma_f\tilde{\Lambda}' + \mathbf{c}')(\tilde{\Lambda}\Sigma_f\tilde{\Lambda}' + \Omega)^{-1}(\tilde{\Lambda}\Sigma_f\lambda_1 + \mathbf{c}). \tag{9.25}$$

The l_2-norm weight vector for β is estimated by

$$\hat{\beta} = \left(\sum_{t=1}^{T} \tilde{\mathbf{y}}_t \tilde{\mathbf{y}}_t'\right)^{-1} \sum_{t=1}^{T} \tilde{\mathbf{y}}_t y_{1t} = \left(\mathbf{Y}_0\mathbf{Y}_0'\right)^{-1} \mathbf{Y}_0 \mathbf{y}_1. \tag{9.26}$$

When $T \longrightarrow \infty$, $\hat{\beta} \longrightarrow_p \beta$.

Proof is provided in the Appendix.

Given the results of Lemma 9.1, the HR predictor of $y_{1,T+1}$ is

$$\hat{y}_{1,T+1}^{HR} = \hat{\alpha}'\mathbf{y}_1, \tag{9.27}$$

where $\hat{\alpha}$ is provided in (9.22), and the VR prediction of $y_{1,T+1}$ is

$$\hat{y}_{1,T+1}^{VR} = \hat{\beta}'\tilde{\mathbf{y}}_{T+1}, \tag{9.28}$$

where $\hat{\beta}$ is provided in (9.26).

Lemma 9.2 *Under Assumptions C1 and C2, conditional on $(\mathbf{Y}_0, \mathbf{y}_1, \tilde{\mathbf{y}}_{T+1})$, when u_{it} is i.i.d over i and t,*

(a) The MSPE for VR is smaller than that of the HR when $T \longrightarrow \infty$ and N is finite or when $(N,T) \longrightarrow \infty$ and $\frac{N}{T} \longrightarrow 0$.

(b) When $N \longrightarrow \infty$ and T is fixed or when $(N,T) \longrightarrow \infty$ and $\frac{T}{N} \longrightarrow 0$, the MSPE for HR is smaller than that of the VR.

Proof is provided in the Appendix.

9.4 More than One Period Ahead Measurement of Treatment Effects

When $y_{1,T+1},\ldots,y_{1,T+m}$ and $\mathbf{Y}_1$ are unknown, Bai and Ng (2006) suggest to predict $y_{1,T+h}$ by the diffusion model approach, namely, predicting $y_{1,T+h}$ by

$$\hat{y}_{1,T+h} = \hat{\lambda}_1'\hat{\mathbf{f}}_T, \tag{9.29}$$

which in the case of HR or VR model, would imply predicting $y_{1,T+h}$ by (9.27) or (9.28), where $\hat{\alpha}$ and $\hat{\beta}$ are given by (9.22) and (9.26), respectively. In other words, the diffusion model prediction of $y_{1,T+h}$ is a constant for $h = 1,\ldots,m$. However, in the case of panel treatment effects measurement, there are post-treatment control units information such as the data in the form of (9.6) that can be exploited to get more accurate prediction of $y_{1,T+h}$. Here, we suggest three approaches:

Approach 1: VR prediction

$$\hat{y}^{VR}_{1,T+h} = \hat{\beta}'\tilde{\mathbf{y}}_{T+h}, \tag{9.30}$$

where $\hat{\beta}$ is estimated by (9.26) and $\tilde{\mathbf{y}}_{T+h} = \left(y_{2,T+h}, y_{3,T+h},\ldots,y_{N,T+h}\right)'$.

Approach 2: HR prediction conditional on $\mathbf{Y}_0$, $\tilde{\mathbf{y}}_{T+h}$ and $\mathbf{y}_1$,

$$y_{1,T+h} = E\left(y_{1,T+h}|\mathbf{Y}_0, \tilde{\mathbf{y}}_{T+h}, \mathbf{y}_1\right) + \eta_{T+h} = \alpha^{(h)\prime}\mathbf{y}_1 + \eta_{T+h}, \tag{9.31}$$

where $\alpha^{(h)} = \left(E\left(\mathbf{y}_i\mathbf{y}_i'\right)\right)^{-1} E\left(\mathbf{y}_i y_{i,T+h}\right)$. Then the l_2-norm estimate of $\alpha^{(h)}$ is given by

$$\hat{\alpha}^{(h)} = \left(\sum_{i=2}^{N}\mathbf{y}_i\mathbf{y}_i'\right)^{-1}\sum_{i=2}^{N}\mathbf{y}_i y_{i,T+h} = \left(\mathbf{Y}_0'\mathbf{Y}_0\right)^{-1}\mathbf{Y}_0'\tilde{\mathbf{y}}_{T+h}, \tag{9.32}$$

and the associated HR predicted outcome

$$\hat{y}^{HR,1}_{1,T+h} = \hat{\alpha}^{(h)\prime}\mathbf{y}_1, \quad h = 1,\ldots,m. \tag{9.33}$$

Approach 3: HR prediction conditional on $\mathbf{y}_1$, $\mathbf{Y}_0$ and $\mathbf{Y}_1$

$$y_{1,T+h} = E\left(y_{1,T+h}|\mathbf{y}_1, \mathbf{Y}_0, \hat{\mathbf{y}}_1^{T+h-1}\right) + \eta^{**}_{T+h}, \tag{9.34}$$

where $\hat{\mathbf{y}}_1^{T+h-1} = \left(\mathbf{y}_1', \hat{y}_{1,T+1},\ldots,\hat{y}_{1,T+h-1}\right)'$. Since the outcomes of the first unit at period $T+j$, denoted as $\hat{y}_{1,T+j}$ are not available for $j = 1,\ldots,h-1$, we consider replacing unknown $y_{1,T+j}$ by VR predictor

$$\hat{y}_{1,T+j} = \hat{\beta}'\tilde{\mathbf{y}}_{T+j}, \quad j = 1,\ldots,h-1, \tag{9.35}$$

in implementing HR. Then the associated predictor is:

$$\hat{y}^{HR,2}_{1,T+h} = \hat{\alpha}'\hat{\mathbf{y}}_1^{T+h-1}, \quad h = 1,\ldots,m, \tag{9.36}$$

where $\hat{\mathbf{y}}_1^{T+h-1} = (\mathbf{y}_1', \hat{y}_{1,T+1}, \ldots, \hat{y}_{1,T+h-1})$,

$$\hat{\alpha} = \left(\sum_{i=2}^{N} \mathbf{y}_i^{T+h-1} \mathbf{y}_i^{T+h-1\prime}\right)^{-1} \left(\sum_{i=2}^{N} \mathbf{y}_i^{T+h-1} y_{i,T+h}\right)$$
$$= \left(\mathbf{Y}^{T+h-1\prime} \mathbf{Y}^{T+h-1}\right)^{-1} \mathbf{Y}^{T+h-1\prime} \tilde{\mathbf{y}}_{T+h}, \tag{9.37}$$

$\mathbf{Y}^{T+h-1} = (\mathbf{Y}_0, \mathbf{Y}_{1,T+h-1})$, $\mathbf{Y}_{1,T+h-1} = (\tilde{\mathbf{y}}_{T+1}, \tilde{\mathbf{y}}_{T+2}, \ldots, \tilde{\mathbf{y}}_{T+h-1})$ and $\tilde{\mathbf{y}}_{T+s} = (y_{2,T+s}, y_{3,T+s}, \ldots, y_{N,T+s})'$ for $s = 1, 2, \ldots, h-1$.

Lemma 9.3 *Approach 1 (e.g., (9.30)), Approach 2 (e.g., (9.33)) and Approach 3 (e.g., (9.36)) yield identical point prediction for* $y_{1,T+h}$.

Proof is provided in the Appendix.

Lemma 9.4 *Suppose Assumption C1 and C2 hold, h is fixed and u_{it} is i.i.d over i and t. Denote* $\mathbf{y}_1^{T+h-1} = (\mathbf{y}_1', y_{1,T+1}, y_{1,T+2}, \ldots, y_{1,T+h-1})'$.

(i) When $T \longrightarrow \infty$ and N is finite or $(N,T) \longrightarrow \infty$ and $\frac{N}{T} \longrightarrow 0$, the VR prediction of $y_{1,T+h}$ using $\hat{y}_{1,T+h}^{VR}$ has smaller MSPE than that using $\hat{y}_{1,T+h}^{HR,1}$.

(ii) When $N \longrightarrow \infty$ and T is finite or $(N,T) \longrightarrow \infty$ and $\frac{T}{N} \longrightarrow 0$, the HR prediction of $y_{1,T+h}$ using $\hat{y}_{1,T+h}^{HR,1}$ has a smaller MSPE than that using $\hat{y}_{1,T+h}^{VR}$.

(iii) If $\mathbf{y}_1$, $\mathbf{Y}_0$, $\mathbf{Y}_1$, $\mathbf{y}_1^{T+h-1}$ are available, define the VR prediction as

$$\hat{y}_{1,T+h}^{VR,0} = \tilde{\beta}' \tilde{\mathbf{y}}_{T+h}, \tag{9.38}$$

where $\tilde{\beta} = \left(\mathbf{Y}^{T+h-1} \mathbf{Y}^{T+h-1\prime}\right)^{-1} \mathbf{Y}^{T+h-1} \left(\mathbf{y}_1^{T+h}\right)'$, where $\mathbf{Y}^{T+h-1} = \left(\mathbf{Y}_0, \mathbf{Y}_1^{T+h-1}\right)$. Suppose $T \longrightarrow \infty$ and $\frac{N}{T} \longrightarrow 0$, then prediction of $y_{1,T+h}$ using $\hat{y}_{1,T+h}^{VR}$ has the same MSPE as that using $\hat{y}_{1,T+h}^{VR,0}$.

(iv) If $\mathbf{y}_1$, $\mathbf{Y}_0$, $\mathbf{Y}_1$, $\mathbf{y}_1^{T+h-1}$ are available, define the HR prediction as

$$\hat{y}_{1,T+h}^{HR,0} = \hat{\alpha}' \mathbf{y}_1^{T+h-1}, \tag{9.39}$$

where $\hat{\alpha}$ is defined in (9.37). Suppose $N \longrightarrow \infty$ and $\frac{T}{N} \longrightarrow 0$, then the prediction using $\hat{y}_{1,T+h}^{HR,0}$ is smaller than that using $\hat{y}_{1,T+h}^{HR,1}$ in (9.33).

Proof is provided in the Appendix.

Remark 9.1 Under Assumption C1 and C2, Lemma 9.3 implies that the diffusion index prediction of $y_{1,T+h}$ in (9.30) or (9.31) yields identical point prediction. There is no need to consider making use of observed $\tilde{\mathbf{y}}_{T+1}, \ldots, \tilde{\mathbf{y}}_{T+h-1}$ as by (9.36). They do not change the prediction of $y_{1,T+h}$. However, their prediction error variance depends on the sample configuration of N and T. Moreover, when VR prediction is considered and sample size is sufficiently large, the feasible predictor $\hat{y}_{1,T+h}^{VR}$ is shown to be as accurate as the infeasible predictor $\hat{y}^{VR,0}_{1,T+h}$. In the case where HR prediction is used, however, the feasible predictor $\hat{y}^{HR,1}_{1,T+h}$ is shown to be less accurate as the infeasible predictor $\hat{y}^{HR,0}_{1,T+h}$.

9.5 Monte Carlo Simulations

We consider a panel model with only one factor:

$$y_{it} = \lambda_i f_t + u_{it} \tag{9.40}$$

where $\lambda_i \sim i.i.d.\ \mathcal{N}(1,1)$ are independent of the error term $u_{it} \sim i.i.d.\ \mathcal{N}(0,\sigma_u^2)$ with $\sigma_u = 1$ and the factor f_t is generated as

$$\begin{aligned}&(1)\ f_t = v_t,\\ &(2)\ f_t = 0.5 f_{t-1} + v_t,\\ &(3)\ f_t = 0.9 f_{t-1} + v_t,\end{aligned} \tag{9.41}$$

where $v_t \sim i.i.d.\ \mathcal{N}(0,\sigma_v^2)$ with $\sigma_v = 1$. We set the number of control units to be $N = 30, 50, 100$, the pre-treatment time $T = 30, 50, 100$, and the post-treatment periods $h = 1,2,3,4,5$. The number of replications is set at $R = 1000$.

Though there is algebraic equivalence among Approach 1, 2 , and 3, we include all of them to verify our Lemma 9.3. We also compare the multi-period ahead prediction accuracy of feasible horizontal regression (Approach 3 defined in (9.36)), infeasible horizontal regression[7] (9.39), and the diffusion model prediction (Bai & Ng, 2006) under the assumption that the factor dimension r is known (here $r = 1$), but the factor and factor loading are unknown. We consider two versions of the diffusion model method.

Diffusion model method A:

- Step 1: Use PCA approach to estimate $\hat{f}_t$ for $t = 1,\cdots,T$ based on the data $\{y_{it}\}_{t=1,\cdots,T}^{i=1,\cdots,N+1}$;
- Step 2: Regress $y_{1,t}$ on $\hat{f}_t$ for $t = 1,...,T$ to get estimated factor loading $\hat{\lambda}_1 = \frac{\sum_{t=1}^T \hat{f}_t y_{1,t}}{\sum_{t=1}^T \hat{f}_t^2}$;
- Step 3: Compute predicted value for $y_{1,T+h}$ by:

$$\hat{y}^{DF,A}_{1,T+h} = \hat{\lambda}_1 \hat{f}_T \tag{9.42}$$

Diffusion model method B:

[7] It is unnecessary to include the infeasible vertical regression (9.38), as it produces identical predictions to the infeasible horizontal regression by Shen et al. (2023).

- Step 1: Use PCA approach to estimate $\hat{f}_t$ for $t = 1, \cdots, T$ based on the data $\{y_{it}\}_{t=1,\cdots,T}^{i=1,\cdots,N+1}$;
- Step 2: Regress $y_{1,t}$ on $\hat{f}_{t-h}$ for $t = h+1, ..., T$ to get estimated coefficients $\hat{c}_h = \frac{\sum_{t=h+1}^{T} \hat{f}_{t-h} y_{1,t}}{\sum_{t=h+1}^{T} \hat{f}_{t-h}^2}$[8];
- Step 3: Compute predicted value for $y_{1,T+h}$ by:

$$\hat{y}_{1,T+h}^{DF,B} = \hat{c}_h \hat{f}_T. \tag{9.43}$$

We consider three criteria for comparisons: Bias, Mean Absolute Bias (MAB), and Root Mean Squared Prediction Error (RMSPE). Bias represents the mean difference between the true value and the predicted value at each post-treatment time point, while MAB is the mean of the absolute bias. RMSPE is the square root of the mean of the squared difference between the true value and the predicted value.

Table 9.1: Simulation Results for 2-period ahead predictions using Approach 1, 2, 3 and the diffusion method under DGP (1).

N	T	Approach 1			Approach 2			Approach 3		
		Bias	MAB	RMSPE	Bias	MAB	RMSPE	Bias	MAB	RMSPE
	30	0.0146	1.0183	1.2925	0.0146	1.0183	1.2925	0.0146	1.0183	1.2925
10	50	-0.0301	0.9186	1.1558	-0.0301	0.9186	1.1558	-0.0301	0.9186	1.1558
	100	-0.0012	0.8635	1.0791	-0.0012	0.8635	1.0791	-0.0012	0.8635	1.0791
	30	-0.0004	1.4036	1.7925	-0.0004	1.4036	1.7925	-0.0004	1.4036	1.7925
20	50	0.0012	1.0433	1.3149	0.0012	1.0433	1.3149	0.0012	1.0433	1.3149
	100	0.0448	0.9175	1.1603	0.0448	0.9175	1.1603	0.0448	0.9175	1.1603
30		0.0508	1.0386	1.3261	0.0508	1.0386	1.3261	0.0508	1.0386	1.3261
50	10	-0.0393	0.9035	1.1489	-0.0393	0.9035	1.1489	-0.0393	0.9035	1.1489
100		0.0011	0.8746	1.0861	0.0011	0.8746	1.0861	0.0011	0.8746	1.0861

[8] Our data is generated with $E(f_t) = 0$ in the current simulation. If $E(f_t) = \mu_f \neq 0$, then we regress $y_{1,t}$ on the intercept and $\hat{f}_{t-h}$ and generate the prediction for $y_{1,T+h}$ by the estimated intercept plus $\hat{c}_h \hat{f}_T$.

Table 9.2: Simulation Results for 2-period ahead predictions using Approach 1, 2, 3 and the diffusion method under DGP (2).

N	T	Approach 1			Approach 2			Approach 3		
		Bias	MAB	RMSPE	Bias	MAB	RMSPE	Bias	MAB	RMSPE
	30	-0.0148	1.0269	1.2883	-0.0148	1.0269	1.2883	-0.0148	1.0269	1.2883
10	50	0.0221	0.9766	1.2178	0.0221	0.9766	1.2178	0.0221	0.9766	1.2178
	100	-0.0361	0.8715	1.0958	-0.0361	0.8715	1.0958	-0.0361	0.8715	1.0958
	30	0.0197	1.4870	1.8576	0.0197	1.4870	1.8576	0.0197	1.4870	1.8576
20	50	-0.0730	1.0697	1.3421	-0.0730	1.0697	1.3421	-0.0730	1.0697	1.3421
	100	-0.0455	0.9572	1.2063	-0.0455	0.9572	1.2063	-0.0455	0.9572	1.2063
30		-0.0427	0.9788	1.2483	-0.0427	0.9788	1.2483	-0.0427	0.9788	1.2483
50	10	-0.0620	0.9056	1.1545	-0.0620	0.9056	1.1545	-0.0620	0.9056	1.1545
100		-0.0045	0.9007	1.1299	-0.0045	0.9007	1.1299	-0.0045	0.9007	1.1299

Table 9.3: Simulation Results for 5-period ahead predictions using Approach 1, 2, 3 and the diffusion method under DGP (3).

N	T	Approach 1			Approach 2			Approach 3		
		Bias	MAB	RMSPE	Bias	MAB	RMSPE	Bias	MAB	RMSPE
	30	-0.0391	1.0682	1.3359	-0.0391	1.0682	1.3359	-0.0391	1.0682	1.3359
10	50	-0.0064	0.9561	1.2048	-0.0064	0.9561	1.2048	-0.0064	0.9561	1.2048
	100	-0.0096	0.8722	1.0954	-0.0096	0.8722	1.0954	-0.0096	0.8722	1.0954
	30	-0.0505	1.4530	1.8572	-0.0505	1.4530	1.8572	-0.0505	1.4530	1.8572
20	50	0.0068	1.0723	1.3506	0.0068	1.0723	1.3506	0.0068	1.0723	1.3506
	100	-0.0113	0.8984	1.1354	-0.0113	0.8984	1.1354	-0.0113	0.8984	1.1354
30		-0.0047	1.0710	1.3443	-0.0047	1.0710	1.3443	-0.0047	1.0710	1.3443
50	10	-0.0483	0.9615	1.2174	-0.0483	0.9615	1.2174	-0.0483	0.9615	1.2174
100		0.0093	0.8956	1.1317	0.0093	0.8956	1.1317	0.0093	0.8956	1.1317

Table 9.4: Simulation Results for 1-period ahead predictions using infeasible horizontal method (9.39), Approach 3, and the diffusion methods (9.42), (9.43) under DGP 1-3.

N	T	Infeasible HR			Approach 3			DiffusionA			DiffusionB		
		Bias	MAB	RMSPE	Bias	MAB	RMSPE	Bias	MAB	RMSPE	Bias	MAB	RMSPE
DGP (1)													
10	50	-0.0120	0.9240	1.1642	-0.0120	0.9240	1.1642	-0.0823	1.5900	2.1726	-0.1054	1.2581	1.6746
10	100	0.0007	0.8841	1.1219	0.0007	0.8841	1.1219	-0.1309	1.6457	2.2551	-0.0400	1.3373	1.7847
10	200	0.0416	0.8173	1.0269	0.0416	0.8173	1.0269	0.0583	1.5768	2.2154	0.0707	1.2896	1.7507
500	20	0.0269	0.8166	1.0174	0.0269	0.8166	1.0174	-0.0118	1.5920	2.2089	0.0052	1.3360	1.7979
700	20	0.0143	0.8372	1.0393	0.0143	0.8372	1.0393	0.0462	1.5707	2.1943	0.0071	1.3004	1.7556
1000	20	0.0390	0.8167	1.0335	0.0390	0.8167	1.0335	0.1781	1.5891	2.2087	0.0695	1.2596	1.7118
DGP (2)													
10	50	-0.0224	0.9303	1.1691	-0.0224	0.9303	1.1691	-0.0229	1.3545	1.8484	-0.0055	1.2839	1.7455
10	100	-0.0736	0.8689	1.1076	-0.0736	0.8689	1.1076	-0.0725	1.3811	1.8696	-0.0392	1.2841	1.7270
10	200	-0.0346	0.8592	1.0875	-0.0346	0.8592	1.0875	-0.0265	1.4030	1.9315	-0.0407	1.2888	1.7297
500	20	-0.0019	0.8495	1.0475	-0.0019	0.8495	1.0475	-0.0603	1.4844	2.0122	-0.0441	1.3526	1.7819
700	20	0.0010	0.8510	1.0621	0.0010	0.8510	1.0621	0.0731	1.4960	2.0053	0.0720	1.3923	1.8465
1000	20	0.0162	0.8415	1.0333	0.0162	0.8415	1.0333	0.0215	1.4609	2.0344	0.0223	1.3668	1.8265
DGP (3)													
10	50	-0.0217	0.9354	1.1724	-0.0217	0.9354	1.1724	-0.0287	1.2874	1.7343	-0.0203	1.2811	1.7412
10	100	-0.0715	0.8707	1.1029	-0.0715	0.8707	1.1029	-0.0448	1.3246	1.7743	-0.0468	1.3126	1.7628
10	200	-0.0305	0.8613	1.0884	-0.0305	0.8613	1.0884	-0.0336	1.3297	1.8035	-0.0371	1.3032	1.7609
500	20	-0.0023	0.8438	1.0456	-0.0023	0.8438	1.0456	-0.0567	1.3608	1.8252	-0.0632	1.3571	1.8033
700	20	0.0002	0.8501	1.0656	0.0002	0.8501	1.0656	0.0750	1.4004	1.8772	0.0681	1.4032	1.8631
1000	20	0.0093	0.8306	1.0314	0.0093	0.8306	1.0314	0.0186	1.3648	1.8336	0.0026	1.3667	1.8145

Table 9.5: Simulation Results for 2-period ahead predictions using infeasible horizontal method (9.39), Approach 3, and the diffusion methods (9.42), (9.43) under DGP 1-3.

N	T	Infeasible HR			Approach 3			DiffusionA			DiffusionB		
		Bias	MAB	RMSPE	Bias	MAB	RMSPE	Bias	MAB	RMSPE	Bias	MAB	RMSPE
DGP (1)													
	50	0.0028	0.9349	1.1701	0.0026	0.9372	1.1754	-0.1393	1.6082	2.1775	-0.0390	1.2641	1.6763
10	100	-0.0088	0.8761	1.1252	-0.0089	0.8766	1.1261	0.0055	1.6386	2.3981	-0.0100	1.3065	1.7775
	200	-0.0629	0.9114	1.1290	-0.0629	0.9113	1.1295	-0.0447	1.6896	2.3160	-0.0697	1.3203	1.7755
500		-0.0136	0.8445	1.0581	-0.0124	0.8441	1.0578	0.0125	1.6642	2.2295	0.0148	1.3586	1.7619
700	20	0.0276	0.8155	1.0299	0.0259	0.8149	1.0306	-0.0932	1.5292	2.0718	-0.0304	1.3165	1.7328
1000		-0.0522	0.8134	1.0212	-0.0541	0.8125	1.0200	-0.0978	1.6266	2.2929	-0.0991	1.3201	1.7740
DGP (2)													
	50	-0.0354	0.9104	1.1469	-0.0322	0.9123	1.1503	0.0477	1.5771	2.1909	0.0142	1.3070	1.7813
10	100	-0.0181	0.8829	1.1074	-0.0183	0.8838	1.1091	0.0390	1.7132	2.4040	-0.0267	1.4467	2.0194
	200	-0.0395	0.8559	1.0728	-0.0390	0.8559	1.0725	-0.0158	1.6585	2.2874	-0.0126	1.4081	1.8793
500		0.0307	0.8405	1.0773	0.0359	0.8401	1.0774	0.0595	1.5945	2.2163	0.0170	1.4291	1.9067
700	20	0.0277	0.8266	1.0395	0.0273	0.8271	1.0388	0.0326	1.5639	2.1350	0.0795	1.4220	1.9086
1000		0.0089	0.8310	1.0443	0.0132	0.8301	1.0450	0.0891	1.6463	2.2449	0.0940	1.4510	1.9237
DGP (3)													
	50	-0.0341	0.9231	1.1567	-0.0322	0.9259	1.1626	0.0358	1.5947	2.2056	0.0325	1.5214	2.1030
10	100	-0.0122	0.8780	1.1029	-0.0124	0.8790	1.1041	-0.0059	1.6989	2.4129	-0.0248	1.6683	2.3586
	200	-0.0362	0.8634	1.0770	-0.0354	0.8629	1.0765	0.0096	1.6875	2.3306	0.0165	1.6231	2.2210
500		0.0377	0.8517	1.0875	0.0462	0.8506	1.0873	0.0200	1.6407	2.2488	0.0016	1.6598	2.2800
700	20	0.0318	0.8213	1.0367	0.0299	0.8257	1.0383	0.0667	1.6010	2.2022	0.1052	1.6147	2.2407
1000		0.0181	0.8328	1.0454	0.0207	0.8358	1.0523	0.1224	1.6509	2.2167	0.1285	1.6880	2.2960

Table 9.6: Simulation Results for 3-period ahead predictions using infeasible horizontal method (9.39), Approach 3, and the diffusion methods (9.42), (9.43) under DGP 1-3.

N	T	Infeasible HR			Approach 3			DiffusionA			DiffusionB		
		Bias	MAB	RMSPE	Bias	MAB	RMSPE	Bias	MAB	RMSPE	Bias	MAB	RMSPE
DGP (1)													
10	50	0.0210	0.9686	1.2158	0.0246	0.9763	1.2234	0.0341	1.6888	2.3420	-0.0435	1.3222	1.7397
	100	0.0225	0.9011	1.1249	0.0219	0.9047	1.1291	-0.0503	1.5504	2.1339	-0.0566	1.2972	1.7532
	200	-0.0042	0.8750	1.0962	-0.0040	0.8757	1.0964	-0.0833	1.6316	2.3081	0.0018	1.2922	1.7201
500	20	0.0089	0.8149	1.0172	0.0120	0.8127	1.0172	0.0947	1.6050	2.2517	0.0368	1.3255	1.7911
700		-0.0217	0.8182	1.0180	-0.0230	0.8193	1.0203	-0.0989	1.5808	2.2137	-0.0322	1.2792	1.6977
1000		0.0103	0.8375	1.0366	0.0078	0.8374	1.0360	-0.0990	1.6593	2.2967	-0.0843	1.3897	1.8376
DGP (2)													
10	50	0.0177	0.9032	1.1434	0.0178	0.9018	1.1442	-0.0030	1.7456	2.4549	-0.0009	1.4764	2.0111
	100	0.0167	0.8732	1.0944	0.0167	0.8757	1.0963	0.0740	1.6687	2.2955	0.0592	1.4055	1.8923
	200	0.0239	0.8816	1.1021	0.0239	0.8827	1.1029	-0.0240	1.7642	2.3747	0.0300	1.4504	1.9515
500	20	-0.0008	0.8265	1.0142	0.0067	0.8304	1.0149	0.1705	1.7326	2.4055	0.0955	1.4787	2.0337
700		-0.0704	0.8157	1.0195	-0.0721	0.8215	1.0229	-0.0715	1.7479	2.4310	-0.0774	1.4649	2.0060
1000		-0.0592	0.8445	1.0662	-0.0560	0.8458	1.0687	0.0668	1.6746	2.3368	-0.0049	1.4368	1.9507
DGP (3)													
10	50	0.0184	0.8971	1.1381	0.0195	0.8975	1.1410	-0.0536	1.9073	2.7191	-0.0444	1.8639	2.6126
	100	0.0108	0.8771	1.1035	0.0120	0.8794	1.1054	0.0317	1.8278	2.5562	0.0419	1.7643	2.5191
	200	0.0247	0.8820	1.1020	0.0246	0.8832	1.1030	0.0595	1.9593	2.6865	0.0826	1.8275	2.5160
500	20	-0.0027	0.8335	1.0244	0.0073	0.8375	1.0264	0.1749	1.8344	2.6083	0.1451	1.9268	2.7513
700		-0.0724	0.8241	1.0290	-0.0695	0.8348	1.0369	-0.0608	1.8422	2.6216	-0.0988	1.8518	2.5878
1000		-0.0437	0.8453	1.0620	-0.0386	0.8509	1.0693	-0.0243	1.7240	2.4129	-0.0366	1.7542	2.5283

Table 9.7: Simulation Results for 4-period ahead predictions using infeasible horizontal method (9.39), Approach 3, and the diffusion methods (9.42), (9.43) under DGP 1-3.

N	T	Infeasible HR			Approach 3			DiffusionA			DiffusionB		
		Bias	MAB	RMSPE	Bias	MAB	RMSPE	Bias	MAB	RMSPE	Bias	MAB	RMSPE
DGP (1)													
	50	-0.0570	0.8833	1.1361	-0.0639	0.8944	1.1515	-0.0056	1.5612	2.1453	-0.0062	1.2599	1.6748
10	100	0.0051	0.8941	1.1039	0.0047	0.8965	1.1066	-0.0550	1.5588	2.1522	-0.0227	1.2815	1.7317
	200	0.0112	0.8615	1.0791	0.0119	0.8619	1.0796	0.0011	1.5618	2.1286	-0.0054	1.2342	1.6325
500		0.0455	0.8154	1.0209	0.0475	0.8126	1.0216	0.0264	1.6306	2.2648	-0.0006	1.3350	1.7921
700	20	-0.0284	0.8283	1.0404	-0.0303	0.8261	1.0382	-0.0741	1.6081	2.1977	-0.0638	1.2847	1.7814
1000		0.0153	0.8577	1.0651	0.0179	0.8576	1.0646	-0.0209	1.6444	2.2859	-0.0003	1.3651	1.7934
DGP (2)													
	50	-0.0256	0.9256	1.1717	-0.0163	0.9315	1.1803	0.0225	1.8126	2.4738	-0.0256	1.4606	1.9524
10	100	0.0285	0.8905	1.1135	0.0255	0.8926	1.1167	0.0176	1.6676	2.3864	0.0026	1.4222	1.9107
	200	-0.0892	0.8530	1.0642	-0.0903	0.8531	1.0653	-0.1187	1.8413	2.5983	-0.0955	1.4674	1.9796
500		0.0414	0.8932	1.0882	0.0399	0.8880	1.0882	-0.0742	1.7823	2.4569	-0.0447	1.4717	1.9509
700	20	0.0164	0.8266	1.0224	0.0195	0.8257	1.0245	-0.0052	1.7395	2.4224	-0.0329	1.4459	1.9598
1000		0.0114	0.7920	0.9962	0.0065	0.7911	0.9990	0.0281	1.6988	2.4471	0.0131	1.4394	1.9688
DGP (3)													
	50	-0.0150	0.9313	1.1822	-0.0049	0.9349	1.1883	0.0112	2.0804	2.8881	-0.0237	1.9380	2.6881
10	100	0.0281	0.8887	1.1121	0.0268	0.8902	1.1152	0.0235	1.9925	2.8312	0.0324	1.9130	2.6913
	200	-0.0910	0.8471	1.0597	-0.0918	0.8467	1.0607	-0.1112	2.1363	2.9827	-0.1211	1.9943	2.7940
500		0.0352	0.9023	1.0899	0.0282	0.9045	1.0958	-0.0432	1.9667	2.7216	-0.0301	1.9759	2.8129
700	20	0.0169	0.8290	1.0295	0.0216	0.8308	1.0330	-0.1180	2.0734	2.9439	-0.0788	2.1114	3.1093
1000		0.0170	0.7992	1.0007	0.0095	0.7999	1.0070	0.0801	1.9517	2.7888	0.0530	1.9414	2.7877

Table 9.8: Simulation Results for 5-period ahead predictions using infeasible horizontal method (9.39), Approach 3, and the diffusion methods (9.42), (9.43) under DGP 1-3.

N	T	Infeasible HR			Approach 3			DiffusionA			DiffusionB		
		Bias	MAB	RMSPE	Bias	MAB	RMSPE	Bias	MAB	RMSPE	Bias	MAB	RMSPE
DGP (1)													
	50	-0.0225	0.9376	1.1797	-0.0236	0.9494	1.1913	-0.0898	1.6287	2.2567	-0.0288	1.2936	1.7259
10	100	0.0363	0.9185	1.1456	0.0374	0.9199	1.1477	0.0684	1.6358	2.2589	0.0797	1.3004	1.7656
	200	0.0008	0.8467	1.0706	0.0012	0.8473	1.0714	-0.0162	1.5875	2.2103	0.0168	1.3489	1.7908
500		-0.0014	0.8599	1.0837	0.0017	0.8567	1.0850	-0.0243	1.5803	2.1376	0.0111	1.2732	1.6769
700	20	-0.0142	0.8395	1.0459	-0.0183	0.8398	1.0464	0.0671	1.6555	2.2610	0.0420	1.3797	1.8761
1000		0.0382	0.8267	1.0453	0.0396	0.8284	1.0463	0.1067	1.6365	2.2102	0.0375	1.3309	1.7825
DGP (2)													
	50	0.0414	0.9305	1.1681	0.0494	0.9422	1.1829	0.1074	1.7647	2.5001	0.1255	1.4395	1.9625
10	100	-0.0335	0.8859	1.1074	-0.0301	0.8903	1.1140	0.0609	1.8624	2.5789	0.0172	1.5021	2.0621
	200	0.0279	0.8348	1.0588	0.0267	0.8351	1.0600	0.1182	1.7576	2.4780	0.0707	1.3962	1.9376
500		-0.0004	0.8339	1.0424	0.0037	0.8303	1.0419	0.0183	1.7513	2.3943	0.0088	1.4165	1.9125
700	20	-0.0168	0.8084	1.0177	-0.0256	0.8115	1.0215	-0.0120	1.7867	2.4914	-0.0216	1.4809	2.0695
1000		-0.0025	0.8332	1.0564	-0.0077	0.8450	1.0640	-0.1912	1.8315	2.5033	-0.0120	1.5449	2.0829
DGP (3)													
	50	0.0378	0.9336	1.1690	0.0477	0.9485	1.1867	0.0960	2.1489	3.0661	0.1002	2.0316	2.8772
10	100	-0.0349	0.8869	1.1140	-0.0318	0.8892	1.1196	0.0227	2.2797	3.2564	0.0278	2.1557	3.0786
	200	0.0308	0.8396	1.0624	0.0296	0.8402	1.0633	0.1186	2.1182	3.1232	0.0815	1.9396	2.8433
500		-0.0032	0.8320	1.0440	-0.0036	0.8433	1.0609	0.0146	2.1130	2.9300	-0.0135	2.1872	3.1165
700	20	-0.0127	0.7993	1.0111	-0.0204	0.8155	1.0372	-0.1559	2.2054	3.1280	-0.1012	2.2508	3.2677
1000		-0.0062	0.8380	1.0592	-0.0083	0.8508	1.0739	-0.0819	2.2090	3.0388	-0.0418	2.2219	3.0722

Tables 9.1, 9.2, and 9.3 report the results for 2, 3 and 5 periods ahead prediction under DGP 1-3, respectively. The results for other periods are available upon request. Tables 9.4-9.8 report the results for 1-5 periods ahead prediction under DGP 1-3, respectively.

To compare the prediction accuracy between the diffusion method and HR or VR, we only report the results in terms of Approach 3 for $T+1,\ldots,T+5$ because as shown by Lemma 9.3 and confirmed by our simulations all three approaches yield identical results.

The simulation results confirm that Approach 1 (9.30), Approach 2 (9.33) and Approach 3 (9.36) yield identical point prediction. Additionally, the Infeasible HR and Approach 3 yield identical results for $T+1$ in Table 9.4 because both methods use the same set of information for a prediction in one period. The simulation results also show that HR (both infeasible and feasible) yields a more accurate prediction than the diffusion model method. Moreover, if T is large and h is small, although knowing

$y_{1,T+1},\ldots y_{1,T+h-1}$ to predict $y_{1,T+h}$ is more accurate than (9.36), the difference is negligible. However, when N is large, knowing $y_{1,T+1},\ldots y_{1,T+h-1}$ not only produces smaller RMSPE but also maintains a notable difference from (9.36) even as N grows. This finding echoes (iii) and (iv) in Lemma 9.4.

9.6 Concluding Remarks

We argued that the measurement of treatment effects using panel data is essentially a prediction issue. There are multiple ways to construct counterfactuals based on a hypothetical data generating process (DGP) of the observed data. Under the unconfoundedness assumption, the linear projection approach is widely applied (e.g., Hsiao, 2022). Shen et al. (2023) showed that the projection based on a unit's own past values or based on the contemporary outcomes of other units in the panel data, respectively referred as horizontal regression and vertical regression by Athey et al. (2021), yield identical numerical values. However, Shen et al. (2023) results only hold for the construction of counterfactual first post-treatment period and their statistical inferences are based on different assumptions about the underlying data generating process. We suggested using a factor approach as a unified framework to link the HR and VR approach to counterfactuals and derived their statistical distributions. We showed that the multi-period ahead construction of counterfactuals between the HR and VR are different and their prediction accuracy depends on whether the time series dimension T is fixed and the cross-sectional dimensions N is large, or N is fixed T is large, or both N and T are large. We also suggested different ways to construct multi-period ahead predictions based on the HR regression. We showed that for multi-period post-treatment measurement of treatment effects, in general, the linear projection approach based on the VR yields more accurate measurement than those based on the HR or the diffusion model approach suggested by Bai and Ng (2006).

Appendix: Mathematical Proofs and Further Discussions

This appendix provides the proofs and heuristic arguments that are omitted in the paper.

Proof for Lemma 9.1

Proof For (a), under Assumption C1, C2 and C4, we have

$$
\begin{aligned}
\hat{\alpha} &= \left(\frac{1}{N}\sum_{i=2}^{N}\mathbf{y}_i\mathbf{y}_i'\right)^{-1}\frac{1}{N}\sum_{i=2}^{N}\mathbf{y}_i y_{i,T+1} \\
&= \left(\frac{1}{N}\sum_{i=2}^{N}(\mathbf{F}\lambda_i+\mathbf{u}_i)(\mathbf{F}\lambda_i+\mathbf{u}_i)'\right)^{-1}\frac{1}{N}\sum_{i=2}^{N}(\mathbf{F}\lambda_i+\mathbf{u}_i)\left(\lambda_i'\mathbf{f}_{T+1}+u_{i,T+1}\right) \\
&\to{}_p\alpha = (\mathbf{F}\Sigma_\lambda\mathbf{F}'+\Omega^*)^{-1}(\mathbf{F}\Sigma_\lambda\mathbf{f}_{T+1}+\mathbf{c}^*),
\end{aligned}
$$

because $\frac{1}{N}\sum_{i=2}^{N}\lambda_i\lambda_i' \to \Sigma_\lambda$, $\frac{1}{N}\sum_{i=2}^{N}\mathbf{u}_i\mathbf{u}_i' \to_p \Omega^*$, $\frac{1}{N}\sum_{i=2}^{N}\mathbf{u}_i u_{i,T+1} \to_p \mathbf{c}^*$ and $\frac{1}{N}\sum_{i=2}^{N}\mathbf{u}_i\lambda_i' \to_p \mathbf{0}$.

For σ_η^2, under Assumption C1, C2 and C4, we have

$$
\begin{aligned}
\sigma_\eta^2 &= \text{plim}_{N\to\infty}\frac{1}{N}\sum_{i=2}^{N}\left(y_{i,T+1}-\alpha'\mathbf{y}_i\right)^2 \\
&= \text{plim}_{N\to\infty}\frac{1}{N}\sum_{i=2}^{N}\left(\lambda_i'\mathbf{f}_{T+1}+u_{i,T+1}-\alpha'\mathbf{F}\lambda_i-\alpha'\mathbf{u}_i\right)^2 \\
&= \text{plim}_{N\to\infty}\frac{1}{N}\sum_{i=2}^{N}u_{i,T+1}^2+\mathbf{f}_{T+1}'\left(\text{plim}_{N\to\infty}\frac{1}{N}\sum_{i=2}^{N}\lambda_i\lambda_i'\right)\mathbf{f}_{T+1} \\
&\quad +\alpha'\mathbf{F}\left(\text{plim}_{N\to\infty}\frac{1}{N}\sum_{i=2}^{N}\lambda_i\lambda_i'\right)\mathbf{F}'\alpha+\alpha'\left(\text{plim}_{N\to\infty}\frac{1}{N}\sum_{i=2}^{N}\mathbf{u}_i\mathbf{u}_i'\right)\alpha \\
&\quad -2\mathbf{f}_{T+1}'\left(\text{plim}_{N\to\infty}\frac{1}{N}\sum_{i=2}^{N}\lambda_i\lambda_i'\right)\mathbf{F}'\alpha-2\left(\text{plim}_{N\to\infty}\frac{1}{N}\sum_{i=2}^{N}u_{i,T+1}\mathbf{u}_i'\right)\alpha \\
&= \sigma_1^2+\mathbf{f}_{T+1}'\Sigma_\lambda\mathbf{f}_{T+1}+\alpha'\mathbf{F}\Sigma_\lambda\mathbf{F}'\alpha+\alpha'\Omega^*\alpha-2\mathbf{f}_{T+1}'\Sigma_\lambda\mathbf{F}'\alpha-2\mathbf{c}^{*\prime}\alpha \\
&= \sigma_1^2+\mathbf{f}_{T+1}'\Sigma_\lambda\mathbf{f}_{T+1}-(\mathbf{f}_{T+1}'\Sigma_\lambda\mathbf{F}'+\mathbf{c}^{*\prime})(\mathbf{F}\Sigma_\lambda\mathbf{F}'+\Omega^*)^{-1}(\mathbf{F}\Sigma_\lambda\mathbf{f}_{T+1}+\mathbf{c}^*)
\end{aligned}
$$

where the last identity holds since

$$
\begin{aligned}
&\alpha'\mathbf{F}\Sigma_\lambda\mathbf{F}'\alpha+\alpha'\Omega^*\alpha-2\mathbf{f}_{T+1}'\Sigma_\lambda\mathbf{F}'\alpha-2\mathbf{c}^{*\prime}\alpha \\
&= \left(\alpha'\mathbf{F}\Sigma_\lambda\mathbf{F}'+\alpha'\Omega^*-2\mathbf{f}_{T+1}'\Sigma_\lambda\mathbf{F}'-2\mathbf{c}^{*\prime}\right)\alpha \\
&= \left(\alpha'(\mathbf{F}\Sigma_\lambda\mathbf{F}'+\Omega^*)-2\mathbf{f}_{T+1}'\Sigma_\lambda\mathbf{F}'-2\mathbf{c}^{*\prime}\right)\alpha \\
&= \left((\mathbf{f}_{T+1}'\Sigma_\lambda\mathbf{F}'+\mathbf{c}^{*\prime})(\mathbf{F}\Sigma_\lambda\mathbf{F}'+\Omega^*)^{-1}(\mathbf{F}\Sigma_\lambda\mathbf{F}'+\Omega^*)-2\mathbf{f}_{T+1}'\Sigma_\lambda\mathbf{F}'-2\mathbf{c}^{*\prime}\right)\alpha \\
&= \left((\mathbf{f}_{T+1}'\Sigma_\lambda\mathbf{F}'+\mathbf{c}^{*\prime})-2\mathbf{f}_{T+1}'\Sigma_\lambda\mathbf{F}'-2\mathbf{c}^{*\prime}\right)\alpha \\
&= -\left(\mathbf{f}_{T+1}'\Sigma_\lambda\mathbf{F}'+\mathbf{c}^{*\prime}\right)(\mathbf{F}\Sigma_\lambda\mathbf{F}'+\Omega^*)^{-1}(\mathbf{F}\Sigma_\lambda\mathbf{f}_{T+1}+\mathbf{c}^*).
\end{aligned}
$$

For (b), it can be derived similarly. □

Proof for Lemma 9.2

Proof When u_{it} is i.i.d over i and t, we have $\Omega^* = \mathbf{I}_T$ and $\mathbf{c}^* = 0$, if $\Sigma_\lambda = \mathbf{I}_r$, then by Woodbury matrix identity,

$$\begin{aligned}\sigma_\eta^2 &= \sigma_1^2 + \mathbf{f}'_{T+1}\mathbf{f}_{T+1} - \mathbf{f}'_{T+1}\mathbf{F}'\left(\mathbf{F}\mathbf{F}' + \mathbf{I}_T\right)^{-1}\mathbf{F}\mathbf{f}_{T+1} \\ &= \sigma_1^2 + \mathbf{f}'_{T+1}\left(\mathbf{I}_r - \mathbf{F}'\left(\mathbf{F}\mathbf{F}' + \mathbf{I}_T\right)^{-1}\mathbf{F}\right)\mathbf{f}_{T+1} \\ &= \sigma_1^2 + \mathbf{f}'_{T+1}\left(\mathbf{I}_r + \mathbf{F}'\mathbf{F}\right)^{-1}\mathbf{f}_{T+1}. \end{aligned} \tag{9.44}$$

Similarly, if $\Sigma_f = \mathbf{I}_r$, then

$$\sigma_{\eta^*}^2 = \sigma_1^2 + \lambda_1'\left(\mathbf{I}_r + \tilde{\Lambda}'\tilde{\Lambda}\right)^{-1}\lambda_1. \tag{9.45}$$

Let $\hat{e}_{1,T+1}^{HR} = y_{1,T+1} - \hat{y}_{1,T+1}^{HR}$ be the prediction error, under $\Sigma_\lambda = \mathbf{I}_r$, then

$$\hat{e}_{1,T+1}^{HR} = \alpha'\mathbf{y}_1 + \eta_1 - \hat{\alpha}'\mathbf{y}_1 = \eta_1 + (\alpha' - \hat{\alpha}')\mathbf{y}_1, \tag{9.46}$$

with the mean square prediction error equals, conditional on $\mathbf{y}_1$,

$$\begin{aligned} MSPE\left(\hat{y}_{1,T+1}^{HR}\right) &= Var\left(\hat{e}_{1,T+1}^{HR}\right) = \sigma_\eta^2 + \mathbf{y}_1' Var(\hat{\alpha})\mathbf{y}_1 \\ &= \sigma_\eta^2\left(1 + \mathbf{y}_1'\left(\mathbf{Y}_0'\mathbf{Y}_0\right)^{-1}\mathbf{y}_1\right). \end{aligned} \tag{9.47}$$

On the other hand, we have

$$\frac{1}{N}\mathbf{Y}_0'\mathbf{Y}_0 = \frac{1}{N}\sum_{i=2}^{N}\mathbf{y}_i\mathbf{y}_i',$$

where $\mathbf{y}_i = (y_{i1}, \ldots, y_{iT})'$ is a $T \times 1$ vector. Also, from (9.17), we have $\mathbf{y}_i = \mathbf{F}\lambda_i + \mathbf{u}_i$, then as $N \to \infty$,

$$\begin{aligned} \frac{1}{N}\mathbf{Y}_0'\mathbf{Y}_0 &= \frac{1}{N}\sum_{i=2}^{N}\left(\mathbf{F}\lambda_i + \mathbf{u}_i\right)\left(\mathbf{F}\lambda_i + \mathbf{u}_i\right)' \\ &\to_p \left(\mathbf{F}\Sigma_\lambda\mathbf{F}' + \Omega^*\right)_{T\times T}. \end{aligned} \tag{9.48}$$

Since Ω^* is a $T \times T$ p.d. matrix with bounded eigenvalues, and $\mathbf{F}\Sigma_\lambda\mathbf{F}'$ is a $T \times T$ matrix p.s.d matrix of rank r, whose largest eigenvalue is of order T since $\frac{1}{T}\mathbf{F}'\mathbf{F} \to \Sigma_f$, which is a $r \times r$ matrix with bounded eigenvalues under Assumption 1. The above results show that the minimum eigenvalue of $\mathbf{F}\Sigma_\lambda\mathbf{F}' + \Omega^*$ is $O(1)$ and the maximum eigenvalue of $\mathbf{F}\Sigma_\lambda\mathbf{F}' + \Omega^*$ is $O(T)$. Then the minimum eigenvalue of $(\mathbf{F}\Sigma_\lambda\mathbf{F}' + \Omega^*)^{-1}$ is $O\left(\frac{1}{T}\right)$ and the maximum eigenvalue of $(\mathbf{F}\Sigma_\lambda\mathbf{F}' + \Omega^*)^{-1}$ is $O(1)$.

Consequently, we have

$$MSPE\left(\hat{y}_{1,T+1}^{HR}\right) = \sigma_{\eta}^{2}\left(1+\frac{1}{N}\mathbf{y}_{1}'\left(\frac{1}{N}\mathbf{Y}_{0}'\mathbf{Y}_{0}\right)^{-1}\mathbf{y}_{1}\right)$$
$$\leq \sigma_{\eta}^{2}\left(1+C\frac{1}{N}\mathbf{y}_{1}'\mathbf{y}_{1}\right) \leq \sigma_{\eta}^{2}\left(1+C\frac{T}{N}\right), \tag{9.49}$$

since $\frac{1}{T}\mathbf{y}_1'\mathbf{y}_1 = \frac{1}{T}\sum_{t=1}^{T} y_{1t}^2$ will converge to a finite constant under Assumption C1 and C2 as $T \to \infty$.

In sum, we have

$$MSPE\left(\hat{y}_{1,T+1}^{HR}\right) \to \sigma_{\eta}^{2} \text{ as } \frac{T}{N} \to 0. \tag{9.50}$$

Similarly, let $\hat{e}_{1,T+1}^{VR} = y_{1,T+1} - \hat{y}_{1,T+1}^{VR}$, then

$$\hat{e}_{1,T+1}^{VR} = \beta'\tilde{\mathbf{y}}_{T+1} + \eta_1^* - \hat{\beta}'\tilde{\mathbf{y}}_{T+1} = \eta_1^* + \left(\beta' - \hat{\beta}'\right)\tilde{\mathbf{y}}_{T+1}, \tag{9.51}$$

with the mean square prediction error equals and conditional on $\tilde{\mathbf{y}}_{T+1}$,

$$MSPE\left(\hat{y}_{1,T+1}^{VR}\right) = Var\left(\hat{e}_{1,T+1}^{VR}\right) = \sigma_{\eta^*}^{2} + \tilde{\mathbf{y}}_{T+1}'Var\left(\hat{\beta}\right)\tilde{\mathbf{y}}_{T+1}$$
$$= \sigma_{\eta^*}^{2}\left(1+\tilde{\mathbf{y}}_{T+1}'\left(\mathbf{Y}_0\mathbf{Y}_0'\right)^{-1}\tilde{\mathbf{y}}_{T+1}\right) = \sigma_{\eta^*}^{2}\left(1+\frac{1}{T}\tilde{\mathbf{y}}_{T+1}'\left(\frac{1}{T}\mathbf{Y}_0\mathbf{Y}_0'\right)^{-1}\tilde{\mathbf{y}}_{T+1}\right). \tag{9.52}$$

As shown above, we can obtain

$$MSPE\left(\hat{y}_{1,T+1}^{VR}\right) \to \sigma_{\eta^*}^{2} \text{ as } \frac{N}{T} \to 0. \tag{9.53}$$

For (a), when $(N,T) \to \infty$ and $\frac{T}{N} \to 0$, from (9.44) and (9.50), we have

$$\sigma_{\eta}^{2} \to \sigma_{1}^{2}, \tag{9.54}$$

since $\mathbf{F}'\mathbf{F}$ is of order T. Moreover,

$$MSPE\left(\hat{y}_{1,T+1}^{VR}\right) = \sigma_{\eta^*}^{2}\left(1+\frac{1}{T}\tilde{\mathbf{y}}_{T+1}'\left(\frac{1}{T}\mathbf{Y}_0\mathbf{Y}_0'\right)^{-1}\tilde{\mathbf{y}}_{T+1}\right) \geq \sigma_{\eta^*}^{2} > \sigma_{1}^{2}. \tag{9.55}$$

Thus, we can claim

$$MSPE\left(\hat{y}_{1,T+1}^{VR}\right) > MSPE\left(\hat{y}_{1,T+1}^{HR}\right) \text{ as } (N,T) \to \infty \text{ and } \frac{T}{N} \to 0.$$

Similar result can be obtained when $T \to \infty$ only since $\frac{N}{T} \to 0$ when N is finite.

For (b), when $(N,T) \to \infty$ and $\frac{N}{T} \to 0$, from (9.45) and (9.53), we have

$$\sigma_{\eta^*}^{2} \to \sigma_{1}^{2}, \tag{9.56}$$

since $\tilde{\Lambda}'\tilde{\Lambda}$ is of order N. Moreover,

$$MSPE\left(\hat{y}_{1,T+1}^{HR}\right) = \sigma_{\eta}^{2}\left(1+\frac{1}{N}\mathbf{y}_1'\left(\frac{1}{N}\mathbf{Y}_0'\mathbf{Y}_0\right)^{-1}\mathbf{y}_1\right) \geq \sigma_{\eta}^{2} > \sigma_{1}^{2}. \tag{9.57}$$

Consequently, we have

$$MSPE\left(\hat{y}_{1,T+1}^{HR}\right) > MSPE\left(\hat{y}_{1,T+1}^{VR}\right) \text{ as } (N,T)\to\infty \text{ and } \frac{N}{T}\to 0.$$

Similar result can be obtained when $N \to \infty$ only, because $\frac{T}{N} \to 0$ when T is finite.□

Proof for Lemma 9.3

Proof The equivalence between Approach 1 and Approach 2 is straightforward, as it extends directly from one-period to multi-period ahead predictions. The main task is to show the equivalence between Approach 2 and Approach 3. The predictor by Approach 3, as shown in (9.36), can be written as

$$\begin{aligned}\hat{y}_{1,T+h}^{HR,2} &= \mathbf{y}_{T+h}'\mathbf{Y}^{T+h-1}\left(\mathbf{Y}^{T+h-1\prime}\mathbf{Y}^{T+h-1}\right)^{-1}\hat{\mathbf{y}}_1^{T+h-1}\\ &= \hat{\alpha}_1'\mathbf{y}_1+\hat{\alpha}_2'\hat{\mathbf{y}}_1^{h-1}\end{aligned}$$

where $(\hat{\alpha}_1'\ \hat{\alpha}_2') = \tilde{\mathbf{y}}_{T+h}'[\mathbf{Y}_0\ \ \mathbf{Y}_{1,T+h-1}]\left([\mathbf{Y}_0\ \ \mathbf{Y}_{1,T+h-1}]'[\mathbf{Y}_0\ \ \mathbf{Y}_{1,T+h-1}]\right)^{-1}$ and $\hat{\mathbf{y}}_1^{h-1\prime} = (\hat{y}_{1,T+1},\ldots,\hat{y}_{1,T+h-1})$. By the block matrix inversion formula, we have

$$\left([\mathbf{Y}_0\ \ \mathbf{Y}_{1,T+h-1}]'[\mathbf{Y}_0\ \ \mathbf{Y}_{1,T+h-1}]\right)^{-1} = \begin{bmatrix} B_{11} & B_{12}\\ B_{21} & B_{22}\end{bmatrix} \tag{9.58}$$

where $B_{11} = (\mathbf{Y}_0'\mathbf{Y}_0)^{-1} + (\mathbf{Y}_0'\mathbf{Y}_0)^{-1}\mathbf{Y}_0'\mathbf{Y}_{1,T+h-1}B_{22}\mathbf{Y}_{1,T+h-1}'\mathbf{Y}_0(\mathbf{Y}_0'\mathbf{Y}_0)^{-1}$, $B_{22} = (\mathbf{Y}_{1,T+h-1}'\mathbf{Y}_{1,T+h-1} - \mathbf{Y}_{1,T+h-1}'\mathbf{Y}_0(\mathbf{Y}_0'\mathbf{Y}_0)^{-1}\mathbf{Y}_0'\mathbf{Y}_{1,T+h-1})^{-1}$, $B_{12} = -(\mathbf{Y}_0'\mathbf{Y}_0)^{-1}\mathbf{Y}_0'\mathbf{Y}_{1,T+h-1}B_{22}$, $B_{21} = -B_{22}\mathbf{Y}_{1,T+h-1}'\mathbf{Y}_0(\mathbf{Y}_0'\mathbf{Y}_0)^{-1}$. It follows that

$$\begin{aligned}\hat{\alpha}_1' &= \tilde{\mathbf{y}}_{T+h}'\mathbf{Y}_0B_{11}+\tilde{\mathbf{y}}_{T+h}'\mathbf{Y}_{1,T+h-1}B_{21}\\ &= \tilde{\mathbf{y}}_{T+h}'\mathbf{Y}_0(\mathbf{Y}_0'\mathbf{Y}_0)^{-1}-\tilde{\mathbf{y}}_{T+h}'M_{\mathbf{Y}_0}C,\end{aligned}$$

where $C = \mathbf{Y}_1B_{22}\mathbf{Y}_{1,T+h-1}'\mathbf{Y}_0(\mathbf{Y}_0'\mathbf{Y}_0)^{-1}$, $M_{\mathbf{Y}_0} = \mathbf{I}_T - \mathbf{Y}_0(\mathbf{Y}_0'\mathbf{Y}_0)^{-1}\mathbf{Y}_0'$, and

$$\begin{aligned}\hat{\alpha}_2' &= \tilde{\mathbf{y}}_{T+h}'\mathbf{Y}_0B_{12}+\tilde{\mathbf{y}}_{T+h}'\mathbf{Y}_{1,T+h-1}B_{22}\\ &= \tilde{\mathbf{y}}_{T+h}'M_{\mathbf{Y}_0}\mathbf{Y}_{1,T+h-1}B_{22}.\end{aligned}$$

Recall that predicted values for $y_{1,T+1}, \ldots, y_{1,T+h-1}$ are $\hat{\mathbf{y}}_1^{h-1} = \mathbf{Y}'_{1,T+h-1}\left(\mathbf{Y}_0\mathbf{Y}'_0\right)^{-}\mathbf{Y}_0\mathbf{y}_1$, and note $\left(\mathbf{Y}_0\mathbf{Y}'_0\right)^{-}\mathbf{Y}_0 = \left(\mathbf{Y}'_0\right)^{-}$ by part (xxvi) of Proposition 6.16 in Bernstein (2009), then we have

$$\hat{\mathbf{y}}_1^{h-1} = \mathbf{Y}'_{1,T+h-1}(\mathbf{Y}'_0)^{-}\mathbf{y}_1 = (\mathbf{y}'_1(\mathbf{Y}_0)^{-}\mathbf{Y}_{1,T+h-1})' = (\mathbf{y}'_1(\mathbf{Y}'_0\mathbf{Y}_0)^{-1}\mathbf{Y}'_0\mathbf{Y}_{1,T+h-1})', \tag{9.59}$$

and

$$\begin{aligned}\hat{\alpha}'_1\mathbf{y}_1 + \hat{\alpha}'_2\hat{\mathbf{y}}_1^{h-1} &= \mathbf{y}'_1\hat{\alpha}_1 + \hat{\mathbf{y}}_1^{h-1\prime}\hat{\alpha}_2 \\ &= \mathbf{y}'_1\left(\mathbf{Y}'_0\mathbf{Y}_0\right)^{-1}\mathbf{Y}'_0\tilde{\mathbf{y}}_{T+h} - \mathbf{y}'_1C'M_{\mathbf{Y}_0}\tilde{\mathbf{y}}_{T+h} + \mathbf{y}'_1C'M_{\mathbf{Y}_0}\tilde{\mathbf{y}}_{T+h} \\ &= \mathbf{y}'_1\left(\mathbf{Y}'_0\mathbf{Y}_0\right)^{-1}\mathbf{Y}'_0\tilde{\mathbf{y}}_{T+h}.\end{aligned}$$

This is identical to the predictor by Approach 2, as shown in (9.33). □

Proof for Lemma 9.4

Proof We first note that under Assumption 1-2, for $\hat{\alpha}^{(h)}$, following the derivation of Lemma 9.2, we have

$$\begin{aligned}\hat{\alpha}^{(h)} &= \left(\frac{1}{N}\sum_{i=2}^{N}\mathbf{y}_i\mathbf{y}'_i\right)^{-1}\frac{1}{N}\sum_{i=2}^{N}\mathbf{y}_i y_{i,T+h} \\ &\to_p \alpha^{(h)} = (\mathbf{F}\Sigma_\lambda\mathbf{F}' + \Omega^*)^{-1}\left(\mathbf{F}\Sigma_\lambda\mathbf{f}_{T+1} + \mathbf{c}^*_h\right),\end{aligned} \tag{9.60}$$

where $\mathbf{c}^*_h = E\left(\mathbf{u}_i u_{i,T+h}\right)$. Moreover, noting $\eta_{T+h} = y_{1,T+h} - \alpha^{(h)\prime}\mathbf{y}_1$ by (9.31) and the data generating processes of $y_{i,T+h}$ as well as $\mathbf{y}_i$, we can generate the variance of η_{T+h} as follows,

$$\begin{aligned}\sigma^2_{\eta(h)} &= \text{plim}_{N\to\infty}\frac{1}{N}\sum_{i=2}^{N}\left(y_{i,T+h} - \alpha^{(h)\prime}\mathbf{y}_i\right)^2 \\ &= \text{plim}_{N\to\infty}\frac{1}{N}\sum_{i=2}^{N}\left(\lambda'_i\mathbf{f}_{T+h} + u_{i,T+h} - \alpha^{(h)\prime}\mathbf{F}\lambda_i - \alpha^{(h)\prime}\mathbf{u}_i\right)^2 \\ &= \sigma^2_1 + \mathbf{f}'_{T+h}\Sigma_\lambda\mathbf{f}_{T+h} - (\mathbf{f}'_{T+h}\Sigma_\lambda\mathbf{F}' + \mathbf{c}^{*\prime}_h)(\mathbf{F}\Sigma_\lambda\mathbf{F}' + \Omega^*)^{-1}\left(\mathbf{F}\Sigma_\lambda\mathbf{f}_{T+h} + \mathbf{c}^*_h\right)\end{aligned} \tag{9.61}$$

Similarly, noting $\eta^*_{T+h} = y_{1,T+h} - \beta'\tilde{\mathbf{y}}_{T+h}$ by (9.23) where β is defined in (9.24), then the variance of η^*_{T+h} can be shown as

$$\sigma^{*2}_{\eta(h)} = \sigma^2_1 + \lambda'_1\Sigma_f\lambda_1 - (\lambda'_1\Sigma_f\tilde{\Lambda}' + \mathbf{c}'_h)(\tilde{\Lambda}\Sigma_f\tilde{\Lambda}' + \Omega)^{-1}(\tilde{\Lambda}\Sigma_f\lambda_1 + \mathbf{c}_h), \tag{9.62}$$

where $\mathbf{c}_h = E(\tilde{\mathbf{u}}_{t+h}u_{1,t+h})$. When u_{it} is i.i.d over i and t, following the proof of Lemma 9.2, we have

$$MSPE\left(\hat{y}_{1,T+h}^{HR,1}\right) = \sigma_{\eta(h)}^2 \left(1 + \frac{1}{N}\mathbf{y}_1' \left(\frac{1}{N}\mathbf{Y}_0'\mathbf{Y}_0\right)^{-1} \mathbf{y}_1\right), \tag{9.63}$$

and

$$MSPE\left(\hat{y}_{1,T+h}^{VR}\right) = \sigma_{\eta(h)}^{*2} \left(1 + \tilde{\mathbf{y}}_{T+h}' \left(\mathbf{Y}_0\mathbf{Y}_0'\right)^{-1} \tilde{\mathbf{y}}_{T+h}\right). \tag{9.64}$$

For (i), when $(N,T) \longrightarrow \infty$ and $\frac{N}{T} \longrightarrow 0$, we have

$$MSPE\left(\hat{y}_{1,T+h}^{VR}\right) \longrightarrow \sigma_1^2, \tag{9.65}$$

and

$$MSPE\left(\hat{y}_{1,T+h}^{HR,1}\right) = \sigma_{\eta(h)}^2 \left(1 + \frac{1}{N}\mathbf{y}_1' \left(\frac{1}{N}\mathbf{Y}_0'\mathbf{Y}_0\right)^{-1} \mathbf{y}_1\right) \geq \sigma_{\eta(h)}^2 > \sigma_1^2. \tag{9.66}$$

Consequently, we have

$$MSPE\left(\hat{y}_{1,T+h}^{HR,1}\right) > MSPE\left(\hat{y}_{1,T+h}^{VR}\right) \text{ as } (N,T) \longrightarrow \infty \text{ and } \frac{N}{T} \longrightarrow 0. \tag{9.67}$$

Similar result can be obtained when $T \longrightarrow \infty$ only since $\frac{N}{T} \longrightarrow 0$ when N is finite.

For (ii), when $(N,T) \longrightarrow \infty$ and $\frac{T}{N} \longrightarrow 0$, we have

$$MSPE\left(\hat{y}_{1,T+h}^{HR,1}\right) \longrightarrow \sigma_1^2, \tag{9.68}$$

while

$$MSPE\left(\hat{y}_{1,T+h}^{VR}\right) = \sigma_{\eta(h)}^{*2} \left(1 + \tilde{\mathbf{y}}_{T+h}' \left(\mathbf{Y}_0\mathbf{Y}_0'\right)^{-1} \tilde{\mathbf{y}}_{T+h}\right) \geq \sigma_{\eta(h)}^{*2} > \sigma_1^2. \tag{9.69}$$

Thus, we have

$$MSPE\left(\hat{y}_{1,T+1}^{VR}\right) > MSPE\left(\hat{y}_{1,T+h}^{HR,1}\right) \text{ as } (N,T) \longrightarrow \infty \text{ and } \frac{T}{N} \longrightarrow 0. \tag{9.70}$$

Similar result can be obtained when $N \longrightarrow \infty$ only, because $\frac{T}{N} \longrightarrow 0$ when T is finite.

For (iii), consider $\hat{y}_{1,T+h}^{VR}$ and note since u_{it} is i.i.d over i and t, $\Omega^* = \mathbf{I}_T$ and $\mathbf{c}_h^* = \mathbf{0}$, $\Sigma_f = \mathbf{I}_r$, then using (9.61) it follows

$$\begin{aligned}
\sigma_{\eta(h)}^{*2} &= \sigma_1^2 + \lambda_1'\lambda_1 - (\lambda_1'\tilde{\Lambda}' + \mathbf{c}')(\tilde{\Lambda}\tilde{\Lambda}' + \Omega)^{-1}(\tilde{\Lambda}\lambda_1 + \mathbf{c}) \\
&= \sigma_1^2 + \lambda_1' \left(\mathbf{I}_r + \tilde{\Lambda}'\tilde{\Lambda}\right)^{-1} \lambda_1
\end{aligned}$$

where the second line holds by Woodbury matrix identity. Invoking $\sigma_{\eta(h)}^{*2}$ in (9.64) then yields

$$MSPE\left(\hat{y}_{1,T+h}^{VR}\right) = \left[\sigma_1^2 + \lambda_1' \left(\mathbf{I}_r + \tilde{\Lambda}'\tilde{\Lambda}\right)^{-1} \lambda_1\right] \left[1 + \frac{1}{T}\tilde{\mathbf{y}}_{T+h}' \left(\frac{1}{T}\mathbf{Y}_0\mathbf{Y}_0'\right)^{-1} \tilde{\mathbf{y}}_{T+h}\right]. \tag{9.71}$$

Now consider $\hat{y}_{1,T+h}^{VR,0}$ and note by the proof of Lemma 9.2, we have

$$MSPE\left(\hat{y}_{1,T+h}^{VR,0}\right) = \left[\sigma_1^2 + \lambda_1' \left(\mathbf{I}_r + \tilde{\Lambda}'\tilde{\Lambda}\right)^{-1} \lambda_1\right] \times \left[1 + \frac{1}{T}\tilde{\mathbf{y}}_{T+h}' \left(\frac{1}{T}\mathbf{Y}^{T+h-1}\mathbf{Y}^{T+h-1\prime}\right)^{-1} \tilde{\mathbf{y}}_{T+h}\right]. \tag{9.72}$$

Following the analysis after (9.48) in proof of Lemma 9.2, as $T \to \infty$, we further have

$$\frac{1}{T}\tilde{\mathbf{y}}_{T+h}' \left(\frac{1}{T}\mathbf{Y}_0\mathbf{Y}_0'\right)^{-1} \tilde{\mathbf{y}}_{T+h} = O_p\left(\frac{N}{T}\right),$$

$$\frac{1}{T}\tilde{\mathbf{y}}_{T+h}' \left(\frac{1}{T}\mathbf{Y}^{T+h-1}\mathbf{Y}^{T+h-1\prime}\right)^{-1} \tilde{\mathbf{y}}_{T+h} = O_p\left(\frac{N}{T}\right).$$

By (9.71) and (9.72), it can be shown $MSPE\left(\hat{y}_{1,T+h}^{VR}\right) = MSPE\left(\hat{y}_{1,T+h}^{VR,0}\right)$ as long as $N/T \to 0$.

For (iv), consider $\hat{y}_{1,T+h}^{HR,1}$ and note since u_{it} is i.i.d over i and t, $\Omega^* = \mathbf{I}_T$ and $\mathbf{c}_h^* = \mathbf{0}$, $\Sigma_\lambda = \mathbf{I}_r$, then using (9.61) it follows

$$\begin{aligned}\sigma_{\eta(h)}^2 &= \sigma_1^2 + \mathbf{f}_{T+h}'\mathbf{f}_{T+h} - \left(\mathbf{f}_{T+h}'\mathbf{F}'\right)\left(\mathbf{F}\mathbf{F}' + \mathbf{I}_T\right)^{-1}\left(\mathbf{F}\mathbf{f}_{T+h}\right)\\ &= \sigma_1^2 + \mathbf{f}_{T+h}'\left(\mathbf{I}_r + \mathbf{F}'\mathbf{F}\right)^{-1}\mathbf{f}_{T+h},\end{aligned}$$

where the second line holds by Woodbury matrix identity. Substituting $\sigma_{\eta(h)}^2$ in (9.63) then yields

$$MSPE\left(\hat{y}_{1,T+h}^{HR,1}\right) = \left[\sigma_1^2 + \mathbf{f}_{T+h}'\left(\mathbf{I}_r + \mathbf{F}'\mathbf{F}\right)^{-1}\mathbf{f}_{T+h}\right]\left[1 + \frac{1}{N}\mathbf{y}_1'\left(\frac{1}{N}\mathbf{Y}_0'\mathbf{Y}_0\right)^{-1}\mathbf{y}_1\right]. \tag{9.73}$$

Now for $\hat{y}_{1,T+h}^{HR,0}$, using the proof of Lemma 9.2 yields

$$MSPE\left(\hat{y}_{1,T+h}^{HR,0}\right) = \left[\sigma_1^2 + \mathbf{f}_{T+h}'\left(\mathbf{I}_r + \mathbf{F}_{T+h-1}'\mathbf{F}_{T+h-1}\right)^{-1}\mathbf{f}_{T+h}\right] \times \left[1 + \frac{1}{N}\mathbf{y}_1^{T+h-1\prime}\left(\frac{1}{N}\mathbf{Y}_{T+h-1}'\mathbf{Y}_{T+h-1}\right)^{-1}\mathbf{y}_1^{T+h-1}\right], \tag{9.74}$$

where $\mathbf{F}_{T+h-1} = (\mathbf{f}_1, \mathbf{f}_2, \ldots, \mathbf{f}_T, \mathbf{f}_{T+1}, \ldots, \mathbf{f}_{T+h-1})'$ and $\mathbf{Y}_{T+h-1} = (\mathbf{Y}_0, \mathbf{Y}_1)$. As $N \to \infty$ and following the analysis after (9.48) in proof of Lemma 9.2, we further have

$$\frac{1}{N}\mathbf{y}_1'\left(\frac{1}{N}\mathbf{Y}_0'\mathbf{Y}_0\right)^{-1}\mathbf{y}_1 = O_p\left(\frac{T}{N}\right),$$

$$\frac{1}{N}\mathbf{y}_1^{T+h-1\prime}\left(\frac{1}{N}\mathbf{Y}_{T+h-1}'\mathbf{Y}_{T+h-1}\right)^{-1}\mathbf{y}_1^{T+h-1} = O_p\left(\frac{T}{N}\right).$$

For the case where T is fixed, note

$$\begin{aligned}&\mathbf{f}_{T+h}'\left(\mathbf{I}_r+\mathbf{F}'\mathbf{F}\right)^{-1}\mathbf{f}_{T+h}-\mathbf{f}_{T+h}'\left(\mathbf{I}_r+\mathbf{F}_{T+h-1}'\mathbf{F}_{T+h-1}\right)^{-1}\mathbf{f}_{T+h}\\&=\mathbf{f}_{T+h}'\left[\left(\mathbf{I}_r+\mathbf{F}'\mathbf{F}\right)^{-1}-\left(\mathbf{I}_r+\mathbf{F}_{T+h-1}'\mathbf{F}_{T+h-1}\right)^{-1}\right]\mathbf{f}_{T+h}.\end{aligned} \tag{9.75}$$

Let $A=\mathbf{I}_r+\mathbf{F}'\mathbf{F}$ and $\mathbf{U}\mathbf{U}'=\sum_{s=1}^{h-1}\mathbf{f}_{T+s}\mathbf{f}_{T+s}'$ where $\mathbf{U}=(\mathbf{f}_{T+1},\ldots,\mathbf{f}_{T+h-1})$. By Woodbury matrix identity formula, we have

$$\begin{aligned}&\mathbf{f}_{T+h}'\left[\left(\mathbf{I}_r+\mathbf{F}'\mathbf{F}\right)^{-1}-\left(\mathbf{I}_r+\mathbf{F}_{T+h-1}'\mathbf{F}_{T+h-1}\right)^{-1}\right]\mathbf{f}_{T+h}\\&=\mathbf{f}_{T+h}'\left[A^{-1}-\left(A+\mathbf{U}'\mathbf{U}\right)^{-1}\right]\mathbf{f}_{T+h}\\&=\left(\mathbf{f}_{T+h}'A^{-1}\mathbf{U}\right)\left[\mathbf{I}_{h-1}+\mathbf{U}'A^{-1}\mathbf{U}\right]^{-1}\left(\mathbf{U}'A^{-1}\mathbf{f}_{T+h}\right).\end{aligned}$$

For any nonzero vector $z\in\mathbb{R}^{h-1}$, we can write:

$$z'\left(\mathbf{I}_{h-1}+\mathbf{U}'^{-1}A^{-1}\mathbf{U}\right)z=z'z+\left(z'\mathbf{U}'^{-1}\right)A^{-1}(\mathbf{U}z).$$

Since A is a sum of a positive definite (p.d.) matrix and a positive semidefinite (p.s.d.) matrix, A itself is p.d. Thus, $(z'\mathbf{U}'^{-1})A^{-1}(\mathbf{U}z)\geq 0$, which implies that $z'z+(z'\mathbf{U}'^{-1})A^{-1}(\mathbf{U}z)>0$. It shows that $\mathbf{I}_{h-1}+\mathbf{U}'^{-1}A^{-1}\mathbf{U}$ is p.d and so is its inverse. Hence, by (9.75) we conclude:

$$\mathbf{f}_{T+h}'\left(\mathbf{I}_r+\mathbf{F}'\mathbf{F}\right)^{-1}\mathbf{f}_{T+h}\geq\mathbf{f}_{T+h}'\left(\mathbf{I}_r+\mathbf{F}_{T+h-1}'\mathbf{F}_{T+h-1}\right)^{-1}\mathbf{f}_{T+h}>0,$$

which in conjunction with (9.73) and (9.74) shows

$$MSPE\left(\hat{y}_{1,T+h}^{HR,1}\right)\geq MSPE\left(\hat{y}_{1,T+h}^{HR,0}\right).$$

When $(N,T)\to\infty$ and $T/N\to 0$,

$$\mathbf{f}_{T+h}'\left(\mathbf{I}_r+\mathbf{F}'\mathbf{F}\right)^{-1}\mathbf{f}_{T+h}\to 0,\mathbf{f}_{T+h}'\left(\mathbf{I}_r+\mathbf{F}_{T+h-1}'\mathbf{F}_{T+h-1}\right)^{-1}\mathbf{f}_{T+h}\to 0,$$

so by (9.73) and (9.74) it follows $MSPE\left(\hat{y}_{1,T+h}^{HR,1}\right)=MSPE\left(\hat{y}_{1,T+h}^{HR,0}\right)$. □

References

Ahn, S. C. & Horenstein, A. R. (2013). Eigenvalue ratio test for the number of factors. *Econometrica*, *81*(3), 1203–1227.

Athey, S., Bayati, M., Doudchenko, N., Imbens, G. & Khosravi, K. (2021). Matrix completion methods for causal panel data models. *Journal of the American Statistical Association*, *116*(536), 1716–1730.

Bai, J. (2003). Inferential theory for factor models of large dimensions. *Econometrica*, *71*(1), 135–171.

Bai, J. (2009). Panel data models with interactive fixed effects. *Econometrica*, *77*(4), 1229–1279.

Bai, J. & Ng, S. (2002). Determining the number of factors in approximate factor models. *Econometrica*, *70*(1), 191–221. (First published: 12 December 2003) doi: 10.1111/1468-0262.00273

Bai, J. & Ng, S. (2006). Confidence intervals for diffusion index forecasts and inference for factor-augmented regressions. *Econometrica*, *74*(4), 1133–1150.

Bai, J. & Ng, S. (2021). Matrix completion, counterfactuals, and factor analysis of missing data. *Journal of the American Statistical Association*, *116*(536), 1746–1763.

Bernstein, D. S. (2009). *Matrix mathematics: theory, facts, and formulas*. Princeton University Press.

Hsiao, C. (1974). Statistical inference for a model with both random cross-sectional and time effects. *International Economic Review*, *15*(1), 12–30.

Hsiao, C. (1975). Some estimation methods for a random coefficients model. *Econometrica*, *43*(2), 305–325.

Hsiao, C. (2022). *Analysis of panel data* (4th ed.). Cambridge University Press.

Hsiao, C., Ching, S. & Wan, S. K. (2012). A panel data approach for program evaluation: Measuring the benefits of political and economic integration of hong kong with mainland china. *Journal of Applied Econometrics*, *27*(5), 705–740.

Hsiao, C., Shi, Z. & Zhou, Q. (2022). Transformed estimation for panel interactive effects models. *Journal of Business & Economic Statistics*, *40*(4), 1831–1848.

Hsiao, C. & Zhou, Q. (2019). Panel parametric, semiparametric, and nonparametric construction of counterfactuals. *Journal of Applied Econometrics*, *34*(4), 463–481.

Hsiao, C. & Zhou, Q. (2024). Panel treatment effects measurement: Factor or linear projection modelling? *Journal of Applied Econometrics*, *39*(7), 1332–1358.

Jin, S., Lu, X. & Su, L. (2024, June). *Three-dimensional factor models with global and local factors*. (Working paper. Available at http://dx.doi.org/10.2139/ssrn.4867187)

Li, X., Shen, Y. & Zhou, Q. (2024). Confidence intervals of treatment effects in panel data models with interactive fixed effects. *Journal of Econometrics*, *240*, 105684.

Shen, D., Ding, P., Sekhon, J. & Yu, B. (2023). Same root different leaves: Time series and cross-sectional methods in panel data. *Econometrica*, *91*(6), 2125–2154.

Swamy, P. A. (1971). *Statistical inference in random coefficient regression models*. Springer-Verlag.

Chapter 10
Nonparametric Correlated Random-Effects Models

Daniel J. Henderson, Emma Kate Henry and Alexandra Soberon

Abstract This chapter develops methods for estimation and inference in nonparametric panel data models with correlated random-effects. Using the Mundlak specification to control for unobserved heterogeneity, this nonparametric estimation procedure can identify both the nonparametric function and a finite-dimensional parameter associated with (potentially) observed time-invariant regressors. We develop the necessary asymptotic theory for our proposed estimator. To assess the validity of our method in practice, we propose a consistent specification test for whether the model controls for the correlation between the unobserved individual effects and the regressors. Monte Carlo simulations support the asymptotic developments. We illustrate the practical utility of our approach via an empirical application.

10.1 Introduction

The analysis of panel data has a long history in econometrics/statistics (Nerlove, 2005, Chapter 2). Fixed-effects and random-effects models have served as workhorses for panel data analysis (Baltagi, 2021). While random-effects models are easy to employ, fixed-effects models are typically used in economics because the assumption of uncorrelatedness between unobserved factors and explanatory variables often proves unrealistic in practice. Correlated random-effects (CRE) models, pioneered by

Daniel J. Henderson ✉
Department of Economics, Finance and Legal Studies, University of Alabama, Tuscaloosa, AL 35487-0224, USA, e-mail: daniel.henderson@ua.edu

Emma Kate Henry
Department of Economics, Finance and Legal Studies, University of Alabama, Tuscaloosa, AL 35487-0224, USA, e-mail: ekhenry@crimson.ua.edu

Alexandra Soberon
Department of Economics, University of Cantabria, Santander, Cantabria 39005, Spain, e-mail: alexandra.soberon@unican.es

B. H. Baltagi and L. Mátyás (eds.), *Seven Decades of Econometrics and Beyond*, Advanced Studies in Theoretical and Applied Econometrics 57,
https://doi.org/10.1007/978-3-031-92699-0_10

Mundlak (1978) and further developed by Chamberlain (1982), offer a middle ground that combines the simplicity of random-effects with the robustness of fixed-effects approaches (see Chapter 14).

CRE models maintain many advantages of random-effects estimation while accommodating potential correlation between individual-specific effects and explanatory variables (Wooldridge, 2019). Although the traditional CRE framework has proven fruitful, it still relies on potentially restrictive parametric assumptions about the conditional mean function that relate unobserved effects to observed covariates. This chapter proposes semiparametric methods that relax these assumptions and can accommodate non-linear relationships and interactions that standard parametric specifications might miss.

The field of nonparametric and semiparametric analysis in panel data modelling has expanded considerably. Estimators for the random (Henderson & Ullah, 2005) and fixed-effects (Henderson, Carroll & Li, 2008) settings exist and comprehensive reviews are available in the literature.[1] However, the incorporation of fixed-effects in these models presents significant computational challenges. Various estimation approaches have been proposed, ranging from iterative methodologies to profile least-squares to marginal integration techniques. These methods necessitate the estimation of fixed-effects parameters or rely on nontestable assumptions. Our semiparametric approach preserves the core insights of the CRE models and leads to straightforward estimation.[2] We both develop and demonstrate how researchers can implement these methods in practice while maintaining the interpretability and efficiency that have made CRE models valuable in applied work.

A key contribution of this chapter is the development of a specification test. This test provides researchers with a formal tool to evaluate whether the CRE specification controls for correlation between the composite error term and the explanatory variables. Our test compares the performance of our estimator, which is consistent under the null and alternative hypothesis versus the commonly used local-constant least-squares estimator, which is only consistent under the null hypothesis.

The remainder of this chapter is organized as follows. Section 10.2 introduces our semiparametric estimator and establishes its asymptotic properties. Section 10.3 develops the specification test and examines its properties under the null and alternative hypotheses. In Section 10.4, we conduct Monte Carlo simulations to investigate the finite sample performance of our estimator and test statistic. Section 10.5 demonstrates the practical utility of our methods through an empirical application that examines the relationship between R&D expenditure and industry-level regulations. Finally, Section 10.6 concludes.

---

[1] See the surveys of Ai and Li (2008), Henderson and Parmeter (2015), Sun, Zhang and Li (2015), Parmeter and Racine (2019), Rodriguez-Poo and Soberon (2017) and/or Su and Ullah (2011) and the references within.

[2] Bester and Hansen (2009) are able to identify average marginal effects in CRE models using sufficient statistics and index restrictions.

10.2 Estimation

Consider the standard nonparametric one-way error component model where the outcome variable, y_{it}, is related to the regressors, x_{it}, through the following regression

$$y_{it} = m(x_{it}) + \mu_i + u_{it}, \quad i = 1, \ldots, N, \quad t = 1, \ldots, T \tag{10.1}$$

where $x_{it} \in \mathbb{R}^d$ is a vector of explanatory variables, μ_i is the unobserved heterogeneity that may be correlated with x_{it}, $m(\cdot)$ is an unknown smooth function to be estimated, and u_{it} is the idiosyncratic error term with $u_{it} \sim IID(0, \sigma_u^2)$. The exogeneity condition $E(u_{it}|x_{i1}, \ldots, x_{iT}, \mu_i) = 0$ for each i is assumed throughout the paper.

To relax the strict exogeneity condition and allow for correlation between the regressors of the model and the unobserved heterogeneity, we follow Mundlak (1978) and model the correlated random-effects as a linear function of all the explanatory variables averaged across time or time-invariant regressors as

$$\mu_i = \bar{x}_i^\top \psi + z_i^\top \gamma + v_i, \tag{10.2}$$

where $\bar{x}_i = T^{-1} \sum_{t=1}^T x_{it}$ is the temporal average of x_{it}, $z_i \in R^q$ is a vector of time-invariant regressors outside the model, and v_i is an error term that is assumed to be independent of (x_{it}, z_i, u_{it}) and $v_i \sim IID(0, \sigma_v^2)$.

Plugging (10.2) in (10.1) gives the partially linear model:

$$y_{it} = m(x_{it}) + \omega_i^\top \theta + \varepsilon_{it}, \tag{10.3}$$

where $\omega_i^\top = (\bar{x}_i^\top, z_i^\top)$, $\theta = (\psi^\top, \gamma^\top)^\top$, and $\varepsilon_{it} = v_i + u_{it}$.

Let $\mathcal{J}_m(\mathbf{x}) = \frac{\partial m(\cdot)}{\partial \mathbf{x}}$ be the $(d \times 1)$ vector of first-order derivatives of $m(\cdot)$. Using a first-order Taylor expansion of $m(\cdot)$, the objective function (with known θ) is

$$\arg\min_{a,b} \sum_{i=1}^N \sum_{t=1}^T [y_{it} - a - (x_{it} - \mathbf{x})^\top b - \omega_i^\top \theta]^2 K_h(x_{it} - \mathbf{x}),$$

where $K_h(\cdot)$ is a kernel function. For multivariate x_{it}, we use a product kernel $K_h(v) = \prod_{l=1}^d k_h(v_l)$ with $v = (v_1, \ldots, v_d)^\top$ and $k_h(x_{it} - \mathbf{x}) = h^{-1} k((x_{it} - \mathbf{x})/h)$, where h is the smoothing (bandwidth) parameter. Let $\widehat{a}(\mathbf{x})$ and $\widehat{b}(\mathbf{x})$ be the resulting nonparametric estimators of $m(\mathbf{x})$ and $\mathcal{J}_m(\mathbf{x})$, respectively.

In this paper, we propose to use a local-constant approach to obtain the estimator of the unknown function $m(\cdot)$. Following Robinson (1988), we take the conditional expectation over x_{it} on both sides of (10.3) to obtain

$$E(y_{it}|x_{it}) = m(x_{it}) + E(\omega_i^\top | x_{it})\theta + E(\varepsilon_{it}|x_{it}), \tag{10.4}$$

and using the fact that $E(\varepsilon_{it}|x_{it}) = 0$ and subtracting (10.4) from (10.3), we obtain

$$y_{it} - E(y_{it}|x_{it}) = \{\omega_i - E(\omega_i|x_{it})\}^\top \theta + \varepsilon_{it}. \tag{10.5}$$

In order to avoid the random denominator problem which is quite common in the nonparametric kernel estimation (see Powell, Stock & Stoker, 1989), we premultiply both sides of (10.5) by the density function of x_{it}, i.e., $f(x_{it})$, to obtain

$$\widetilde{y}_{it} = \widetilde{\omega}_{it}\theta + \widetilde{\varepsilon}_{it} \tag{10.6}$$

where $\widetilde{y}_{it} = \{y_{it} - E(y_{it}|x_{it})\}f(x_{it})$, $\widetilde{\omega}_{it} = \{\omega_i - E(\omega_i|x_{it})\}f(x_{it})$, and $\widetilde{\varepsilon}_{it} = \{\varepsilon_{it} - E(\varepsilon_{it}|x_{it})\}f(x_{it})$.

Unfortunately, the resulting least-squares estimator from (10.6) is infeasible as $E(y_{it}|x_{it})f(x_{it})$, $E(\omega_i|x_{it})f(x_{it})$, and $f(x_{it})$ are unknown functions, but they can be consistently estimated using nonparametric techniques. We propose to estimate these via $\widehat{E}(y_{it}|x_{it})\widehat{f}(x_{it}) = (NT)^{-1}\sum_{j=1}^{N}\sum_{s=1}^{T} y_{js}K_h(x_{it},x_{js})$, $\widehat{E}(\omega_i|x_{it})\widehat{f}(x_{it}) = (NT)^{-1}\sum_{j=1}^{N}\sum_{s=1}^{T}\omega_i K_h(x_{it},x_{js})$, and $\widehat{f}(x_{it}) = (NT)^{-1}\sum_{j=1}^{N}\sum_{s=1}^{T} K_h(x_{it},x_{js})$, respectively, where $K_h(x_{it},x_{js}) = h^{-d}K((x_{it}-x_{js})/h)$.

This leads to the feasible estimator of θ in (10.5) as

$$\widehat{\theta} = \left(\sum_{i=1}^{N}\sum_{t=1}^{T}\widetilde{\omega}_{it}\widetilde{\omega}_{it}^{\top}\right)^{-1}\sum_{i=1}^{N}\sum_{t=1}^{T}\widetilde{\omega}_{it}\widetilde{y}_{it}. \tag{10.7}$$

To obtain $m(\cdot)$ in (10.1), we plug (10.7) into (10.1) to obtain

$$\ddot{y}_{it} = m(x_{it}) + e_{it}, \tag{10.8}$$

where $\ddot{y}_{it} = y_{it} - \omega_i^{\top}\widehat{\theta}$ is the new left-hand-side variable and $e_{it} = \varepsilon_{it} - \omega_i^{\top}(\widehat{\theta}-\theta)$ is the new error term.

Using a local-constant approach to estimate $m(\cdot)$ in (10.8), we propose to minimize the following objective function

$$\arg\min_{a_0}\sum_{i=1}^{N}\sum_{t=1}^{T}(\ddot{y}_{it}-a_0)^2K_h(x_{it}-\mathbf{x}),$$

and the resulting nonparametric estimator of $m(\cdot)$ is

$$\widehat{m}(\mathbf{x};h) = \frac{\sum_{i=1}^{N}\sum_{t=1}^{T}\ddot{y}_{it}K_h(x_{it},\mathbf{x})}{\sum_{i=1}^{N}\sum_{t=1}^{T}K_h(x_{it},\mathbf{x})}. \tag{10.9}$$

To derive the large sample properties of the proposed estimators in (10.7) and (10.9), we first provide a definition and state some assumptions by extending what is assumed in Li and Stengos (1996) and Soberon, Rodriguez-Poo and Robinson (2021). We shall use $\mathcal{G}_{\upsilon}^{r}$ to denote the class of smooth functions such that if $g \in \mathcal{G}_{\upsilon}^{r}$, then g is bounded and υ times differentiable; g and its partial derivative functions (up to order υ) all satisfy some Lipschitz-type conditions such as $|g(\mathbf{x}) - g(\mathbf{x}')| \le \mathcal{H}_g(\mathbf{x})||\mathbf{x}'-\mathbf{x}||$, where $\mathcal{H}_g(\mathbf{x})$ is a continuous function having rth moment, and where $||\cdot||$ denotes the Euclidean norm, i.e., $||\mathbf{x}|| = \sqrt{\sum_{l=1}^{d} \mathbf{x}_j^2}$ (see Definition 2 of Robinson, 1988, pp. 939). In addition, r controls the moment properties of the remainder term.

***Assumption* A1:** For $t = 1,\ldots,T$, $(x_{it}, z_i, v_i, u_{it})$ are $i.i.d.$ in the subscript i and strict stationarity in t for fixed i. x_{it} admits a probability density function $f \in \mathcal{G}_{\upsilon-1}^{\infty}$ (i.e., f is bounded), $m(\cdot) \in \mathcal{G}_{\upsilon}^{4}$, $E(z_i|x_{it}) \in \mathcal{G}_{\upsilon}^{4}$, and $E(\bar{x}_{iA}|x_{it}) \in \mathcal{G}_{\upsilon}^{4}$, where $\upsilon \geq 2$ is an integer. □

***Assumption* A2:** $E(u_{it}|x_{it}) = 0$, $E(v_i|x_{it}) = 0$. For $\varepsilon_{it} = v_i + u_{it}$, $E(\varepsilon_{it}^2|\mathbf{x}) = \sigma_{\varepsilon}^2 \equiv \sigma_v^2 + \sigma_u^2$. $(x_{it}, z_i, v_i, u_{it})$ have finite fourth moments. □

***Assumption* A3:** $K(\cdot)$ is a product kernel, the univariate kernel $k(\cdot)$ is a bounded υth order kernel, and $k(v) = O(1/(1+|v|)^{\upsilon+1})$. □

***Assumption* A4:** As $N \longrightarrow \infty$ for T fixed, $h \longrightarrow 0$, $Nh^{2d} \longrightarrow \infty$ and $Nh^{4\upsilon} \longrightarrow 0$. □

***Assumption* A5:** $E[(\omega_i - E(\omega_{it}|x_{it}))(\omega_i - E(\omega_{it}|x_{it}))^{\top} f^2(x_{it})]$ is a $(d+q)\times(d+1)$ non-singular matrix function. □

***Assumption* A6:** For some $\delta > 0$, $E[|v_i|^{(4+\delta)}|x_{it}] < \infty$ and $E[|u_{it}|^{(4+\delta)}|x_{it}] < \infty$.□

These assumptions are fairly standard in the semiparametric literature, but some remarks are required. Assumption (A1) is stronger than what it is in Li and Stengos (1996). The density function is assumed to be bounded and at least first-order partially differentiable with a Lipschitz-continuous remainder as this is required for estimating the nonlinear portion. This stronger condition is not necessary for the asymptotic properties of $\widehat{\theta}$, only for $\widehat{m}(\cdot)$. The strict stationarity condition is imposed for the simplicity of the mathematical proofs, but can be relaxed. Assumption (A4) implies that $(2\upsilon > d)$ or $(2\upsilon \geq d+1)$ (because υ is an integer), which in turn is equivalent to $(\upsilon \geq (d+1)/2)$ as in Robinson (1988) or Li and Stengos (1996). This assumption is stronger than what is assumed in Li (1996), but it implies that a standard second-order kernel $(\upsilon = 2)$ can be used if $(d \leq 3)$ and the proofs of the asymptotic properties of the proposed estimators are considerably simpler. Assumption (A6) is required to obtain the asymptotic distribution of the nonparametric estimator, $\widehat{m}(\mathbf{x};h)$.

We are now ready to proceed with the asymptotic results for our parametric and nonparametric components, respectively:

Theorem 10.1 *Under assumptions (A1)-(A5), as $N \longrightarrow \infty$ and T is fixed*

$$\sqrt{NT}(\widehat{\theta} - \theta) \xrightarrow{d} N\left(0, \Phi^{-1}\Psi\Phi^{-1}\right)$$

where $\eta_{it} = \omega_i - E(\omega_i|x_{it})$, $\Phi = E[\eta_{it}\eta_{it}^{\top} f^2(x_{it})]$ *is positive definite, and* $\Psi = \sigma_{\varepsilon}^2 E[\eta_{it}\eta_{it}^{\top} f^4(x_{it})]$. *Moreover,* $\widehat{\Phi} = (NT)^{-1}\sum_{i=1}^{N}\sum_{t=1}^{T}\widehat{\eta}_{it}\widehat{\eta}_{it}^{\top}\widehat{f}^2(x_{it})$ *and* $\widehat{\Psi} = (1/NT^2)\sum_{i=1}^{N}\sum_{t=1}^{T}\sum_{s=1}^{T}\widehat{\eta}_{it}\widehat{\eta}_{is}^{\top}\widehat{\varepsilon}_{it}\widehat{\varepsilon}_{is}\widehat{f}(x_{it})\widehat{f}(x_{is})$ *are the consistent estimators of* Ψ *and* Φ^{-1}, *respectively, where* $\widehat{\eta}_{it} = \omega_i - \widehat{E}(\omega_i|x_{it})$.

The proof of Theorem 10.1 is a straightforward extension of Theorem 1 in Li and Stengos (1996). The detailed proof is available upon request.

We now turn to the asymptotic properties of the nonparametric estimator proposed for $m(\cdot)$ in (10.9). Denote $\mu_2 = \int v^2 K(v)dv$ and $\mathcal{R} = \int k^2(v)dv$, where μ_2 and $\mathcal{R}$ are scalars different from zero.

Theorem 10.2 *Under Assumptions (A1)-(A6), as $N \to \infty$ and T is fixed*

$$\sqrt{NTh^d}\left\{\widehat{m}(\mathbf{x};h) - m(\mathbf{x}) + \frac{h^2\mu_2^d}{2} tr\{\mathcal{H}_m(\mathbf{x})\} + o_p(h^2)\right\} \xrightarrow{d} N\left(0, \frac{\sigma_\varepsilon^2 \mathcal{R}^d}{f(\mathbf{x})}\right),$$

where $\mathcal{H}_m(\mathbf{x}) = \partial m(\mathbf{x})/\partial\mathbf{x}\partial\mathbf{x}^\top$ is the Hessian matrix of $m(\cdot)$.

The proof of Theorem 10.2 is obtained via a similar scheme as in Fan and Gijbels (1995) or Soberon et al. (2021), among others. The detailed proof is available upon request.

10.3 Inference

In this section, we discuss how to test for whether the estimation procedure proposed in this chapter accounts for correlation between the individual effects v_i and the regressor vector x_{it}. The null and alternative hypothesis can be written as

$$H_0 : Pr[E(v_i|x_{it}) = 0] = 1 \text{ for all } i$$
$$H_1 : Pr[E(v_i|x_{it}) \neq 0] > 0 \text{ for some } i.$$

The commonly used local-constant estimator (i.e., Nadaraya-Watson estimator) is consistent when $E(v_i|x_{it}) = 0$, but is inconsistent when $E(v_i|x_{it}) \neq 0$. Our nonparametric correlated random-effects estimator is consistent in both cases.

Motivated by Sun, Carroll and Li (2009) in a different context, we propose a test statistic based on

$$I = \int [\widehat{m}(\mathbf{x};h) - \widetilde{m}(\mathbf{x};h)]^2 d\mathbf{x}, \tag{10.10}$$

where $\widehat{m}(\mathbf{x})$ is our nonparametric correlated random-effects local-constant estimator and $\widetilde{m}(\mathbf{x})$ is the Nadaraya-Watson estimator of the form

$$\widetilde{m}(\mathbf{x};h) = \frac{\sum_{i=1}^{N}\sum_{t=1}^{T} y_{it} K_h(x_{it}, \mathbf{x})}{\sum_{i=1}^{N}\sum_{t=1}^{T} K_h(x_{it}, \mathbf{x})}.$$

Using $\sum_{i=1}^{N}\sum_{t=1}^{T} K_h(x_{it}, \mathbf{x})$ to remove the random denominators and by defining $\widehat{\epsilon}_{it} = y_{it} - \omega_i^\top \widehat{\theta} - \widetilde{m}(\mathbf{x};h)$, the test statistic becomes

$$I_n = \frac{1}{N^2T^2h^d}\sum_{i=1}^{N}\sum_{j=1}^{N}\sum_{t=1}^{T}\sum_{s=1}^{T}\int \widehat{\epsilon}_{it}\widehat{\epsilon}_{js}K_h(x_{it},\mathbf{x})K_h(x_{js},\mathbf{x})d\mathbf{x}, \tag{10.11}$$

because by rearranging terms, we can write

$$\sum_{i=1}^{N}\sum_{t=1}^{T}K_h(x_{it},\mathbf{x})\left[\widehat{m}(\mathbf{x};h)-\widetilde{m}(\mathbf{x};h)\right] = \sum_{i=1}^{N}\sum_{t=1}^{T}K_h(x_{it},\mathbf{x})\widehat{\epsilon}_{it}.$$

To simplify the integration in (10.11), we use the fact that the typical element of the integration term can be written as a twofold convolution kernel that acts as a weighting function to select the observations such that only those (x_{it},x_{js}) close to each other are used. Finally, to avoid the asymptotic bias term of this type of double summation test, we remove the $i=j$ terms of the kernel matrix of I_n, and obtain the following test statistic

$$\widehat{I}_n = \frac{1}{N^2T^2h^d}\sum_{i=1}^{N}\sum_{j\neq i}^{N}\sum_{t=1}^{T}\sum_{s=1}^{T}\widehat{\epsilon}_{it}\widehat{\epsilon}_{js}K_h(x_{it},x_{js}). \tag{10.12}$$

To derive the asymptotic properties of the test statistic proposed in (10.12), the following assumptions are required:

***Assumption* A7:** For all $t\neq s$, (x_{it},x_{is}) has a joint probability density function, $f_{ts}(\mathbf{x}_1,\mathbf{x}_2)$, that is continuously differentiable and $\sup_{t\neq s}\int f_{ts}(\mathbf{x},\mathbf{x})d\mathbf{x} < M < \infty$. $E(x_{l,it}^{v}|x_{is})$, $f(\mathbf{x})$, $f_{ts}(\mathbf{x}_1,\mathbf{x}_2)$, and their first-order partial derivatives are all uniformly bounded for $l=1,\ldots,d$, $v=1,\ldots,4$, and for all i and $t\neq s$. □

***Assumption* A8:** As $N\longrightarrow\infty$ for T fixed, $h\longrightarrow 0$, and $Nh^d\longrightarrow\infty$. □

The asymptotic distribution under the null and the power of our test are given by the following theorems. They are obtained following similar reasoning as to that in Zheng (1996) and/or Henderson and Soberon (2024), among others. The detailed proofs are available upon request.

Theorem 10.3 *Under Assumptions (A1)-(A8) and assuming $f(x)>0$ for each x in the support of x_{it}, under H_0, as $N\longrightarrow\infty$ and T is fixed,*

$$J_n = NTh^{d/2}\frac{\widehat{I}_n}{\sqrt{\widehat{\Sigma}}} \xrightarrow{d} N(0,1)$$

where $\widehat{\Sigma} = \frac{1}{N^2T^2h^d}\sum_{i=1}^{N}\sum_{j\neq i}^{N}\sum_{t=1}^{T}\sum_{s=1}^{T}\widehat{\epsilon}_{it}^2\widehat{\epsilon}_{js}^2K_h^2(x_{it},x_{js})$ is a consistent estimator of the asymptotic variance of $NTh^{d/2}\widehat{I}_{NT}$ where

$$\Sigma = \frac{2\sigma_{\varepsilon}^2\mathcal{R}^d(K)}{T^2}E[f(x_{it})].$$

Theorem 10.4 *Under Assumptions (A1)-(A8) and assuming $f(x) > 0$ for each x in the support of x_{it}, under H_1, we have $Pr(J_n \geq c) \to 1$ as $N \to \infty$, where (c) is any positive constant.*

In practice, the asymptotic distribution of the test statistic is not useful for finite samples. As is standard in nonparametric kernel based tests, we resort to a bootstrap procedure. The steps for the wild bootstrap are as follows:

1. Compute the test statistic $\widehat{J}_n$ for the original sample of data $\{y_{it}, x_{it}, \omega_i\}$.
2. For each observation i, for each time period t, draw a wild residual bootstrap v_{it}^* and construct the bootstrapped left-hand-side variable as $y_{it}^* = \widehat{m}(x_{it}) + \omega_i^\top \widehat{\theta} + v_{it}^*$ and call $\{y_{it}^*, x_{it}, \omega_i\}$ the bootstrap sample.[3]
3. Calculate $\widehat{T}_n^*$ where $\widehat{T}_n^*$ is calculated the same way as $\widehat{T}_n$ except that y_{it} is replaced by y_{it}^*.
4. Repeat steps 2 and 3 a large number (B) of times and then construct the sampling distribution of the bootstrapped test statistics. Reject the null if the estimated statistic $\widehat{T}_n$ is greater than the upper $\alpha-$percentile of the bootstrapped test statistics.

10.4 Simulations

We investigate the finite sample performance of our proposed estimators and test via Monte Carlo simulations. We begin with Equation (10.1) and consider various forms for $m(\cdot)$ and μ_i. Our study considers two distinct functional forms for $m(\cdot)$:

$$m(x_{it}) = x_{it}^\top \beta$$
$$m(x_{it}) = \sin(x_{it}),$$

where x_{it} is generated as a standard random normal variable. We examine three specifications for the unobserved heterogeneity, μ_i:

$$\mu_i = \bar{x}_i^\top \psi + v_i$$
$$\mu_i = z_i^\top \gamma + v_i$$
$$\mu_i = z_i^\top \gamma + \bar{x}_i^\top \psi + v_i,$$

and when we study the size of our test, we set $\mu_i = v_i$.

Both error terms (v_i and ε_{it}) are generated as i.i.d. random variables, normally distributed with zero mean and variance $1/2$. Our simulation design varies both the cross-sectional dimension $N \in \{100, 200, 400\}$ and the time dimension $T \in \{3, 5\}$. For each configuration, we conduct 999 simulations, and use Gaussian kernel functions with bandwidths selected according to Silverman (1986). For our testing procedure, we use 399 bootstrap replications for each simulation.

[3] $v_{it}^* = \widehat{\epsilon}_{it} * \varphi_i^b$, where $\varphi_i^b \overset{i.i.d.}{\sim} N(0,1)$. It is easy to see that $E[\widehat{\epsilon}_{it}\varphi_i^b] = 0$, $Cov[\widehat{\epsilon}_{it}\varphi_i^b, \widehat{\epsilon}_{js}\varphi_j^b] = Cov[\widehat{\epsilon}_{it}, \widehat{\epsilon}_{js}]$, for all $i, j = 1, \ldots, N$, $t, s = 1, \ldots, T$.

10.4.1 Estimation

To assess estimation accuracy, we compute the average mean squared error (AMSE) for the nonparametric function $m(\cdot)$

$$AMSE[\widehat{m}(\cdot)] = \frac{1}{999}\sum_{j=1}^{999}\left[\frac{1}{NT}\sum_{i=1}^{N}\sum_{t=1}^{T}(\widehat{m}_j(x_{it}) - m(x_{it}))^2\right],$$

and for specifications involving γ, we also analyze its behavior via $MSE(\widehat{\gamma}) = \frac{1}{999}\sum_{j=1}^{999}(\widehat{\gamma}_j - \gamma)^2$.

The results of the simulations are summarized in Figures 10.1 and 10.2.[4] The results reveal several key patterns. For the linear specification $m(\cdot) = x_{it}^{\top}\beta$, as expected, the least-squares estimator consistently outperforms the nonparametric approach across all specifications, exhibiting lower AMSE values and reduced variance. This advantage is particularly pronounced for smaller sample sizes. However, the performance gap narrows as the sample size increases, with $N = 400$ showing markedly improved precision for both methods.

The nonparametric estimation of the nonlinear function $m(\cdot) = \sin(x_{it})$ demonstrates consistent patterns across all three μ_i specifications. The AMSE values decrease with larger sample sizes, and estimation precision improves when $T = 5$ compared to $T = 3$. The consistency properties shown in the theorems appear to be validated here.

The estimation results appear relatively robust across the different specifications of unobserved heterogeneity, though the more complex specification $\mu_i = z_i^{\top}\gamma + \bar{x}_i^{\top}\psi + v_i$ shows slightly higher MSE values, particularly in smaller samples. This suggests that the additional complexity of controlling for both time-invariant regressors and time averages of x_{it} introduces modest efficiency costs in finite samples.

The results for the estimation of the parameter γ can be found in Figures 10.3 and 10.4. The estimates here come from the same simulations runs in panels (c-f) in Figure 10.1. What is obvious from the comparison between the two figures is that in Figure 10.3, we can see that the parametric components from the semiparametric procedure are estimated as precisely as those from the least-squares estimator. As expected, the results for Figure 10.4 for the semiparametric estimator look nearly identical to those from the previous figure. Again, the least-squares estimates are not given in Figure 10.4 as the estimates of $m(\cdot)$ are inconsistent.

[4] We show the results for both the least-squares and semiparametric estimates for the linear specification, but only the semiparametric results for the nonlinear specification as our least-squares estimators are inconsistent in this setting.

Fig. 10.1: AMSE for the linear specification of $m(x) = x_{it}\beta$ for least-squares and semiparametric estimators: $N \in \{100, 200, 400\}$ and $T \in \{3, 5\}$

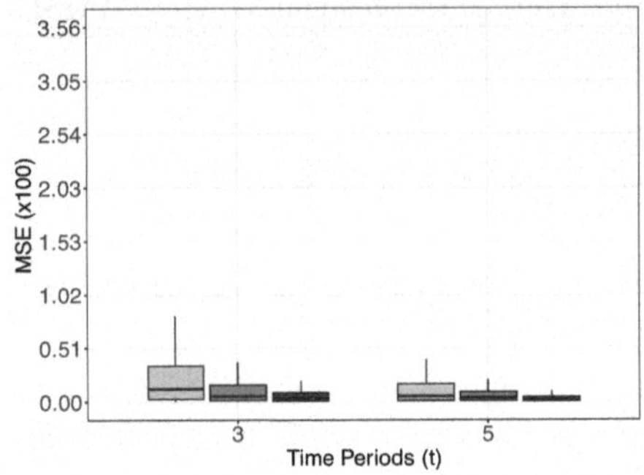

(a) $\mu_i = \bar{x}_i^\top \psi + v_i$: Least-Squares

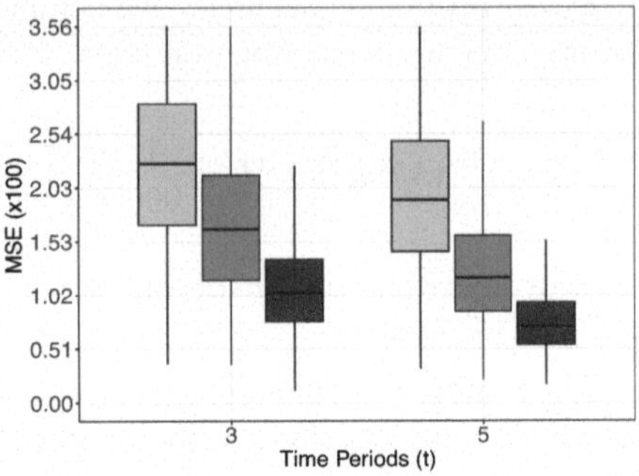

(b) $\mu_i = \bar{x}_i^\top \psi + v_i$: Semiparametric

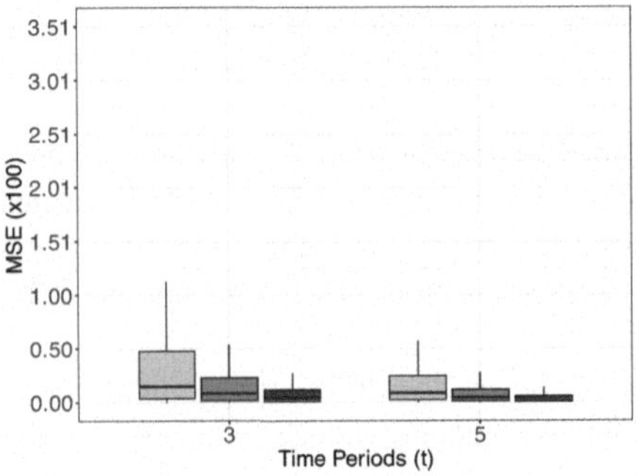

(c) $\mu_i = Z_i \gamma + v_i$: Least-Squares

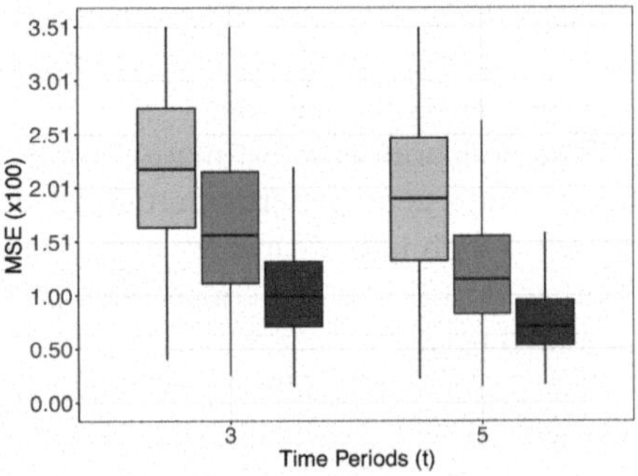

(d) $\mu_i = z_i^\top \gamma + v_i$: Semiparametric

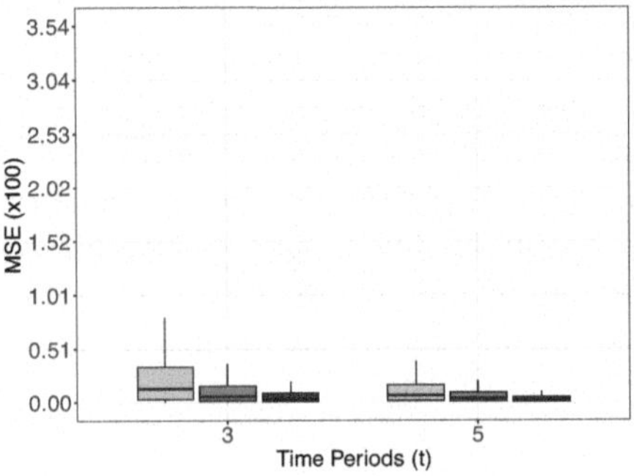

(e) $\mu_i = z_i^\top \gamma + \bar{x}_i^\top \psi + v_i$: Least-Squares

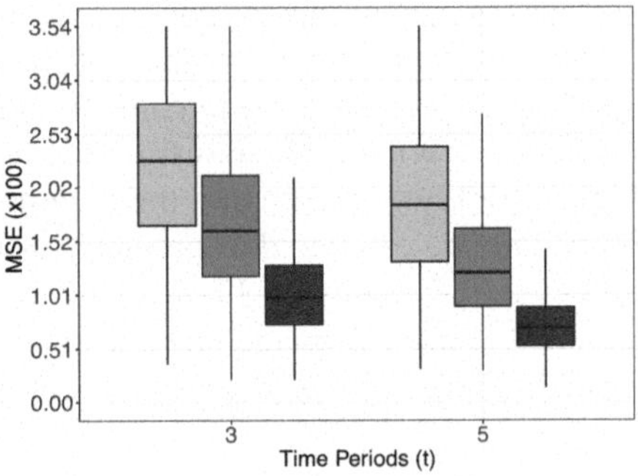

(f) $\mu_i = z_i^\top \gamma + \bar{x}_i^\top \psi + v_i$: Semiparametric

Fig. 10.2: AMSE for the nonlinear specification of $m(x) = \sin(x_{it})$ for the semiparametric estimator: $N \in \{100, 200, 400\}$ and $T \in \{3, 5\}$

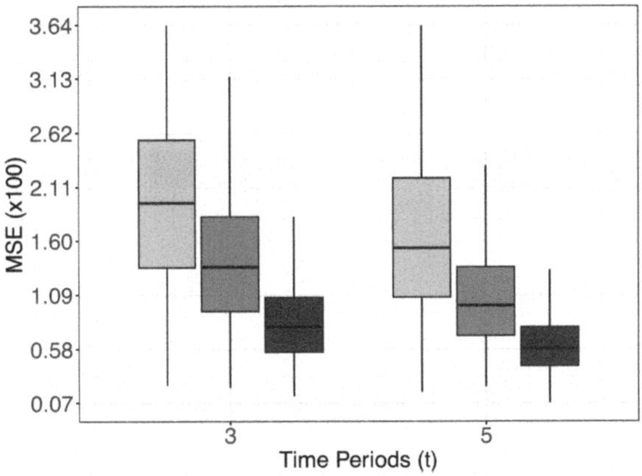

(a) $\mu_i = \bar{x}_i^\top \psi + v_i$

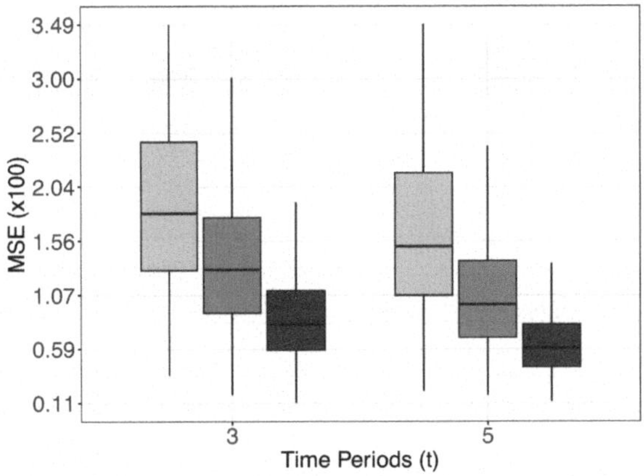

(b) $\mu_i = z_i^\top \gamma + v_i$

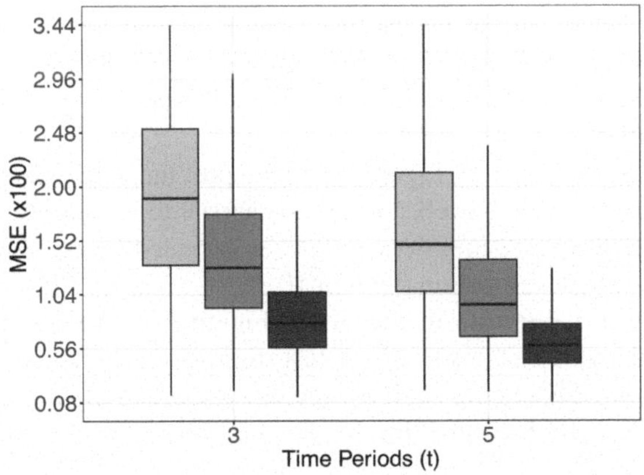

(c) $\mu_i = z_i^\top \gamma + \bar{x}_i^\top \psi + v_i$

Fig. 10.3: Estimates of $\widehat{\gamma}$ for the linear specification of $m(x) = x_{it}^{\top}\beta$ for least-squares and semiparametric estimators: $N \in \{100, 200, 400\}$ and $T \in \{3, 5\}$

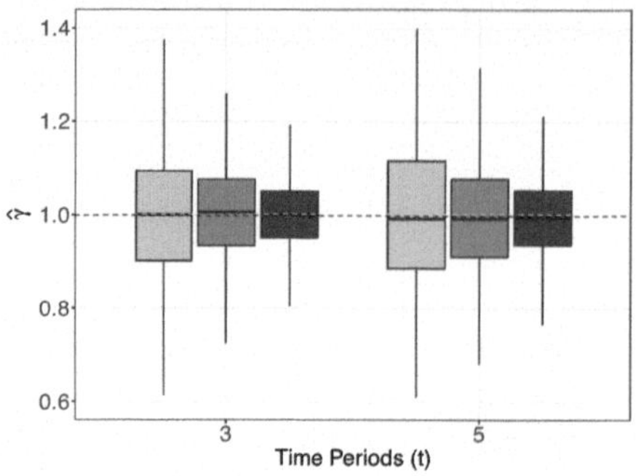

(a) $\mu_i = Z_i\gamma + v_i$: Least-Squares

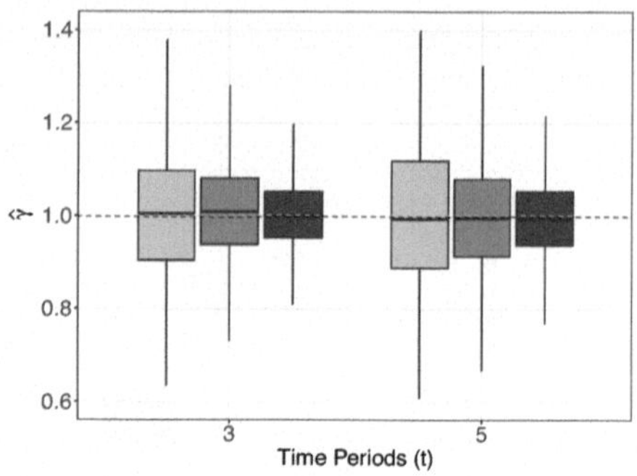

(b) $\mu_i = z_i^{\top}\gamma + v_i$: Semiparametric

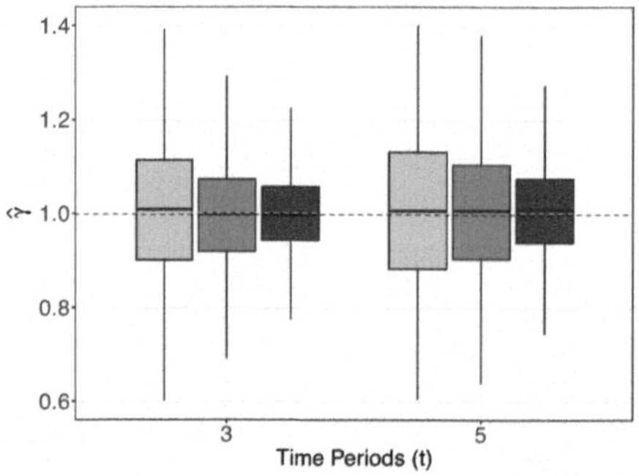

(c) $\mu_i = z_i^{\top}\gamma + \bar{x}_i^{\top}\psi + v_i$: Least-Squares

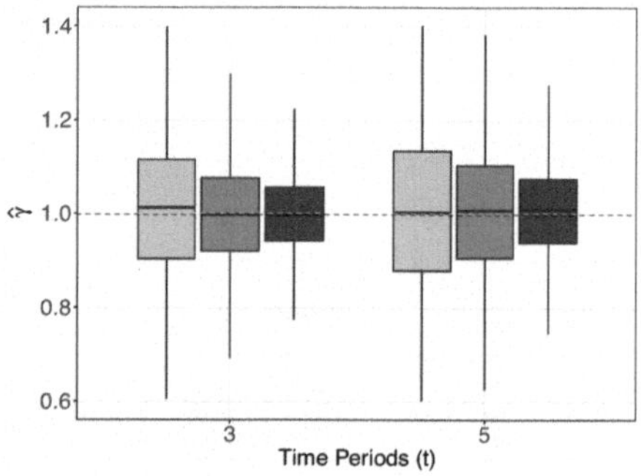

(d) $\mu_i = z_i^{\top}\gamma + \bar{x}_i^{\top}\psi + v_i$: Semiparametric

10.4.2 Inference

To study the performance of our testing procedure, we first study its finite sample size properties. The results are similar for the linear and nonlinear functions of $m(\cdot)$ and so we only consider the linear function. Everything remains the same as in the previous subsection except to study size, we set $\mu_i = v_i$. 399 bootstrap replications for each of our 999 simulations.

The results for size can be found in Figure 10.5. We plot the actual level versus the nominal level. Each of the actual levels for the tests appear to be near the nominal levels, but note that the performance improves with N, as expected. These simulations appear to support the theory developed in Section 10.3.

The results for the power of our test can be found in Figure 10.6. We plot the actual level versus the nominal level. The power of the test improves with N, as expected, for each specification of μ_i. That being said, we see that the probability of correctly rejecting the null is higher when controlling for $\bar{x}_i$ as compared to z_i, and is even

Fig. 10.4: Estimates of $\widehat{\gamma}$ for the nonlinear specification of $m(x) = \sin(x_{it})$ for the semiparametric estimator: $N \in \{100, 200, 400\}$ and $T \in \{3, 5\}$

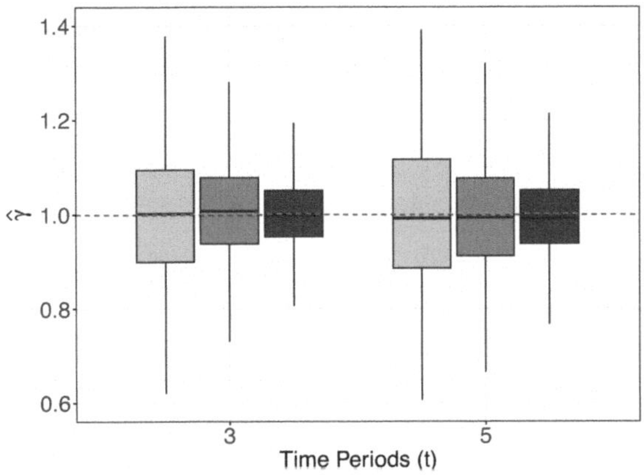

(a) $\mu_i = z_i^\top \gamma + v_i$

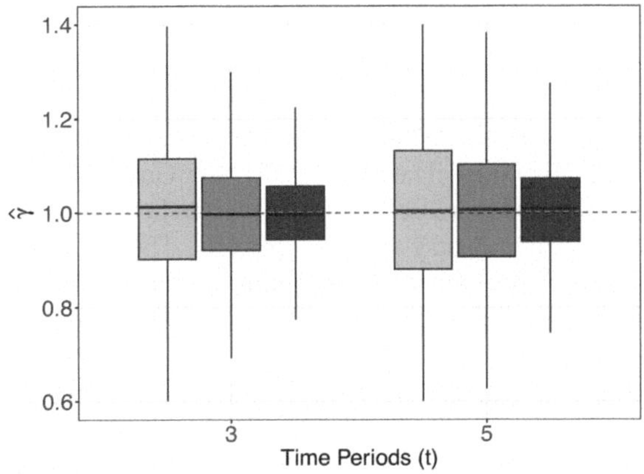

(b) $\mu_i = z_i^\top \gamma + \bar{x}_i^\top \psi + v_i$

higher when controlling for both. These simulations appear to support the theory developed in Section 10.3.

10.5 Empirical Illustration

To demonstrate the empirical relevance of our methodological innovations, we analyze the relationship between firms' research and development expenditures, current assets, and regulatory restrictions across different industries. Our data combines firm-level

Fig. 10.5: Nominal size (vs actual size) of our specification test: 999 simulations each with 399 bootstrap replications: $N \in \{100, 200, 400\}$ and $T \in \{3, 5\}$

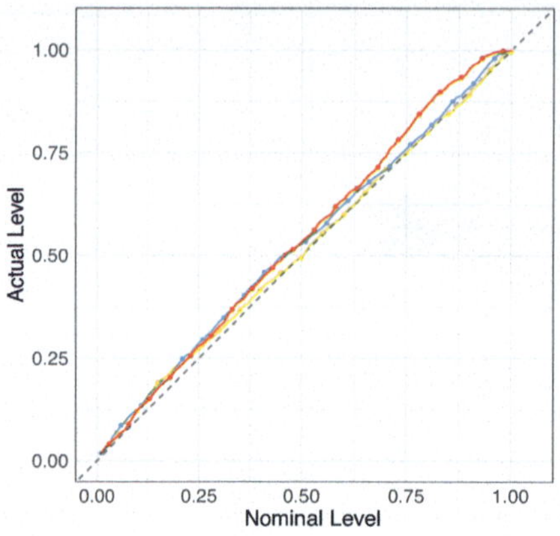

(a) $\mu_i = \nu_i, T = 3$

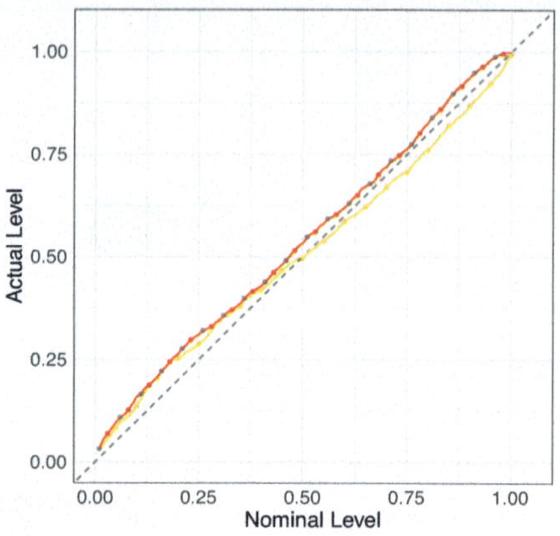

(b) $\mu_i = \nu_i, T = 5$

financial information from Compustat (Wharton Research Data Services, 2024) with industry-level regulatory data from RegData U.S. 4.1 (QuantGov, 2024).

RegData U.S. 4.1, released in March 2022, provides comprehensive measurements of federal regulations and their industry-specific impacts. The dataset quantifies regulatory restrictions through algorithmic identification of prohibited or required activities in the Code of Federal Regulations, offering a systematic measure of regulatory burden at the industry level.

Instead of a traditional panel data model (firms measured over time), in our application, we consider a repeated measure problem. Our unit of observation will be the industry and the repeated measure will be observing different firms in each industry. We will repeat this analysis over three different time periods: 2019, 2020 and 2021.

More formally, we examine the relationship between research and development expenditures (XRD) as our left-hand-side variable, current assets (ACT) as our firm-varying explanatory variable, and industry-specific regulatory restrictions (RESTRICTIONS) as our firm-invariant variable. Our specification follows the form:

$$y_{ij} = x_{ij}^\top \beta + \bar{x}_i^\top \psi + z_i^\top \gamma + \varepsilon_{ij}, \quad i = 1, \ldots, N, \quad j = 1, \ldots, J, \tag{10.13}$$

where y_{ij} represents (the log of) research and development expenditures of firm j in industry i, x_{ij} denotes current (the log of) assets of firm j in industry i, $\bar{x}_i$ is the average value of x_{ij} over the firms (j) in industry i, and z_i captures regulatory restrictions in industry i. To ensure sufficient within-industry variation, we select the top five firms ($J = 5$) from each industry and exclude industries with fewer than five

Fig. 10.6: Power of our specification test: 999 simulations each with 399 bootstrap replications: $N \in \{100, 200, 400\}$ and $T \in \{3, 5\}$

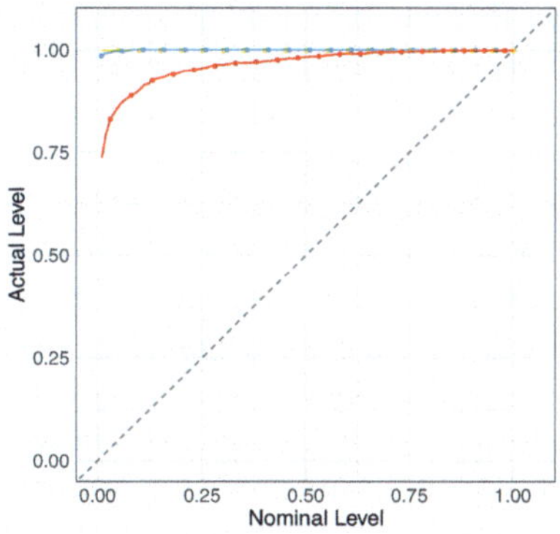

(a) $\mu_i = \overline{x}_i^{\top}\psi + v_i$, $T = 3$

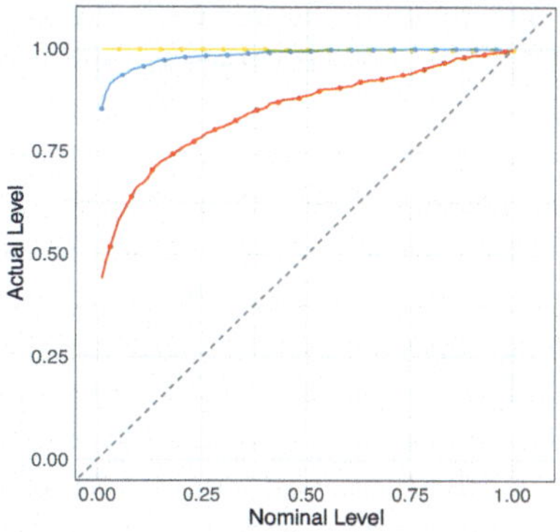

(b) $\mu_i = \overline{x}_i^{\top}\psi + v_i$, $T = 5$

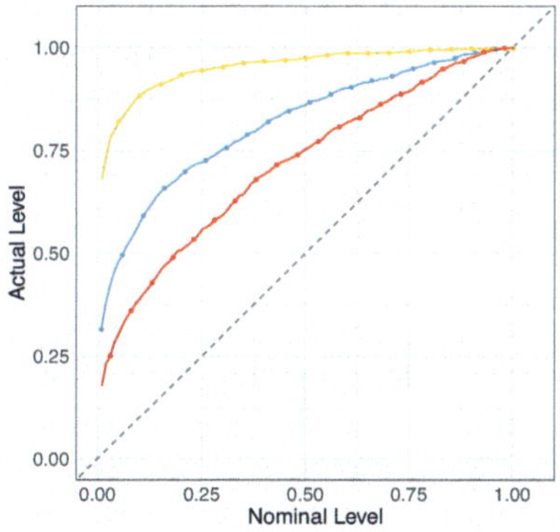

(c) $\mu_i = z_i^{\top}\gamma + v_i$, $T = 3$

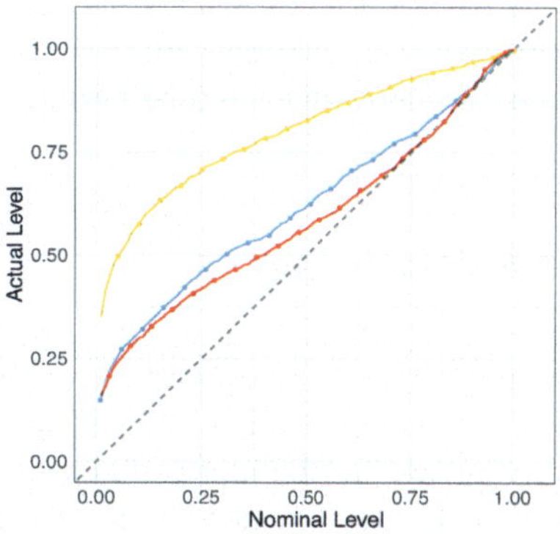

(d) $\mu_i = z_i^{\top}\gamma + v_i$, $T = 5$

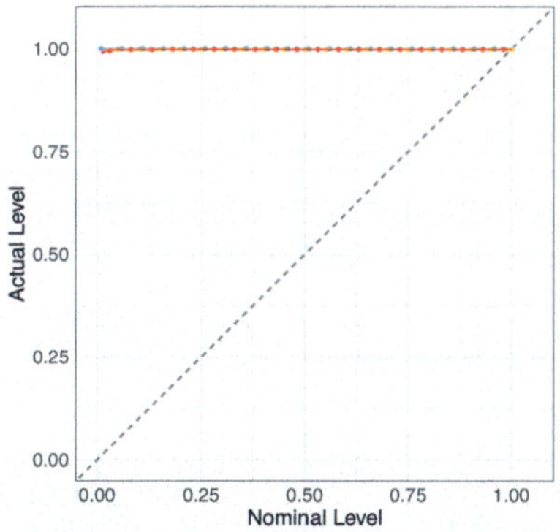

(e) $\mu_i = \overline{x}_i^{\top}\psi + z_i^{\top}\gamma + v_i$, $T = 3$

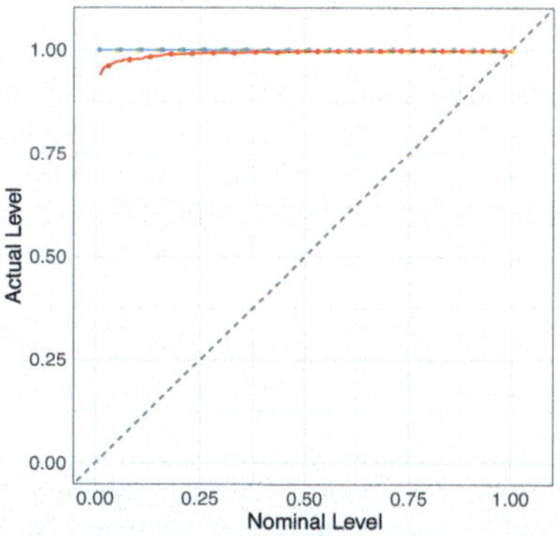

(f) $\mu_i = \overline{x}_i^{\top}\psi + z_i^{\top}\gamma + v_i$, $T = 5$

firms. This filtering process yields $N = 47$ industries in 2019, 48 industries in 2020, and 50 industries in 2021.

The results for both the parametric and semiparametric estimation of Equation 10.13 can be found in Figures 10.7 and 10.8. The former (Figure 10.7) displays scatterplots of $\ddot{y}_{ij} = y_{ij} - \omega_i^\top \widehat{\theta}$ versus x_{ij} and lays the parametric $(x_{ij}^\top \widehat{\beta})$ and nonparametric fits $(\widehat{m}(x_{ij}))$ on top of the scatterplots for each year (rows) in the first and second columns, respectively. The parametric and nonparametric fits are difficult to distinguish from one another. Any functional form test for these sample sizes would likely lead to a failure to reject the parametric model. That being said, the semiparametric fit appears to go through the center of the points whereas the parametric fit appears to be impacted by a few large (outlier) values for $\ddot{y}_{ij}$ for smaller values of x_{ij}. The semiparametric (local-estimator) is less impacted by these values. The slightly better fit is summarized via the relative values of pseudo-R^2 (squared correlation between y and the fitted value of y).[5] Figure 10.8 gives the point estimates of γ, for each method, for each year, for each estimation method (ψ is a nuisance parameter and is not reported). We can see that the point estimates for γ are negative for each method in each year.[6] If we are to take these estimates literally, it suggests that increased regulation leads to less expenditure on research and development. That being said, these models are very simple and the estimates of $\widehat{\gamma}$ are insignificant (for each estimator) in each time period.

10.6 Conclusion

In this chapter, we proposed a semiparametric procedure for estimating CRE models. Our estimators result in closed-form solutions and achieve the optimal rates of convergence for the nonparametric and parametric components of our models. We further develop a test to check if the CRE specification captures the correlation between the unobserved effects and the regressors. Our finite sample simulations support our asymptotic theory. Finally, we provided an empirical illustration to show how the methods work with real data.

Acknowledgements The authors thank the editors, Badi Baltagi and László Mátyás, as well as Soroush Ghazi, Chris Parmeter and Mike Price for useful comments and suggestions. Henderson and Soberon acknowledge financial support from the I+D+i project Ref. TED2021-131763A-I00 financed by MCIN/AEI/10.13039/501100011033 and by the European Union NextGenerationEU/PRTR.

[5] Pseudo-R^2 values; 2019: least-squares = 0.7450, semiparametric = 0.7880; 2020: least-squares = 0.7342, semiparametric = 0.7893; and 2021: least-squares = 0.7195, semiparametric = 0.7847.

[6] The small values for $\widehat{\gamma}$ should not be surprising as a one-unit increase in regulation should lead to only small changes in the log of research and development expenditure.

Fig. 10.7: Application: $x_{ij}^{\top}\widehat{\beta}$ and $\widehat{m}(x_{ij})$ versus x_{ij} for the least-squares and semiparametric estimators

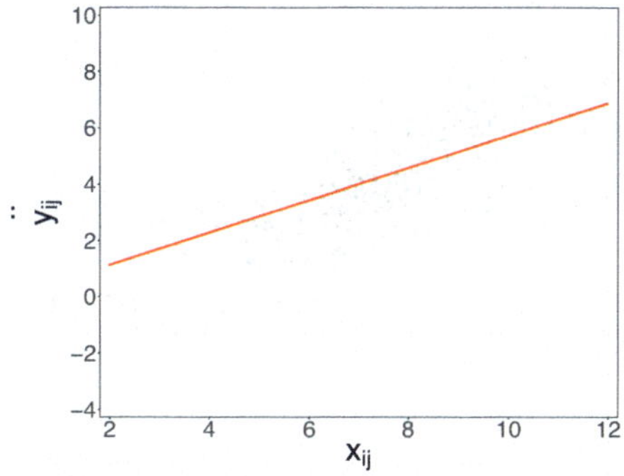

(**a**) 2019: Least-Squares

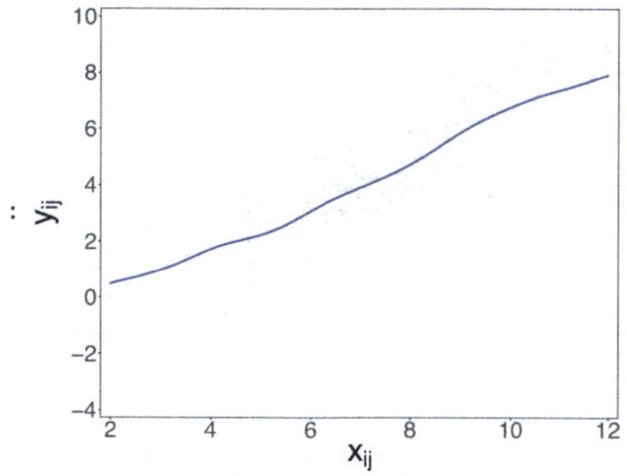

(**b**) 2019: Semiparametric

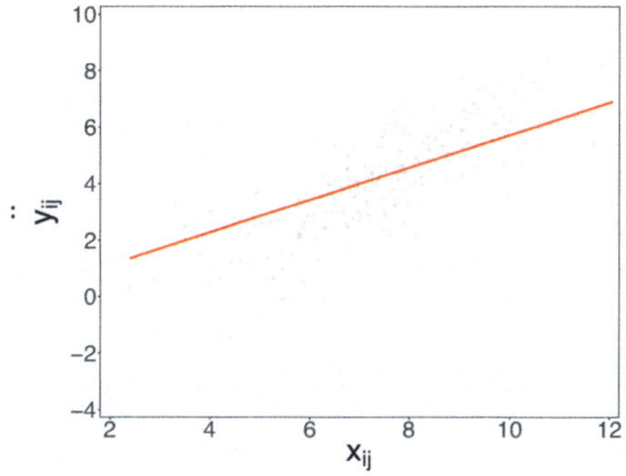

(**c**) 2020: Least-Squares

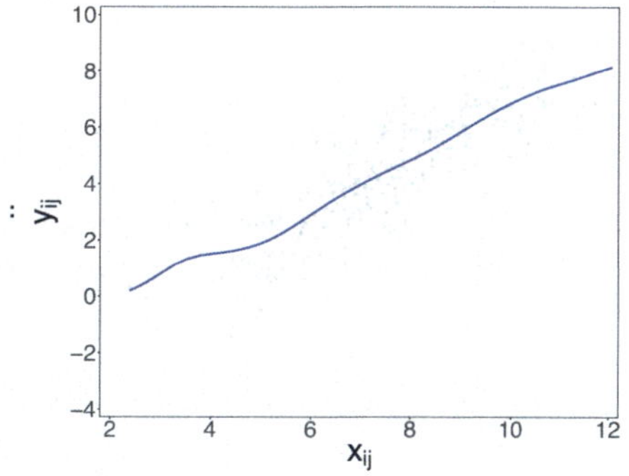

(**d**) 2020: Semiparametric

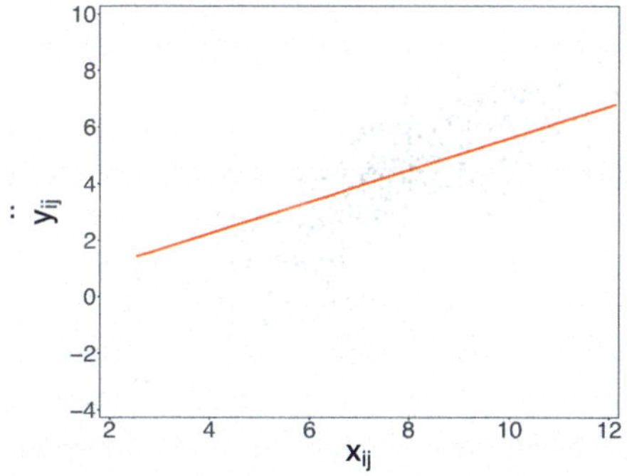

(**e**) 2021: Least-Squares

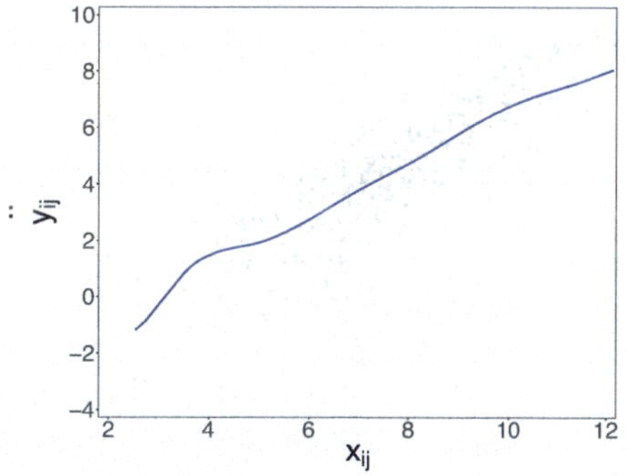

(**f**) 2021: Semiparametric

Fig. 10.8: Application: point estimates of γ and 90% confidence bounds for the years 2019, 2020, and 2021 for both the least-squares and semiparametric estimators

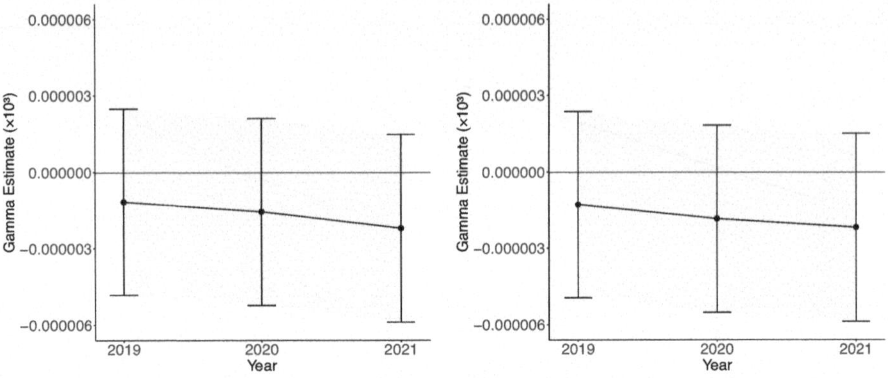

(a) Least-Squares **(b)** Semiparametric

References

Ai, C. & Li, Q. (2008). Semi-parametric and non-parametric methods in panel data models. In *The Econometrics of Panel Data: Fundamentals and Recent Developments in Theory and Practice* (pp. 451–478). Springer.

Baltagi, B. (2021). *Econometric Analysis of Panel Data*. Springer.

Bester, C. A. & Hansen, C. (2009). Identification of marginal effects in a nonparametric correlated random effects model. *Journal of Business & Economic Statistics*, *27*(2), 235–250.

Chamberlain, G. (1982). Multivariate regression models for panel data. *Journal of Econometrics*, *18*(1), 5–46.

Fan, J. & Gijbels, I. (1995). *Local Polynomial Modelling and its Applications*. Chapman & Hall.

Henderson, D. J., Carroll, R. J. & Li, Q. (2008). Nonparametric estimation and testing of fixed effects panel data models. *Journal of Econometrics*, *144*(1), 257–275.

Henderson, D. J. & Parmeter, C. F. (2015). *Applied Nonparametric Econometrics*. Cambridge University Press.

Henderson, D. J. & Soberon, A. (2024). Nonparametric models with fixed effects. In L. Matyas (Ed.), *The Econometrics of Multi-dimensional Panels, Advanced Studies in Theoretical and Applied Econometrics 54* (pp. 285–323). Springer Nature Switzerland AG.

Henderson, D. J. & Ullah, A. (2005). A nonparametric random effects estimator. *Economics Letters*, *88*(3), 403–407.

Li, Q. (1996). On the root-n-consistent semiparametric estimation of partially linear models. *Economics Letters*, *51*(3), 277–285.

Li, Q. & Stengos, T. (1996). Semiparametric estimation of partially linear panel data models. *Journal of Econometrics*, *71*(1), 389–397.

Mundlak, Y. (1978). On the pooling of time series and cross section data. *Econometrica*, *46*(1), 69–85.

Nerlove, M. (2005). *Essays in Panel Data Econometrics*. Cambridge University Press.

Parmeter, C. F. & Racine, J. S. (2019). Nonparametric estimation and inference for panel data models. *Panel Data Econometrics*, 97–129.

Powell, J. L., Stock, J. H. & Stoker, T. M. (1989). Semiparametric estimation of index coefficients. *Econometrica*, *57*(6), 1403–1430.

QuantGov. (2024). *RegData United States 4-1.* https://www.quantgov.org/csv-download. (Accessed: 12-2024)

Robinson, P. M. (1988). Root-n-consistent semiparametric regression. *Econometrica*, *56*(4), 931–954.

Rodriguez-Poo, J. M. & Soberon, A. (2017). Nonparametric and semiparametric panel data models: Recent developments. *Journal of Economic Surveys*, *31*(4), 923–960.

Silverman, B. W. (1986). *Density Estimation for Statistics and Data Analysis*. Routledge.

Soberon, A., Rodriguez-Poo, J. M. & Robinson, P. M. (2021). Nonparametric panel data regression with parametric cross-sectional dependence. *Econometrics Journal*, *25*(1), 114–133.

Su, L. & Ullah, A. (2011). Nonparametric and semiparametric panel econometric models: estimation and testing. *Handbook of Empirical Economics and Finance*, 455–497.

Sun, Y., Carroll, R. J. & Li, D. (2009). Semiparametric estimation of fixed effects panel data varying coefficient models. *Advances in Econometrics*, *25*, 101–130.

Sun, Y., Zhang, Y. Y. & Li, Q. (2015). Nonparametric panel data regression models. In B. Baltagi (Ed.), *The Oxford Handbook of Panel Data* (pp. 285–324). Oxford University Press.

Wharton Research Data Services. (2024). *Compustat Annual Fundamentals Database*. https://wrds-www.wharton.upenn.edu/pages/grid-items/compustat-annual-updates-fundamentals-annual-demo/. (Accessed: 12-2024)

Wooldridge, J. M. (2019). Correlated random effects models with unbalanced panels. *Journal of Econometrics*, *211*(1), 137–150.

Zheng, J. X. (1996). A consistent test of functional form via nonparametric estimation techniques. *Journal of Econometrics*, *75*(2), 263–289.

Chapter 11
The Correlated Random Effects GMM-Level Estimation: Monte Carlo Evidence and Empirical Applications

Maria Elena Bontempi and Jan Ditzen

Abstract We introduce CRE-GMM, a new estimator that exploits correlated random effects (CRE) within the generalised method of moments on level equations (GMM-lev) in a dynamic (but also static) model on panel data. Unlike GMM-dif, it allows the estimation of the effects of measurable time-invariant covariates and, compared to GMM-sys, makes efficient use of all available information. CRE-GMM considers explanatory variables that may be affected by double endogeneity (correlation with individual heterogeneity and idiosyncratic shocks), models initial conditions and improves inference. Monte Carlo simulations validate CRE-GMM across panel types and endogeneity scenarios. Empirical applications to R&D, production, and wage functions illustrate the advantages of CRE-GMM.

11.1 Introduction

Many economic relationships involve dynamic adjustment processes due to, e.g., habits, adjustment costs, gestation lags, and the wait and see role of uncertainty. The generalised method of moments (GMM) is widely used in applied economic research to estimate linear dynamic panel data models, mainly as GMM-dif in which the model in first differences is instrumented by lagged levels (Arellano & Bond, 1991; Alvarez & Arellano, 2003) and GMM-sys which adds to the model in first differences the model in levels only for one period, instrumented by lagged first differences (Arellano & Bover, 1995; Blundell & Bond, 1998). In the words of Kiviet (2007), the issue is whether it is possible to remove unobservable individual heterogeneity from the regressors (GMM-dif case), or from any variable that may be used as an instrument

Maria Elena Bontempi ✉
Department of Economics, University of Bologna, Bologna, Italy, e-mail: mariaelena.bontempi@unibo.it

Jan Ditzen
Facoltà di Economia, Free University of Bozen-Bolzano, Bolzano, Italy, e-mail: jan.ditzen@unibz.it

B. H. Baltagi and L. Mátyás (eds.), *Seven Decades of Econometrics and Beyond*, Advanced Studies in Theoretical and Applied Econometrics 57,
https://doi.org/10.1007/978-3-031-92699-0_11

(GMM-sys case in the not redundant moment condition in levels), or from the model, which is our proposal.

Indeed, we investigate the effect of maintaining the model in levels (GMM-lev) combined with the Correlated Random Effects (CRE) specification into a unique stage framework, the CRE-GMM estimator. Our approach explicitly extends the equation to be estimated so to capture the initial endowment of each unit, as measured by the pre-sample realisations of the variables of the model. Hence, it can handle a mix of units whose differences are not simply captured by the initial observation of the dependent variable, but by a set of observations of all variables concurring to the dynamic process. In our extended level regressions, we treat individual heterogeneity as random, to be consistent with Haavelmo's view (Haavelmo, 1944) that the population of interest consists of an infinity of decisions made by individuals who are different from each other and who may change their behavior over time (Nerlove, Sevestre & Balestra, 2008). We consider the random effects similar in spirit to latent variables driving the distribution of the correlated explanatory variables.

We show that our proposed methodology is effective in addressing the double endogeneity of the explanatory variables, which originates from two sources. The first source of endogeneity, termed *endogeneity due to heterogeneity*, arises from the correlation between covariates and unobservable, unit-specific characteristics, that may vary over time, thereby invalidating the GMM-dif and, partially, the GMM-sys. The second source of endogeneity, termed *standard endogeneity*, is contingent on the correlation of covariates with idiosyncratic shocks that vary with units and over time. It is managed by GMM-lev where the instruments are defined as lagged first differences or levels depending on the presence or absence of correlation between the explanatory variables and individual heterogeneity; the selection of lags hinges on the classification of the explanatory variables as exogenous, predetermined and endogenous in terms of their correlation with the idiosyncratic shock. We implement Monte Carlo simulations under alternative settings, with the driving schedule being macro panels (small N and long T, e.g., N=25 and T=40), multilevel panels (the number of groups is large relative to the number of observations per group, for example N=100 and T=20), and longitudinal panels (N much larger than T, e.g., N=1000 and T=10).

The primary objective of our CRE-GMM method is to maintain the levels of the equation of interest, thereby enabling the efficient estimation of the effects of measurable and time-invariant explanatory variables, while controlling for unmeasurable individual heterogeneity. In macro panels, our method controls for the effects of measurable institutional traits that drive time-invariant heterogeneity along with unobserved country-specific characteristics, avoiding bias in the estimation of, for example, the 'resource curse' (Haber & Menaldo, 2011). By also considering cross-sectional variation, CRE-GMM avoids using only within variation and the critique given by Kropko and Kubinec (2020) of Acemoğlu, Johnson, Robinson and Yared's (Acemoğlu et al., 2008) counter-intuitive finding that GDP exhibits no relationship with democratisation. In longitudinal panels, CRE-GMM can estimate innovative investments (Gormley & Matsa, 2014) as a function of measurable individual characteristics that are of great interest to researchers (like industries and technological opportunities, location, and market power) while controlling for unobservable factors (like managerial quality, ownership motivation and cost of capital). In fields such as education, psychology, sociology, and political science, based on multilevel or multidimensional panels, GMM-CRE can estimate the effects of higher-level, time-constant variables (e.g., educational system, family background, and social norms) while controlling for endogeneity due to the heterogeneity of lower-level predictors (Mátyás, 2017; Yang & Schmidt, 2021; Hill & In Song, 2020; Imai, Davis, Roos & French, 2019).

Another aim of CRE-GMM is to tackle a common challenge in applied empirical studies. When implementing the GMM-lev estimator on dynamic panel data models, a high estimated autoregressive parameter is observed, akin to the upward-biased pooled OLS estimation that disregards individual heterogeneity. We love to quote Nerlove and Balestra (1966): "The presence of lagged endogenous variables may make it difficult, if not impossible, to separate the individual [...] effects from the effect induced by the lagged variable". This suggests that neglecting or not capturing individual heterogeneity can result in a combination of 'spurious' persistence, due to unobserved unit-specific permanent characteristics, and 'true' persistence, defined as the causal effect of past realizations on the current realization of the dependent variable (Heckman, 1991). The innovation literature frequently points out the difficulty of estimating the causal effect of past R&D activities on current R&D investment due to the path-dependent nature of technical changes (Atkinson & Stiglitz, 1969), if individual heterogeneity that generates spurious persistence is not controlled for (Peters, 2009).

Our approach can be related to that of Riju and Wooldridge (2019), who consider CRE with instrumental variables on a static panel, a situation to which our CRE-GMM approach can also be applied. Our aim of identifying the effects of time-invariant variables in the presence of unobserved heterogeneity is similar to that of the sequential approaches suggested by Hausman and Taylor (1981) and Pesaran and Zhou (2018) for static models, and by Kripfganz and Schwarz (2019) for dynamic models. However, in the latter the explanatory variables are assumed to be strictly exogenous with respect to the idiosyncratic error term, an assumption we relax. Our CRE-GMM approach avoids the use of a two-stage standard error correction and the bias due to time-invariant variables omitted in the first stage (if relevant and related with unit- and time-varying covariates). The use of GMM-lev is taken from Arellano and Bover (1995) and, especially, Bun and Kiviet (2006), who compare various GMM implementations under the assumption that the model includes a predetermined unit- and time-varying explanatory variable, possibly correlated with individual heterogeneity. For GMM-lev, the leading term of the bias is strongly influenced by the magnitude of the individual effects and any correlation between regressors and the effects, which is something we address with our CRE-GMM estimator.

Compared to maximum likelihood, GMM is less restrictive in its assumptions and more useful in modeling complex economic relationships in a world with limited information (Bera & Bilias, 2002). GMM can also be considered as encompassing almost all common estimation methods (Imbens, 2002).

Studies exploiting maximum likelihood dynamic models (Bhargava & Sargan, 1983; Phillips, 2010, 2015; Hsiao & Zhou, 2018; Hsiao, 2020; Alvarez & Arellano, 2022) assume a simple autoregressive specification or, in more general dynamic models, that the explanatory variable other than the lagged dependent variable is uncorrelated with idiosyncratic shocks. GMM allows this assumption to be relaxed, an advantage for most applications where many or all explanatory variables are affected by both endogeneity due to heterogeneity, and standard endogeneity. Accordingly, our research extends the literature focused on the comparative analysis of GMM estimation on dynamic panel data models, Brown and Newey (2002); Hayakawa (2007, 2009, 2015); Hayakawa and Nagata (2016); Bun and Kiviet (2006); Bun and Windmeijer (2010); Kiviet (2007); Kiviet, Pleus and Poldermans (2017); Kiviet (2020); Alvarez and Arellano (2003); Jin, Lee and J. (2021).

The chapter is organized as follows. Section 11.2 introduces the model and Section 11.3 presents the CRE-GMM and its motivations for the applied researcher. Section 11.4 presents results of our Monte Carlo experiments. Section 11.5 reports an empirical example. Section 11.6 concludes. Monte Carlo setup and further results are in the *Online Appendix* (Bontempi & Ditzen, 2025).

11.2 The Model

We consider a dynamic panel model in the form of an ARDL(1,0), or partial adjustment model (PAM):

$$\begin{aligned} y_{it} &= \alpha + \boldsymbol{\beta}'\mathbf{x}_{it} + \boldsymbol{\theta}'\mathbf{w}_i + \rho y_{it-1} + \nu_{it} \\ \nu_{it} &= \mu_i + \upsilon_{it}, \end{aligned} \tag{11.1}$$

where $|\rho| < 1$, $\mathbf{x}_{it}$ is a $1 \times K$ vector of measurable explanatory variables changing with $i = 1, \ldots, N$ and $t = 1, \ldots, T_i$,[1] and $\mathbf{w}_i$ is a $1 \times D$ vector of measurable time-invariant explanatory variables changing only with i. The composite error term is $\nu_{it} = \mu_i + \upsilon_{it}$, where $\mu_i \sim i.i.d.(0, \sigma_\mu^2)$ represents randomly drawn individual-specific unobserved effects, possibly correlated with $\mathbf{x}_{it}$ and $\mathbf{w}_i$ and by definition correlated with y_{it-1}. The component $\upsilon_{it} \sim i.i.d.(0, \sigma_\upsilon^2)$ represents the idiosyncratic errors. The individual heterogeneity is uncorrelated with the random noise, i.e. $Cov(\mu_i, \upsilon_{it}) = 0$ as this is needed for the validity of moment conditions in GMM-dif and GMM-sys (Chudik & Pesaran, 2022).

The assumptions regarding the idiosyncratic shocks are:

(i) $\mathbb{E}(\upsilon_{it}\upsilon_{jt}) = 0 \ \forall j$ and $i = 1, \ldots, N$, $t = 1, \ldots, T$ with $i \neq j$, the errors are uncorrelated across units;

(ii) $\mathbb{E}(\upsilon_{it}\upsilon_{il}) = 0 \ \forall i = 1, \ldots, N$, l and $t = 1, \ldots, T$ with $t \neq l$, the errors are serially uncorrelated over time.

[1] We explicitly allow for unbalanced panels. In balanced panel $T_i = T, \forall i = 1, \ldots, N$.

The easiest way to ensure validity of (i) is to assume $\upsilon_{it} = \tau_t + \varepsilon_{it}$ and $\mathbb{E}(\varepsilon_{it}\varepsilon_{jt}) = 0\ \forall j, i = 1,...,N, t = 1,...,T$ with $i \neq j$. Time dummies τ_t aim to explicitly capture CCE, common correlated effects or period-specific factors of 'aggregate influence' on micro units, such as business cycle, neighbourhood effects, herd behaviour and social norms. If not accounted for, these unobservable common factors may generate weak cross-sectional dependence (Chudik, Pesaran & Tosetti, 2011). Another simple way to account for common correlated effects is to use cross-sectional demeaned data as in Moral-Benito (2013); Alvarez and Arellano (2022).[2]

To guarantee assumption (ii), which underlies the appropriate setting of the moment conditions exploited by CRE-GMM, the dynamics of Equation (11.1) can be extended, e.g., to an ARDL(1,1) and the corresponding equilibrium correction model (ECM):

$$y_{it} = \alpha + \boldsymbol{\beta}_1'\mathbf{x}_{it} + \boldsymbol{\beta}_2'\mathbf{x}_{it-1} + \boldsymbol{\theta}'\mathbf{w}_i + \rho y_{it-1} + \nu_{it}. \tag{11.2}$$

In the case of $\rho = 0$ and $\boldsymbol{\beta}_2 = 0$, we have a static model, for which our CRE-GMM approach is interesting as it provides 'internal' instruments (information within the model), thus avoiding the difficulties of finding good 'external' instruments.

The error components μ_i and υ_{it} are sometimes referred to as 'permanent' and 'transitory' components.[3] They imply that the moment conditions exploited to estimate Equation (11.1) in levels must tackle the possible *double* endogeneity of the explatory variables. Specifically:

1. Endogeneity *due to heterogeneity*:

$$\mathbb{E}(y_{it-p}\mu_i) \neq 0 \qquad p \geq 1, \forall t = 1,\ldots,T_i;$$
$$\mathbb{E}(\mathbf{x}_{it-q}\mu_i) \neq 0 \qquad q \geq 0, \forall t = 1,\ldots,T_i.$$

2. *Standard* endogeneity

$$\mathbb{E}(\mathbf{x}_{it-q}\upsilon_{it}) \neq 0 \qquad q \geq 0, \forall t = 1,\ldots,T_i.$$

The lagged dependent variable is predetermined (uncorrelated with $\{\upsilon_{it}, \upsilon_{it+1},\ldots,\upsilon_{iT_i}\}$), but, including by definition μ_i, it is endogenous due to *individual heterogeneity*. The $\mathbf{x}_{it}$ variables could be correlated with both the individual heterogeneity μ_i (*endogeneity due to heterogeneity*) and the shock υ_{it} (*standard endogeneity*). Indeed, a variable in $\mathbf{x}_{it}$ could be predetermined rather than strictly exogenous; for example, in Vella and Verbeek (1998)'s model explaining workers' wages, a reduction in t of the dependent variable wages could lead to union

[2] Strong cross-sectional dependence is beyond our setting. It is captured by the interactive fixed effects models, $\upsilon_{it} = \varphi_i\lambda_t + \varepsilon_{it}$, where λ_t indicates factors and φ_i individual-specific loadings, meaning that the regression is augmented with cross-sectional averages Pesaran (2006), or principal components Bai (2009).

[3] In the words of Crowder and Hand (1990), the term α is the "immutable constant of the universe", μ_i represents the "lasting characteristics of individuals" and thus captures the unobserved, and omitted, time-constant variables representing individual specificities, while the idiosyncratic shocks, υ_{it}, are the "fleeting aberration of the moment".

membership in $t+1$. Often the dynamic model in Equation (11.1) could be affected by omitted regressors correlated with $\mathbf{x}_{it}$, by measurement errors in $\mathbf{x}_{it}$, by simultaneity, thus producing endogeneity of the variables $\mathbf{x}_{it}$.

11.3 The CRE-GMM Estimation

One of the goals of our Correlated Random Effects GMM-lev (CRE-GMM) method is to keep Equation (11.1) in levels, which allow us to estimate the effects of the measurable time-invariant explanatory variables, $\mathbf{w}_i$, while also considering the role of unmeasurable individual heterogeneity and controlling for both types of endogeneity. If μ_i were omitted or not appropriately captured, it would generate an upward bias of the autoregressive parameter and bias of all other parameters through the smearing effect. The individual effects μ_i capture an additive and linear combination of all time-invariant unit-specific unobservable variables, e.g., the differences of each individual with respect to the benchmark α. GMM-dif could solve the upward bias due to endogeneity because of heterogeneity, but it does not allow estimating the parameters associated with $\mathbf{w}_i$. This problem also affects GMM-sys, at least in terms of efficiency, as most of the equations are first-differenced while the level equation is only retained for non-redundant moment conditions, actually for only one more period per panel unit, the T_i period (Kiviet et al., 2017). Thus, the GMM-sys estimation of $\boldsymbol{\theta}$ does not exploit all available information.

Indeed, GMM-dif and GMM-sys are uniquely (GMM-dif) or mostly (GMM-sys) based on first differences of Equation (11.1):

$$\Delta y_{it} = \boldsymbol{\beta}_1' \Delta \mathbf{x}_{it} + \rho \Delta y_{it-1} + \Delta \upsilon_{it}. \tag{11.3}$$

First differencing removes μ_i, one source of endogeneity, under the condition that individual characteristics are constant over time, an assumption that is not tested; another drawback is that the estimation of $\boldsymbol{\theta}$ is not possible (GMM-dif) or not fully informed (GMM-sys).

One more reason why GMM-lev is an attractive alternative to the GMM-dif estimator is that its performance does not deteriorate when ρ is high (Binder, Hsiao & Pesaran, 2005). Interestingly, Bun and Windmeijer (2010) interpret the GMM-sys as a weighted average of GMM-dif and GMM-lev where the weight on the moment conditions in levels increases with increasing persistence of the series. The higher the autoregressive parameter, the weaker the relationship between lagged levels and the first-differenced variables. In contrast, in GMM-lev, a large autoregressive parameter implies a strong link between lagged first differences and level variables (Bewley, 1979) and an even stronger link between lagged levels and level variables. We therefore believe that GMM-lev provides more informative and relevant estimates than GMM-dif and GMM-sys. Our GMM-CRE estimator can also be combined with the first differences and extended to GMM-sys, as tested in Section 11.4.

The intuition of our CRE-GMM estimator comes from iterating Equation (11.1) backwards for an arbitrary $\tilde{t}$:

$$
\begin{aligned}
y_{i\tilde{t}} &= \rho^{\tilde{t}} y_{i0} + \sum_{\tau=0}^{\tilde{t}-1} \rho^{\tau}\alpha_i + \sum_{\tau=0}^{\tilde{t}-1} \rho^{\tau}\boldsymbol{\theta}'\mathbf{w}_i + \sum_{\tau=0}^{\tilde{t}-1} \rho^{\tau}\boldsymbol{\beta}'\mathbf{x}_{i\tilde{t}-\tau} + \sum_{\tau=0}^{\tilde{t}-1} \rho^{\tau} \upsilon_{i\tilde{t}-\tau} \\
&= \rho^{\tilde{t}} y_{i0} + \frac{1-\rho^{\tilde{t}-1}}{1-\rho}\alpha_i + \frac{1-\rho^{\tilde{t}-1}}{1-\rho}\boldsymbol{\theta}'\mathbf{w}_i + \sum_{\tau=0}^{\tilde{t}-1} \rho^{\tau}\boldsymbol{\beta}'\mathbf{x}_{i\tilde{t}-\tau} + \sum_{\tau=0}^{\tilde{t}-1} \rho^{\tau} \upsilon_{i\tilde{t}-\tau},
\end{aligned} \tag{11.4}
$$

where $\alpha_i = \alpha + \mu_i$ and $\tilde{t}$ captures the sample splitting, as will be discussed later. The dependent variable can be separated into four components. The first component, $\rho^{\tilde{t}} y_{i0}$, is the term that depends on the initial observations, y_{i0}, and influences the behaviour of any estimators as long as T_i is finite; its effect does not vanish and is reflected in each subsequent period when the time dimension is short, particularly for some units in unbalanced panels. Instead, its relevance decreases when T_i is large, under the weak stationarity condition $|\rho| < 1$. The effect of the initial conditions does not vanish with T_i when ρ is close to unity. The starting values may be seen as representing the initial individual endowments. Particularly in longitudinal panels where T_i is rather small and asymptotic concerns $N \to \infty$, the effects of the initial conditions are not asymptotically diminishing, and hence the assumptions on initial observations play an important role in determining the properties of the level equations used by GMM-sys and GMM-lev. Hahn (1999) argues that in estimating an AR(1) model on panel data it is fairly common to disregard the potentially informative role of the distribution of initial conditions y_{i0} for the estimation of the autoregressive parameter ρ. This practice is understandable because misspecification of the distribution of y_{i0} would result in the inconsistency of the resultant estimator. Perhaps because of this concern, efficiency in the dynamic panel literature has been discussed in the framework where y_{i0} was assumed to be ancillary for the parameter of interest. Hahn (1999) shows that the marginal information contained in the initial condition is substantially even when T_i is relatively large, and the efficiency gain tends to be larger for ρ close to one, as the coefficient $\rho^{\tilde{t}}$ of y_{i0} indicates that the importance of initial condition in $y_{i\tilde{t}}$ is an increasing function of $|\rho|$.

The second term, $\left[(1-\rho^{\tilde{t}-1})/(1-\rho)\right]\alpha_i + \left[(1-\rho^{\tilde{t}-1})/(1-\rho)\right]\boldsymbol{\theta}'\mathbf{w}_i$, is the equilibrium that depends on the unmeasurable, μ_i, and potentially measurable individual characteristics, $\mathbf{w}_i$; they interact with the autocorrelation coefficient, to determine the unit-specific limiting distribution of the series $y_{i\tilde{t}}$. The third term, $\sum_{\tau=0}^{\tilde{t}-1} \rho^{\tau}\boldsymbol{\beta}'\mathbf{x}_{i\tilde{t}-\tau}$, is a component that depends on the current and past values of $\mathbf{x}_{i\tilde{t}}$, and is related to the dynamics of the model; it condenses the forces producing path dependence (Page, 2006), such as increasing returns, intertemporal spillovers, and externalities. Finally, the last term is a moving average of the disturbances $\upsilon_{i\tilde{t}}$ (considered by the weighting matrix of GMM).

Let us split the temporal observations $\tilde{t}$ as:

pre-sample	**estimation sample**
$s = 1,2,3,\ldots,S_i$	$t = S_{i+1},\ldots,T_i$
$s = \ldots,-2,-1,0$	$t = 1,2,3,\ldots,T_i$
$\tau = q+1,\ldots,\infty$	$\tau = 0,1,\ldots\ldots,q,$

and assume that the process for $y_{i\tilde{t}}$ has been going on for some time, i.e. that $\tilde{t} \rightarrow \infty$. Then, Equation (11.4) can be re-formulated as:

$$
\begin{aligned}
y_{i\tilde{t}} &= \sum_{\tau=0}^{\infty} \rho^{\tau}\alpha_i + \sum_{\tau=0}^{\infty} \rho^{\tau}\boldsymbol{\theta}'\mathbf{w}_i + \sum_{\tau=0}^{\infty} \rho^{\tau}\boldsymbol{\beta}'\mathbf{x}_{i\tilde{t}-\tau} + \sum_{\tau=0}^{\infty} \rho^{\tau}\upsilon_{i\tilde{t}-\tau} = \\
&= \frac{1}{1-\rho}\alpha_i + \frac{1}{1-\rho}\boldsymbol{\theta}'\mathbf{w}_i + \sum_{\tau=0}^{\infty} \rho^{\tau}\boldsymbol{\beta}'\mathbf{x}_{i\tilde{t}-\tau} + \sum_{\tau=0}^{\infty} \rho^{\tau}\upsilon_{i\tilde{t}-\tau} = \\
&= \frac{1}{1-\rho}\alpha_i + \frac{1}{1-\rho}\boldsymbol{\theta}'\mathbf{w}_i + \sum_{\tau=0}^{S_i-1} \rho^{\tau}\boldsymbol{\beta}'\mathbf{x}_{it-\tau} + \sum_{\tau=S_i}^{\infty} \rho^{\tau}\boldsymbol{\beta}'\mathbf{x}_{it-\tau} + \\
&+ \sum_{\tau=0}^{S_i-1} \rho^{\tau}\upsilon_{it-\tau} + \sum_{\tau=S_i}^{\infty} \rho^{\tau}\upsilon_{it-\tau}.
\end{aligned} \tag{11.5}
$$

The available sample with finite T_i does not allow for the estimation of ARDL(∞, ∞) models. Instead, the dynamics are usually truncated to some lag (p,q) lags implying, for example, that Equation (11.1) omits $\sum_{\tau=q+1}^{\infty} \rho^{\tau}\boldsymbol{\beta}'_{\tau}\mathbf{x}_{i\tilde{t}-\tau}$.[4]

We thus estimate:

$$
y_{it} = \frac{1}{1-\rho}\left(\boldsymbol{\pi}'_x\breve{\mathbf{x}}_{i.} + \pi_y\breve{y}^1_{i.}\right) + \frac{1}{1-\rho}\boldsymbol{\theta}'\mathbf{w}_i + \sum_{\tau=0}^{S_i-1} \rho^{\tau}\boldsymbol{\beta}'\mathbf{x}_{it-\tau} + \sum_{\tau=0}^{S_i-1} \rho^{\tau}\upsilon_{it-\tau} + \eta_{it}, \tag{11.6}
$$

where the term $\sum_{\tau=0}^{S_i-1} \rho^{\tau}\boldsymbol{\beta}'\mathbf{x}_{it-\tau}$ includes the observations of the estimation sample $\{\mathbf{x}_{it}, y_{it-1}\}$ for $t = S_i+1,\ldots,T_i$.

The term $\sum_{\tau=S_i}^{\infty} \rho^{\tau}\boldsymbol{\beta}'\mathbf{x}_{it-\tau}$ of Equation (11.5) is proxied by a CRE approach based on pre-sample observations $\{\mathbf{x}_{is}, y_{is-1}\}$ for $s = 1,2,3,\ldots,S_i$ and the auxiliary Equation $\alpha_i = \alpha + \mu_i = \boldsymbol{\pi}'_x\breve{\mathbf{x}}_{i.} + \pi_y\breve{y}^1_{i.} + e_i$, where $\breve{\mathbf{x}}_{i.} = S_i^{-1}\sum_{s=1}^{S_i}\mathbf{x}_{is}$, $\breve{y}^1_{i.} = S_i^{-1}\sum_{s=1}^{S_i} y_{is-1}$ and $S_i < T_i$ is the pre-sample period; $\eta_{it} \approx e_i + \sum_{\tau=S_i}^{\infty}\rho^{\tau}\upsilon_{it-\tau}$.

In words, we compute the averages of the explanatory variables for the periods $s = 1,\ldots,S_i$ to capture the initial conditions and estimate the dynamic model over the periods $t = S_i+1,\ldots,T_i$. The individual effects are considered as random and functions of past histories of the stochastic variables concurring to the path dependence process but omitted due to lag truncation, where $\breve{\mathbf{x}}_{i.}$ and $\breve{y}^1_{i.}$ represent the *systematic*

[4] The lag length of the ARDL(p,q) must be chosen to imply uncorrelated errors and accordingly to the frequency of the data.

component capturing the permanent differences between units, and the *unsystematic component* is treated as an additional random term, e_i.[5]

For example, using the average innovative activity carried out by firms in the period prior to the estimation allows us to capture the unobservable differences in accumulated knowledge that determine the initial conditions of R&D activity. Indeed, as technological knowledge is an economic good characterised by cumulability and non-exhaustibility, companies can rely on it to generate additional new knowledge at a lower cost (the 'learning to learn' and 'learning to do' effects). Once research has started, the opportunity cost of interrupting it is rather high due to high start-up costs for research facilities and staff training, and long-term investment commitments ('sunk-costs' effect generating barriers and negative externalities). Firms may also have accumulated cash flow to finance new research projects ('success-breads-success' effect). We will return to this in the empirical example in Section 11.5.

Our idea to capture initial conditions is sufficiently general and encompasses several other specifications of the initial values considered in the literature as special cases. Using values dated before the estimation sample to compute proxies for unobserved heterogeneity avoids correlation with any later shock in the equation of interest. This has the distinct advantage of producing weakly exogenous (predetermined) regressors, as the measurement of individual effects is based solely on pre-sample information.

For stationary stochastic processes, such as ARs, the pre-sample mean is a more informed estimate of the steady state solution than the first sample observation. Averages condense past informations and therefore are better suited to represent the initial conditions in comparison to first observations in a sample. A further advantage is that they mitigate possible large variations in the time series and measurement errors (being divided by S_i, the statistical averaging effect is achieved). In contrast, the single initial observation can be strongly influenced by the short-term cyclical position of the variables and/or the occurrence of random shocks. Apparently the use of the ith individual's time series mean (Mundlak, 1978's approach) is more restrictive than using each observed variable at all the different time periods for each unit i, $\tilde{y} = (y_{i1}, \ldots, y_{iS_i})$ and $\tilde{\mathbf{x}} = (\mathbf{x}_{i1}, \ldots, \mathbf{x}_{iS_i})$, as in the Chamberlain (1980)'s approach. However, the simulation results in Hsiao and Zhou (2018) suggest that the averages tend to perform better when the temporal dimension is large (possibly larger than N); also for smaller than N temporal dimensions, the Mundlak (1978)'s approach yields asymptotically unbiased inference with smaller RMSE.

---

[5] Take the example in Nerlove et al. (2008) in which $\mathbf{x}_{it} = \gamma_i'\mathbf{x}_{it-1} + \omega_{it}$ with $\omega_{it} \sim i.i.d.(0, \sigma^2_{\omega_i})$, cross-sectionally and serially unrelated. For $t, j \in \{0, \ldots, Q\}$ (the set of indices for which $\mathbf{x}_{it}$ is observed, with q used to specify the dynamics chosen much less than Q), the j-order autocorrelation is $\mathbb{E}(\mathbf{x}_{it}\mathbf{x}_{it-j}) = [\gamma_i^j/(1-\gamma_i^2)]\sigma^2_{\omega_i}$. It follows that $\mathbf{x}_{it}$ and α_i are correlated, $\mathbb{E}(\mathbf{x}_{ij}\alpha_i) = \sum_{\tau=q+1}^{\infty} \boldsymbol{\beta}_\tau' \mathbb{E}(\mathbf{x}_{ij}\mathbf{x}_{it-\tau}) = [\sigma^2_{\omega_i}/(1-\gamma_i^2)] \sum_{\tau=q+1}^{\infty} \boldsymbol{\beta}_\tau \gamma_i^{|j-\tau|}$, with a correlation depending on how close to the beginning of the sample period the observation on $\mathbf{x}_{it}$ is taken. This introduces additional '$\boldsymbol{\beta}_\tau$' parameters in Equation (11.1) capturing the relationship between the individual effects and the observed past values of the explanatory variables $\mathbf{x}_{it}$; the greater $\sigma^2_{\omega_i}$ the greater is the signal to noise ratio on one side, but the greater the dependence between $\mathbf{x}_{it}$ and α_i on the other side (especially for j near the beginning of the observation period). A beautiful discussion is in Nerlove (1999).

The selection of S_i is research specific; indeed, Bhargava (1987) suggests using a length necessary to ensure that the systematic part of the initial observations is well approximated; Kuchibhotla, Kolassa and Kuffner (2022) suggest sample splitting to assess uncertainty in model selection and state that there is no clear guidance. In Monte Carlo of Section 11.4 we set S_i at 10% of T; in the empirical example of Section 11.5 we used 41% of the theoretical T to smooth the effect of a temporary fiscal incentive. Thus, applied economists should evaluate the choice of S_i based on the research question, the events that occurred during the sample, and the pattern of the variables: as T_i and/or ρ and within-cluster variability increase, the length of S_i must be increased (Grilli & Rampichini, 2011).

Since panels are often unbalanced and each unit has its own pre-sample S_i, our approach requires a sufficient number of pre-sample observations for each unit i. If we set S_i at a fixed date, initial conditions for units entering the sample after S_i could be estimated by computing averages of units available in the pre-sample and characterised by 'similar' features (same size, same industry, same geographical area, etc.). Our approach is valid under the assumption that (11.6) is not misspecified; it is suitable if the π_x and π_y parameters of Equation (11.6) do not vary between i, otherwise the incidental parameters problem occurs. However, the CRE approach does not lead to incidental-parameters bias when T and N are of comparable size (Bai, 2009).

11.3.1 The CRE-GMM Estimation – Advantages

Our CRE-GMM approach augments the dynamic panel data regression with the systematic part of the individual effects considered as random and yields a model representation that includes the random and fixed effect specifications as special cases:

$$y_{it} = \boldsymbol{\beta}'\mathbf{x}_{it} + \boldsymbol{\theta}'\mathbf{w}_i + \rho y_{it-1} + \boldsymbol{\pi}_x'\breve{\mathbf{x}}_{i.} + \pi_y \breve{y}_{i.}^1 + v_{it} + e_i. \tag{11.7}$$

In our random effects framework, the initial conditions give rise to the 'between' Equation $\alpha_i = \alpha + \mu_i = \boldsymbol{\pi}_x'\breve{\mathbf{x}}_{i.} + \pi_y \breve{y}_{i.}^1 + e_i$ which captures the sample variation across units together with the parameters $\boldsymbol{\theta}$ of time-invariant regressors $\mathbf{w}_i$, while the estimated ρ and $\boldsymbol{\beta}$ of Equation (11.7) are 'within' parameters which capture the sample variation within each unit over time.[6] The CRE-GMM approach based on Equation (11.7) has a number of advantages. The first is that it avoids the omission of, or the inability to capture, individual heterogeneity and the mixture of 'spurious' persistence, due to the serial correlation generated by the unobservable permanent component μ_i, and 'true' (path/behavioural) persistence measured by ρ which is the causal effect of past values of the dependent variable on its current realization.

The second advantage is that the inclusion of individual averages avoids omitted variables bias and endogeneity bias due to heterogeneity, while standard endogeneity

[6] A similar idea in the maximum likelihood framework is in Lee and Yu (2020). The RE estimator under the hypothesis of exogenous $\mathbf{x}_{it}$ is investigated also by Hsiao and Zhou (2018); Hsiao (2020).

is treated by using 'internal' instruments (even external ones, if they exist). In general, it is rather difficult to exclude a statistical dependence between individual-specific effects and explanatory variables. GMM-lev estimation applied to the Equation (11.7) extended with CRE treats endogeneity due to heterogeneity as a *substantive phenomenon* without necessarily requiring instrumental variables estimation to address this endogeneity.

The third advantage is to avoid exploiting only within variability (as in GMM-dif) or a weighted average of within and between variability (as in GMM-lev). Instead, CRE-GMM estimates the within and between effects of the covariates separately and assesses whether the between effect is relevant and possibly opposite to the within effect. It then simultaneously estimates the within effect of unit-specific temporary deviations from the individual averages of the covariates and the between effect of permanent differences among individuals in the covariates. For example, firm-specific technical efficiency and economies of scale are between effects, while technological changes over time are within effects. Permanent unemployment is a between measure, while temporary unemployment is a within measure. When studying the birth weight of newborns (Abrevaya, 2006), weight may be more related to maternal smoking behaviour than smoking cessation. In psychological research, the *ecological fallacy* arises from confounding within- and between-group differences (Robinson, 1950).

The fourth advantage is that obtaining the within estimate of each covariate while controlling for systematic differences in the levels of the covariates between i leads to a more convincing analysis. CRE-GMM has greater variability and exploits more informative data transformations due to the combination of variation among cross-sectional units (between) and variation over time (within). As a result, it allows more efficient estimations and mitigates multicollinearity problems, particularly in the weighting matrix used by the GMM to handle moment conditions. Our approach permits estimating the $\boldsymbol{\theta}$ coefficients of time-constant variables and at the same time obtaining 'fixed-effects' estimates of the ρ and $\boldsymbol{\beta}$ parameters of the time-varying variables (Wooldridge, 2019, 2021).

Finally, the CRE-GMM approach can estimate the additional effects of measurable time-constant covariates. If $\mathbf{w}_i$ is correlated with individual heterogeneity, its effect can be identified in the spirit of Hausman and Taylor (1981) who use, as instruments in a static model, both the within and between transformations of the components of $\mathbf{x}$ uncorrelated with individual heterogeneity. Our CRE-GMM approach, instead, directly adds the individual averages of y_{it-1} and $\mathbf{x}_{it}$ to the model, along with any measurable time-invariant variables useful for capturing individual heterogeneity. Indeed, Equation (11.7) allows for a robust version, based on variable addition, of the Hausman (1978) test on $H_0 : \boldsymbol{\pi}_x = 0$ for individual effects uncorrelated with $\mathbf{x}_{it}$ covariates, Arellano (1993). This version of the test can also be implemented for subsets of $\mathbf{x}_{it}$, avoids computational problems and can be robust, preventing the severe size distortion on inference related to the non-robust Hausman (1978) pretest (Guggenberger, 2010).

11.3.2 The CRE-GMM Estimation – Moment Conditions

We propose six versions of the CRE-GMM that are based on the moment conditions presented comparatively to those exploited by the GMM-dif and GMM-sys in Table 11.1. When the model is first-differenced, as in the GMM-dif, the moment conditions are based on the y_{it-m} and $\mathbf{x}_{it-n}$ levels; the idea is that first differencing the equation is sufficient to consider individual heterogeneity; the opposite is the case in the GMM-sys level component. In the CRE-GMM approach we leave the equations in levels and, differently from GMM-sys based on the moment conditions that are not redundant, we exploit all the available moment conditions.[7] We comparatively explore alternative identification strategies. In the six CRE-GMM-CRE-GMM5 methods y_{it-1} is, by definition, correlated with the individual heterogeneity, μ_i, so we use GMM-sys-style moment conditions based on first differences Δy_{it-m}[8] In the three CRE-GMM-CRE-GMM2 estimations, we assume that $\mathbf{x}_{it}$ is correlated with the individual heterogeneity, μ_i, so we use GMM-sys-style moment conditions based on first differences $\Delta\mathbf{x}_{it-n}$. In the three CRE-GMM3-CRE-GMM5 estimations we exploit GMM-dif-style lags of the levels $\mathbf{x}_{it-n}$ which should increase the efficiency of CRE-GMM as they are more correlated with endogenous variables. The inclusion of unit-specific averages in Equation (11.7) allows for the comparison of the two sets of estimations, CRE-GMM-CRE-GMM2 and CRE-GMM3-CRE-GMM5, as it explicitly models μ_i which is thus removed from the error term, making the instruments in levels valid for level equations. This can be an advantage, as level instruments improve the performance of the estimator as the autocorrelation coefficient ρ increases (promising Monte Carlo results). Furthermore, this inclusion is useful when there is no guarantee that the first-differenced instruments for the untransformed equations are uncorrelated with the unit-specific error component. For example, Macher, Miller and Osborne (2021) examine the adoption of fuel-efficient precalciner kilns in the cement industry and have a region-specific term that affects all cement plants in the same geographic region. The first differences are only valid instruments if the region-specific effect is constant over time, a process that could only occur in practice if regional differences were due to factors at state level, e.g., in trade union policies or tax rates.

Regarding the assumptions on individual averages, we compare three alternative cases: the individual pre-sample averages $\check{\mathbf{x}}_{i.}$ and $\check{y}^{1}_{i.}$ are exogenous (CRE-GMM and CRE-GMM3 cases); only the individual pre-sample averages of $\mathbf{x}_{it}$ are exogenous (CRE-GMM1 and CRE-GMM4 cases); the individual pre-sample averages of y_{it-1} and x_{it} are endogenous CRE-GMM2 and CRE-GMM5 cases).[9] The individual

---

[7] In GMM-sys the moment conditions available for $t = 2, \dots, T_i - 1$ are redundant because they can be expressed as a linear combination of the moment conditions used in GMM-dif, Kiviet et al. (2017).

[8] Robustness checks on the use of lagged levels y_{it-m} as instruments show no improvement in the results, which means that the inclusion of individual pre-sample averages is useful for capturing a possibly non-constant correlation with individual heterogeneity over time.

[9] The idea behind these comparisons is to understand what happens when using suspect moment conditions. DiTraglia (2016) suggests that, in finite samples, the addition of a slightly endogenous but highly relevant instruments can reduce estimator variance much more than it increases the bias.

averages assumed to be endogenous are instrumented as the explanatory variables of the model, y_{it-1} and $\mathbf{x}_{it}$. Instrumenting the averages obtained from the pre-sample with lags belonging to the estimation sample resembles the forward orthogonal deviations suggested by Arellano and Bover (1995).[10]

An interesting aspect that emerges from comparing the CRE-GMM-CRE-GMM1 and CRE-GMM3-CRE-GMM4 estimations with the CRE-GMM2 and CRE-GMM5 estimations is whether individual characteristics have evolved over time. If agents maintain their personal characteristics unchanged over time, it clearly follows that $\mathbb{E}[\breve{y}^1_{i.}\mu_i] \neq 0$ and $\mathbb{E}[\breve{\mathbf{x}}_{i.}\mu_i] \neq 0$ and the moment conditions under CRE-GMM-CRE-GMM1 and CRE-GMM3-CRE-GMM4 are invalid. If, instead, the behaviour of the agents evolves over time or, even better, if there is a structural and status change in the individual characteristics such that $\alpha_i = \boldsymbol{\pi}'_x\breve{\mathbf{x}}_{i.} + \pi_y\breve{y}^1_{i.} + e_i$, where $\breve{\mathbf{x}}_{i.} = S_i^{-1}\sum_{s=1}^{S_i}\mathbf{x}_{is}$ and $\breve{y}^1_{i.} = S_i^{-1}\sum_{s=1}^{S_i}y_{is-1}$ for $S_i < T_i$, but $\alpha_i \neq \boldsymbol{\pi}'_x\tilde{\mathbf{x}}_{i.} + \pi_y\tilde{y}^1_{i.} + e_i$, where $\tilde{\mathbf{x}}_{i.} = T_i^{-1}\sum_{t=S-i+1}^{T_i}\mathbf{x}_{it}$ and $\tilde{y}^1_{i.} = T_i^{-1}\sum_{t=S_i+1}^{T_i}y_{it-1}$ for $t = S_i + 1, \ldots, T_i$, hence the moment conditions used in CRE-GMM-CRE-GMM1 and CRE-GMM3-CRE-GMM4 are valid.[11] Initial conditions are important when T_i small/ρ high (Hahn, 1999) and we exploit a general formulation able to capture the heterogeneous starting points of the units, without the need to assume that the correlation between y_{it-1}, x_{it} and μ_i is constant over time or that individuals must be close to their steady state (a function of μ_i) because deviations from long-term values are assumed to be systematically uncorrelated with μ_i (effect stationarity, Kiviet, 2007; Bun & Sarafidis, 2015; Alvarez & Arellano, 2022). Indeed, by conditioning on initial observations, CRE-GMM can handle a mix of units in which, for example, younger firms, still far from their steady state compared to mature firms, grow faster at the beginning of the sample period (skewed distributions of firms, (Blundell & Smith, 1991; Barbosa & Moreira, 2021)). Another example concerns educational experience, which has an effect on the earning structure that is not loosely captured by years of schooling (a between effect), but also depends on on-the-job training (a within effect). During the early stages of their careers, high-skilled workers may accept lower earnings because they expect that, as they accumulate more experience, they will develop the necessary skills to compensate them with higher future earnings: adding the individual average of work experience and wages (our CRE-GMM approach) helps to capture the relationship over time between unobservable skills, experience and wages.

In the Monte Carlo simulations in the next Section, we include parameters measuring the possible correlation between individual heterogeneity, μ_i, and $\breve{y}^1_{i.}$, $\breve{\mathbf{x}}_{i.}$ and w_i.

[10] In Alvarez and Arellano (2003) for fixed T the IV estimators in orthogonal deviations and in first differences are both consistent, whereas as T increases the former remains consistent but the latter is inconsistent. The use of past observations has its antecedent in the long lags of Chamberlain (1982).

[11] We have that $\mathbb{E}[\breve{y}^1_{i.}v_{it}] = 0$ and $\mathbb{E}[\breve{\mathbf{x}}_{i.}v_{it}] = 0$ by definition, as the individual averages are computed by exploiting the pre-estimation period $s = 1, \ldots, S_i$.

Table 11.1: Moment conditions - comparison with GMM-dif & GMM-sys

Classification		Method		
		GMM-dif	Additional in GMM-sys	CRE-GMM-CRE-GMM5
Correlat.	$\mathbb{E}[y_{it-p}\mu_i] \neq 0, p > 0$			
Predet.	$\mathbb{E}[y_{it-p}\upsilon_{it}] = 0, p \geq 1$	$\mathbb{E}[y_{it-m}\Delta\upsilon_{it}] = 0$	$\mathbb{E}[\Delta y_{it-m}(\mu_i + \upsilon_{it})] = 0$	$\mathbb{E}[\Delta y_{it-m}(\mu_i + \upsilon_{it})] = 0$
		$m \geq 2; \quad t = 3, \ldots, T_i$	$m = 1, 2, 3; \quad t = T_i$	$m = 1, 2, 3; \quad t = S_i + 2, \ldots, T_i$
				CRE-GMM-CRE-GMM2
Correlat.	$\mathbb{E}[\mathbf{x}_{it-q}\mu_i] \neq 0, q \geq 0$			
Predet.	$\mathbb{E}[\mathbf{x}_{it-q}\upsilon_{it}] = 0, q \geq 1$	$\mathbb{E}[x_{it-n}\Delta\upsilon_{it}] = 0$		
		$n \geq 1 \quad t = 3, \ldots, T_i$	$\mathbb{E}[\Delta x_{it-n}(\mu_i + \upsilon_{it})] = 0$	$\mathbb{E}[\Delta x_{it-n}(\mu_i + \upsilon_{it})] = 0$
Endog.	$\mathbb{E}[\mathbf{x}_{it-q}\upsilon_{it}] \neq 0, q = 0$	$\mathbb{E}[x_{it-n}\Delta\upsilon_{it}] = 0$	$n = 1, 2, 3 \quad t = T_i$	$n = 1, 2, 3 \quad t = S_i + 2, \ldots, T_i$
		$m \geq 2 \quad t = 3, \ldots, T_i$		
				CRE-GMM3-CRE-GMM5
Uncorrelat.	$\mathbb{E}[\mathbf{x}_{it-q}\mu_i] = 0, q \geq 0$			
Predet.	$\mathbb{E}[\mathbf{x}_{it-q}\upsilon_{it}] = 0, q \geq 1$	$\mathbb{E}[x_{it-n}\Delta\upsilon_{it}] = 0$		
		$n \geq 1 \quad t = 3, \ldots, T_i$	$\mathbb{E}[\Delta x_{it-n}(\mu_i + \upsilon_{it})] = 0$	$\mathbb{E}[x_{it-n}(\mu_i + \upsilon_{it})] = 0$
Endog.	$\mathbb{E}[\mathbf{x}_{it-q}\upsilon_{it}] \neq 0, q = 0$	$\mathbb{E}[x_{it-n}\Delta\upsilon_{it}] = 0$	$n = 1, 2, 3 \quad t = T_i$	$n = 1, 2, 3 \quad t = S_i + 2, \ldots, T_i$
		$m \geq 2 \quad t = 3, \ldots, T_i$		
				CRE-GMM and CRE-GMM3 potential MCs
				$\mathbb{E}[\breve{y}^1_{i.}\mu_i] = 0, \mathbb{E}[\breve{\mathbf{x}}^1_{i.}\mu_i] = 0, \mathbb{E}[\mathbf{w}_i\mu_i] = 0$
				CRE-GMM1 and CRE-GMM4 potential MCs
				$\mathbb{E}[\breve{\mathbf{x}}^1_{i.}\mu_i] = 0, \mathbb{E}[\mathbf{w}_i\mu_i] = 0$
				CRE-GMM2 and CRE-GMM5 potential MCs
				$\mathbb{E}[\mathbf{w}_i\mu_i] = 0$

Note: Panels often have moderate N and long T, and applied econometricians tend in practice to use fewer GMM-style instruments than available if their total number (a quadratic function of T for each variable to be instrumented) is not deemed sufficiently small relative to N. To combine this practice with a more structured lag selection, we follow Ziliak (1997); Bun and Kiviet (2006) and restrict the moment conditions to lags $t-1$ to $t-3$. The selected lags are valid for dealing with predetermined and endogenous explanatory variables in the equation in levels. Under the strict exogeneity assumption of $\mathbf{x}_{it}$ variables, we can exploit $n = 0, 1, 2, 3$ moment conditions.

11.4 Monte Carlo Simulations

To asses the performance of CRE-GMM, we employ a Monte Carlo simulation with the DGP based on an ARDL(1,1) model:

$$
\begin{aligned}
y_{it} &= \beta_0 + \rho y_{it-1} + \beta_1 x_{it} + \beta_2 x_{it-1} + \beta_3 w_i + u_{it}, \\
x_{it} &= \gamma_1 \mu_i + \vartheta x_{it-1} + \gamma_2 \epsilon_{it} + \gamma_5 w_i + \xi_{it}, \\
u_{it} &= \mu_i + e_{it}, \\
e_{it} &= \gamma_3 \mu_i + \gamma_4 w_i + \epsilon_{it}.
\end{aligned}
\tag{11.8}
$$

The error term u_{it} is composed into individual effects μ_i and idiosyncratic shocks ϵ_{it}. The parameters γ_1 and γ_2 control the degree of endogeneity by specifying the correlation of $\mathbf{x}_{it}$ with the individual effects μ_i and the random noise ϵ_{it}, respectively; hence γ_1 is important to assess the validity of the potential moment condition $\mathbb{E}[\breve{\mathbf{x}}^1_{i.}\mu_i] = 0$ of Table 11.1. The parameter γ_3 sets the variance of the individual effects μ_i equal or higher than the variance of the shocks ϵ_{it}; it also influences the validity of the potential moment condition $\mathbb{E}[\breve{y}^1_{i.}\mu_i] = 0$ of Table 11.1 and allows assessing how the performance of GMM-dif/-sys is adversely affected when variability of μ_i is larger than variability of e_{it} (Bun & Kiviet, 2006; Hayakawa, 2007; Kiviet, 2007). The parameter γ_4, in combination with γ_3, defines the relative importance of measurable and non-measurable individual heterogeneity (hence the validity of potential moment condition $\mathbb{E}[\mathbf{w}_i\mu_i] = 0$ of Table 11.1), and γ_5 sets the correlation between $\mathbf{x}_{it}$ and w_i (hence possible collinearity issues).

We explore $N = [25, 100, 1000]$, $T = [10, 20, 40]$, $\gamma_1 = [0, 0.25, 0.8]$, $\gamma_2 = [0, 0.25, 0.8]$, $\gamma_3 = [0, 0.25, 0.8]$, $\gamma_4 = [0, 0.3]$, $\gamma_5 = [0, 0.3]$ yielding 756 combinations that, with $\rho = [0.5, 0.8]$, $\beta_3 = [0, 0.3]$ and different setups of the variances (see the *Online Appendix*, Bontempi & Ditzen, 2025), produce 12,096 experiments for the PAM-ARDL(1,0) model, $\beta_2 = 0$ in Equation (11.8), and as many for the ECM-ARDL(1,1) model. We perform many Monte Carlo simulations because the robustness of our CRE-GMM estimator is an important advantage, as in practice, on real data, it is not known whether and which restrictions are satisfied (Chudik & Pesaran, 2022). We summarize here the results for PAM with $\rho = 0.5$, $\vartheta = 0.5$, $\beta_1 = 1$, $\beta_2 = 0$, $\beta_3 = 0.1$ and the variances are in the *Online Appendix* (Bontempi & Ditzen, 2025). We use the nested loop plots in Figures 11.1, 11.2 and 11.3, which allow a direct comparison of the percentage bias of the estimators across parametrisations (on the horizontal axis).[12]

Under the assumption that x_{it} is uncorrelated with the idiosyncratic shock ($\gamma_2 = 0$), the Random Effects (RE) and Fixed Effects (FE) estimators serve as benchmarks for the consistent estimation of the parameter ρ. The upper bound is provided by the RE which assumes no correlation between the regressor x_{it} and the individual effects μ_i ($\gamma_1 = 0$).[13] The lower bound is represented by the FE which removes the influence of any time-invariant variable from the model by exploiting only the within-transformation of the data, producing a consistent estimate of the autoregressive parameter for $T \to \infty$. In the two correlated random effects models, we add only the individual pre-sample average of the lagged dependent variable (CRE1, addition of $\breve{y}^1_{i.}$) and also the individual pre-sample average of x_{it} (CRE2, addition of $\breve{y}^1_{i.}$ and $\breve{x}_{i.}$). Especially CRE2, which has the advantage over the FE and RE estimators to exploit the within and between effects separately, gives indications of the bias due to endogeneity because of the individual heterogeneity of the RE estimator and thus performs well for β_1 and, especially, β_3.

The presence of standard endogeneity, $\gamma_2 > 0$, produces a bias in the estimates increasing with the γ_2 parameter capturing the correlation between x_{it} and ϵ_{it}.

[12] See Rücker and Schwarzer (2014); the plots were produced using the `siman` suite in Stata (Marley-Zagar, White & Morris, 2022).

[13] By definition, y_{it-1} is correlated with the individual effects.

The bias is greatly reduced by IVs in the GMM framework. The standard GMM-lev (GL), CRE-GMM2 (IVs in first differences and individual averages, $\breve{y}^1_{i.}$ and $\breve{x}_{i.}$, instrumented), CRE-GMM5 (IVs in levels and individual averages, $\breve{y}^1_{i.}$ and $\breve{x}_{i.}$, instrumented), standard GMM-sys (GS, only not redundant moment conditions exploited for the level equation), our CRE applied to GMM-sys (IVs in first differences for the level equation and individual averages, $\breve{y}^1_{i.}$ and $\breve{x}_{i.}$, added and instrumented) and Kripfganz and Schwarz (2019) combining GMM-dif at the first step and GMM-lev at the second step (KS2) are the most appropriate.[14]

Monte Carlo results can be summarised as follows.

- In instances involving longitudinal panels and the estimation of the ρ and β_1 parameters, the GMM-sys often demonstrates superior performance. However, the inclusion of individual pre-sample averages proves advantageous in capturing individual heterogeneity. As GSC exhibits a performance comparable to that of GMM-sys, our approach can be useful in longitudinal panels when γ_1 is high. It is also noteworthy that both CRE-GMM2 (use IVs in first differences) and CRE-GMM5 (use IVs in levels) maintain adequate performance, with a negligible preference of CRE-GMM5 for the ρ parameter and CRE-GMM2 for the β_1 parameter under case $\gamma_1 > 0$.
- When we switch to panels with large N and T (multilevel panels) and $T > N$ (macro panels), the CRE-GMM2 and CRE-GMM5 estimators are generally the least biased and more efficient for the ρ and β_1 parameters compared to GL/GS/KS2. The preference for CRE-GMM5 over CRE-GMM2 tends to hold regardless of the prevalence of correlation with individual heterogeneity or correlation with idiosyncratic shocks (cases $0 < \gamma_2 < \gamma_1$ or $0 < \gamma_1 < \gamma_2$). A longer T improves the possibility to better approximate the initial conditions and the CRE-GMM approach is less affected by spurious persistence (note that for long T the FE tends to the true ρ parameter for $\gamma_2 = 0$).
- The advantage of CRE-GMM5 over GL/GS/KS2 is even more evident when $\gamma_3 > 0$, passing from the case $\sigma^2_\mu/\sigma^2_e = 1$, where σ^2_μ is the variance of individual heterogeneity, μ_i, and σ^2_e is the variance of shocks e_{it}, to the case $\sigma^2_\mu/\sigma^2_e > 1$.
- It is noteworthy that CRE-GMM5 has the best performance in the estimation of the β_3 parameter.
- Considering different panel types, and γ_1, γ_2 and γ_3 combinations, the bias for the parameter ρ is 0.12 (GL), -0.06 (KS2), 0.06 (GS), 0.04 (CRE-GMM2), 0.03 (CRE-GMM5), 0.00 (GSC); for the parameter β_1 the bias is -0.08 (GL), 0.08 (KS2), -0.01 (GS), 0.00 (CRE-GMM2 and CRE-GMM5), 0.04 (GSC); for the

[14] The Figures show the cases with individual averages instrumented, CRE-GMM2 with instruments in first differences and CRE-GMM5 with instruments in level for the variable x_{it}, as having better performance in terms of bias and standard errors even in longitudinal panels. Details of the six versions of CRE-GMM are given in the *Online Appendix* (Bontempi & Ditzen, 2025). In addition, the figures present the GSC estimator that implements the CRE-GMM2 approach in the GMM-sys, i.e. the equation in levels of the system is instrumented by the first differences. Although it performs worse than the CRE-GMM5 approach applied to GMM-sys, it is still interesting from the comparative point of view as it is more similar to the GMM-sys estimator.

parameter β_3 the bias is 0.18 (GL), 0.36 (KS2), 0.23 (GS), 0.02(CRE-GMM2), 0.00 (CRE-GMM5), 0.02 (GSC).

- Thus, looking for a compromise capable of capturing different possible situations that a researcher may face, CRE-GMM5 appears to be the best choice for all parameters and is characterised by good coverage, size and power.
- Some preliminary results, which need further investigation, show that CRE-GMM5 performs well in the case of persistence ($\rho = 0.8$): the bias is reduced thanks to the additional moment conditions based on the pre-sample averages of y_{it-1} and x_{it} in the case of correlation with individual heterogeneity and only x_{it} in the case of no such correlation.

Fig. 11.1: Nested loop Plot for ρ, PAM
Bias for $\rho = 0.5$ across different specifications. Parameters shown on horizontal axis.

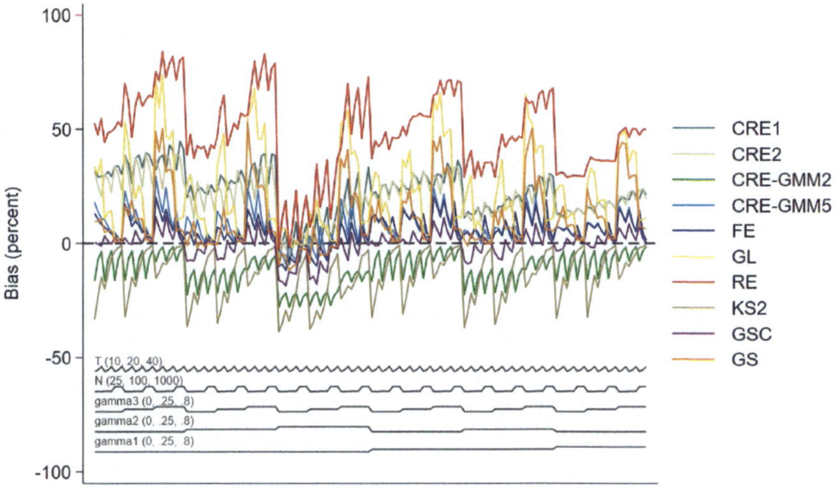

11.5 Empirical Application - Persistence of R&D and Market Power

We estimate R&D investment on an unbalanced panel of 3,971 Italian companies over the period 1984-2012.[15] Individual averages are computed over the years 1984-1995,

[15] Two other examples, estimating a production function and estimating the role of education on workers' wages in two balanced panels for the US, are given in the *Online Appendix* (Bontempi & Ditzen, 2025). The comparison also add maximum likelihood estimates and the Hausman and Taylor (1981) estimator.

Fig. 11.2: Nested loop plot for β_1, PAM
Bias for $\beta_1 = 1$ across different specifications. Parameters shown on horizontal axis.

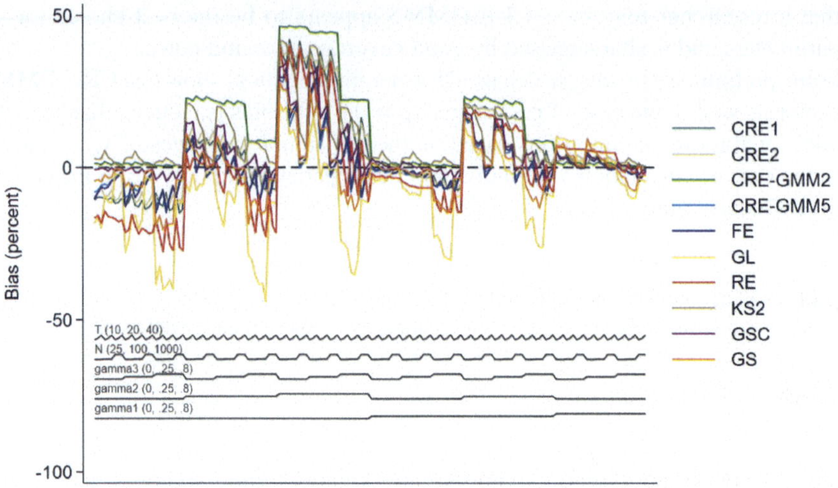

Fig. 11.3: Nested loop plot for β_3, PAM
Bias for $\beta_3 = 0.1$ across different specifications. Parameters shown on horizontal axis.

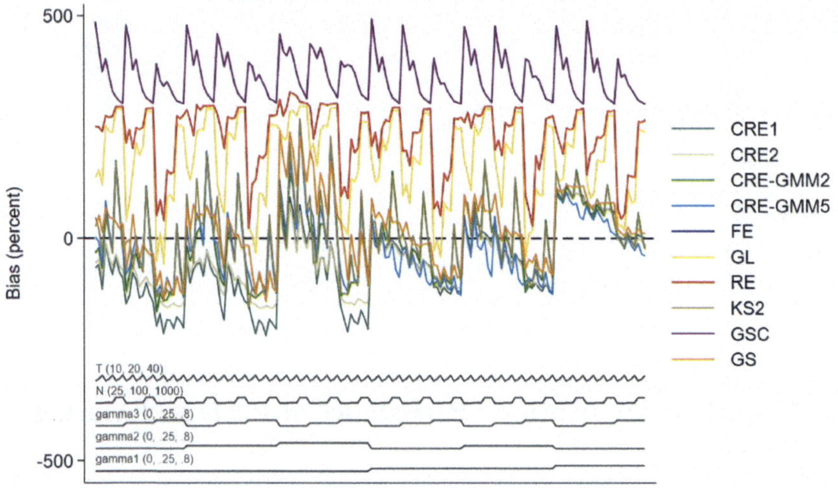

while the years 1996-2012 are used to estimate the constrained $ARDL(1,1)$ model:

$$R\&D_{it} = \alpha + \rho R\&D_{it-1} + \boldsymbol{\beta}' \mathbf{x}_{it-1} + \boldsymbol{\theta}' \mathbf{w}_i + \lambda_t + \mu_i + \epsilon_{it}.$$

The equation is derived from standard IO methods based on the intensive margin, where R&D, the input in levels, is measured as the R&D investment over employees. Bontempi, Lambertini and Parigi (2024) argue that the lagged amount of R&D investment is not necessarily a meaningful measure of the accumulated knowledge, because even after a discovery has been made, companies must continue to invest in R&D, as it can take a long time to convert innovation into economic results. Therefore, we estimate a second dynamic model in which the lagged R&D investment is replaced by the lagged R&D stock, $R\&D^{stock}$, measured as the logarithm of the innovation stock and better able to capture inter-temporal externalities and temporal spillovers between subsequent R&D investments:

$$R\&D_{it} = \alpha + \rho_1 R\&D^{stock}_{it-1} + \boldsymbol{\beta}' \mathbf{x}_{it-1} + \boldsymbol{\theta}' \mathbf{w}_i + \lambda_t + \mu_i + \epsilon_{it}.$$

In both specifications, $\mathbf{x}_{it-1}$ includes a set of controls supposed to be endogenous and instrumented in GMM methods: size, financing, planned investments, firm-specific uncertainty, firm openness and international competition; λ_t is the macroeconomic uncertainty measured by Baker, Bloom and Davis (2016) which is an alternative to the demeaning of the variables by means of time dummies. Among the measurable individual characteristics $\mathbf{w}_i$ (like age, type of ownership, geographical localization, industry), we are particularly interested in the role of market power, as the theoretical literature is still debated between a Schumpeterian positive effect (monopolistic firms can appropriate the returns from innovation) and an Arrovian negative effect (competition positively affects innovation). Market power is measured by the firm-specific elasticity of demand which is a constant characteristic of companies over the observed temporal span (see the discussion in Bontempi et al., 2024).

The total variability of the dependent variable is dominated, in Table 11.2, by the variation over the units, with the exception of the pre-sample period in which a temporary tax incentive was granted to investments in the years 1994-95.[16] The within variability is mostly firm-specific. From Figure 11.4 it is clearly visible the effect of the 'Tremonti Law' (the vertical bar) which was temporary and can be smoothed by using the average over 12 years to capture initial conditions of firms relative to R&D investment and variables included in $\mathbf{x}_{it-1}$.

This is an example that provocatively highlights the loss of observations in unbalanced panels where the dependent variable is characterized by many firm-specific discontinuities (which is why Bontempi et al. (2024) prefer to rely on duration models). Despite this, the example is fitting for those researchers who prefer to use standard dynamic models to estimate R&D investment. The estimates for the lagged dependent variable, $R\&D_{it-1}$, reported in Table 11.3, are characterised by the typical bias in a dynamic panel that is inherent to RE and FE, as well as the non-use of instruments also in CRE1 and CRE2. In contrast, the ρ estimates are in a similar range, irrespective of the estimation method employed, GL, CRE-GMM2 or CRE-GMM5. The same applies to the use of the common sample resulting from

[16] It was the first 'Tremonti Law' with a tax benefit consisting in the exclusion from the formation of the company's income of 50% of the increase in investments made in the current tax period compared to the average of those made in the previous five years.

Table 11.2: Variance decomposition for the dependent variable R&D investment

Period	1984-2012	1984-95	1996-2012
Between variability	65.54 %	47.49%	72.08%
Within variability	34.46%	52.51%	27.92%
common to all the units	(0.48%)	(1.92%)	(0.15%)
unit-specific	(33.99%)	(50.60%)	(27.77%)

Note: Computations implemented by the author-written procedure `xtsum3`.

the availability of the pre-sample individual averages or one-step cluster standard errors or two-step standard errors (the latter are biased in small samples, say $T \leq 5$ and N/T high, Windmeijer, 2005). A completely different situation emerges when we estimate the effect of the true accumulated knowledge (the stock of past R&D, $R\&D_{it-1}^{stock}$); when comparing one-step, two-step and samples, the GL method is not robust.[17] In contrast, the decomposition of the effects into the within and between components exploited by the CRE-GMM method improves the estimation of the weighting matrix used in the two-step. The Hausman tests confirm the correlation between certain explanatory variables (mainly financing and firm-specific uncertainty) and unobservable individual effects.

The most interesting result concerns the estimation of market power, which is of course not available in the FE estimation (and thus would not be available in the GMM-dif estimation either) and is not robust in the GL method, particularly when the true accumulated knowledge, $R\&D_{it-1}^{stock}$, is used in the model. In contrast, the CRE-GMM5 method shows robustness irrespective of the use of one-step or two-step and the way of measuring accumulated knowledge ($R\&D_{it-1}$ or $R\&D_{it-1}^{stock}$); among other things, CRE-GMM5 reveals a Schumpeterian effect, confirming results in Bontempi et al. (2024). The CRE-GMM approach is able to address the difficulties of the empirical literature on innovation in consistently estimating the role of market power while considering the causal effect of past R&D activities on current R&D investment due to the 'true' path-dependent nature of technical changes. The CRE-GMM results are in line with the few other results available in the empirical literature. It is also interesting to note that GMM-sys tends to produce results in line with CRE-GMM5, despite not being robust in estimating the effect of market power, while KS2 is in line with GL.[18]

[17] The GMM-sys produces estimates of 0.546, one-step, and 0.393, two-step, in the full estimation sample, and 0.308, one-step, and 0.309, two-step, in the common sample, indicating greater robustness than GL.

[18] Results available on request.

Table 11.3: Estimation of R&D for Italian Firms

	RE	FE	CRE1	CRE2	GL	GLt	CRE-GMM2	CRE-GMM2t	CRE-GMM5	CRE-GMM5t
Panel A: Lagged R&D - full estimation sample										
$R\&D_{it-1}$	0.743***	0.335***	0.735***		0.626***	0.632***				
	(0.0510)	(0.0675)	(0.0647)		(0.0616)	(0.0029)				
Market power	0.201***	-	0.122		0.261***	0.114**				
	(0.0686)		(0.1993)		(0.1006)	(0.0495)				
Panel A: Lagged R&D - common sample										
$R\&D_{it-1}$	0.739***	0.473***	0.735***	0.735***	0.673***	0.674***	0.649***	0.650***	0.696***	0.696***
	(0.0641)	(0.0956)	(0.0646)	(0.0631)	(0.0683)	(0.0001)	(0.0649)	(0.0003)	(0.0764)	(0.0002)
Market power	0.152	-	0.111	0.159	0.200	0.203***	0.430	0.424***	0.422*	0.418***
	(0.2068)		(0.2121)	(0.1943)	(0.2561)	(0.0031)	(0.2887)	(0.0124)	(0.2446)	(0.0093)
NT	1596	1596	1596	1596	1596	1596	1596	1596	1596	1596
N	284	284	284	284	284	284	284	284	284	284
$\bar{T}$	6	6	6	6	6	6	6	6	6	6
ar1 pval.					0.00	0.00	0.00	0.00	0.00	0.00
ar2 pval.					0.24	0.25	0.23	0.24	0.26	0.27
ar3 pval.					0.33	0.33	0.33	0.33	0.32	0.32
Hansen pval.					0.968	0.971	0.949	0.938	0.947	0.945
Hansen df.					329	329	322	322	322	322
Hausman pval.			0.146	0.210			0.058	0.000	0.355	0.000
Hausman df.			1	7			7	7	7	7
R^2	0.63	0.43	0.63	0.63	0.62	0.62	0.61	0.61	0.62	0.62
Panel B: Accumulated knowledge - full estimation sample										
$R\&D^{stock}_{it-1}$	0.199***	0.023	0.093***		0.981***	0.670***				
	(0.0291)	(0.0254)	(0.0221)		(0.2044)	(0.0371)				
Market power	0.750**	-	-0.144		-0.073	0.016				
	(0.3390)		(0.3542)		(0.3805)	(0.1992)				
Panel B: Accumulated knowledge - common sample										
$R\&D^{stock}_{it-1}$	0.106***	0.018	0.093***	0.094***	0.269***	0.270***	0.309***	0.309***	0.255***	0.255***
	(0.0221)	(0.0233)	(0.0227)	(0.0235)	(0.0659)	(0.0010)	(0.0833)	(0.0019)	(0.0815)	(0.0012)
Market power	-0.005	-	-0.066	0.281	-0.574	-0.580***	0.243	0.212***	0.503	0.494***
	(0.3737)		(0.3780)	(0.4175)	(0.4644)	(0.0125)	(0.6095)	(0.0315)	(0.6875)	(0.0233)
NT	1967	1967	1967	1967	1967	1967	1967	1967	1967	1967
N	334	334	334	334	334	334	334	334	334	334
Tavg	6	6	6	6	6	6	6	6	6	6
ar1 pval.					0.07	0.05	0.06	0.04	0.04	0.02
ar2 pval.					0.40	0.40	0.36	0.38	0.49	0.53
ar3 pval.					0.52	0.50	0.43	0.41	0.39	0.40
Hansen pval.					0.874	0.874	0.930	0.951	0.786	0.795
Hansen df.					329	329	322	322	322	322
Hausman pval.			0.038	0.232			0.119	0.000	0.224	0.000
Hausman df.			1	7			7	7	7	7
R^2	0.37	0.05	0.37	0.38	0.36	0.36	0.34	0.34	0.37	0.37

Note: We report only estimates of persistence and market power. The full estimation sample is composed by 8,109 observations, 1,415 firms in Panel A (1,629 observations and 288 firms in CRE1)) and 6,414 observations, 1,136 firms in Panel B (2,068 observations and 349 firms in CRE1). The common sample is derived from CRE2, CRE-GMM2 and CRE-GMM5, consisting of companies with available 1984-1995 pre-sample averages. Cluster standard errors in RE, FE, CRE1-CRE2; one-step cluster standard errors in GMM, unless the label includes 't' to indicate two-step standard errors. Estimates are implemented using `xtdpdgmm`, (Kripfganz, 2019) in Stata; the Arellano and Bond (1991) test for autocorrelation is ar#; Hausman is Hausman (1978)'s test; Hansen is Hansen (1982)'s test.

Fig. 11.4: The temporal pattern of the dependent variable R&D investment

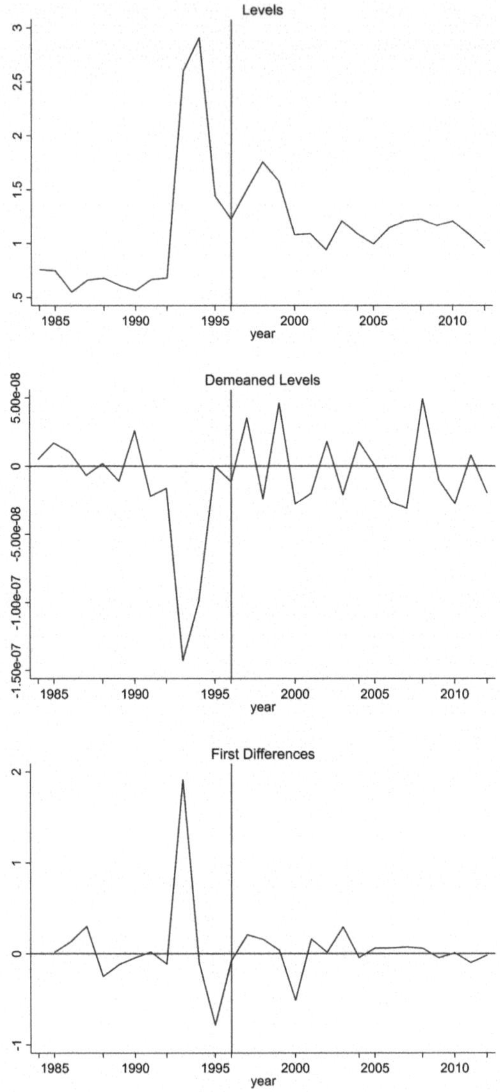

11.6 Conclusions

We present a new approach to dynamic panel data models, also suitable for static models, which merges the GMM applied to level equations with the CRE approach. The levels allow the estimation of the effect of measurable time-invariant covariates. Individual averages computed in the pre-estimation period capture the initial conditions of the units and help manage endogeneity due to heterogeneity that may not be removed by the use of instruments in first differences. Our method works well in case of double endogeneity due to correlation with idiosyncratic shocks and individual heterogeneity, and reduces the bias that characterises the GMM-lev when T is large and the variance of individual heterogeneity is greater than the variance of idiosyncratic shocks. It is more efficient than GMM-sys. The inclusion of individual averages makes level instruments valid, another positive feature of our approach, as instruments in level are preferable when series tend to be persistent; level instruments produce similar results as instruments in first differences when the autoregressive parameter is small.

Acknowledgements We are grateful to the participants of the International Panel Data Conference for their helpful comments. Special thanks to Badi H. Baltagi, Alok Bhargava, Maurice Bun, Jaya Krishnakumar, Jacques Mairesse, Patrick Sevestre, and Tom Wansbeek. Thanks to Annapaola Zenato for her initial research help. Last but not least, thanks to László Mátyás for his valuable suggestions and to Ramzi Chariag for technical help. Bontempi and Ditzen acknowledge financial support from Italian Ministry MIUR under the PRIN project 2022H2STF2, Finanziamento dell'Unione Europea - NextGenerationEU - PNRR M4.C2.1.1.

References

Abrevaya, J. (2006). Estimating the effect of smoking on birth outcomes using a matched panel data approach. *Journal of Applied Econometrics*, *21*(2), 489–519.

Acemoğlu, D., Johnson, S., Robinson, J. A. & Yared, P. (2008). Income and Democracy. *American Economic Review*, *98*(3), 808-842.

Alvarez, J. & Arellano, M. (2003). The time series and cross-section asymptotics of dynamic panel data estimators. *Econometrica*, *71*(4), 1121–1159.

Alvarez, J. & Arellano, M. (2022). Robust likelihood estimation of dynamic panel data models. *Journal of Econometrics*, *226*(1), 21-61.

Arellano, M. (1993). On the testing of correlated effects with panel data. *Journal of Econometrics*, *59*, 87–97.

Arellano, M. & Bond, S. R. (1991). Some tests of specification for panel data: Monte Carlo Evidence and an application to employment equations. *Review of Economic Studies*, *58*(2), 277–297.

Arellano, M. & Bover, O. (1995). Another look at the instrumental variable estimation of error-components models. *Journal of Econometrics*, *68*, 29–52.

Atkinson, A. B. & Stiglitz, J. E. (1969). A New View of Technological Change. *Economic Journal*, *79*, 573–578.

Bai, J. (2009). Panel Data Models With Interactive Fixed Effects. *Econometrica*, *77*(4), 1229–1279.

Baker, S. R., Bloom, N. & Davis, S. J. (2016). Measuring Economic Policy Uncertainty. *The Quarterly Journal of Economics*, *131*(4), 1593–1636.

Barbosa, J. D. & Moreira, M. J. (2021). Likelihood inference and the role of intial conditions for the dynamic panel data model. *Journal of Econometrics*, *221*, 160–179.

Bera, A. K. & Bilias, Y. (2002). The MM, ME, ML, EL, EF and GMM Approaches to Estimation: a Synthesis. *Journal of Econometrics*, *107*, 51–86.

Bewley, R. A. (1979). The Direct Estimation of the Equilibrium Response in a Linear Model. *Economics Letters*, *3*, 357–361.

Bhargava, A. (1987). Wald tests and systems of stochastic equations. *International Economic Review*, *28*(3), 789–808.

Bhargava, A. & Sargan, J. D. (1983). Estimating dynamic random effects models from panel data covering short time periods. *Econometrica*, *51*, 1635–1660.

Binder, M., Hsiao, C. & Pesaran, M. H. (2005). Estimation and inference in short panel vector autoregressions with unit roots and cointegration. *Econometric Theory*, *21*, 795–837.

Blundell, R. W. & Bond, S. R. (1998). Initial conditions and moment restrictions in dynamic panel data models. *Journal of Econometrics*, *87*, 115–143.

Blundell, R. W. & Smith, R. J. (1991). Conditions initiales et estimation efficace dans les modeles dynamiques sur donnees de panel: Une application au comportement d'investissement des entreprises. *Annales d'Economie et de Statistique*, *20/21*, 109–123.

Bontempi, M. E. & Ditzen, J. (2025). Online Supplement to Chapter 11 of the volume: Seven Decades of Economentrics and Beyond. In B. H. Baltagi & L. Matyas (Eds.), *Seven decades of economentrics and beyond.* Springer. https://www.dropbox.com/scl/fi/hrgts0gigsex871bt5wn1/Online_Appendix_March1.pdf?rlkey=gcahc7vnrzgys6rhzeffzm7cr&e=1&st=vhebmo0p&dl=0.

Bontempi, M. E., Lambertini, L. & Parigi, G. (2024). Exploring the innovative effort: Duration Models and Heterogeneity. *Eurasian Business Review*, *14*, 587–656.

Brown, B. W. & Newey, W. K. (2002). Generalized Method of Moments, Efficient Bootstrapping, and Improved Inference. *Journal of Business & Economic Statistics - Twentieth Anniversary Issue on the Generalized Method of Moments*, *20:4*, 507–517.

Bun, M. J. G. & Kiviet, J. F. (2006). The effects of dynamic feedbacks on LS and MM estimator accuracy in panel data models. *Journal of Econometrics*, *132*, 409–444.

Bun, M. J. G. & Sarafidis, V. (2015). Dynamic Panel Data Data Models. In B. H. Baltagi (Ed.), *The Oxford Handbook of Panel Data* (pp. 76–110). Oxford University Press.

Bun, M. J. G. & Windmeijer, F. (2010). The weak instrument problem of the system GMM estimator in dynamic panel data models. *Econometrics Journal*, *13*, 95–126.

Chamberlain, G. (1980). Analysis of covariance with qualitative data. *Review of Economic Studies*, *47*, 225–238.

Chamberlain, G. (1982). Multivariate Regression Models for Panel Data. *Journal of Econometrics*, *18*, 5–46.

Chudik, A. & Pesaran, M. H. (2022). An Augmented Anderson-Hsiao Estimator for Dynamic Short-T panels. *Econometric Reviews*, *41:4*, 416–447.

Chudik, A., Pesaran, M. H. & Tosetti, E. (2011, feb). Weak and strong cross-section dependence and estimation of large panels. *The Econometrics Journal*, *14*(1), C45–C90.

Crowder, M. J. & Hand, D. (1990). *Analysis of repeated measures*. Chapman & Hall CRC.

DiTraglia, F. J. (2016). Using Invalid Instruments on Purpose: Focused Moment Selection and Averaging for GMM. *Journal of Econometrics*, *195*(2), 187–208.

Gormley, T. A. & Matsa, D. A. (2014). Common errors: How to (and Not to) Control for Unobserved Heterogeneity. *The Review of Financial Studies*, *27*(2), 617–661.

Grilli, L. & Rampichini, C. (2011). The role of sample cluster means in multilevel models: A view on endogeneity and measurment error issues. *Methodology: European Journal of Research Methods for the Behavioral and Social Sciences*, *7:4*, 121–133.

Guggenberger, P. (2010). The impact of a Hausman pretest on the size of a hypothesis test: The panel data case. *Journal of Econometrics*, *156*(2), 337–343.

Haavelmo, T. (1944). The probability approach in econometrics. *Econometrica, Supplement*, *12*, iii–vi+1–115.

Haber, S. & Menaldo, V. (2011). Do Natural Resources Fuel Authoritarianism? A Reappraisal of the Resource Curse. *The American Political Science Review*, *105*(1), 1-26.

Hahn, J. (1999). How Informative Is the Initial Condition in the Dynamic Panel Model with Fixed Effects? *Journal of Econometrics*, *93*, 309–326.

Hansen, L. P. (1982). Large Sample Properties of Generalized Method of Moments Estimators. *Econometrica*, *50*, 1029–1054.

Hausman, J. A. (1978). Specification Tests in Econometrics. *Econometrica*, *46*, 1251–1271.

Hausman, J. A. & Taylor, W. E. (1981). Panel data and unobservable individual effects. *Econometrica*, *49*, 1377–1398.

Hayakawa, K. (2007). Small sample bias properties of the system GMM estimator in dynamic panel data models. *Economics Letters*, *95*, 32–38.

Hayakawa, K. (2009). On the effect of mean-nonstationarity in dynamic panel data models. *Journal of Econometrics*, *153*, 133–135.

Hayakawa, K. (2015). The asymptotic properties of the system GMM estimator in dynamic panel data models when both N and T are large. *Econometric Theory*, *31:3*, 647–667.

Hayakawa, K. & Nagata, S. (2016). On the behaviour of the GMM estimator in persistent dynamic panel data models with unrestricted initial conditions. *Computational Statistics and Data Analysis*, *100*, 265–303.

Heckman, J. J. (1991). Identifying the hand of past: Distinguishing state dependence from heterogeneity. *The American Economic Review*, *81*(2), 75–79.

Hill, T. D. & In Song, K. (2020). Limitations of fixed-effects models for panel data. *Sociological Perspectives*, *63*(3), 357–369.

Hsiao, C. (2020). Estimation of Fixed Effects Dynamic Panel Data Models: Linear Differencing or Conditional Expectation. *Econometric Reviews*, *39*(8), 858–874.

Hsiao, C. & Zhou, Q. (2018). Incidental Parameters, Initial Conditions and Sample Size in Statistical Inference for Dynamic Panel Data Models. *Journal of Econometrics*, *207*, 114–128.

Imai, K., Davis, A. P., Roos, J. M. & French, M. T. (2019). When Should we use unit fixed effects regression models for causal inference with longitudinal data? *American Journal of Political Science*, *63*(2), 467–490.

Imbens, G. W. (2002). Generalized Method of Moments and Empirical Likelihood. *Journal of Business & Economic Statistics*, *20:4*, 493–506.

Jin, F., Lee, L.-F. & J., Y. (2021). Sequential and efficient GMM estimation of dynamic short panel data models. *Econometric Reviews*, *40:10*, 1007–1037.

Kiviet, J. F. (2007). Judging Contending Estimators by Simulation: Tournaments in Dynamic Panel Data Models. In G. Phillips & E. Tzavalis (Eds.), *The refinement of econometric estimation and test procedures; finite sample and asymptotic analysis* (pp. 282–318). Cambridge University Press: Cambridge, UK.

Kiviet, J. F. (2020). Microeconometric dynamic panel data methods: Model specification and selection issues. *Econometrics and Statistics*, *13*, 16–45.

Kiviet, J. F., Pleus, M. & Poldermans, R. W. (2017). Accuracy and Efficiency of Various GMM Inference Techniques in Dynamic Micro Panel Data Models. *Econometrics*, *5*(14), 1–54.

Kripfganz, S. (2019). *Generalized Method of Moments Estimation of Linear Dynamic Panel Data Models.* (Proceedings of the 2019 London Stata Conference)

Kripfganz, S. & Schwarz, C. (2019). Estimation of linear dynamic panel data models with time-invariant regressors. *Journal of Applied Econometrics*, *34*(4), 526–546.

Kropko, J. & Kubinec, R. (2020). Interpretation and identification of within-unit and cross-sectional variation in panel data models. *PLoS ONE*, *15*(4), 1–22.

Kuchibhotla, A. K., Kolassa, J. E. & Kuffner, T. A. (2022). Post-Selection Inference. *Annual Review of Statistics and Its Application*, *9:1*, 505–527.

Lee, L.-F. & Yu, J. (2020). Initial conditions of dynamic panel data models: on within and between equations. *The Econometrics Journal*, *23:1*, 115–136.

Macher, J. T., Miller, N. H. & Osborne, M. (2021). Finding Mr. Schumpeter: Technology Adoption in the Cement Industry. *Rand Journal of Economics*, *52*, 78–99.

Marley-Zagar, E., White, I. & Morris, T. (2022). `siman`: A suite of Stata programs for analysing simulation studies. *MRC Clinical Trials Unit at UCL, London Stata Conference, 8 September 2022.*

Moral-Benito, E. (2013). Likelihood-based estimation of dynamic panels with predetermined regressors. *Journal of Business and Economic Statistics*, *31*, 451–472.

Mundlak, Y. (1978). On the pooling of time series and cross-sectional data. *Econometrica*, *46*, 69–86.

Mátyás, L. (2017). *The econometrics of multi-dimensional panels: theory and applications*. Springer-Verlag Berlin Heidelberg.

Nerlove, M. (1999). Likelihood inference for dynamic panel data models. *Annales d'Economie et de Statistique, 55/56*, 370–410.

Nerlove, M. & Balestra, P. (1966). Pooling cross-section and time-series data in the estimation of a dynamic economic model: the demand for natural gas. *Econometrica, 34*, 585–612.

Nerlove, M., Sevestre, P. & Balestra, P. (2008). Introduction. In L. Mátyás & P. Sevestre (Eds.), *The econometrics of panel data. fundamentals and recent developments in theory and practice* (pp. 3–21). Springer-Verlag Berlin Heidelberg, Third edition.

Page, S. E. (2006). Path dependence. *Quarterly Journal of Political Science, 1*, 87–115.

Pesaran, M. H. (2006). Estimation and inference in large heterogeneous panels with a multifactor error structure. *Econometrica, 74*(4), 967–1012.

Pesaran, M. H. & Zhou, Q. (2018). Estimation of time-invariant effects in static panel data models. *Econometric Reviews, 37*(10), 1137–1171.

Peters, B. (2009). Persistence of Innovation: Stylised Facts and Panel Data Evidence. *The Journal of Technology Transfer, 34*, 226-243.

Phillips, R. F. (2010). Iterated Feasible Generalized Least-Squares Estimation of Augmented Dynamic Panel Data Models. *Journal of Business & Economic Statistics, 28*(3), 410–422.

Phillips, R. F. (2015). On quasi maximum-likelihood estimation of dynamic Panel Data Models. *Economic Letters, 137*, 91–94.

Riju, J. & Wooldridge, J. M. (2019). Correlated Random effects models with endogenous variables and unbalanced panels. *Annals of Economics and Statistics*(134), 243–268.

Robinson, W. S. (1950). Ecological correlation and the behaviour of individuals. *American Sociological Review, 15*, 351–357.

Rücker, G. & Schwarzer, G. (2014). Presenting simulation results in a nested loop plot. *BMC Medical Research Methodology, 14*(1), 1–8.

Vella, F. & Verbeek, M. J. C. (1998). Whose wages do unions raise? A dynamic model of unionism and wage rate determination for young men. *Journal of Applied Econometrics, 13*, 163–183.

Windmeijer, F. (2005). A finite sample correction for the variance of linear efficient two-step GMM estimators. *Journal of Econometrics, 126*(1), 25–51.

Wooldridge, J. M. (2019). Correlated Random effects models with unbalanced panels. *Journal of Econometrics, 211*, 137-150.

Wooldridge, J. M. (2021). *Two-way Fixed-Effects, the Two-way Mundlak regression, and difference-in-difference Estimators.* https://papers.ssrn.com/sol3/papers.cfm?abstract_id=3906345. SSRN 3906345.

Yang, Y. & Schmidt, P. (2021). An econometric approach to the estimation of multi-level models. *Journal of Econometrics, 220*, 532–543.

Ziliak, J. P. (1997). Efficient Estimation with panel data when instruments are predetermined. An empirical comparison of moment-condition estimators. *Journal of Business & Economic Statistics, 15:4*, 419-431.

Chapter 12
Estimation of Serially Correlated Error Components Models Using Whittle's Approximate Maximum Likelihood Method

Badi H. Baltagi, Georges Bresson and Jean-Michel Etienne

Abstract This chapter studies the estimation of error components models with serial correlation of the $ARMA(p,q)$ type using Whittle's (Whittle, 1953) approximate maximum likelihood method. This is done for the one-way and two-way error components panel data model. Monte Carlo simulations are performed that investigate the small sample performance of this method.

12.1 Introduction

This chapter is a tribute to Marc Nerlove's contributions to panel data and spectral analysis in time series. In addition to his seminal contributions to panel data, most notably Balestra and Nerlove (1966), Marc Nerlove carried out groundbreaking work in spectral analysis of time series and seasonal adjustment (see Nerlove, 1964, Grether & Nerlove, 1970 and Nerlove, Grether & Carvalho, 2014). Marc left his imprint as a scholar on many diverse and different areas, see the Econometric Theory interview of Marc by Eric Ghysels, Ghysels (1993).

In his book titled 'Essays in Panel Data Econometrics', Nerlove (2005) reviews serial correlation in the one-way and two-way error components models, presenting generalized least squares (GLS) matrix transformations proposed in the literature between the 70s and 90s. For the one-way error components model, many estimators were proposed that take into account serial correlation on the remainder disturbances. These include the autoregressive $AR(p)$, moving average $MA(q)$ or mixed $ARMA(p,q)$ type (see Baltagi, 2021 for a textbook treatment of this subject). These estimators have been associated with increasingly complicated GLS transformations, of which

Badi H. Baltagi ✉
Department of Economics and Center for Policy Research, Syracuse University, Syracuse, NY, USA. e-mail: bbaltagi@syr.edu

Georges Bresson
Department of Economics, Université Paris Panthéon-Assas, Paris, France, e-mail: georges.bresson@u-paris2.fr

Jean-Michel Etienne
Department of Economics, Université Paris-Saclay, Sceaux, France, e-mail: jean-michel.etienne@u-psud.fr

B. H. Baltagi and L. Mátyás (eds.), *Seven Decades of Econometrics and Beyond*, Advanced Studies in Theoretical and Applied Econometrics 57,
https://doi.org/10.1007/978-3-031-92699-0_12

the most advanced is that of Galbraith and Zinde-Walsh (1992, 1995).[1] For the two-way error components model with serial correlation of the $AR(1)$ type, few estimators have been proposed. Some of these estimators include Revankar (1979) and Karlsson and Skoglund (2004). These papers allow for serial correlation, either on the remainder disturbances or the time effects, but not on both. Brou, Kouassi and Kymn (2011), De Porres and Krishnakumar (2013) focused on serial correlation in both the remainder disturbances and the time effects but again with no Monte Carlo simulations or applications. Baltagi, Bresson and Etienne (2024) proposed a feasible generalized least squares (FGLS) estimator with serial correlation on both the remainder disturbances and the time effects, but only in the case of an $AR(1)$ process. Monte Carlo simulations show the good performance of the proposed estimator.

Given the complexity and cumbersome implementation of the usual GLS transformations in a general serial correlation framework of the $ARMA(p,q)$ type, see Galbraith and Zinde-Walsh (1992, 1995), we propose to use Whittle's (Whittle, 1953) approximate maximum likelihood method, which is much more flexible to implement.[2] In fact, the approximate likelihood function introduced by Whittle was used to estimate the spectral density and parameters of a wide range of time series models. Faced with complicated variance-covariance matrix structures of the disturbances, estimating the model's parameters by maximum likelihood poses computational problems and can be costly in terms of CPU time. An alternative to solving the exact maximum likelihood equations is to maximize an approximation to the likelihood function. Whittle (1953) proposed a log-likelihood formulation based on the discrete Fourier transform and its power spectral density to manage such complicated variance-covariance matrices. While the literature on Whittle likelihood estimation in time series analysis is huge, only few extensions to panel data have been proposed (*e.g.*, Chen, 2006, Chen, 2008 and Wei, Zhang, Jiang & Huang, 2022).

This chapter studies the estimation of error components models with serial correlation of the $ARMA(p,q)$ type using Whittle's (Whittle, 1953) approximate maximum likelihood method. This is done for the one-way and two-way error components panel data model. Section 12.2 focuses on the one-way error components model. Following Wei et al. (2022), we propose an estimator using Whittle approximate maximum likelihood with serial correlation on the remainder disturbances of the $ARMA(p,q)$ type. Section 12.3 focuses on the two-way error components model. We extend the proposed Whittle approximate MLE in Section 12.2 to the case of serial correlation of the $ARMA(p,q)$ type both in the remainder disturbances and in the time effects. In Section 12.4, Monte Carlo simulations for the one-way and two-way error components models with serial correlation of the $ARMA(p,q)$ type are performed. These simulations confirm the suitability of the Whittle estimator for solving the problems associated with serial correlation. Section 12.5 concludes.

---

[1] Surprisingly, to our knowledge, this transformation has never been empirically tested by simulation or used in an application.

[2] We do not cover the literature on nonparametric random effects models which differ from standard (parametric) random effects models in that no assumptions are made about the distribution of the random effects. Actually, this is a form of latent class analysis: the mixing distribution is modelled by means of a finite mixture structure (see for instance Laird, 1978, Heckman & Singer, 1982, Bester & Hansen, 2009 and Chapter 10 in this book to mention a few). A whole body of literature has also focused on arbitrary serial correlation (see Cameron, Gelbach & Miller, 2011, Thompson, 2011, Davezies, D'Haultfoeuille & Guyonvarch, 2021, Menzel, 2021, Chiang, Hansen & Sasaki, 2024 to mention a few). This is also an area that we will not tackle, since we are focusing on parametric serial correlation with known distributions.

12.2 Serial Correlation in the One-way Error Components of the $ARMA(p,q)$ Type

Consider the following one-way random effects (OW-RE) model:

$$\begin{aligned} y_{it} &= X_{it}'\beta + u_{it}, \ , i = 1, \cdots, N \ , t = 1, \cdots, T, \\ \text{with} \quad & u_{it} = \mu_i + \nu_{it}, \end{aligned} \tag{12.1}$$

with i denoting individuals, t denoting time and where X_{it} is a $(K \times 1)$ vector of exogenous covariates, $\beta = (\beta_1, \cdots, \beta_K)'$ is a $(K \times 1)$ vector of parameters and $\mu_i \sim N(0, \sigma_\mu^2)$. We assume that the remainder disturbances ν_{it} follow an $ARMA(p_\nu, q_\nu)$ process:

$$\left(1 - \phi_{\nu,1} B - \cdots - \phi_{\nu,p} B^{p_\nu}\right) \nu_{it} = \left(1 - \theta_{\nu,1} B - \cdots - \theta_{\nu,q} B^{q_\nu}\right) e_{it}, \tag{12.2}$$

$$\nu_{it} = \phi_\nu^{-1}(B) \theta_\nu(B) e_{it}, \tag{12.3}$$

where e_{it} is $N(0, \sigma_e^2)$. $\phi_\nu(B)$ and $\theta_\nu(B)$ are polynomials of the backward-shift operator B such that $B^k e_{it} = e_{it-k}$.

In vector form, (12.1) can be also written as

$$\begin{aligned} y &= X\beta + u, \\ \text{with } u &= (I_N \otimes \iota_T)\mu + \nu, \end{aligned} \tag{12.4}$$

$y = \left(y_1', \cdots, y_N'\right)', y_i = (y_{i1}, \cdots, y_{iT})', X = \left(X_1', \cdots, X_N'\right)', X_i = \left(X_{i1}', \cdots, X_{iT}'\right)'$, ι_T is a $(T \times 1)$ vector of ones, I_N is an identity matrix of dimension N, $\otimes$ is the Kronecker product, $\mu = (\mu_1, \cdots, \mu_N)'$ and $\nu = \left(\nu_1', \cdots, \nu_N'\right)'$. By convention, $X_{i1} (= \iota_T, \forall i)$ and β_1 denotes the intercept.

The general variance-covariance structure of the error components setting is given by

$$Var(u) \equiv \Omega_u(\Psi_\nu) = \Omega_\mu \otimes J_T + I_N \otimes \Omega_\nu(\Psi_\nu), \tag{12.5}$$

with $J_T = \iota_T \iota_T'$. Ω_μ and $\Omega_\nu(\Psi_\nu)$ are respectively the variance-covariance matrices of μ and ν. In our specific case: $\Omega_\mu = \sigma_\mu^2 I_N$. $\Omega_\nu(\Psi_\nu) = \sigma_e^2 \Gamma_\nu(\Psi_\nu)$ where $\Gamma_\nu(\Psi_\nu)$ is a Toeplitz matrix of standardized autocovariances depending on the parameters $\Psi_\nu = \left(\phi_{\nu,1}, \cdots, \phi_{\nu,p}, \theta_{\nu,1}, \cdots, \theta_{\nu,q}\right)$.

For the general $ARMA(p,q)$ case, using (12.2) to (12.5), the log-likelihood function can be written as

$$\begin{aligned} \ln L(\Phi) &= -\frac{NT}{2} \ln 2\pi - \frac{1}{2} \ln |\Omega_u(\Psi_\nu)| - \frac{1}{2} (y - X\beta)' \Omega_u^{-1}(\Psi_\nu) (y - X\beta), \\ &= -\frac{NT}{2} \ln 2\pi - \frac{N}{2} \ln |V_u(\Psi_\nu)| - \frac{1}{2} \sum_{i=1}^{N} (y_i - X_i\beta)' V_u^{-1}(\Psi_\nu) (y_i - X_i\beta), \end{aligned} \tag{12.6}$$

where $\Omega_u(\Psi_\nu) = I_N \otimes V_u(\Psi_\nu)$ where $V_u(\Psi_\nu) = \sigma_\mu^2 J_T + \sigma_e^2 \Gamma_\nu(\Psi_\nu)$ and where Φ is an L-dimensional vector of the whole set of parameters of the model. Since $\Gamma_\nu(\Psi_\nu)$ and $V_u^{-1}(\Psi_\nu)$ are complicated functions of parameters, maximizing the log-likelihood (12.6) is no easy task. Under mild regularity assumptions, this maximization problem can be reformulated in terms of the first partial derivatives. The MLE $\widehat{\Phi}$ is the solution of the system of L equations

$$\frac{\partial}{\partial \Phi_l} \ln L(\Phi) = 0, l = 1, \cdots, L, \tag{12.7}$$

$$\text{where } \frac{\partial}{\partial \Phi_l} \ln L(\Phi) = -\frac{1}{2} \frac{\partial}{\partial \Phi_l} \ln |\Omega_u(\Psi_\nu)| - \frac{1}{2} \frac{\partial}{\partial \Phi_l} \left[(y - X\beta)' \Omega_u^{-1}(\Psi_\nu) (y - X\beta)\right]. \tag{12.8}$$

As underlined by Beran (1994), the problem of estimating Φ by the maximum likelihood poses computational problems. To obtain the solution of (12.7), (12.8) has to be evaluated for many trial values of Φ. This can be costly in terms of CPU time, in particular if the dimension of Φ is high. Also, evaluation of the inverse of the covariance matrix $\Omega_u(\Psi_\nu)$ may be numerically unstable. An alternative to solving the exact maximum likelihood equations is to maximize an approximation to the likelihood function. This is why some authors have suggested using Whittle's (Whittle, 1953) approximation. The literature on Whittle likelihood estimation is abundant in time series analysis, since the seminal works of Whittle (1953) and many others (*e.g.*, Grenander & Szegö, 1958, Hannan, 1970, Priestley, 1981, Dzhaparidze & Yaglom, 1983, Dahlhaus, 1988, Beran, 1994, and more recently Wang & Xia, 2015, Huang, Xia & Qin, 2016 or Huang, Jiang & Wang, 2019 to mention a few). Unfortunately, only few extensions to panel data have been proposed. Examples include Chen (2006), Chen (2008) and Wei et al. (2022).

Let us consider the demeaned model for individual i

$$\widetilde{y}_i = y_i - X_i\beta.$$

By definition

$$\nu_i^* = \widetilde{y}_i - \mu_i \iota_T.$$

is an estimator of ν_i with known β and μ_i. For the error process $\nu_{it} = \phi_\nu^{-1}(B)\theta_\nu(B)e_{it}$ in model (12.1), a set of Fourier frequencies is needed to derive the original Whittle likelihood estimator (WLE),

$$\omega_m = \frac{2\pi m}{T}, \; m = 0, \cdots, M,$$

where M is the bandwidth number, *i.e.*, an integer smaller than T which defines the bandwidth M/T. In general, $\max(M) = T-1$ and $\omega_m \in \Omega_T = \left\{0, \frac{2\pi}{T}, \cdots, \frac{2\pi(T-1)}{T}\right\}$. The periodogram of ν_{it}^* for $t = 1, \cdots, T$ at frequency ω_m, denoted by $I\left(\omega_m, \nu_i^*\right)$, has the following explicit form[3,4]

$$I\left(\omega_m, \nu_i^*\right) = \frac{1}{2\pi T}\left|\sum_{t=1}^{T} \nu_{it}^* e^{-jt\omega_m}\right|^2, \; j = \sqrt{-1}, \tag{12.9}$$

and the spectral density function is given by

$$\begin{aligned} f\left(\omega_m, \Psi_\nu\right) &= \sigma_e^2 f_*\left(\omega_m, \Psi_\nu\right), \\ \text{where } f_*\left(\omega_m, \Psi_\nu\right) &= \frac{1}{2\pi}\frac{\left|1 - \sum_{k=1}^{q} \theta_{\nu,k} e^{-jk\omega_m}\right|^2}{\left|1 - \sum_{k=1}^{p} \phi_{\nu,k} e^{-jk\omega_m}\right|^2}. \end{aligned}$$

$f_*\left(\omega_m, \Psi_\nu\right)$ is the standardized spectral density function. Following Huang et al. (2016), Huang et al. (2019) and Wei et al. (2022), we can estimate μ_i by minimizing the '-ln' WLE (see appendix 12.5), which is equivalent to minimizing

$$Q_i^* = \frac{1}{T}\sum_{m\in\Omega_T} \frac{I\left(\omega_m, \nu_i^*\right)}{f_*\left(\omega_m, \Psi_\nu\right)}. \tag{12.10}$$

It is the minimization of the information divergence between $f_*\left(\omega_m, \Psi_\nu\right)$ and $I\left(\omega_m, \nu_i^*\right)$ (see Parzen, 1983, Dahlhaus, 1988), *i.e.*, the search for the function $f_*\left(\omega_m, \Psi_\nu\right)$ that best approximates

[3] To avoid confusion with the letter (i) associated with individuals ($i = 1, \ldots, N$), we denote the imaginary number by $j(=\sqrt{-1})$ instead of the usual i.

[4] If $z = a + jb$, then $|z|^2 = z\bar{z} = a^2 + b^2$ where $\bar{z}(= a - jb)$ is the corresponding conjugate.

the nonparametric estimate $I\left(\omega_m, v_i^*\right)$. Then, conditional on β, Ψ_v and σ_e^2,

$$\widehat{\mu}_i = \arg\min_{\mu_i} Q_i^*.$$

According to (12.9), the periodogram in (12.10) can be written as

$$I\left(\omega_m, v_i^*\right) = \frac{1}{2\pi T} v_i^{*'} d_T(\omega_m)\bar{d}_T'(\omega_m) v_i^*, \tag{12.11}$$

where $d_T(\omega_m) = \left(e^{-j\omega_m}, \cdots, e^{-jT\omega_m}\right)'$ and $\bar{d}_T(\omega_m)$ is the corresponding conjugate.[5] Let D_T be a Toeplitz matrix given by

$$\begin{aligned} D_T &= \frac{1}{2\pi T} \sum_{m\in\Omega_T} \frac{d_T(\omega_m)\bar{d}_T'(\omega_m)}{f_*\left(\omega_m, \Psi_v\right)}, \qquad (12.12)\\ &= \frac{1}{2\pi T} \begin{pmatrix} \sum_{m\in\Omega_T} \frac{1}{f_*(\omega_m,\Psi_v)} & \cdots & \sum_{m\in\Omega_T} \frac{\cos((T-1)\omega_m)}{f_*(\omega_m,\Psi_v)} \\ \vdots & & \vdots \\ \sum_{m\in\Omega_T} \frac{\cos((T-1)\omega_m)}{f_*(\omega_m,\Psi_v)} & \cdots & \sum_{m\in\Omega_T} \frac{1}{f_*(\omega_m,\Psi_v)} \end{pmatrix}. \end{aligned}$$

Then, $\widehat{\mu}_i$ can be written as

$$\widehat{\mu}_i = \left(\iota_T' D_T \iota_T\right)^{-1} \iota_T' D_T \widetilde{y}_i, \tag{12.13}$$

and

$$\widehat{\sigma}_\mu^2 = \frac{1}{N-1} \sum_{i=1}^N \left(\widehat{\mu}_i - \frac{1}{N}\sum_{i=1}^N \widehat{\mu}_i\right)^2. \tag{12.14}$$

The estimated residuals are given by

$$\widehat{v}_i^* = P_T \widetilde{y}_i \text{ with } P_T = I_T - \iota_T \left(\iota_T' D_T \iota_T\right)^{-1} \iota_T' D_T. \tag{12.15}$$

As β, Ψ_v and σ_e^2 are invariant across all the individuals $\{i = 1, \cdots, N\}$, we can use an estimation method through minimizing a weighted sum of all WLE Q_i, $\{i = 1, \cdots, N\}$,

$$Q = \frac{1}{N}\sum_{i=1}^N \frac{1}{\sigma_e^2} Q_i^* = \frac{1}{NT}\sum_{i=1}^N \sum_{m\in\Omega_T} \frac{I\left(\omega_m, \widehat{v}_i^*\right)}{f\left(\omega_m, \Psi_v\right)}. \tag{12.16}$$

If we know σ_e^2, β and Ψ_v can be estimated by minimizing the profile based Whittle likelihood

$$\begin{aligned} Q^* &= \frac{1}{N}\sum_{i=1}^N Q_i^* = \frac{1}{NT}\sum_{i=1}^N \sum_{m\in\Omega_T} \frac{I\left(\omega_m, \widehat{v}_i^*\right)}{f_*\left(\omega_m, \Psi_v\right)}, \qquad (12.17)\\ &= \frac{1}{N}\sum_{i=1}^N \left(\frac{1}{T}\widehat{v}_i^{*'} D_T \widehat{v}_i^*\right) = \frac{1}{N}\sum_{i=1}^N \left(\frac{1}{T}\widetilde{y}_i' D_{NT} \widetilde{y}_i\right), \\ &= \frac{1}{NT}\left(y - X\beta\right)' \left(I_N \otimes D_{NT}\right)\left(y - X\beta\right), \end{aligned}$$

[5] $d_T(\omega_m)\bar{d}_T'(\omega_m)$ is a $(T\times T)$ Toeplitz matrix with first row given by $(1, \cos(\omega_m), \cdots, \cos((T-1)\omega_m))$, see Baltagi, Bresson and Etienne (2025).

where $D_{NT} = P_T' D_T P_T$. Then, the estimate of σ_e^2 is given by

$$\widehat{\sigma}_e^2 = \widehat{Q}^* = \frac{1}{NT}\left(y - X\widehat{\beta}\right)' \left(I_N \otimes \widehat{D}_{NT}\right)\left(y - X\widehat{\beta}\right), \tag{12.18}$$

Through a nonlinear optimization algorithm, we can get estimates of β, Ψ_ν and σ_e^2. From (12.13) and (12.14), we get an estimate of μ and σ_μ^2. Likewise, from the estimate of the transfer function $\Theta_\nu(B) = \phi_\nu^{-1}(B)\theta_\nu(B) = (\Theta_{\nu,0} + \Theta_{\nu,1}B + \Theta_{\nu,2}B^2 + \cdots)$ and the estimate of σ_e^2, we get an estimate of σ_ν^2

$$\widehat{\sigma}_\nu^2 = \widehat{\sigma}_e^2 \sum_{k=0}^{\infty} \widehat{\Theta}_{\nu,k}^2 \approx \widehat{\sigma}_e^2 \sum_{k=0}^{T-1} \widehat{\Theta}_{\nu,k}^2,$$

where (see Box & Jenkins, 1976 and Karanasos, 1998)

$$\begin{cases} \widehat{\Theta}_{\nu,0} &= 1 \\ \widehat{\Theta}_{\nu,1} &= \widehat{\phi}_{\nu,1} - \widehat{\theta}_{\nu,1} \\ \widehat{\Theta}_{\nu,2} &= \widehat{\phi}_{\nu,1}\widehat{\Theta}_{\nu,1} + \widehat{\phi}_{\nu,2} - \widehat{\theta}_{\nu,2} \\ \vdots & \vdots\ \vdots \\ \widehat{\Theta}_{\nu,\tau} &= \sum_{j=1}^{p_\nu} \widehat{\phi}_{\nu,j}\widehat{\Theta}_{\nu,\tau-j} - \widehat{\theta}_{\nu,\tau}\ ,\ \tau \leq q_\nu \\ \widehat{\Theta}_{\nu,\tau} &= \sum_{j=1}^{p_\nu} \widehat{\phi}_{\nu,j}\widehat{\Theta}_{\nu,\tau-j}\ ,\ \tau > \max(p_\nu - 1, q_\nu) \end{cases} \tag{12.19}$$

$\widehat{\sigma}_\nu^2$ can also be directly obtained using $Var(\widehat{\nu}^*)$ from (12.15): $\widehat{\sigma}_\nu^2 = Var\left[(I_N \otimes P_T)\widetilde{y}\right]$. From the estimated weights of the transfer function $\widehat{\Theta}_{\nu,\tau}$, we can estimate the covariance generating function

$$\widehat{\gamma}_{\nu,\tau} = \widehat{\sigma}_e^2 \widehat{\gamma}_{\nu,\tau}^* \text{ with } \widehat{\gamma}_{\nu,\tau}^* = \sum_{k=0}^{\infty} \widehat{\Theta}_{\nu,k+|\tau|}\widehat{\Theta}_{\nu,k} \approx \sum_{k=0}^{T-\tau-1} \widehat{\Theta}_{\nu,k+|\tau|}\widehat{\Theta}_{\nu,k},$$

and $\Gamma_\nu(\widehat{\Psi}_\nu)$ is a Toeplitz matrix obtained from $\widehat{\gamma}_{\nu,\tau}^*$ for $\tau = 0, \cdots, T-1$. Next, we get the estimated variance-covariance matrix of the residuals u, $\Omega_u(\widehat{\Psi}_\nu) = \widehat{\sigma}_\mu^2 I_{NT} + \widehat{\sigma}_e^2 \Gamma_\nu(\widehat{\Psi}_\nu)$, and the FGLS estimates

$$\widehat{\beta} = \left(X'\Omega_u^{-1}(\widehat{\Psi}_\nu)X\right)^{-1} X'\Omega_u^{-1}(\widehat{\Psi}_\nu)y,$$

$$\text{and } Var\left(\widehat{\beta}\right) = \left(X'\Omega_u^{-1}(\widehat{\Psi}_\nu)X\right)^{-1}.$$

The Whittle likelihood estimation of the one-way random effects model with serial correlation of the $ARMA(p,q)$ type can therefore be summarized by Algorithm 1.

Algorithm 1: Whittle likelihood estimation of a one-way random effects model with serial correlation of the $ARMA(p,q)$ type

1. First step.

   a. Let $\widetilde{y}_i = y_i - X_i\beta$ and $\widetilde{y} = y - X\beta$ with $\widetilde{y} = (\widetilde{y}_1, \cdots, \widetilde{y}_N)'$, $\nu_i^* = \widetilde{y}_i - \mu_i \iota_T$ and $\nu^* = \left(\nu_1^*, \cdots, \nu_N^*\right)'$.

   b. nonlinear optimization algorithm

      i. solve $\arg\min_{\beta,\Psi_\nu,\mu} Q^* = \arg\min_{\beta,\Psi_\nu,\mu} \frac{1}{NT}\left(y - X\beta\right)' (I_N \otimes D_{NT}) \left(y - X\beta\right)$ where $D_{NT} = P_T' D_T P_T$ with $P_T = I_T - \iota_T \left(\iota_T' D_T \iota_T\right)^{-1} \iota_T' D_T$ and

$$D_T = \frac{1}{2\pi T}\begin{pmatrix} \sum_{m\in\Omega_T} \frac{1}{f_*(\omega_m,\Psi_\nu)} & \cdots & \sum_{m\in\Omega_T} \frac{\cos((T-1)\omega_m)}{f_*(\omega_m,\Psi_\nu)} \\ \vdots & & \vdots \\ \sum_{m\in\Omega_T} \frac{\cos((T-1)\omega_m)}{f_*(\omega_m,\Psi_\nu)} & \cdots & \sum_{m\in\Omega_T} \frac{1}{f_*(\omega_m,\Psi_\nu)} \end{pmatrix},$$

with

$$f_*(\omega_m, \Psi_\nu) = \frac{1}{2\pi}\frac{\left|1 - \sum_{k=1}^{q} \theta_{\nu,k} e^{-jk\omega_m}\right|^2}{\left|1 - \sum_{k=1}^{p} \phi_{\nu,k} e^{-jk\omega_m}\right|^2}$$

$$\omega_m \in \Omega_T = \left\{0, \frac{2\pi}{T}, \cdots, \frac{2\pi(T-1)}{T}\right\}.$$

      ii. $\widehat{\mu}_i = \left(\iota_T' D_T \iota_T\right)^{-1} \iota_T' D_T \widetilde{y}_i$, $\widehat{\mu} = (\widehat{\mu}_1, \cdots, \widehat{\mu}_N)'$, $\widehat{\sigma}_\mu^2 = \frac{1}{N-1}\sum_{i=1}^N \left(\widehat{\mu}_i - \frac{1}{N}\sum_{i=1}^N \widehat{\mu}_i\right)^2$.

      iii. $\widehat{\sigma}_e^2 = \frac{1}{NT}\left(y - X\widehat{\beta}\right)' \left(I_N \otimes \widehat{D}_{NT}\right)\left(y - X\widehat{\beta}\right)$.

      iv. $\widehat{\sigma}_\nu^2 = Var\left[(I_N \otimes P_T)\widetilde{y}\right]$ or $\widehat{\sigma}_\nu^2 = \widehat{\sigma}_e^2 \sum_{k=0}^{T-1} \widehat{\Theta}_{\nu,k}^2$ with

$$\begin{cases} \widehat{\Theta}_{\nu,0} &= 1 \\ \widehat{\Theta}_{\nu,1} &= \widehat{\phi}_{\nu,1} - \widehat{\theta}_{\nu,1} \\ \widehat{\Theta}_{\nu,2} &= \widehat{\phi}_{\nu,1}\widehat{\Theta}_{\nu,1} + \widehat{\phi}_{\nu,2} - \widehat{\theta}_{\nu,2} \\ \vdots & \vdots \\ \widehat{\Theta}_{\nu,\tau} &= \sum_{j=1}^{p_\nu} \widehat{\phi}_{\nu,j}\widehat{\Theta}_{\nu,\tau-j} - \widehat{\theta}_{\nu,\tau}\,,\ \tau \le q_\nu \\ \widehat{\Theta}_{\nu,\tau} &= \sum_{j=1}^{p_\nu} \widehat{\phi}_{\nu,j}\widehat{\Theta}_{\nu,\tau-j}\,,\ \tau > \max(p_\nu - 1, q_\nu) \end{cases}$$

2. Second step.

   a. $\Gamma_\nu(\widehat{\Psi}_\nu)$: Toeplitz matrix obtained from $\widehat{\gamma}_{\nu,\tau}^* = \sum_{k=0}^{T-\tau-1} \widehat{\Theta}_{\nu,k+|\tau|}\widehat{\Theta}_{\nu,k}$ for $\tau = 0, \cdots, T-1$.

   b. $\Omega_u(\widehat{\Psi}_\nu) = \widehat{\sigma}_\mu^2 I_{NT} + \widehat{\sigma}_e^2 \Gamma_\nu(\widehat{\Psi}_\nu)$

   c. $\widehat{\beta} = \left(X'\Omega_u^{-1}(\widehat{\Psi}_\nu)X\right)^{-1} X'\Omega_u^{-1}(\widehat{\Psi}_\nu)y$ and $Var\left(\widehat{\beta}\right) = \left(X'\Omega_u^{-1}(\widehat{\Psi}_\nu)X\right)^{-1}$.

Of course, time series specialists have long shown that Whittle's approximation works well in the asymptotic case ($T \longrightarrow \infty$). In the case of small samples in the time dimension, the small sample behavior may be poor if the spectrum of the process contains peaks or if characteristic roots of the $ARMA(p,q)$ process are close to unity. In fact, the expected value of the periodogram is the convolution of the standardized spectral density and the Fejér kernel. This convolution smooths out the peaks in the spectral density function due to the sidelobes in the Fejér kernel. This is the leakage effect which is greater when the spectral density has a large peak and the sample size is small.

Since the Whittle approximation of the log-likelihood function can be considered as the information divergence between the periodogram and the spectral density, we get a parametric leakage effect due to the fact that the leakage effect is also transferred to the parametric estimation procedure. Some advocate the use of tapered data and tapered (or modified) periodograms to reduce variance. The tapered periodogram multiplies the input time series by a taper sequence or window h_t

$$I\left(\omega_m, \nu_i^*\right) = \frac{1}{2\pi \sum_{t=1}^{T} h_t^2} \left| \sum_{t=1}^{T} h_t \nu_{it}^* e^{-jt\omega_m} \right|^2, \tag{12.20}$$

and the number of points that are tapered will impact the bias. The usual periodogram has $h_t = 1$, $\forall t$. Generally, one would prefer a window that leaves the bulk of the data unmodified and just tapers the ends. Tapering can reduce the leakage effect of the periodogram as an estimate of the true spectrum (see Tukey, 1967, Jenkins & Watts, 1969, Hatanaka, 1972, Brillinger, 1981, Priestley, 1981, Dahlhaus, 1988, Zhang, 1992, Ginovyan & Sahakyan, 2021).[6] In panel data, we deal very often with small T as compared to large N (except with monthly or quarterly macroeconomic and daily financial datasets). The introduction of tapered periodogram may be appropriate to reduce potential biases due to the use of an unmodified periodogram, although we have no knowledge of the use of modified periodogram in the case of panel data.[7] Moreover, to our knowledge, there are no Monte Carlo simulation studies or applications for a one-way error components model with $ARMA$ errors using Whittle's approximation with large N and small T.

The h_t window in (12.20) is both data and frequency independent, *i.e.*, h_t is the same at any frequency of the spectrum and for any data sequence. For this non-adaptive window, the consequence of this restriction is twofold: on the one hand, reducing the leakage effect may not be effective and, on the other, any attempt to reduce the leakage effect leads to a reduction of the resolution and *vice versa*. A possible solution is the apodization approach[8], defining a data and frequency dependent temporal window, which mitigates the leakage problem of the periodogram without compromising its resolution (see DeGraaf, 1994, Stankwitz, Dallaire & Fienup, 1994, Thomas, Flores & Sok-Son, 2000 and Stoica & Moses, 2005).

In the apodization literature, the non-adaptive Hanning window $h_t = 1 - \delta \cos(2\pi t/T), (|\delta| \leq 1)$ is replaced by its adaptive version. So, in the panel data case, we have

$$h_{i,m,t} = 1 - \delta_{i,m} \cos(\frac{2\pi t}{T}),$$

with $|\delta_{i,m}| \leq 1$ and the apodized-windowed periodogram is given by

---

[6] In time series analysis, there are many attempts to find the optimum tapering method, either by selecting different taper sequences or windows (Hanning, Tukey, Parzen, polynomial, etc), or by using other approaches (boxed (modified) periodogram, complete periodogram, etc) and there is currently no single, optimal method that yields unbiased estimates of the periodogram when T is small, (see Welch, 1967, Hatanaka, 1972, Priestley, 1981, Dahlhaus, 1988, Hurvich & Ray, 1995, Robinson, 1995, Montanari, Taqqu & Teverovsky, 1999, Velasco, 1999, Velasco & Robinson, 2000, Das, Subba Rao & Yang, 2021, Ginovyan & Sahakyan, 2021, Subba Rao & Yang, 2021 to mention a few).

[7] Chen (2006) used $N = 20$ and $T = 50, 100, 200$ for a one-way error components model with serial correlation of the $ARMA(1,1)$ type. Chen (2008) used $N = 20, 30$ and $T \geq 350$ for the same model to estimate empirical size and power, while Wei et al. (2022) chose $N = T$ $(= 30, 60, 90)$ for a linear dynamic model with serial correlation of the $MA(1)$ type. These authors do not use a tapered periodogram. In their simulation, Wei et al. (2022) obtain relatively large biases for the $MA(1)$ parameter, in excess of 30%.

[8] Apodization is a term borrowed from optics where it has been used to mean a reduction of the sidelobes induced by diffraction. This is a method for increasing contrast and at least partially eliminating diffraction rings produced by an optical instrument, in order to improve the definition of the elements to be studied.

$$
\begin{aligned}
I^*\left(\omega_m, \nu_i^*\right) &= \frac{1}{2\pi T}\left|D_{i,m}^*\right|^2 = \frac{1}{2\pi T}\left|D_{i,m} - \frac{\delta_{i,m}}{2}\left(D_{i,m-1} + D_{i,m+1}\right)\right|^2 \\
&= \frac{1}{2\pi T}\nu_i^{*\prime}\nabla_{i,m}\nu_i^*,
\end{aligned}
\tag{12.21}
$$

where $D_{i,m} = \sum_{t=1}^{T}\nu_{it}^* e^{-jt\omega_m}$ is the discrete Fourier transform (see Baltagi et al. (2025) for derivations). $\nabla_{i,m}$ is a $(T \times T)$ matrix with (t,l) element given by

$$
\nabla_{i,m}(t,l) = \cos\left((t-l)\,\omega_m\right).\left\{\begin{array}{l} 1 - \delta_{i,m}\left[\cos(\frac{2\pi t}{T}) + \cos(\frac{2\pi l}{T})\right] \\ +\delta_{i,m}^2\left[\cos(\frac{2\pi t}{T})\cos(\frac{2\pi l}{T})\right] \end{array}\right\},
$$

for $t, l = 1, \cdots, T$. Of course, for $m = 0$ and $m = T-1$, $\delta_{i,m} \equiv 0$ and we find the expression for $d_T(\omega_m)\bar{d}_T'(\omega_m)$ in (12.11). The filter coefficient $\delta_{i,m}$ is defined as

$$
\delta_{i,m} = \begin{cases} 0 & \text{if } \delta_{0,i,m} < 0 \\ \delta_{0,i,m} & \text{if } 0 \leq \delta_{0,i,m} \leq 1 \\ 1 & \text{if } \delta_{0,i,m} > 1 \end{cases},
$$

where

$$
\delta_{0,i,m} = \frac{2\left(a_{i,m}c_{i,m} + b_{i,m}d_{i,m}\right)}{c_{i,m}^2 + d_{i,m}^2},
$$

with

$$
\begin{aligned}
a_{i,m} &= \sum_{t=1}^{T}\nu_{it}^*\cos(\omega_m t),\ b_{i,m} = \sum_{t=1}^{T}\nu_{it}^*\sin(\omega_m t), \\
c_{i,m} &= \sum_{t=1}^{T}\nu_{it}^*\left(\cos(\omega_{m-1}t) + \cos(\omega_{m+1}t)\right), \\
d_{i,m} &= \sum_{t=1}^{T}\nu_{it}^*\left(\sin(\omega_{m-1}t) + \sin(\omega_{m+1}t)\right),
\end{aligned}
$$

see Baltagi et al. (2025) for derivations.

The introduction of an apodized-windowed periodogram as in (12.21) requires us to rewrite relations (12.12) to (12.18) and to modify step 1.b of Algorithm 1.[9] Similar changes must also be made to steps 1.b and 2.b of Algorithm 2 for the two-way Whittle ML estimate in Section 12.3.

12.3 Serial Correlation in the Two-way Error Components of the $ARMA(p,q)$ Type

Consider now the two-way random effects (TW-RE) model with serial correlation of the $ARMA(p,q)$ type in both $\{\lambda_t\}$ and $\{\nu_{it}\}$:

$$
\begin{aligned}
y_{it} &= X_{it}'\beta + u_{it},\ ,\ i = 1, \cdots, N\ ,\ t = 1, \cdots, T, \\
\text{with} \quad & u_{it} = \mu_i + \lambda_t + \nu_{it}.
\end{aligned}
\tag{12.22}
$$

[9] To save space, the modified Algorithm is given in Baltagi et al. (2025).

We assume that the remainder disturbances ν_{it} follow an $ARMA(p_\nu, q_\nu)$ process:

$$\begin{aligned}\left(1-\phi_{\nu,1}B-\cdots-\phi_{\nu,p}B^{p_\nu}\right)\nu_{it} &= \left(1-\theta_{\nu,1}B-\cdots-\theta_{\nu,q}B^{q_\nu}\right)e_{it},\\ \nu_{it} &= \phi_\nu^{-1}(B)\theta_\nu(B)e_{it},\end{aligned}$$

where e_{it} is $N(0,\sigma_e^2)$. Likewise, the time effects λ_t follow an $ARMA(p_\lambda, q_\lambda)$ process:

$$\begin{aligned}\left(1-\phi_{\lambda,1}B-\cdots-\phi_{\lambda,p}B^{p_\lambda}\right)\lambda_t &= \left(1-\theta_{\lambda,1}B-\cdots-\theta_{\lambda,q}B^{q_\lambda}\right)\varepsilon_t,\\ \lambda_t &= \phi_\lambda^{-1}(B)\theta_\lambda(B)\varepsilon_t,\end{aligned}$$

where ε_t is $N(0,\sigma_\varepsilon^2)$. In vector form, (12.22) can be also written as

$$\begin{aligned}y &= X\beta+u,\\ \text{with } u &= (I_N\otimes\iota_T)\mu+(\iota_N\otimes I_T)\lambda+\nu,\end{aligned}$$

where $\lambda=(\lambda_1,\cdots,\lambda_T)'$. The general variance-covariance structure of the error components setting is given by

$$Var(u)\equiv\Omega_u(\Psi)=\sigma_\mu^2(I_N\otimes J_T)+J_N\otimes\Omega_\lambda(\Psi_\lambda)+I_N\otimes\Omega_\nu(\Psi_\nu),$$

where $\Psi=(\Psi_\lambda',\Psi_\nu')'$, with $\Psi_\lambda'=(\phi_{\lambda,1},\cdots,\phi_{\lambda,p_\lambda},\theta_{\lambda,1},\cdots,\theta_{\lambda,q_\lambda})$ and $\Psi_\nu'=(\phi_{\nu,1},\cdots,\phi_{\nu,p_\nu},\theta_{\nu,1},\cdots,\theta_{\nu,q_\nu})$. $\Omega_\lambda(\Psi_\lambda)$ and $\Omega_\nu(\Psi_\nu)$ are respectively the variance-covariance matrices of λ and ν with $\Omega_\nu(\Psi_\nu)=\sigma_e^2\Gamma_\nu(\Psi_\nu)$ and $\Omega_\lambda(\Psi_\lambda)=\sigma_\varepsilon^2\Gamma_\lambda(\Psi_\lambda)$. $\Gamma_\lambda(\Psi_\lambda)$ and $\Gamma_\nu(\Psi_\nu)$ are $(T\times T)$ Toeplitz matrices of standardized autocovariances depending on the parameters Ψ_λ and Ψ_ν, respectively.

Similar to (12.6), the log-likelihood function is

$$\ln L(\Phi) = -\frac{NT}{2}-\frac{1}{2}\ln|\Omega_u(\Psi)|-\frac{1}{2}(y-X\beta)'\Omega_u^{-1}(\Psi)(y-X\beta).$$

Here again, the problem of estimating Φ by the maximum likelihood poses computational problems and an alternative to solving the exact maximum likelihood equations is to use the Whittle's (Whittle, 1953) approximation. However, to the best of our knowledge, this has never been considered in the case of a two-way random effects (TW-RE) model with serial correlation of the $ARMA(p,q)$ type in both $\{\lambda_t\}$ and $\{\nu_{it}\}$. As the approach is a little more complex than in Section 12.2, we will break it down into 3 steps.

1. In the first step, we use the within-time transformation $(E_N\otimes I_T)$ where $E_N=I_N-\bar{J}_N$ with $\bar{J}_N=\iota_N\iota_N'/N$, leading to the within-time error components[10]

$$u^\bullet=(E_N\otimes I_T)u=(E_N\otimes\iota_T)\mu+(E_N\otimes I_T)\nu=\mu^\bullet+\nu^\bullet.$$

The within-time transformation wipes out the intercept β_1 and the time effects λ_t and leads to the within-time equation $y^\bullet=X^\bullet\beta^\circ+u^\bullet$ with $X^\bullet=(E_N\otimes I_T)X^\circ$. X° (resp. β°) is the set of covariates (resp. coefficients) excluding $X_1(=\iota_{NT})$ (resp. the intercept β_1). With this model, we apply the first step of Algorithm 1 where y is replaced by $y^\bullet$ and X is replaced by $X^\bullet$. This gives us estimates for β°, Ψ_ν, σ_ν^2, σ_e^2, σ_μ^2, μ and $\nu^\bullet$.
2. In a second step, we define

$$\begin{aligned}\delta_{it} &= y_{it}-X_{it}^{\circ\prime}\widehat{\beta}^\circ-\widehat{\mu}_i-\widehat{\nu}_{it}^\bullet \leftrightarrow \delta=y-X^\circ\widehat{\beta}^\circ-(I_N\otimes\iota_T)\widehat{\mu}-\widehat{\nu}^\bullet\\ \text{and } \delta^\star &= (\bar{J}_N\otimes I_T)\delta \text{ with } \delta=(\delta_{11},\cdots,\delta_{1T},\cdots,\delta_{N1},\cdots,\delta_{NT})',\end{aligned}$$

[10] $(E_N\otimes I_T)=I_{NT}-(\bar{J}_N\otimes I_T)$ has a typical element $u_{it}-\bar{u}_{.t}$ where $(\bar{J}_N\otimes I_T)$ is the between-time transformation which averages the data over individuals and has a typical element $\bar{u}_{.t}=\sum_{i=1}^N u_{it}/N$.

where $(\bar{J}_N \otimes I_T)$ is the between-time transformation. We apply again the first step of Algorithm 1 where y is replaced by $\delta^\star$ and X is replaced by a constant ι_{NT}. This gives us estimates for β_1, Ψ_λ, σ_λ^2 and σ_ε^2.

3. In a third step, as in Section 12.2, we use the transfer functions $\Theta_\nu(B) = \phi_\nu^{-1}(B)\theta_\nu(B) = (\Theta_{\nu,0}+\Theta_{\nu,1}B+\Theta_{\nu,2}B^2+\cdots)$ and $\Theta_\lambda(B) = \phi_\lambda^{-1}(B)\theta_\lambda(B) = (\Theta_{\lambda,0}+\Theta_{\lambda,1}B+\Theta_{\lambda,2}B^2+\cdots)$ where the weights are estimated as in (12.19). From the estimated weights of the transfer functions, we can estimate the covariance generating functions

$$\widehat{\gamma}_{\nu,\tau} = \widehat{\sigma}_e^2\widehat{\gamma}_{\nu,\tau}^* \text{ with } \widehat{\gamma}_{\nu,\tau}^* = \sum_{k=0}^{\infty}\widehat{\Theta}_{\nu,k+|\tau|}\widehat{\Theta}_{\nu,k} \approx \sum_{k=0}^{T-\tau-1}\widehat{\Theta}_{\nu,k+|\tau|}\widehat{\Theta}_{\nu,k},$$

$$\widehat{\gamma}_{\lambda,\tau} = \widehat{\sigma}_\varepsilon^2\widehat{\gamma}_{\lambda,\tau}^* \text{ with } \widehat{\gamma}_{\lambda,\tau}^* = \sum_{k=0}^{\infty}\widehat{\Theta}_{\lambda,k+|\tau|}\widehat{\Theta}_{\lambda,k} \approx \sum_{k=0}^{T-\tau-1}\widehat{\Theta}_{\lambda,k+|\tau|}\widehat{\Theta}_{\lambda,k}.$$

$\Gamma_\nu(\widehat{\Psi}_\nu)$ and $\Gamma_\lambda(\widehat{\Psi}_\lambda)$ are Toeplitz matrices obtained from $\widehat{\gamma}_{\nu,\tau}^*$ and $\widehat{\gamma}_{\lambda,\tau}^*$, respectively for $\tau = 0,\cdots,T-1$. As in Section 12.2, $\widehat{\sigma}_\nu^2$ and $\widehat{\sigma}_\lambda^2$ are estimated using the transfer function weights (see step 3 in Algorithm 2) and they can also be directly obtained using the estimated variance of the residuals in steps 1 and 2. The estimated variance-covariance matrix of the residuals u is given by

$$\Omega_u(\widehat{\Psi}) = \widehat{\sigma}_\mu^2(I_N \otimes J_T)+\widehat{\sigma}_\varepsilon^2(J_N \otimes \Gamma_\lambda(\widehat{\Psi}_\lambda))+\widehat{\sigma}_e^2(I_N \otimes \Gamma_\nu(\widehat{\Psi}_\nu)),$$

and the corresponding FGLS estimates are

$$\widehat{\beta} = \left(X'\Omega_u^{-1}(\widehat{\Psi})X\right)^{-1}X'\Omega_u^{-1}(\widehat{\Psi})y,$$
$$\text{and } Var\left(\widehat{\beta}\right) = \left(X'\Omega_u^{-1}(\widehat{\Psi})X\right)^{-1}.$$

In order to speed up the computation of $\Omega_u^{-1}(\widehat{\Psi})$, especially when N and T are large, we can use the Woodbury matrix identity (see Appendix 12.5), *i.e.*,

$$\Omega_u^{-1}(\Psi) = I_N \otimes A^{\star^{-1}} - \left(\iota_N \otimes A^{\star^{-1}}\right)\left[\sigma_\varepsilon^{-2}\Gamma_\lambda^{-1}(\Psi_\lambda)+NA^{\star^{-1}}\right]^{-1}\left(\iota_N' \otimes A^{\star^{-1}}\right),$$

where $A^{\star^{-1}} = \left[\sigma_\mu^2 J_T + \sigma_e^2\Gamma_\nu(\Psi_\nu)\right]^{-1}$.

Algorithm 2: Whittle likelihood estimation of a two-way random effects model with serial correlation of the $ARMA(p,q)$ type

1. First step.

   a. Let $y_{it}^{\bullet} = y_{it} - \bar{y}_{.t}$ with $\bar{y}_{.t} = \sum_{i=1}^{N} y_{it}/N$, $X_{it}^{\bullet} = X_{it}^{\circ} - \bar{X}_{.t}^{\circ}$, $y^{\bullet} = (E_N \otimes I_T)\, y$, $X^{\bullet} = (E_N \otimes I_T)\, X^{\circ}$. Let $\widetilde{y}_i^{\bullet} = y_i^{\bullet} - X_i^{\bullet}\beta^{\circ}$ and $\widetilde{y}^{\bullet} = y^{\bullet} - X^{\bullet}\beta^{\circ}$ with $\widetilde{y}^{\bullet} = \left(\widetilde{y}_1^{\bullet}, \cdots, \widetilde{y}_N^{\bullet}\right)'$, $\nu_i^{*\bullet} = \widetilde{y}_i^{\bullet} - \mu_i^{\bullet}\iota_T$ with $\mu_i^{\bullet} = \mu_i$ and $\nu^{*\bullet} = \left(\nu_1^{*\bullet}, \cdots, \nu_N^{*\bullet}\right)'$ where X_i° is the set of covariates excluding the intercept.

   b. nonlinear optimization algorithm

      i. solve $\arg\min_{\beta^{\circ},\Psi_{\nu},\mu^{\bullet}} Q^{*} = \arg\min_{\beta^{\circ},\Psi_{\nu},\mu^{\bullet}} \frac{1}{NT}\left(y^{\bullet} - X^{\bullet}\beta^{\circ}\right)' (I_N \otimes D_{NT}) \left(y^{\bullet} - X^{\bullet}\beta^{\circ}\right)$, where $D_{NT} = P_T' D_T P_T$ with $P_T = I_T - \iota_T\left(\iota_T' D_T \iota_T\right)^{-1}\iota_T' D_T$ and

$$D_T = \frac{1}{2\pi T}\begin{pmatrix} \sum_{m\in\Omega_T}\frac{1}{f_*(\omega_m,\Psi_\nu)} & \cdots & \sum_{m\in\Omega_T}\frac{\cos((T-1)\omega_m)}{f_*(\omega_m,\Psi_\nu)} \\ \vdots & & \vdots \\ \sum_{m\in\Omega_T}\frac{\cos((T-1)\omega_m)}{f_*(\omega_m,\Psi_\nu)} & \cdots & \sum_{m\in\Omega_T}\frac{1}{f_*(\omega_m,\Psi_\nu)} \end{pmatrix}$$

      with

$$f_*(\omega_m,\Psi_\nu) = \frac{1}{2\pi}\frac{\left|1-\sum_{k=1}^{q}\theta_{\nu,k}e^{-jk\omega_m}\right|^2}{\left|1-\sum_{k=1}^{p}\phi_{\nu,k}e^{-jk\omega_m}\right|^2}$$
$$\omega_m \in \Omega_T = \left\{0, \frac{2\pi}{T}, \cdots, \frac{2\pi(T-1)}{T}\right\}.$$

      ii. $\widehat{\mu}_i^{\bullet} (\equiv \widehat{\mu}_i) = \left(\iota_T' D_T \iota_T\right)^{-1}\iota_T' D_T \widetilde{y}_i^{\bullet}$, $\widehat{\mu} = (\widehat{\mu}_1, \cdots, \widehat{\mu}_N)'$, $\widehat{\sigma}_{\mu}^2 = \frac{1}{N-1}\sum_{i=1}^{N}\left(\widehat{\mu}_i - \frac{1}{N}\sum_{i=1}^{N}\widehat{\mu}_i\right)^2$.

      iii. $\widehat{\sigma}_e^2 = \frac{1}{NT}\left(y^{\bullet} - X^{\bullet}\widehat{\beta}^{\circ}\right)'\left(I_N \otimes \widehat{D}_{NT}\right)\left(y^{\bullet} - X^{\bullet}\widehat{\beta}^{\circ}\right)$.

2. Second step.

   a. Let $\delta_{it} = y_{it} - X_{it}^{\circ\prime}\widehat{\beta}^{\circ} - \widehat{\mu}_i - \widehat{\nu}_{it}^{*\bullet} \leftrightarrow \delta = y - X^{\circ}\widehat{\beta}^{\circ} - (I_N \otimes \iota_T)\,\widehat{\mu} - \widehat{\nu}^{*\bullet}$, and $\delta^{\star} = (\bar{J}_N \otimes I_T)\,\delta$ with $\delta = (\delta_{11}, \cdots, \delta_{1T}, \cdots, \delta_{N1}, \cdots, \delta_{NT})'$ and $T_t = \iota_T$.

   b. nonlinear optimization algorithm

      i. $\arg\min_{\eta,\Psi_\lambda} Q^{*} = \arg\min_{\eta,\Psi_\lambda} \frac{1}{NT}\left(\delta^{\star} - (\iota_N \otimes T_t)\,\eta\right)' (I_N \otimes D_{NT}) \left(\delta^{\star} - (\iota_N \otimes T_t)\,\eta\right)$, where $D_{NT} = P_T' D_T P_T$ with $P_T = I_T - \iota_T\left(\iota_T' D_T \iota_T\right)^{-1}\iota_T' D_T$ and

$$D_T = \frac{1}{2\pi T}\begin{pmatrix} \sum_{m\in\Omega_T}\frac{1}{f_*(\omega_m,\Psi_\lambda)} & \cdots & \sum_{m\in\Omega_T}\frac{\cos((T-1)\omega_m)}{f_*(\omega_m,\Psi_\lambda)} \\ \vdots & & \vdots \\ \sum_{m\in\Omega_T}\frac{\cos((T-1)\omega_m)}{f_*(\omega_m,\Psi_\lambda)} & \cdots & \sum_{m\in\Omega_T}\frac{1}{f_*(\omega_m,\Psi_\lambda)} \end{pmatrix},$$

      with

$$f_*(\omega_m,\Psi_\lambda) = \frac{1}{2\pi}\frac{\left|1-\sum_{k=1}^{q}\theta_{\lambda,k}e^{-jk\omega_m}\right|^2}{\left|1-\sum_{k=1}^{p}\phi_{\lambda,k}e^{-jk\omega_m}\right|^2}$$
$$\omega_m \in \Omega_T = \left\{0, \frac{2\pi}{T}, \cdots, \frac{2\pi(T-1)}{T}\right\}.$$

      ii. $\widehat{\sigma}_{\varepsilon}^2 = \frac{1}{NT}\left(\delta^{\star} - (\iota_N \otimes T_t)\,\widehat{\eta}\right)'\left(I_N \otimes \widehat{D}_{NT}\right)\left(\delta^{\star} - (\iota_N \otimes T_t)\,\widehat{\eta}\right)$.

Algorithm 2: Cont'd — Whittle likelihood estimation of a two-way random effects model with serial correlation of the $ARMA(p,q)$ type

3. Third step.

a. $\widehat{\sigma}_{\nu}^2 = \widehat{\sigma}_e^2 \sum_{k=0}^{T-1} \widehat{\Theta}_{\nu,k}^2$ and $\widehat{\sigma}_{\lambda}^2 = \widehat{\sigma}_{\varepsilon}^2 \sum_{k=0}^{T-1} \widehat{\Theta}_{\lambda,k}^2$ with

$$\begin{cases} \widehat{\Theta}_{\nu,0} &= 1 \\ \widehat{\Theta}_{\nu,1} &= \widehat{\phi}_{\nu,1} - \widehat{\theta}_{\nu,1} \\ \widehat{\Theta}_{\nu,2} &= \widehat{\phi}_{\nu,1}\widehat{\Theta}_{\nu,1} + \widehat{\phi}_{\nu,2} - \widehat{\theta}_{\nu,2} \\ \vdots & \vdots \\ \widehat{\Theta}_{\nu,\tau} &= \sum_{j=1}^{p_\nu} \widehat{\phi}_{\nu,j}\widehat{\Theta}_{\nu,\tau-j} - \widehat{\theta}_{\nu,\tau}\,,\ \tau \le q_\nu \\ \widehat{\Theta}_{\nu,\tau} &= \sum_{j=1}^{p_\nu} \widehat{\phi}_{\nu,j}\widehat{\Theta}_{\nu,\tau-j}\,,\ \tau > \max(p_\nu - 1, q_\nu) \end{cases},\quad \begin{cases} \widehat{\Theta}_{\lambda,0} &= 1 \\ \widehat{\Theta}_{\lambda,1} &= \widehat{\phi}_{\lambda,1} - \widehat{\theta}_{\lambda,1} \\ \widehat{\Theta}_{\lambda,2} &= \widehat{\phi}_{\lambda,1}\widehat{\Theta}_{\lambda,1} + \widehat{\phi}_{\lambda,2} - \widehat{\theta}_{\lambda,2} \\ \vdots & \vdots \\ \widehat{\Theta}_{\lambda,\tau} &= \sum_{j=1}^{p_\lambda} \widehat{\phi}_{\lambda,j}\widehat{\Theta}_{\lambda,\tau-j} - \widehat{\theta}_{\lambda,\tau}\,, \\ & \tau \le q_\lambda \\ \widehat{\Theta}_{\lambda,\tau} &= \sum_{j=1}^{p_\lambda} \widehat{\phi}_{\lambda,j}\widehat{\Theta}_{\lambda,\tau-j}\,, \\ & \tau > \max(p_\lambda - 1, q_\lambda) \end{cases}$$

b. $\Gamma_\nu(\widehat{\Psi}_\nu)$: Toeplitz matrix obtained from $\widehat{\gamma}_{\nu,\tau}^* = \sum_{k=0}^{T-\tau-1} \widehat{\Theta}_{\nu,k+|\tau|}\widehat{\Theta}_{\nu,k}$ for $\tau = 0,\cdots,T-1$.

c. $\Gamma_\lambda(\widehat{\Psi}_\lambda)$: Toeplitz matrix obtained from $\widehat{\gamma}_{\lambda,\tau}^* = \sum_{k=0}^{T-\tau-1} \widehat{\Theta}_{\lambda,k+|\tau|}\widehat{\Theta}_{\lambda,k}$ for $\tau = 0,\cdots,T-1$.

d. $\Omega_u(\widehat{\Psi}) = \widehat{\sigma}_\mu^2(I_N \otimes J_T) + \widehat{\sigma}_\varepsilon^2(J_N \otimes \Gamma_\lambda(\widehat{\Psi}_\lambda)) + \widehat{\sigma}_e^2(I_N \otimes \Gamma_\nu(\widehat{\Psi}_\nu))$.

e. $\widehat{\beta} = \left(X'\Omega_u^{-1}(\widehat{\Psi})X\right)^{-1} X'\Omega_u^{-1}(\widehat{\Psi})y$ and $Var\left(\widehat{\beta}\right) = \left(X'\Omega_u^{-1}(\widehat{\Psi})X\right)^{-1}$.

12.4 Monte Carlo Simulations for the One-way and Two-way Error Components Models with Serial Correlation of the $ARMA(p,q)$ Type

12.4.1 Monte Carlo Simulations for the One-way Error Components Models with Serial Correlation of the $ARMA(p,q)$ Type

We consider the following one-way random effects (OW-RE) model with a time-varying exogenous covariate and remainder disturbances that follow an $ARMA(1,1)$:

$$\begin{aligned} y_{it} &= \beta_1 + \beta_2 X_{it} + u_{it},\ i = 1,\cdots,N,\ t = 1,\cdots,T \\ \text{with} \quad u_{it} &= \mu_i + \nu_{it}, \\ \nu_{it} &= \phi_{\nu,1}\nu_{i,t-1} + e_{it} - \theta_{\nu,1}e_{it-1}, \end{aligned}$$

where $\beta_1 = 5$, $\beta_2 = 1$. Following Nerlove (1971) p. 367, the variable X_{it} is generated by

$$X_{it} = 0.1t + 0.5X_{i,t-1} + w_{it},$$

where w_{it} is uniform on $[-0.5, 0.5]$ and $X_{i0} = 5 + 10w_{i0}$. $\mu_i \sim N(0,\sigma_\mu^2)$, $\nu_{it} \sim N(0,\sigma_\nu^2)$, $e_{it} \sim N(0,\sigma_e^2)$ and $\sigma_\nu^2 = \sigma_e^2 \frac{(1+\theta_{\nu,1}^2 - 2\phi_{\nu,1}\theta_{\nu,1})}{(1-\phi_{\nu,1}^2)}$. We vary the duration of the panel. We choose three

(N,T) pairs with $N = 200$ and $T = 200$, 100, 50 and the number of replications is 100. All the variables were generated over $T + T_0$ time periods and we drop the first $T_0 (= 20)$ observations to reduce the dependence on the initial values. In total we have 4 experiments for each (N,T) pair:[11]

Table 12.1: Parameters for the 4 experiments in the one-way error components model

	$\phi_{\nu,1}$	$\theta_{\nu,1}$	σ_μ^2	σ_ν^2
1	0.8	0.6	10	6
2	0.4	-0.7	10	6
3	0.8	0.6	10	4
4	0.4	-0.7	10	4

The choices of $\phi_{\nu,1}$ and $\theta_{\nu,1}$ parameters are guided by different dynamic responses summarized by the transfer function weights $\Theta_\nu(B) = (1 + \Theta_{\nu,1}B + \Theta_{\nu,2}B^2 + \cdots)$ in $\nu_{it} = \Theta_\nu(B)e_{it}$. Indeed, for experiment 1 (or 3), the $\Theta_{\nu,\tau}$ weights are initially small and decrease slightly: $\Theta_{\nu,1} = 0.2$, $\Theta_{\nu,2} = 0.16$, $\Theta_{\nu,5} = 0.082$, $\Theta_{\nu,10} = 0.027$, whereas for experiment 2 (or 4), the weights are initially large and decrease quickly: $\Theta_{\nu,1} = 1.1$, $\Theta_{\nu,2} = 0.44$, $\Theta_{\nu,5} = 0.028$, $\Theta_{\nu,10} = 0.0003$.

We compare the Monte Carlo performance with two estimators that ignore serial correlation: OLS (labelled `OLS`) and one-way FGLS (labelled `FGLS`) and two estimators that correct for serial correlation: the Whittle estimator (labelled `Whittle`) and the true one-way GLS (labelled `true GLS`)[12]. Estimates from the `true OW-GLS` allow us to compute the relative efficiency of any estimator $\widehat{\Psi}_{\nu,k}$ by $RMSE(\widehat{\Psi}_{\nu,k})/RMSE(\Psi_{\nu,k,true})$.[13] So, we give Tables of the relative RMSE of the intercept β_1 and the slope coefficient β_2 with respect to true GLS and Tables of the mean absolute percentage error (MAPE) of the parameters $(\frac{1}{rep}\sum_{j=1}^{rep} |\frac{\widehat{\Psi}_{\nu,k,j} - \Psi_{\nu,k,j,true}}{\Psi_{\nu,k,j,true}}|)$. MAPE measures the average magnitude of error in the estimation of the parameter. A MAPE value of 10% means that the average absolute percentage difference between the estimated parameter and the true GLS parameter is 10%. For large T ($T = 200$), there is no need to use a modified periodogram. On the other hand, for smaller T, ($T = 100$, $T = 50$), we use an apodized-windowed periodogram to reduce the leakage effect (see the supplementary material).[14,15]

Table 12.2 gives the relative RMSE with respect to true GLS for the intercept β_1 and the slope coefficient β_2 of 3 estimators for three (N,T) pairs ($N = 200$, $T = 200$, 100 and 50). When $T = 200$, `Whittle` provides the best relative RMSE followed by `FGLS` for both β_1 and β_2 for experiments 1, 3 and 4. Only in experiment 2 does `FGLS` give slightly better relative RMSE than `Whittle`, but the differences are very small. However, as soon as T decreases ($T = 100$ or $T = 50$), `Whittle`

[11] We fix σ_μ^2 at 10 greater than that of σ_ν^2 at 6 and 4 by analogy with the estimated variances on well-known applications such as those of Grunfeld (1958) for an investment equation, Baltagi and Griffin (1983) for a gasoline demand equation and Munnell (1990) for the productivity of public capital in the private sector (see also Baltagi, 2021).

[12] $\beta_{true,GLS} = \left(X'\Omega_u^{-1}(\Psi_\nu)X\right)^{-1} X'\Omega_u^{-1}(\Psi_\nu)y$ and $Var\left(\beta_{true,GLS}\right) = \left(X'\Omega_u^{-1}(\Psi_\nu)X\right)^{-1}$.

[13] Where $RMSE(\widehat{\Psi}_{\nu,k}) = \sqrt{(1/rep)\sum_{j=1}^{rep}(\widehat{\Psi}_{\nu,k,j} - \Psi_{\nu,k})^2}$ with $rep = 100$.

[14] The simulation study was conducted with a MacBook Pro featuring a 3.58 GHz Apple M3 Max chip, a 16-core CPU, 48GB of unified memory and a 1TB SSD.

[15] We have used the Matlab `fminsearch` function based on the Nelder-Mead simplex direct search algorithm. The code was executed within our `R` code using the `matlabr` library, an interface that allows system call to Matlab. We would like to thank Honglei Wei for providing us with part of his Matlab code for the application developed in Wei et al. (2022).

unquestionably delivers better relative RMSEs except for experiment 1 when $T = 100$, where again the differences are very small. When T is small ($T = 50$) and for all experiments, `Whittle` always gives the best relative RMSE. As the values of the relative RMSEs are very close to unity meaning that the Whittle MLE does almost as well as the true GLS, even when T is small. Moreover, the non linear optimization Algorithm 1 is not that computationally intensive even for large T (see Table 12.2).[16]

Table 12.3 gives the mean absolute percentage error (MAPE) of all the parameters for the 4 experiments. The same rankings with MAPE in Table 12.3 emerge as in the relative RMSE rankings in Table 12.2. MAPE of β_1 and β_2 are always lower for `Whittle` than for `FGLS` whatever the sample size T. MAPE of σ_μ^2 for `Whittle` and `FGLS` are small and very close to each other for all experiments and all T. On the other hand, the gap in the MAPEs of σ_ν^2 increases to the detriment of `Whittle` (relative to `FGLS`) as T decreases in all experiments. The coefficients $\phi_{\nu,1}$ and $\theta_{\nu,1}$ for serial correlation have very small MAPEs for experiments 1 and 3 but increase slightly as T decreases. On the other hand, for experiments 2 and 4, MAPEs of the order of 15% are observed for $\phi_{\nu,1}$, while they are much lower (5%) for $\theta_{\nu,1}$ whatever the size of T. Fortunately, the MAPEs for the residual variance σ_e^2 of the $ARMA(1,1)$ process are very small for all experiments and for all T and will neutralize the bias observed for σ_ν^2 since this is σ_e^2 and σ_μ^2 associated with the coefficients $\phi_{\nu,1}$ and $\theta_{\nu,1}$ that define the variance-covariance matrix $\Omega_u(\widehat{\Psi})$. Furthermore, there is no change when the weight of the residual effect variance (σ_ν^2) in the total variance (σ_u^2) is $37.5\%(=6/16)$ or $28.5\%(=4/14)$. But the main lesson to be learned from this Table 12.3 concerns the MAPEs of the standard errors se_{β_1}, se_{β_2} of the coefficients for `Whittle`. We can see that they are significantly smaller than those for `FGLS` (not to mention those for `OLS`) for all experiments and all sample sizes. For example, `Whittle`'s MAPEs of se_{β_2} are 2 to 9 times smaller than those of `FGLS`. This is an extremely clear indication of the better estimation of the variances of β_1 and β_2 using the Whittle MLE. This allows better inference for these parameters. The very good results in terms of standard errors underline the fact that, even if there are biases in the estimation of σ_ν^2 with `Whittle`, these are largely compensated for by taking into account the $ARMA$ structure of the variance-covariance matrix $\Omega_u(\widehat{\Psi})$ when calculating the variances of β_1 and β_2. These MAPEs confirm the very good estimation of the coefficients and their variances for the one-way error components model using Whittle's approximate maximum likelihood method in the presence of serial correlation of the $ARMA(p,q)$ type. Finally, when T is small, the use of an apodized-windowed periodogram, by reducing the leakage effect, seems to provide quite satisfactory results.

[16] In the case of $(N = 200, T = 200)$ with 4×10^4 elements in the (NT, NT) matrices, it takes around 215 seconds per replication for one experiment. This computing time decreases quickly as T decreases. For $T = 100$ (respectively $T = 50$), with 2×10^4 (respectively 10^4) elements in the (NT, NT) matrices, it takes 30 seconds (respectively 18 seconds) per replication for one experiment, and all this after setting the maximum number of iterations to 10^8 in the Nelder-Mead simplex direct search algorithm.

Table 12.2: Relative RMSE with respect to true GLS for β_1 and β_2 - $N = 200$, $T = 200$, $T = 100$, $T = 50$, 100 replications. One-way error components model with serial correlation of the $ARMA(1,1)$ type.

	OLS	FGLS	Whittle	OLS	FGLS	Whittle	OLS	FGLS	Whittle
experiment 1 : $\phi_{\nu,1} = 0.8$, $\theta_{\nu,1} = 0.6$, $\sigma_\mu^2 = 10$, $\sigma_\nu^2 = 6$.									
	$T = 200$			$T = 100$			$T = 50$		
β_1	1.0007	1.0050	1.0002	1.0122	1.0012	1.0048	1.0269	1.0446	1.0151
β_2	1.0030	1.0028	1.0004	1.0085	1.0025	1.0032	1.0269	1.0447	1.0145
experiment 2 : $\phi_{\nu,1} = 0.4$, $\theta_{\nu,1} = -0.7$, $\sigma_\mu^2 = 10$, $\sigma_\nu^2 = 6$.									
	$T = 200$			$T = 100$			$T = 50$		
β_1	1.0057	1.0002	1.0030	1.0057	1.0057	1.0031	1.0212	1.0414	1.0103
β_2	1.0059	1.0001	1.0028	1.0012	1.0101	1.0008	1.0236	1.0441	1.0113
experiment 3 : $\phi_{\nu,1} = 0.8$, $\theta_{\nu,1} = 0.6$, $\sigma_\mu^2 = 10$, $\sigma_\nu^2 = 4$.									
	$T = 200$			$T = 100$			$T = 50$		
β_1	1.0007	1.0064	1.0001	1.0107	1.0046	1.0046	1.0209	1.0304	1.0128
β_2	1.0026	1.0046	1.0004	1.0076	1.0037	1.0032	1.0209	1.0305	1.0124
experiment 4 : $\phi_{\nu,1} = 0.4$, $\theta_{\nu,1} = -0.7$, $\sigma_\mu^2 = 10$, $\sigma_\nu^2 = 4$.									
	$T = 200$			$T = 100$			$T = 50$		
β_1	1.0042	1.0032	1.0023	1.0034	1.0024	1.0018	1.0135	1.0273	1.0069
β_2	1.0046	1.0028	1.0023	1.0002	1.0159	1.0001	1.0155	1.0295	1.0078

`OLS`: OLS, `FGLS`: one-way FGLS, `Whittle`: Whittle MLE.

Computing time per replication for one experiment:

$(N = 200, T = 200)$: 214.88 sec.

$(N = 200, T = 100)$: 30.04 sec.

$(N = 200, T = 50)$: 17.94 sec.

Table 12.3: Mean absolute percentage error (MAPE), $N = 200$, $T = 200$, $T = 100$, $T = 50$, 100 replications. One-way error components model with serial correlation of the $ARMA(1,1)$ type.

	OLS	FGLS	Whittle	OLS	FGLS	Whittle	OLS	FGLS	Whittle
experiment 1 : $\phi_{\nu,1} = 0.8,\ \theta_{\nu,1} = 0.6,\ \sigma_\mu^2 = 10,\ \sigma_\nu^2 = 6.$									
	$T = 200$			$T = 100$			$T = 50$		
β_1	0.0004	0.0006	0.0001	0.0008	0.0013	0.0003	0.0017	0.0040	0.0010
β_2	0.0001	0.0001	$< 10^{-4}$	0.0004	0.0006	0.0001	0.0016	0.0040	0.0010
σ_μ^2		0.0144	0.0131		0.0215	0.0253		0.0351	0.0449
$\phi_{\nu,1}$			0.0374			0.0783			0.1804
$\theta_{\nu,1}$			0.0443			0.0935			0.2148
σ_ν^2		0.0129	0.0178		0.0248	0.0569		0.0482	0.1059
σ_e^2			0.0043			0.0317			0.0628
se_{β_1}	0.3034	2.9268	0.0781	0.2917	1.8113	0.1120	0.2625	1.0225	0.1460
se_{β_2}	0.3005	0.5736	0.0771	0.2853	0.5670	0.1085	0.2451	0.5483	0.1331
experiment 2 : $\phi_{\nu,1} = 0.4,\ \theta_{\nu,1} = -0.7,\ \sigma_\mu^2 = 10,\ \sigma_\nu^2 = 6.$									
	$T = 200$			$T = 100$			$T = 50$		
β_1	0.0003	0.0004	0.0001	0.0006	0.0013	0.0003	0.0015	0.0042	0.0008
β_2	0.0001	0.0001	$< 10^{-4}$	0.0003	0.0006	0.0002	0.0017	0.0042	0.0009
σ_μ^2		0.0138	0.0123		0.0206	0.0242		0.0342	0.0438
$\phi_{\nu,1}$			0.1646			0.1524			0.1408
$\theta_{\nu,1}$			0.0494			0.0560			0.0627
σ_ν^2		0.0116	0.1130		0.0228	0.1256		0.0466	0.1306
σ_e^2			0.0185			0.0078			0.0124
se_{β_1}	0.4730	1.9695	0.3325	0.4698	1.1035	0.3308	0.4582	0.4851	0.3204
se_{β_2}	0.4720	0.6779	0.3319	0.4670	0.6768	0.3291	0.4487	0.6698	0.3144

OLS: OLS, FGLS: one-way FGLS, Whittle: Whittle MLE.

Table 12.3: Cont'd — Mean absolute percentage error (MAPE), $N = 200$, $T = 200$, $T = 100$, $T = 50$, 100 replications. One-way error components model with serial correlation of the $ARMA(1,1)$ type.

	OLS	FGLS	Whittle	OLS	FGLS	Whittle	OLS	FGLS	Whittle
experiment 3 : $\phi_{\nu,1} = 0.8$, $\theta_{\nu,1} = 0.6$, $\sigma_\mu^2 = 10$, $\sigma_\nu^2 = 4$.									
	$T = 200$			$T = 100$			$T = 50$		
β_1	0.0003	0.0004	0.0001	0.0005	0.0012	0.0002	0.0011	0.0038	0.0007
β_2	0.0001	0.0001	$< 10^{-4}$	0.0002	0.0006	0.0001	0.0011	0.0037	0.0006
σ_μ^2		0.0120	0.0100		0.0169	0.0186		0.0257	0.0314
$\phi_{\nu,1}$			0.0374			0.0786			0.1811
$\theta_{\nu,1}$			0.0443			0.0943			0.2163
σ_ν^2		0.0129	0.0178		0.0248	0.0562		0.0482	0.1046
σ_e^2			0.0043			0.0312			0.0616
se_{β_1}	0.2569	3.4650	0.0676	0.2463	2.1786	0.0960	0.2203	1.2574	0.1238
se_{β_2}	0.2544	0.6031	0.0667	0.2406	0.5984	0.0929	0.2053	0.5848	0.1127
experiment 4 : $\phi_{\nu,1} = 0.4$, $\theta_{\nu,1} = -0.7$, $\sigma_\mu^2 = 10$, $\sigma_\nu^2 = 4$.									
	$T = 200$			$T = 100$			$T = 50$		
β_1	0.0002	0.0004	0.0001	0.0004	0.0012	0.0002	0.0011	0.0040	0.0006
β_2	0.0001	0.0001	$< 10^{-4}$	0.0002	0.0006	0.0001	0.0012	0.0040	0.0006
σ_μ^2		0.0115	0.0094		0.0163	0.0178		0.0251	0.0307
$\phi_{\nu,1}$			0.1647			0.1517			0.1385
$\theta_{\nu,1}$			0.0493			0.0564			0.0641
σ_ν^2		0.0116	0.1130		0.0228	0.1248		0.0466	0.1284
σ_e^2			0.0185			0.0080			0.0120
se_{β_1}	0.4212	2.4768	0.3002	0.4181	1.4533	0.2982	0.4075	0.7149	0.2883
se_{β_2}	0.4203	0.6912	0.2996	0.4156	0.6906	0.2966	0.3987	0.6856	0.2826

OLS: OLS, FGLS: one-way FGLS, Whittle: Whittle MLE.

12.4.2 Monte Carlo Simulations for the Two-way Error Components Models with Serial Correlation of the $ARMA(p,q)$ Type

We extend the DGP of Section 12.4.1 to the case of a two-way random effects (TW-RE) model with serial correlation of the $ARMA(1,1)$ type in both $\{\lambda_t\}$ and $\{\nu_{it}\}$:

$$
\begin{aligned}
y_{it} &= \beta_1 + \beta_2 X_{it} + u_{it},\ i = 1, \cdots, N,\ t = 1, \cdots, T \\
\text{with} \quad u_{it} &= \mu_i + \lambda_t + \nu_{it}, \\
\nu_{it} &= \phi_{\nu,1} \nu_{i,t-1} + e_{it} - \theta_{\nu,1} e_{it-1}, \\
\lambda_t &= \phi_{\lambda,1} \lambda_{t-1} + \varepsilon_t - \theta_{\lambda,1} \varepsilon_{t-1},
\end{aligned}
$$

where $\lambda_t \sim N(0, \sigma_\lambda^2)$, $\varepsilon_{it} \sim N(0, \sigma_\varepsilon^2)$ and $\sigma_\lambda^2 = \sigma_\varepsilon^2 \frac{(1+\theta_{\lambda,1}^2 - 2\phi_{\lambda,1}\theta_{\lambda,1})}{(1-\phi_{\lambda,1}^2)}$. For each (N, T) pair, we have 4 experiments :

Table 12.4: Parameters for the 4 experiments in the two-way error components model

	$\phi_{\nu,1}$	$\theta_{\nu,1}$	$\phi_{\lambda,1}$	$\theta_{\lambda,1}$	σ_μ^2	σ_ν^2	σ_λ^2
1	0.8	0.6	0.9	0.7	10	6	4
2	0.4	-0.7	0.2	-0.8	10	6	4
3	0.8	0.6	0.9	0.7	10	4	6
4	0.4	-0.7	0.2	-0.8	10	4	6

Here too, the values chosen for $\phi_{\lambda,1}$ and $\theta_{\lambda,1}$ allow us to have different dynamic responses summarized by the transfer function weights $\Theta_\lambda(B) = (1 + \Theta_{\lambda,1} B + \Theta_{\lambda,2} B^2 + \cdots)$ in $\lambda_t = \Theta_\lambda(B)\varepsilon_t$. Indeed, for experiment 1 (or 3), the $\Theta_{\lambda,\tau}$ weights are initially small and decrease slightly: $\Theta_{\lambda,1} = 0.2$, $\Theta_{\lambda,2} = 0.18$, $\Theta_{\lambda,5} = 0.1312$, $\Theta_{\lambda,10} = 0.0775$, whereas for experiment 2 (or 4), the weights are initially large and decrease quickly: $\Theta_{\lambda,1} = 1$, $\Theta_{\lambda,2} = 0.2$, $\Theta_{\lambda,5} = 0.0016$, $\Theta_{\lambda,10} = 5.10^{-7}$.

We compare the Monte Carlo performance with two estimators that ignore serial correlation: OLS (labelled `OLS`) and two-way FGLS (labelled `FGLS`) and two estimators that allow for serial correlation: the Whittle estimator (labelled `Whittle`) and the true two-way GLS (labelled `true GLS`). As in Section 12.4.1, we provide Tables of relative RMSEs of the intercept β_1 and the slope coefficient β_2 with respect to true GLS and Tables of MAPEs of all the parameters.

Table 12.5 gives the relative RMSE with respect to true two-way GLS for the intercept β_1 and the slope coefficient β_2 for three estimators and three (N, T) pairs ($N = 200$, $T = 200$, 100 and 50). This time, `Whittle` provides the best RMSEs for β_1 and β_2 for all T and all experiments. Whittle's ML estimate is clearly superior to the others and almost identical to the true two-way GLS, given that the relative RMSE values are very close to unity. Table 12.6 gives the mean absolute percentage error (MAPE) of all the parameters for the 4 experiments. The same rankings with MAPE in Table 12.6 emerge as in the relative RMSE rankings in Table 12.5. MAPE of β_1 and β_2 are always lower for `Whittle` than for `FGLS` whatever the sample size T and MAPE of σ_μ^2 for `Whittle` and `FGLS` are small and very close to each other for all T and all experiments.

As in Table 12.3, the MAPE of σ_ν^2 is slightly greater for `Whittle` than for `FGLS`. The coefficients $\phi_{\nu,1}$ and $\theta_{\nu,1}$ of the serial correlation have MAPEs less than 20% for all experiments but increase slightly as T decreases. Fortunately, and as in Table 12.3, the MAPEs for the residual variance σ_e^2 of the $ARMA(1,1)$ process are very small for all experiments and for all T and will neutralize the bias observed for σ_ν^2 since this is σ_e^2 and σ_μ^2 associated with the coefficients $\phi_{\nu,1}$ and $\theta_{\nu,1}$ that define part of the variance-covariance matrix $\Omega_u(\widehat{\Psi})$. On the other hand, for the $ARMA(1,1)$ coefficients associated with the serial correlation of time effects, we note higher MAPEs than those associated with the serial correlation of the remainder error. While they are quite acceptable for $T = 200$ or $T = 100$, they are worse for $T = 50$. However, this depends on the experiment. For experiments 1 and 3, both $\phi_{\lambda,1}$ and $\theta_{\lambda,1}$ MAPEs are large (> 30%), while for experiments 2 and 4, only $\theta_{\lambda,1}$ MAPEs are large. These biases can be explained by the small time dimension and therefore the small number of observations available in step 2 of Algorithm 2 with the between-time transformation. Here again, the MAPEs for the residual variance σ_ε^2 of the $ARMA(1,1)$ process are small for all experiments and for all T (less than 15%). This holds also for σ_{λ}^{2} and will neutralize the impact of biases observed for $\phi_{\lambda,1}$ and $\theta_{\lambda,1}$ in the computation of the variance-covariance matrix $\Omega_u(\widehat{\Psi})$. As in Table 12.3, and as for the one-way error components model, one of the main lessons to be learned from this Table 12.6 concerns the MAPEs of the standard errors se_{β_1}, se_{β_2} of the coefficients for `Whittle`. We can see that they are significantly smaller than those for `FGLS` (not to mention those for `OLS`) for all experiments and all sample sizes. One more time, this is an extremely clear indication of the better estimation of the variances of β_1 and β_2 using the Whittle MLE and this allows better inference for these parameters.

Table 12.5: Relative RMSE with respect to true GLS for β_1 and β_2 - $N = 200$, $T = 200$, $T = 100$, $T = 50$, 100 replications. Two-way error components model with serial correlation of the $ARMA(1,1)$ type.

	OLS	FGLS	Whittle	OLS	FGLS	Whittle	OLS	FGLS	Whittle
exp. 1: $\phi_{\nu,1} = 0.8$, $\theta_{\nu,1} = 0.6$, $\phi_{\lambda,1} = 0.9$, $\theta_{\lambda,1} = 0.7$, $\sigma_\mu^2 = 10$, $\sigma_\nu^2 = 6$, $\sigma_\lambda^2 = 4$.									
	$T = 200$			$T = 100$			$T = 50$		
β_1	1.3744	1.2174	1.0162	1.8951	1.2820	1.0338	2.4711	1.1393	1.0284
β_2	1.4760	1.2875	1.0489	2.5464	1.5285	1.0790	5.0683	1.3608	1.0524
exp. 2: $\phi_{\nu,1} = 0.4$, $\theta_{\nu,1} = -0.7$, $\phi_{\lambda,1} = 0.2$, $\theta_{\lambda,1} = -0.8$, $\sigma_\mu^2 = 10$, $\sigma_\nu^2 = 6$, $\sigma_\lambda^2 = 4$.									
	$T = 200$			$T = 100$			$T = 50$		
β_1	1.5411	1.3757	1.0097	2.7343	1.8806	1.0502	4.4312	1.5474	1.0293
β_2	1.5423	1.3767	1.0100	2.7850	1.8989	1.0537	4.8312	1.6023	1.0456
exp. 3: $\phi_{\nu,1} = 0.8$, $\theta_{\nu,1} = 0.6$, $\phi_{\lambda,1} = 0.9$, $\theta_{\lambda,1} = 0.7$, $\sigma_\mu^2 = 10$, $\sigma_\nu^2 = 4$, $\sigma_\lambda^2 = 6$.									
	$T = 200$			$T = 100$			$T = 50$		
β_1	1.7483	1.3687	1.0246	2.2910	1.2166	1.0178	2.6708	1.0492	1.0017
β_2	1.9362	1.4640	1.0577	3.6434	1.5234	1.0521	7.5791	1.2455	1.0259
exp. 4: $\phi_{\nu,1} = 0.4$, $\theta_{\nu,1} = -0.7$, $\phi_{\lambda,1} = 0.2$, $\theta_{\lambda,1} = -0.8$, $\sigma_\mu^2 = 10$, $\sigma_\nu^2 = 4$, $\sigma_\lambda^2 = 6$.									
	$T = 200$			$T = 100$			$T = 50$		
β_1	2.0024	1.5888	1.0090	3.8306	2.0512	1.0378	5.9013	1.4400	1.0247
β_2	2.0059	1.5911	1.0086	4.0136	2.1174	1.0453	7.1761	1.5863	1.0460

`OLS`: OLS, `FGLS`: two-way FGLS, `Whittle`: Whittle MLE.

Computing time per replication for one experiment:

$(N = 200, T = 200)$: 252.81 sec.

$(N = 200, T = 100)$: 37.55 sec.

$(N = 200, T = 50)$: 22.99 sec.

Table 12.6: Mean absolute percentage error (MAPE), $N = 200$, $T = 200$, $T = 100$, $T = 50$, 100 replications. Two-way error components model with serial correlation of the $ARMA(1,1)$ type.

	OLS	FGLS	Whittle	OLS	FGLS	Whittle	OLS	FGLS	Whittle
exp. 1: $\phi_{\nu,1} = 0.8$, $\theta_{\nu,1} = 0.6$, $\phi_{\lambda,1} = 0.9$, $\theta_{\lambda,1} = 0.7$, $\sigma_\mu^2 = 10$, $\sigma_\nu^2 = 6$, $\sigma_\lambda^2 = 4$.									
	$T = 200$			$T = 100$			$T = 50$		
β_1	0.0722	0.0538	0.0170	0.1252	0.0615	0.0163	0.1900	0.0361	0.0186
β_2	0.0178	0.0133	0.0040	0.0629	0.0300	0.0069	0.1826	0.0297	0.0108
σ_μ^2		0.0156	0.0142		0.0240	0.0259		0.0350	0.0426
$\phi_{\nu,1}$			0.0369			0.0774			0.1806
$\theta_{\nu,1}$			0.0423			0.0917			0.2145
$\phi_{\lambda,1}$			0.0724			0.1460			0.3589
$\theta_{\lambda,1}$			0.1332			0.2638			0.5170
σ_ν^2		0.0127	0.0226		0.0247	0.0390		0.0472	0.0704
σ_λ^2		0.0421	0.0668		0.0580	0.1115		0.1025	0.0767
σ_e^2			0.0077			0.0360			0.0660
σ_ε^2			0.0835			0.1068			0.1041
se_{β_1}	0.9284	0.4426	0.1528	0.9045	0.4137	0.1868	0.8771	0.4662	0.2926
se_{β_2}	0.9128	0.4922	0.1214	0.8455	0.2952	0.0654	0.6483	0.1414	0.0764
exp. 2: $\phi_{\nu,1} = 0.4$, $\theta_{\nu,1} = -0.7$, $\phi_{\lambda,1} = 0.2$, $\theta_{\lambda,1} = -0.8$, $\sigma_\mu^2 = 10$, $\sigma_\nu^2 = 6$, $\sigma_\lambda^2 = 4$.									
	$T = 200$			$T = 100$			$T = 50$		
β_1	0.0442	0.0348	0.0058	0.0739	0.0476	0.0070	0.1116	0.0336	0.0049
β_2	0.0110	0.0086	0.0014	0.0363	0.0229	0.0034	0.1102	0.0329	0.0046
σ_μ^2		0.0149	0.0135		0.0230	0.0248		0.0342	0.0418
$\phi_{\nu,1}$			0.1641			0.1568			0.1451
$\theta_{\nu,1}$			0.0493			0.0535			0.0602
$\phi_{\lambda,1}$			0.2912			0.5258			0.6573
$\theta_{\lambda,1}$			0.0797			0.1344			0.1523
σ_ν^2		0.0114	0.0213		0.0229	0.0372		0.0455	0.0686
σ_λ^2		0.0210	0.0324		0.0322	0.0954		0.0487	0.0541
σ_e^2			0.0149			0.0075			0.0147
σ_ε^2			0.0869			0.1252			0.1603
se_{β_1}	0.8930	0.1633	0.0275	0.8586	0.1266	0.0568	0.8202	0.2107	0.0706
se_{β_2}	0.8581	0.1662	0.0340	0.7640	0.0853	0.0354	0.5121	0.1925	0.0194

Table 12.6: Cont'd — Mean absolute percentage error (MAPE), $N = 200$, $T = 200$, $T = 100$, $T = 50$, 100 replications. Two-way error components model with serial correlation of the $ARMA(1,1)$ type.

	OLS	FGLS	Whittle	OLS	FGLS	Whittle	OLS	FGLS	Whittle
exp. 3: $\phi_{\nu,1} = 0.8$, $\theta_{\nu,1} = 0.6$, $\phi_{\lambda,1} = 0.9$, $\theta_{\lambda,1} = 0.7$, $\sigma_\mu^2 = 10$, $\sigma_\nu^2 = 4$, $\sigma_\lambda^2 = 6$.									
	$T = 200$			$T = 100$			$T = 50$		
β_1	0.1090	0.0692	0.0135	0.1659	0.0557	0.0146	0.2391	0.0345	0.0202
β_2	0.0269	0.0171	0.0033	0.0833	0.0265	0.0048	0.2290	0.0206	0.0074
σ_μ^2		0.0129	0.0108		0.0190	0.0189		0.0262	0.0300
$\phi_{\nu,1}$			0.0369			0.0775			0.1810
$\theta_{\nu,1}$			0.0423			0.0923			0.2156
$\phi_{\lambda,1}$			0.0716			0.1539			0.3567
$\theta_{\lambda,1}$			0.1317			0.2661			0.5001
σ_ν^2		0.0127	0.0226		0.0247	0.0389		0.0472	0.0702
σ_λ^2		0.0390	0.0408		0.0564	0.0781		0.0995	0.0574
σ_e^2			0.0077			0.0354			0.0647
σ_ε^2			0.0793			0.1023			0.0946
se_{β_1}	0.9333	0.4315	0.1386	0.9138	0.4433	0.2053	0.8957	0.5112	0.3179
se_{β_2}	0.9101	0.4083	0.0780	0.8219	0.2069	0.0449	0.5782	0.1151	0.0620
exp. 4: $\phi_{\nu,1} = 0.4$, $\theta_{\nu,1} = -0.7$, $\phi_{\lambda,1} = 0.2$, $\theta_{\lambda,1} = -0.8$, $\sigma_\mu^2 = 10$, $\sigma_\nu^2 = 4$, $\sigma_\lambda^2 = 6$.									
	$T = 200$			$T = 100$			$T = 50$		
β_1	0.0640	0.0452	0.0063	0.0949	0.0488	0.0063	0.1376	0.0280	0.0045
β_2	0.0159	0.0112	0.0016	0.0467	0.0235	0.0028	0.1353	0.0270	0.0036
σ_μ^2		0.0124	0.0103		0.0183	0.0183		0.0257	0.0293
$\phi_{\nu,1}$			0.1640			0.1549			0.1430
$\theta_{\nu,1}$			0.0494			0.0546			0.0615
$\phi_{\lambda,1}$			0.2798			0.5672			0.6845
$\theta_{\lambda,1}$			0.0769			0.1448			0.1554
σ_ν^2		0.0114	0.0213		0.0229	0.0371		0.0455	0.0685
σ_λ^2		0.0168	0.0198		0.0278	0.0634		0.0425	0.0382
σ_e^2			0.0149			0.0075			0.0137
σ_ε^2			0.0823			0.1192			0.1553
se_{β_1}	0.8983	0.1309	0.0224	0.8692	0.1525	0.0763	0.8437	0.2590	0.0863
se_{β_2}	0.8523	0.0218	0.0379	0.7263	0.2209	0.0406	0.4113	0.2285	0.0233

12.5 Conclusion

In this chapter, we proposed the use of the Whittle MLE to account for serial correlation of the $ARMA(p,q)$ type in one-way and two-way error components models. This spectral method, rarely used in panel data econometrics, allows us to take into account the complex structures of the variance-covariance matrices of the residuals. Whether in the one-way case with $ARMA(p,q)$ correlation on the remainder error or in the two-way case with $ARMA(p,q)$ correlation on both the remainder error and the time effects, the Monte Carlo simulations demonstrate the great adaptability and the good estimates obtained with this method. Admittedly, this method is known in time series to work very well in the asymptotic case ($T \to \infty$), but poses problems of leakage effect for small samples. To remedy this problem in the context of panel data where the time dimension is often small, we used the apodization approach, defining a data and frequency dependent temporal window, which mitigates the leakage problem of the periodogram without compromising its resolution. Here again, the results are satisfactory, demonstrating the value of the Whittle MLE for handling serial correlation of the $ARMA(p,q)$ type. To our knowledge, this is the first time that the Whittle MLE method is used in conjunction with the apodization approach for periodogram in panel data econometrics. Of course, throughout this chapter, we have imposed a specific structure on serial correlation for both the one-way and two-way error components models. But we can also imagine a more general arbitrary serial correlation. In that case, an extension of the Whittle MLE to a penalized Whittle MLE approach[17] in the panel data framework can be envisaged. This will be the subject of future research.

Appendix

The Whittle Likelihood Estimator (WLE)

For individual i, ν_i^* is a $(T \times 1)$ time series which follows an $ARMA(p_\nu, q_\nu)$ process with a $(p_\nu + q_\nu)$-dimensional vector of parameters Ψ_ν and a $(T \times T)$ variance-covariance matrix $\Omega_\nu(\Psi_\nu) = \sigma_e^2 \Gamma_\nu(\Psi_\nu)$. Its log-likelihood is given by

$$\ln L(\Psi_\nu) = -\frac{T}{2} \ln 2\pi - \frac{1}{2} \ln |\Omega_\nu(\Psi_\nu)| - \frac{1}{2} \nu_i^{*'} \Omega_\nu^{-1}(\Psi_\nu) \nu_i^*, \tag{12.23}$$

We assume that ν_i^* has a spectral density function $f(\omega_m, \Psi_\nu) = \sigma_e^2 f_*(\omega_m, \Psi_\nu)$ where $f_*(\omega_m, \Psi_\nu)$ is the standardized spectral density function with $\omega_m = 2\pi m/T$, $m = 0, \cdots T-1$. $\Omega_\nu(\Psi_\nu)$ is a symmetric Toeplitz matrix and it has been shown (see Beran, 1994, Hurvich & Ray, 2003, Golub & Van Loan, 2013) that, for large T, all $(T \times T)$ symmetric Toeplitz matrices have complex orthonormal eigenvectors which can be approximated by

$$V_m = \frac{1}{\sqrt{T}} \left\{ \exp(-j\omega_m t) \right\}_{t=1}^{T},$$

and the corresponding eigenvalues of $\Omega_\nu(\Psi_\nu)$ are well approximated by $2\pi f(\omega_m, \Psi_\nu)$. If $V = (V_0, \cdots, V_{T-1})$ and Λ is a $(T \times T)$ diagonal matrix with $\{2\pi f(\omega_m, \Psi_\nu)\}_{m=0}^{T-1}$ on the main diagonal and zero elsewhere, then $\Omega_\nu(\Psi_\nu) \approx V \Lambda \bar{V}'$ and $\Omega_\nu^{-1}(\Psi_\nu) \approx V \Lambda^{-1} \bar{V}'$ where $\bar{V}'$ is the conjugate transpose of V.[18] The log-likelihood (12.23) can be written as

[17] See for instance Pawitan and O'Sullivan (1994), Guo, Dai, Ombao and Von Sachs (2003) or Krafty and Collinge (2013) in the time series litterature.

[18] V is a unitary matrix: $V\bar{V}' = \bar{V}'V = I_T$ and $|V| = 1$.

$$\begin{aligned}\ln L(\Psi_\nu) &\approx -\frac{T}{2}\ln 2\pi - \frac{1}{2}\ln|V\Lambda\bar{V}'| - \frac{1}{2}\nu_i^{*'}V\Lambda^{-1}\bar{V}'\nu_i^*,\\ &\approx -\frac{T}{2}\ln 2\pi - \frac{1}{2}\ln|\Lambda| - \frac{1}{2}\left(\nu_i^{*'}V\Lambda^{-1/2}\right)\overline{\left(\nu_i^{*'}V\Lambda^{-1/2}\right)}.\end{aligned}$$

The m-th entry of $\nu_i^{*'}V$ is

$$\nu_i^{*'}V_m = \frac{1}{\sqrt{T}}\sum_{t=1}^{T}\nu_{it}^* e^{-j\omega_m t} = \sqrt{T}D_{i,m},$$

where $D_{i,m}$ is the discrete Fourier transform and the m-th entry of $\nu_i^{*'}V\Lambda^{-1/2}$ is

$$\nu_i^{*'}V_m\Lambda_m^{-1/2} = \frac{\sqrt{T}D_{i,m}}{\sqrt{2\pi f(\omega_m,\Psi_\nu)}}.$$

The log-likelihood is therefore written as

$$\begin{aligned}\ln L(\Psi_\nu) &\approx -\frac{T}{2}\ln 2\pi - \frac{1}{2}\ln\left(\prod_{m=0}^{T-1}2\pi f(\omega_m,\Psi_\nu)\right) - \frac{1}{2}\sum_{m=0}^{T-1}\frac{T}{2\pi f(\omega_m,\Psi_\nu)}|D_{i,m}|^2,\\ &\approx -T\ln 2\pi - \frac{1}{2}\sum_{m=0}^{T-1}\left[\ln f(\omega_m,\Psi_\nu) + \frac{I(\omega_m,\nu_i^*)}{f(\omega_m,\Psi_\nu)}\right].\end{aligned}$$

with $I(\omega_m,\nu_i^*) = \frac{1}{2\pi T}|\sum_{t=1}^{T}\nu_{it}^*e^{-j\omega_m t}|^2$. The value of Ψ_ν which minimizes the righthand side of $-\ln L(\Psi_\nu)$ is called the Whittle estimator of Ψ_ν.

Inverse of the Variance-covariance Matrix $\Omega_u(\Psi)$

Let

$$\begin{aligned}A &= \sigma_\mu^2(I_N\otimes J_T) + \sigma_e^2(I_N\otimes\Gamma_\nu(\Psi_\nu)),\\ &= I_N\otimes\left[\sigma_\mu^2 J_T + \sigma_e^2\Gamma_\nu(\Psi_\nu)\right] = I_N\otimes A^\star \rightarrow A^{-1} = I_N\otimes A^{\star^{-1}},\end{aligned}$$

then

$$\begin{aligned}\Omega_u(\Psi) &= \sigma_\mu^2(I_N\otimes J_T) + \sigma_\varepsilon^2(J_N\otimes\Gamma_\lambda(\Psi_\lambda)) + \sigma_e^2(I_N\otimes\Gamma_\nu(\Psi_\nu)),\\ &= A + \sigma_\varepsilon^2(\iota_N\otimes I_T)\Gamma_\lambda(\Psi_\lambda)(\iota_N'\otimes I_T).\end{aligned}$$

Using the Woodbury matrix identity,[19] we get

$$\begin{aligned}\Omega_u^{-1}(\Psi) &= A^{-1} - A^{-1}(\iota_N\otimes I_T)B^{-1}(\iota_N'\otimes I_T)A^{-1}\\ &\quad\text{with } B^{-1} = \left[(\iota_N'\otimes I_T)A^{-1}(\iota_N\otimes I_T) + \sigma_\varepsilon^{-2}\Gamma_\lambda^{-1}(\Psi_\lambda)\right]^{-1}\\ &= A^{-1} - A^{-1}(\iota_N\otimes I_T)\left[\sigma_\varepsilon^{-2}\Gamma_\lambda^{-1}(\Psi_\lambda) + NA^{\star^{-1}}\right]^{-1}(\iota_N'\otimes I_T)A^{-1}\\ &= I_N\otimes A^{\star^{-1}} - \left(\iota_N\otimes A^{\star^{-1}}\right)\left[\sigma_\varepsilon^{-2}\Gamma_\lambda^{-1}(\Psi_\lambda) + NA^{\star^{-1}}\right]^{-1}\left(\iota_N'\otimes A^{\star^{-1}}\right).\end{aligned}$$

[19] $(D+EFE')^{-1} = D^{-1} - D^{-1}E(E'D^{-1}E+F^{-1})^{-1}E'D^{-1}$.

with $A^{\star^{-1}} = \left[\sigma_\mu^2 J_T + \sigma_e^2 \Gamma_\nu(\Psi_\nu)\right]^{-1}$.

References

Balestra, P. & Nerlove, M. (1966). Pooling cross section and time series data in the estimation of a dynamic model: The demand for natural gas. *Econometrica, 34*(3), 585–612.

Baltagi, B. H. (2021). *Econometric Analysis of Panel Data* (Sixth ed.). Springer Nature.

Baltagi, B. H., Bresson, G. & Etienne, J.-M. (2024). Two-way random effects model with serial correlation. *Empirical Economics, (*December, 19*)*, 1–32. doi: https://doi.org/10.1007/s00181-024-02695-9

Baltagi, B. H., Bresson, G. & Etienne, J. M. (2025). Online Supplement to Chapter 12 of the volume: Seven Decades of Economentrics and Beyond. In B. H. Baltagi & L. Matyas (Eds.), *Seven decades of economentrics and beyond.* Springer. https://www.dropbox.com/scl/fi/5yjy99hzy9rqkrg41wwht/Online_Appendix_Chapter12.pdf?rlkey=6a3rsa35jzlijybupefvhlhwe&st=7835swgf&dl=0.

Baltagi, B. H. & Griffin, J. M. (1983). Gasoline demand in the OECD: An application of pooling and testing procedures. *European Economic Review*, 22(2), 117–137.

Beran, J. (1994). *Statistics for Long-Memory Processes. Monographs on Statistics and Applied Probability, 61.* Chapman & Hall/Crc.

Bester, C. A. & Hansen, C. (2009). Identification of marginal effects in a nonparametric correlated random effects model. *Journal of Business & Economic Statistics, 27*(2), 235–250.

Box, G. E. & Jenkins, G. M. (1976). *Time Series Analysis: Forecasting and Control.* Holden Day, San Franscisco.

Brillinger, D. R. (1981). *Time Series: Data Analysis and Theory.* SIAM.

Brou, J. M. B., Kouassi, E. & Kymn, K. O. (2011). Double autocorrelation in two way error component models. *Open Journal of Statistics, 1*(3), 185–198.

Cameron, A. C., Gelbach, J. B. & Miller, D. L. (2011). Robust inference with multiway clustering. *Journal of Business & Economic Statistics, 29*(2), 238–249.

Chen, W.-D. (2006). An approximate likelihood function for panel data with a mixed $ARMA(p,q)$ remainder disturbance model. *Journal of Time Series Analysis, 27*(6), 911–921.

Chen, W.-D. (2008). Detecting and identifying interventions with the Whittle spectral approach in a long memory panel data model. *Journal of Applied Statistics, 35*(8), 879–892.

Chiang, H. D., Hansen, B. E. & Sasaki, Y. (2024). Standard errors for two-way clustering with serially correlated time effects. *Review of Economics and Statistics*, 1–40.

Dahlhaus, R. (1988). Small sample effects in time series analysis: a new asymptotic theory and a new estimate. *The Annals of Statistics, 16*(2), 808–841.

Das, S., Subba Rao, S. & Yang, J. (2021). Spectral methods for small sample time series: A complete periodogram approach. *Journal of Time Series Analysis*, *42*(5-6), 597–621.

Davezies, L., D'Haultfoeuille, X. & Guyonvarch, Y. (2021). Empirical process results for exchangeable arrays. *Annals of Statistics*, *49*(2), 845–862.

DeGraaf, S. R. (1994). Sidelobe reduction via adaptive FIR filtering in SAR imagery. *IEEE Transactions on Image Processing*, *3*(3), 292–301.

De Porres, C. & Krishnakumar, J. (2013). General variance covariance structures in two-way random effects models. *Applied Mathematics*, *4*, 614–623.

Dzhaparidze, K. & Yaglom, A. (1983). Spectrum parameter estimation in time series analysis. In P. Krishnaiah (Ed.), *Developments in Statistics* (Vol. 4, pp. 1–96). Elsevier.

Galbraith, J. W. & Zinde-Walsh, V. (1992). The GLS transformation matrix and a semi-recursive estimator for the linear regression model with ARMA errors. *Econometric Theory*, *8*(1), 95–111.

Galbraith, J. W. & Zinde-Walsh, V. (1995). Transforming the error-components model for estimation with general ARMA disturbances. *Journal of Econometrics*, *66*(1-2), 349–355.

Ghysels, E. (1993). The ET interview: Professor Marc Nerlove. *Econometric Theory*, *9*, 117–143.

Ginovyan, M. S. & Sahakyan, A. A. (2021). Statistical inference for stationary linear models with tapered data. *Statistic Surveys*, *15*, 154–194.

Golub, G. H. & Van Loan, C. F. (2013). *Matrix Computations*. JHU press.

Grenander, U. & Szegö, G. (1958). *Toeplitz Forms and their Applications. California Monographs in Mathematical Sciences*. University of California Press.

Grether, D. M. & Nerlove, M. (1970). Some properties of "optimal" seasonal adjustment. *Econometrica*, *38*, 682–703.

Grunfeld, Y. (1958). *The Determinants of Corporate Investment* (Unpublished doctoral dissertation). Department of Economics, University of Chicago.

Guo, W., Dai, M., Ombao, H. C. & Von Sachs, R. (2003). Smoothing spline ANOVA for time-dependent spectral analysis. *Journal of the American Statistical Association*, *98*(463), 643–652.

Hannan, E. J. (1970). *Multiple Time Series*. John Wiley & Sons.

Hatanaka, M. (1972). The estimation of spectra and cross-spectra on short time series data. *International Economic Review*, *13*(3), 679–704.

Heckman, J. J. & Singer, B. (1982). Population heterogeneity in demographic models. In K. Land & A. Rogers (Eds.), *Multidimensional Mathematical Demography* (pp. 567–599). Academic Press.

Huang, L., Jiang, H. & Wang, H. (2019). A novel partial-linear single-index model for time series data. *Computational Statistics & Data Analysis*, *134*, 110–122.

Huang, L., Xia, Y. & Qin, X. (2016). Estimation of semivarying coefficient time series models with *ARMA* errors. *Annals of Statistics*, *44*(4), 1618–1660.

Hurvich, C. M. & Ray, B. K. (1995). Estimation of the memory parameter for nonstationary or noninvertible fractionally integrated processes. *Journal of Time Series Analysis*, *16*(1), 17–41.

Hurvich, C. M. & Ray, B. K. (2003). The local Whittle estimator of long-memory stochastic volatility. *Journal of Financial Econometrics*, *1*(3), 445–470.

Jenkins, G. & Watts, D. G. (1969). *Spectral Analysis and Its Applications*. Holden-Day.

Karanasos, M. (1998). A new method for obtaining the autocovariance of an ARMA model: An exact form solution. *Econometric Theory*, *14*(5), 622–640.

Karlsson, S. & Skoglund, J. (2004). Maximum-likelihood based inference in the two-way random effects model with serially correlated time effects. *Empirical Economics*, *29*, 79–88.

Krafty, R. T. & Collinge, W. O. (2013). Penalized multivariate Whittle likelihood for power spectrum estimation. *Biometrika*, *100*(2), 447–458.

Laird, N. (1978). Nonparametric maximum likelihood estimation of a mixing distribution. *Journal of the American Statistical Association*, *73*(364), 805–811.

Menzel, K. (2021). Bootstrap with cluster-dependence in two or more dimensions. *Econometrica*, *89*(5), 2143–2188.

Montanari, A., Taqqu, M. S. & Teverovsky, V. (1999). Estimating long-range dependence in the presence of periodicity: an empirical study. *Mathematical and Computer Modelling*, *29*(10-12), 217–228.

Munnell, A. H. (1990). Why has productivity growth declined? Productivity and public investment. *New England Economic Review*, 3–22.

Nerlove, M. (1964). Spectral analysis of seasonal adjustment procedures. *Econometrica*, *32*, 241–286.

Nerlove, M. (1971). Further evidence on the estimation of dynamic economic relations from a time series of cross sections. *Econometrica*, *39*, 359–382.

Nerlove, M. (2005). *Essays in Panel Data Econometrics*. Cambridge University Press.

Nerlove, M., Grether, D. M. & Carvalho, J. L. (2014). *Analysis of Economic Time Series: A Synthesis*. Academic Press.

Parzen, E. (1983). Autoregressive spectral estimation. In D. Brillinger & P. Krishnaiah (Eds.), *Handbook of Statistics* (Vol. 3, pp. 221–247). North-Holland.

Pawitan, Y. & O'Sullivan, F. (1994). Nonparametric spectral density estimation using penalized Whittle likelihood. *Journal of the American Statistical Association*, *89*(426), 600–610.

Priestley, M. (1981). *Spectral Analysis and Time Series, Volume 1*. Academic Press, London.

Revankar, N. S. (1979). Error component models with serial correlated time effects. *Journal of the Indian Statistical Association*, *17*, 137–160.

Robinson, P. M. (1995). Log-periodogram regression of time series with long range dependence. *The Annals of Statistics*, *23*(3), 1048–1072.

Stankwitz, H. C., Dallaire, R. J. & Fienup, J. R. (1994). Spatially variant apodization for sidelobe control in SAR imagery. In *Proceedings of 1994 IEEE national radar conference* (pp. 132–137).

Stoica, P. & Moses, R. L. (2005). *Spectral Analysis of Signals* (Vol. 452). Pearson Prentice Hall Upper Saddle River, NJ.

Subba Rao, S. & Yang, J. (2021). Reconciling the Gaussian and Whittle likelihood with an application to estimation in the frequency domain. *The Annals of Statistics*, *49*(5), 2774–2802.

Thomas, G., Flores, B. C. & Sok-Son, J. (2000). SAR sidelobe apodization using the Kaiser window. In *Proceedings 2000 International Conference on Image Processing* (Vol. 1, pp. 709–712).

Thompson, S. B. (2011). Simple formulas for standard errors that cluster by both firm and time. *Journal of Financial Economics*, *99*(1), 1–10.

Tukey, J. (1967). An introduction to the calculations of numerical spectrum analysis. In B. Harris (Ed.), *Advanced Seminar on Spectral Analysis of Time Series* (pp. 25–46). John Wiley & Sons.

Velasco, C. (1999). Non-stationary log-periodogram regression. *Journal of Econometrics*, *91*(2), 325–371.

Velasco, C. & Robinson, P. M. (2000). Whittle pseudo-maximum likelihood estimation for nonstationary time series. *Journal of the American Statistical Association*, *95*(452), 1229–1243.

Wang, T. & Xia, Y. (2015). Whittle likelihood estimation of nonlinear autoregressive models with moving average residuals. *Journal of the American Statistical Association*, *110*(511), 1083–1099.

Wei, H., Zhang, H., Jiang, H. & Huang, L. (2022). On the semi-varying coefficient dynamic panel data model with autocorrelated errors. *Computational Statistics & Data Analysis*, *173*, 107458.

Welch, P. (1967). The use of fast Fourier transform for the estimation of power spectra: a method based on time averaging over short, modified periodograms. *IEEE Transactions on Audio and Electroacoustics*, *15*(2), 70–73.

Whittle, P. (1953). Estimation and information in stationary time series. *Arkiv för Matematik*, *2*(5), 423–434.

Zhang, H.-C. (1992). Reduction of the asymptotic bias of autoregressive and spectral estimators by tapering. *Journal of Time Series Analysis*, *13*(5), 451–469.

Chapter 13
Dynamic Heterogeneous Linear Models for Three-level Panel Data with Short Time Dimension and Stratification

Monika Avila Márquez and Jaya Krishnakumar

Abstract Most national survey data are obtained through a sampling scheme that is stratified at multiple levels such as regions, socioeconomic groups, gender, etc., to ensure adequate representativeness of the underlying population. In such a design, the strata as well as the membership of individuals into a stratum are known. This chapter considers a three-level dynamic panel data model, the levels being group (stratum), individual (household, firm etc.) and time. It specifies a model with additive stratum fixed effects and a mixed coefficients structure composed of stratum-specific fixed effects and random stratum-individual-time specific effects. We examine the identification and estimation of this dynamic heterogeneous three-level linear panel data model under stratification when the time dimension is as short as 3. We propose a Mean Stratum-FGLS estimator and a Mean Stratum-OLS estimator to estimate the mean coefficients. To make the GLS estimation of the Stratum-specific parameters feasible, we introduce a ridge estimator of the variance-covariance matrix of the model. We show consistency and asymptotic normality of the Mean Stratum estimators for short panels, under the assumptions that apply to stratified sampling such as the number of strata (groups) is fixed, all strata are observed, and the number of individuals per stratum is large (growing to infinity). We also show the consistency of the variance parameter estimators. We discuss similarities and differences between our specification and a dynamic two-level panel data model with random coefficients. Finally, we discuss the setup of long time span.

13.1 Introduction

Model-based inference in panel data is usually developed under a random sampling assumption or independently of the sampling design. This chapter is concerned with model-based inference for a stratified random sampling scheme . Stratified sampling designs are commonly used in national surveys, in which the sample is divided into different strata according to some pre-specified criterion such as age, gender, region and so on, in order to ensure representativeness of the underlying population. Our model is suitable for three-level data, for instance regions and individuals within regions, or industries and firms within industries, with repeated observations on these units over time. We assume a short time dimension as is typically the case in such surveys. Heterogeneity of

The original version of the chapter has been revised: The author name has been updated. A correction to this chapter can be found at https://doi.org/10.1007/978-3-031-92699-0_18

Monika Avila Márquez ✉
University of Bristol, Bristol, UK, e-mail: monika.avilamarquez@gmail.com

Jaya Krishnakumar
University of Geneva, Geneva, Switzerland, e-mail: jaya.krishnakumar@unige.ch

B. H. Baltagi and L. Mátyás (eds.), *Seven Decades of Econometrics and Beyond*, Advanced Studies in Theoretical and Applied Econometrics 57,
https://doi.org/10.1007/978-3-031-92699-0_13

behaviour is assumed across strata and individuals by means of a mixed coefficients component structure, consisting of fixed stratum-specific effects and random stratum-individual-time-specific effects.

This mixed coefficients structure is related, but not equal, to the assumption presented by Krishnakumar, Avila Márquez and Balazsi (2017) for a *static* three-level linear panel data model, and it is equivalent to the assumption presented by Avila Marquez (2022). The assumption presented by Krishnakumar et al. (2017) states that the coefficient vector is equal to the sum of a mean coefficient vector plus random specific effects. In contrast, our assumption states that the coefficient vector equals the sum of varying fixed coefficients at stratum level plus stratum-individual-time random components. The difference between Krishnakumar et al. (2017) and this paper, is that they consider random effects that are not nested. This assumption is appropriate for setups with no nesting in the data, such as multi-way clustering or crossed-random effects models. The assumption in this paper is suitable for data with nesting due to stratification and for data with nesting due to known clustering Avila Marquez (2022) .

We propose two Mean Stratum (MS) estimators, which are respectively the means of the FGLS and OLS estimators of parameters obtained using observations belonging to each stratum g. In order to make GLS feasible, we propose a ridge estimation of the variance-covariance components along with a modification suitable for a large sample size. These estimators are consistent under stratified sampling when the number of strata is fixed, and the proportion of observed strata is equal to 1.

In order to test the assumption of stratum heterogeneity , we propose two specification tests that are extensions of the Hausman test (Hausman & Taylor, 1981). First, testing the null hypothesis of stratum additive and multiplicative heterogeneity versus stratum-individual additive and multiplicative heterogeneity is not feasible when the time dimension is as short as 3. But, testing the null hypothesis of complete homogeneity versus stratum additive and multiplicative heterogeneity is possible. In this case, we propose to compare the Mean Stratum estimator with the simple Pooled OLS estimator. In addition, testing the null hypothesis of stratum additive and multiplicative effects versus stratum-individual additive heterogeneity and stratum multiplicative heterogeneity is also feasible. In this case, we propose to compare the Mean Stratum estimator against the Mean Stratum first-difference GMM estimator or the Mean Stratum estimator using a Mundlak approach. The study of the statistical properties of these tests is left for further research.

If stratum-individual fixed effects are present in addition to stratum fixed effects, our estimators become inconsistent. As a possible solution, we extend our baseline model to allow for stratum-individual-specific additive effects. In this setting, we are back to the problems of incidental parameters and dependency on initial conditions. To deal with the incidental parameters, we use the Mundlak approach and propose a Bayesian hierarchical estimator with a prior for the initial conditions. The Bayesian estimator requires the correct specification of the prior of the initial conditions. Thus, the assumption of initial conditions generated from the stationary distribution is critical for consistency of the proposed Bayesian estimator. While we present an alternative prior allowing for initial conditions that are not generated from the stationary distribution, it is not straightforward to decide which is the correct assumption for the initial conditions. As an alternative, we use the Chamberlain or the Mundlak approach conditioning on the initial values (Wooldridge, 2005b), and we propose to estimate the stratum-specific parameters using a factor analytical method following Bai (2013). Another issue is the potential cross-sectional dependence within strata. To address this problem, we extend the baseline model to a setting that includes common factors and propose Mean Stratum estimation using the time-demeaned variables (Sarafidis & Robertson, 2009).

The rest of the paper is organized as follows: Section 13.2 explains the structure of the data, Section 13.3 presents the model with its assumptions, Section 13.4 states the identification strategy of the parameters of interest, Section 13.5 presents the estimation strategy, Section 13.6 exposes the statistical properties of the methods proposed, Section 13.7 discusses consequences of misspecification of the variance-covariance matrix of the disturbance term of the model, 13.8 compares the Mean Stratum estimator with other available estimators, Section 13.9 discusses the relationship between our baseline model and a two-level panel data model, highlighting how the former can be obtained from the later by imposing certain pertinent restrictions. Section Section 13.10 presents specification tests, Section 13.11 relaxes the assumption of additive stratum effects to stratum-individual additive specific effects and presents Bayes estimation and estimation conditioning on the initial values, Section 13.12 presents an extension of the model with cross sectional dependence, Section 13.13 discusses the behaviour of the Mean Stratum estimator in a setting with long time dimension, Section 13.14 describes the Monte Carlo experiments and the results, Section 13.15 gives the conclusions.

Notation: $||\cdot||^2$ is the Euclidean norm. $||\cdot||_F$ is the Frobenius norm. Scalar random variables are collected in column vectors; for instance y_{git} can be collected in the vector $y \in \mathbb{R}^M$ ($y = [y_{111} \quad \dots \quad y_{mN_mT_{im}}]'$). Matrices are denoted by uppercase letters; for instance the matrix $X \in \mathbb{R}^{n\times K}$ that collects the transpose of the column vector $x_{git} \in \mathbb{R}^{K\times 1}$ containing K regressors corresponding to individual i belonging to stratum g at period t. I_A represents the identity matrix with dimension $A \times A$ where A is a positive integer. $\mathbf{0}$ represents a vector of zeros with dimensions $K \times 1$.

13.2 Data Structure

We define the following subscripts:

- g denotes each strata and takes values $g \in \{1, 2, ..., m\}$.
- i_g denotes individual i_g in strata g and takes values $i_g \in \{1, 2, ..., N_g\}$.
- t_{i_g} denotes time observation t of individual i_g in group g and takes values $t_{i_g} \in \{1, 2, ..., T_{i_g}\}$.

Data $\{y_{gi_gt_{i_g}}, x_{gi_gt_{i_g}}\}_{g=1,i_g=1,t_{i_g}=1}^{m,N_g,T_{i_g}}$ are obtained from stratified sampling, from a population that is stratified into m independent known strata that do not overlap. This means that the number of observed strata m is equal to the total number of strata in the population and the dataset can be partitioned into m non-overlapping subsets $\{y_{i_gt_{i_g}}, x_{i_gt_{i_g}}\}_{i_g=1,t_{i_g}=1}^{N_g,T_{i_g}}$. Individuals are independent within a stratum (this is relaxed in Section 13.12). For each stratum g, N_g individuals are sampled over T_{i_g} periods. The total number of individuals across strata is $N = \sum_g^m N_g$. The total number of observations per stratum g is $n_g = \sum_{i_g} T_{i_g}$. The total number of observations in the data set is $n = \sum_g^m n_g$. This data can be seen as an unbalanced three-level panel.

Remark 13.1 For simplicity, we use i and t equivalently to i_g and t_{i_g}. This does not mean that we assume that individual i is part of stratum g.

13.3 The Model

We consider the autoregressive distributed lag ARDL(1,0) heterogeneous panel data model for a random draw i from the population of stratum g:

$$y_{git} = \alpha_g + \rho_g y_{git-1} + x'_{git}\beta_{git} + \varepsilon_{git}, \qquad t = 1, ..., T_{i_g}, \tag{13.1}$$

with:

$$\beta_{git} = \beta_g + \lambda_{git}, \tag{13.2}$$

where y_{git} is the observed outcome variable with support $\mathbb{Y} \subseteq \mathbb{R}$, y_{git-1} is the first lag of the outcome variable, x_{git} is a $K \times 1$ vector of observed explanatory variables for individual i in stratum g for period t with support $\mathbb{X} \subseteq \mathbb{R}^K$ (variables with finite support are also allowed), and ε_{git} is an unobserved idiosyncratic stratum-individual error term in period t.

The unobserved parameters of interest are the stratum-specific parameter (ρ_g), and the stratum-specific slope coefficients (β_g). The model also includes stratum additive specific fixed effects (α_g) as well as multiplicative stratum-individual-time specific effects (λ_{git}). There is also interest in the overall averages of the parameters $E[\rho_g]$, $E[\beta_g]$ which can be seen as average partial effects as explained by Wooldridge (2005a).

We assume that ρ_g is *fixed* and the slope coefficient vector presents a *mixed* structure ($\beta_{git} = \beta_g + \lambda_{git}$) composed of a stratum-specific *fixed* component (β_g) and a *random* stratum-individual-time specific effect λ_{git}. In addition, we assume a full variance-covariance matrix for the random stratum-individual-time specific effect that captures the covariance between marginal effects of the included regressors in the model. This coefficient structure allows for possible stratified endogenous heterogeneity while admitting random deviations of individual time-specific marginal effects from their stratum mean. For instance, one could think that common cultural unobserved characteristics drive the heterogeneous habit formation of individuals in a certain stratum while possible deviations are random and noncorrelated to 'taste-shifters'. [1]

The total number of time observations per individual T_{i_g} is small and considered fixed in the asymptotic analysis. The number of individuals per stratum is N_g and the total number of individuals in the panel N are growing to infinity. The number of strata is fixed under stratified sampling. This setting can be evaluated using an asymptotic sequence framework where we allow N_g to grow and the time dimension T_{i_g} is fixed (Moon, Shum & Weidner, 2018). Model 13.1 is relevant for different empirical applications because it permits accounting for correlated stratum heterogeneity and individual and time heterogeneity. For instance, one could be interested in studying dynamic heterogeneous demand equations, the heterogeneity of habit formation, income persistence, dynamic heterogeneous treatment effects and so on.

In the following lines, we present the assumptions of the model in more detail.

Assumption 1 : Stratum membership is known and fixed over time. □

The researcher knows the strata based on observed characteristics. For instance, stratification can be done by counties, sub-regions, economic activity categories at a detailed level, among others. The membership of individual i into stratum g is denoted by the indicator variable $s_i^{(g)} \in \{0,1\}$ that takes value 1 if the individual belongs to stratum g and 0 otherwise. Thus, each individual has m indicator variables. It is crucial to notice that stratum belonging does not vary with time.
The sum of $s_i^{(g)}$ for all individuals in the panel gives the number of individuals in the stratum g ($\sum_i^N s_i^{(g)} = N_g$).

Assumption 2 : Number of individuals within stratum is growing.

$$N \to \infty \Rightarrow N_g \to \infty, \quad \forall g \in \{1,2,\ldots,m\}.$$

The number of individuals within stratum grows to infinity when the number of individuals in the panel grows to infinity. This could happen for households within sub-region or enterprises in an economic sector.

Assumption 3 : Non vanishing strata.

$$\lim_{N\to\infty} \frac{N_g}{N} \to \pi_g, \quad \forall g \in \{1,2,\ldots,m\},$$
$$\pi_g \in (0,1).$$

The proportion of stratum population to the overall population converges to a fixed number greater than 0 but less than 1 as the number of individuals within stratum and the total number of individuals in the panel grows to infinity. This assumption implies that the number of strata is fixed.

---

[1] Dynan (2000) calls 'taste-shifter' as preference related variables.

Assumption 4 : The proportion of observed strata (q) concerning to the total number of strata in the population is equal to 1.

This assumption is in line with stratified sampling. If the proportion of observed strata (m) with respect to the total number of strata in the population is lower than 1 and the number of strata in the population is small, the sample is not representative of the underlying population. As a result, the Mean Stratum estimator is unfeasible as there is insufficient information.

Assumption 5 : Fixed stratum additive specific effects α_g. □

Assumption 6 : Fixed stratum specific persistence parameter.

$$\rho_g \in (-1, 1).$$

Assumption 7 : Mixed stratum-individual-time specific coefficients.

$$\beta_{git} = \beta_g + \lambda_{git},$$

with

$$E[\lambda_{git}\lambda'_{g'i't'}|x_{gi1}, x_{gi2}, ..., x_{giT}] = \begin{cases} \Delta_{\lambda_g} & \text{if } g = g', i = i' \text{ and } t = t', \\ 0 & \text{otherwise.} \end{cases}$$

The unobserved coefficient vector is composed of a fixed stratum coefficient vector (β_g), and a heteroskedastic random component (λ_{git}) conditional on covariates that captures the multiplicative heterogeneity over time for each individual of stratum g. Specifically, λ_{git} varies across strata (indicated by the sub-index g).

The mixed coefficients structure can have three possible interpretations: i) the data is sampled from a density function with heterogeneous parameters, ii) the correlation of the regressors with unobserved multiplicative individual heterogeneity is equal within stratum, or iii) the regressors are freely correlated to multiplicative stratum unobserved heterogeneity while preserving noncorrelation with multiplicative stratum-individual-time specific unobserved heterogeneity. An example of the second interpretation is that innate ability and the marginal return to education of individuals are equally correlated to education within a city if we believe that individuals with higher ability do not only self-select into education levels but also into the city where they will have the highest return to their education.

Assumption 8: The random stratum-individual-time effects have zero mean conditional on the covariates.

$$E[\lambda_{git}|x_{gi1}, x_{gi2}, ..., x_{giT}, y_{git-1}] = 0.$$

This implies that $E[\beta_{git}|x_{gi1}, x_{gi2}, ..., x_{giT}, y_{git-1}] = \beta_g$. As a consequence of this assumption, $E[\lambda_{git}] = 0$.

Assumption 9 : Strict exogeneity of the covariates with the disturbance term.

$$E[\varepsilon_{git}|x_{gi1}, x_{gi2}, ..., x_{giT}, y_{git-1}] = 0.$$

This assumption is in line with Hsiao, Pesaran and Tahmiscioglu (1998) and it rules out possible feedback of y_{git} with future values of the covariates. It implies the model presents dynamic completeness without conditioning on stratum effects because stratum-specific effects are considered fixed parameters. It is also possible to assume that the stratum effects are random and correlated with the regressors. With correlated stratum effects, the strict exogeneity of the covariates must be conditional on the stratum-specific effects. The orthogonality conditions presented in section 13.4 hold under strict exogeneity of the covariates conditional on the stratum-specific effects. As a consequence of this assumption, $E[\varepsilon_{git}] = 0$.

Remark 13.2 According to Wooldridge (2010), strict exogeneity rules out possible feedback of the past values of the dependent variable to the covariates. Allowing for this feedback requires relaxing this assumption to sequential exogeneity. The assumption of sequential exogeneity is weaker than strict exogeneity since it allows for feedback from y_{git} to $x_{git+1}, ..., x_{giT}$. For instance, consumption in period t can have an effect on taste shifters in periods after t. In order to allow for this possible feedback, it is necessary to modify the first stage of the estimation method proposed in section 13.5 by replacing OLS or GLS with GMM using instrumental variables.

Assumption 10 : The error term ε_{git} is homoskedastic, and uncorrelated within each stratum g but heteroskedastic across strata conditional on regressors.

$$E[\varepsilon_{git}^2 | x_{gi1}, x_{gi2}, ..., x_{giT}] = \sigma_{\varepsilon_g}^2 < \infty.$$

$$E[\varepsilon_{git}, \varepsilon_{g'i't'} | x_{gi1}, x_{gi2}, ..., x_{giT}] = 0, \quad \text{if} \quad g \neq g', i \neq i', t \neq t'.$$

Assumption 11 :
y_{git} are generated from the stationary distribution of the process with initialization values $y_{gi,-h_{i_g}}$ sampled h_{i_g} number of periods before the data collection in period 0.

The initial conditions are given by:

$$y_{gi0} = \rho_g^{h_{i_g}} y_{gi,-h_{i_g}} + \alpha_g \frac{1-\rho_g^{h_{i_g}}}{1-\rho_g} + \sum_{l=0}^{h_{g_i}} \rho_g^l x'_{gi-l} \beta_{gi-l} + \sum_{l=0}^{h_{i_g}} \rho_g^l \varepsilon_{gi-l}. \tag{13.3}$$

with h_{i_g} unrestricted. It is possible to set h_{i_g} free because the initial conditions (y_{gi0}) dependence is controlled under the assumption of fixed stratum additive effects (Assumption 13).
Assumption 19 is not necessary in the presence of fixed stratum-additive effects (Assumption 13). The reason is that the dependence of the initial conditions y_{gi0} is controlled because the stratum-additive effects are assumed to be fixed. As a result, the Mean Stratum estimator (Section 13.5) is consistent without Assumption 19 if the Assumption 13 holds.
In contrast, Assumption 19 is essential if there are stratum-individual additive effects instead of stratum additive effects. The reason is that, under Assumptions 19 and 20, the initial conditions can be projected into all past, present and future values of the regressors. This means that it is possible to estimate the model either using a Bayesian approach or conditioning on the initial value y_{gi0} (Hsiao, Hashem Pesaran & Kamil Tahmiscioglu, 2002).

In addition, if the model presents stratum-individual additive fixed effects instead of stratum additive effects and h_{i_g} is small, the individual initialization values $y_{gi,-h_{i_g}}$ are essential because there exist initial conditions (y_{gi0}) dependence. In that case, there is a need to add an assumption to avoid the incidental parameter problem: $E[y_{gi,-h_{i_g}}] = b_g$. On the other hand, having $h_{i_g} \to \infty$ means that the effect of the initialization value dies (similar to Hsiao et al. (2002)).
Assumption 19 can be relaxed in the presence of stratum-individual additive effects. In this case, the Bayesian estimator requires a prior for the initial conditions not generated from the stationary distribution. A simpler solution is to condition on the initial conditions as suggested by Wooldridge (2005b).

Assumption 12 : x_{git} are generated from:

$$x_{git} = \mu_g + \rho_x x_{git-l} + \omega_{git}, \qquad |\rho_x| < 1$$

μ_g are fixed stratum-specific effects, x_{git} are stationary with ω_{git} i.i.d with variance $\sigma_\omega^2 I_K$. This assumption is similar but not equal to the one presented by Hsiao et al. (2002). Assumption 20 in combination with Assumption 14 states that the dependent variable and the regressors are both integrated of order 0.
It is possible to relax Assumption 20 and allow for the presence of stratum-specific trends as follows:

$$x_{git} = \mu_g + b_g t + \rho_x x_{git-l} + \omega_{git}, \qquad |\rho_x| < 1.$$

The Mean Stratum estimator presented in section 13.5 is consistent with trend stationary regressors if we include a deterministic trend in model (13.1) as above. Otherwise, it is consistent only if the data-generating process started a short time ago (small h_{i_g}). An example could be the wage of young individuals, which means one could include age or experience as regressors in the model.
When the model presents stratum additive specific effects, if we relax Assumption 20 to allow for some non-stationary regressors and assuming non-stationary dependent variable, our stratum-specific estimators remain consistent if the regressors, the dependent variable and its lag are co-integrated per stratum (Hamilton, 1994). Then the Mean Stratum estimator may also be asymptotically normal but this is left for further research. If the regressors and the dependent variable are not co-integrated per stratum, it is not clear if the pooled stratum OLS estimator is consistent even if Phillips and Moon (1999) show that pooled OLS is a consistent estimator of the long-run average regression coefficient if the regressors are non-stationary and there is no co-integration. The reason is that they considered a model that does not present an intercept and the lag of the dependent variable. In order to test for co-integration, one needs to extend the test proposed by Im, Pesaran and Shin (2003) to allow for stratum-specific parameters instead of individual-specific parameters. Concluding that there is co-integration would entail that $u_{git} = x'_{git}\lambda_{git} + \varepsilon_{git}$ is stationary, implying that λ_{git} could be considered as a random co-integrating vector. A study of a co-integration test and the properties of the Mean-stratum estimator when there is no co-integration is outside the scope of this paper, and both issues are left for further research.
When the model presents stratum-individual specific effects instead of stratum-specific effects, the assumption of stationary regressors is important. The reason is that the presence of stratum-individual specific effects causes the problem of initial conditions dependency. This problem can be solved by projecting the initial conditions on the past, the present, and the future values of the regressors. Moreover, the projection of the initial conditions on the regressors is only possible if the regressors are stationary (Hsiao, 2020). Thus, non-stationary regressors cause the failure of the Bayes estimator proposed in section 13.11.1. A solution is conditioning on the initial conditions as proposed by Wooldridge (2005b) because this does not require Assumption 20. In this case, the Mean Stratum estimator is consistent in the presence of non-stationary regressors with or without co-integration (Phillips & Moon, 1999). Alternatively, one can include non-stationary regressors in the model after first differencing them.
Another issue is binary regressors. Under Assumption 20, binary regressors are modeled with a linear probability model. In this case, a more suitable assumption could be a dynamic latent model. Another option could be a Markov chain assumption. These extensions are left for further research.

13.4 Identification

For identification, we can rewrite the model 13.1 as:

$$y_{git} = \rho_g y_{git-1} + \alpha_g + x'_{git}\beta_g + u_{git} = z'_{git}\theta_g + u_{git}, \tag{13.4}$$

where: $z_{git} = [y_{git-1} \quad 1 \quad x'_{git}]'$, $\theta_g = [\rho_g \quad \alpha_g \quad \beta'_g]'$, and $u_{git} = x'_{git}\lambda_{git} + \varepsilon_{git}$ is a composite error term.
Assumptions 16 and 17 imply the following orthogonality conditions:

$$E[u_{git}x_{gis}] = 0, \quad s = 1,2,\dots,T_{i_g}, \quad i = 1,2,\dots,N_g, \quad g = 1,2,\dots,m, \tag{13.5}$$

$$E[u_{git}y_{git-1}] = 0, \quad t = 1,2,\dots,T_{i_g}, \quad i = 1,2,\dots,N_g, \quad g = 1,2,\dots,m. \tag{13.6}$$

Consequently, the moment conditions used for the estimation of the stratum-specific parameters are:

$$E[u_{git}z_{git}] = 0, \quad t = 1,2,...,T_{i_g}, \quad i = 1,2,...,N_g, \quad g = 1,2,...,m. \tag{13.7}$$

Note that we only use contemporaneous exogeneity for estimation of the stratum-specific parameters using stratum-specific data which is in line with Hsiao, Li, Liang and Xie (2019). According to Wooldridge (2010) contemporaneous exogeneity can be exploited when the variance-covariance of the model is diagonal as it is in each stratum.

Additionally, we also assume that the second-order moment matrix of z_{git} is of full rank which means that the regressors vary within stratum.

Assumption 13 :

For OLS: The matrix $E[z_{git}z'_{git}]$ is of full rank.
For GLS: $E[u_g u'_g|Z_g]$ is positive definite and the matrix $E[Z'_g E[u_g u'_g]^{-1} Z_g] = Q_g$ is nonsingular. □

13.5 Estimation

If we re-write model (13.1) using backward substitution, we obtain the following expression of the dependent regressor:

$$y_{git} = \rho_g^t y_{gi0} + \sum_{l=0}^{t} \rho_g^l (\alpha_g + x'_{git-l}(\beta_g + \lambda_{git-l})) + \sum_{l=0}^{t} (\rho_g^l)\varepsilon_{git-l}. \tag{13.8}$$

Using this result, the first lag of the dependent variable can be rewritten as:

$$y_{git-1} = \rho_g^{t-1} y_{gi0} + \sum_{l=0}^{t-1} \rho_g^l (\alpha_g + x'_{git-1-l}(\beta_g + \lambda_{git-1-l})) + \sum_{l=0}^{t-1} (\rho_g^l)\varepsilon_{git-1-l}. \tag{13.9}$$

It is easy to see from (13.9) that first-difference GMM estimation (Arellano & Bond, 1991) ignoring the subpopulation structure of the data leads to inconsistent estimates of the mean parameters. This is caused by the presence of the first lag and the stratum-specific effects in the right-hand side of the model causing endogeneity. Moreover, it is not possible to find an instrument that is uncorrelated with the composite error term and correlated with the regressors. [2]

Similarly, one could argue that the researcher could perform Mean Group estimation per individual within stratum. Mean Group estimation could be used to estimate stratum-specific parameters only if the time dimension is bigger than the number of covariates and growing to infinity or using small sample debiasing techniques (available only if $T > 3$). Thus, when the time dimension is fixed and the number of individuals per stratum is big it would be beneficial to use another estimation strategy.

In order to fill this gap, we propose a methodology that allows for the estimation of the Mean Stratum and the stratum-specific coefficients using a two-stage procedure. This estimation technique is an extension of the Mean Group Estimator presented by Pesaran and Smith (1995). The two-stage procedure is the following:

---

[2] Ignoring stratum effects is equivalent to performing first-difference GMM estimation on the model: $\Delta y_{it} = \rho\Delta y_{it-1} + \Delta x'_{it}\beta + \Delta u_{it}$ with: $\Delta u_{it} = \Delta y_{it-1}\alpha_{2,g} + \Delta x'_{it}\alpha_{3,g} + \Delta x'_{it}\lambda_{git} + \Delta\varepsilon_{it}$, $\alpha_{2,g} = \rho_g - E[\rho_g]$ and $\alpha_{3,g} = \beta_g - E[\beta_g]$. Thus, we would not have available instruments. Another possibility could be first-difference GMM estimation on the model in first differences using multiplicative stratum dummies when $T > 2$. But one could run into the problem of weak instrumental variables (Bun & Windmeijer, 2010).

First stage: In the first stage, one estimates the stratum-specific coefficients by exploiting the population moment condition for individual i within group g:

$$E[u_{git} z_{git}] = 0, \quad t = 1, 2, ..., T_{i_g}. \tag{13.10}$$

The sample moment conditions per stratum g are given by:

$$\frac{1}{N_g} u_g' Z_g = 0, \quad g = 1, 2, ..., m. \tag{13.11}$$

where u_g and Z_g stack u_{git} and z'_{git} respectively.

It is easy to see that using the sample moment conditions (13.11) as estimating equations leads to a simple ordinary least squares estimator :

$$\hat{\theta}_{g,OLS} = (Z_g' Z_g)^{-1} (Z_g' y_g).$$

This estimator is not the most efficient since the model presents a non-i.i.d composite error term u_{git}. A straightforward solution is to set a GLS estimator :

$$\hat{\theta}_{g,GLS} = (Z_g' \Omega_g^{-1} Z_g)^{-1} (Z_g' \Omega_g^{-1} y_g),$$

where $\Omega_g = E[u_g u_g' | Z_g] = diag(X_g)(I_{N_g T} \otimes \Delta_{\lambda_g}) diag(X_g) + \sigma^2_{\varepsilon_g} I_{N_g}$ if $T_g = T$, and $diag(X_g)$ is a block diagonal matrix with blocks equal to x'_{git}. If $T_g \neq T$, one just needs to set up the adequate design matrix to allow unbalancedness in the time dimension.
Since Ω_g is unknown, we propose an estimation procedure for Ω_g in Subsection 13.5.1 which leads to FGLS estimator of θ_g.

The assumptions of unobserved additive and multiplicative stratum fixed effects allow us to estimate the specific parameters by pooling observations within each stratum (Assumptions 13, 14, 15). Additionally, OLS or FGLS estimation is consistent under the assumptions presented in Section 13.3 because the model is dynamic complete conditional on stratum-specific effects. But the FGLS estimator is non-robust to violations of the assumptions 15 and 18 because the variance-covariance matrix is not diagonal if λ_{git} and ϵ_{git} are heteroskedastic, serially correlated and/or present cross-sectional correlation. In this case, it is better to use the OLS estimator with a fully robust variance estimator as explained in Section 13.7.

In the case of endogenous regressors, it is possible to replace the OLS or FGLS first-stage estimation with GMM estimation using instrumental variables. In this case, identification is done using the population moment conditions $E[u_{git} p_{git}] = 0$ with p_{git} a vector of appropriate instruments. Moreover, for identification it is also necessary to assume that the number of instrumental variables is equal or greater than the endogenous regressors.

Second stage: The estimator of $E[\theta_g]$ is equal to the weighted average of the stratum estimated parameters. This is called the Mean Stratum estimator, and it is given by:

$$\hat{\hat{\theta}}_{MS} = \sum_g^m \hat{\pi}_g \hat{\theta}_g,$$

where $\hat{\pi}_g$ is an appropriate estimator of the importance of the stratum in the population, $\hat{\hat{\theta}}_{MS} = [\hat{\hat{\rho}} \quad \hat{\alpha}_g \quad \hat{\hat{\beta}}]'$, $\hat{\theta}_g = [\hat{\rho}_g \quad \hat{\alpha}_g \quad \hat{\beta}_g]'$.
Under stratified sampling, we propose a weighted average of the stratum-specific coefficients where the weights represent the importance of each stratum in the population.

The difference between the Mean Stratum (MS) estimators and the Mean Group (MG) estimator proposed by Pesaran and Smith (1995) is that the MG is obtained by averaging the estimators for each individual in the panel. In contrast, the MS averages stratum pooled estimators.

13.5.1 Variance-Covariance Matrix Estimation

In order to make GLS feasible , we propose a ridge regression estimation method of the variance-covariance components of $\triangle_{\lambda_g}$ and $\sigma^2_{\epsilon_g}$.
First, we derive the linear decomposition of the variance-covariance matrix for each stratum:

$$\Omega_g = \sum_{k=1}^{K}\sum_{k'=1}^{K} \sigma_{\lambda_g,kk'} H_{g,kk',\lambda_g} + \sigma^2_{\epsilon_g} I_{n_g}. \tag{13.12}$$

with the design matrices equal to:

$$H_{g,kk',\lambda_g} = \tilde{X}_{g,k}\tilde{X}'_{g,k'},$$

where $\tilde{X}_{g,k} = diag(x'_{git,k})$.
Now, we obtain a first stage estimator of the residuals for each stratum using OLS estimation $r_{gOLS} = (I_{n_g} - Z_g(Z'_gZ_g)^{-1}Z'_g)y_g = M_g w_g$ where $Z_g \in \mathbb{R}^{n_g\times(K+1)}$ is the matrix stacking up all the observations for $z_{git} = [y_{git-1} \quad 1 \quad x'_{git}]'$. Then, it follows that:

$$E[r_{gOLS} r'_{gOLS}|Z_g] = M_g\Omega_g M_g. \tag{13.13}$$

Replacing expression (13.12) into equation (13.13) and applying the vec operator, we obtain:

$$vec(E[r_{gOLS} r'_{gOLS}|Z_g]) = \sum_{k=1}^{K}\sum_{k'=1}^{K} \sigma_{\lambda_g,kk'} vec(M_g H_{g,kk',\lambda_g} M_g) + \sigma^2_{\epsilon_g} vec(M_g). \tag{13.14}$$

Now, we can rewrite the previous expression in matrix form:

$$vec(E[r_{gOLS} r'_{gOLS}|Z_g]) = B_{\lambda_g} vec(\triangle_{\lambda_g}) + \sigma^2_{\epsilon_g} vec(M_g). \tag{13.15}$$

In order to avoid double estimation of the covariances in the variance-covariance matrix, we use the identity $vec(A) = Dvech(A)$ where A is a square symmetric matrix and we re-express the previous equation as:

$$vec(E[r_{gOLS} r'_{gOLS}|Z_g]) = B_{\lambda_g} Dvech(\triangle_{\lambda_g}) + \sigma^2_{\epsilon_g} vec(M_g). \tag{13.16}$$

The expectation of the outer product of the residuals is replaced by the point estimator of the OLS residuals for each stratum and we add the error ν_g that captures the sampling error.

$$vec(r_{gOLS} r'_{gOLS}) = B_{\lambda_g} Dvech(\triangle_{\lambda_g}) + \sigma^2_{\epsilon_g} vec(M_g) + \nu_g. \tag{13.17}$$

Finally, notice that (13.17) is a simple linear model that can be rewritten as:

$$R_g = C_g\eta_g + \nu_g,$$

where:

$$R_g = vec(r_{gOLS} r'_{gOLS}),$$
$$C_g = [\quad B_{\lambda_g} D \quad vec(M_g)],$$
$$B_{\lambda_g} = [vec(M_g H_{g,11,\lambda_g} M_g) \quad vec(M_g H_{g,12,\lambda_g} M_g) \quad \ldots \quad vec(M_g H_{g,KK,\lambda_g} M_g)],$$
$$\eta_g = [vech(\triangle_{\lambda_g})' \quad \sigma^2_{\epsilon_g}]'.$$

Now, the estimators of the elements of variance-covariance are obtained by minimizing the following penalized loss function:

$$L(\eta_g) = (R_g - C_g\eta_g)'(R_g - C_g\eta_g) + \tau \parallel \eta_g \parallel_2^2,$$

where τ is the penalisation parameter.

Notice, that for identification of η_g we implicitly assume:

Assumption 14 : $E[\nu_g C_g] = 0.$ □

Assumption 22 states that the error term ν_g is orthogonal to the covariates included in C_g. The penalization term using the l_2-norm allows us to tackle the problem of high multicollinearity in the matrix $C_g' C_g$. We follow Hoerl, Kannard and Baldwin (1975), Cule and De Iorio (2012) by estimating τ from the data as follows:

$$\hat{\tau} \geq \frac{\hat{\sigma}^2}{\hat{\beta}'_{OLS}\hat{\beta}_{OLS}},$$

with $\hat{\sigma}^2 = \frac{(y - X\hat{\beta}_{OLS})'(y - X\hat{\beta}_{OLS})}{NT-K-1}$.

Following Hoerl and Kennard (1970), we can prove that the MSE of $\hat{\beta}_{FGLS}$ is monotonically decreasing on τ. Thus, we can choose a $\tau > 0$ that minimizes MSE. The choice $\hat{\tau}$ is heuristic, and we acknowledge that it might be possible to derive an optimal estimator of τ (this is left for further research).

Remark 13.3 We could provide a Bayesian interpretation to the ridge estimation of the variance-covariance matrix under the assumption that the error term in the linearized variance-covariance matrix (13.17) follows a Gaussian distribution, and the parameters in the same equation also follow a Gaussian distribution. This is different from assuming that the random components in the baseline model (13.1) follow a normal distribution along with the coefficients. If we would like to use a Bayesian framework to estimate the variance-covariance matrix using the baseline model (13.1), we would need to assume an appropriate prior distribution for the variance-covariance matrix of the random components of the baseline model (13.1). For example, we could assume that the prior distribution of the inverse of the variance-covariance matrix is an inverse-Whisart. Under this assumption, one needs to use Markov Chain Monte Carlo simulation methods sample from the joint posterior distribution.

A Note on Large and Huge Sample Size

When the sample size is big, there are problems due to memory requirements for storing vectorized matrices. In order to tackle this issue and reduce the computing requirements by half, we modify the method proposed above using the vech operator instead of the vec operator. It is possible to do this replacement since we are dealing with square symmetric matrices.

$$R_g = vech(r_g r_g'),$$

$$C_g = [B_{\lambda,g} \quad vech(M_g)],$$

$$B_{\lambda_g} = [vech(M_g H_{g,11,\lambda_g} M_g) \quad vech(M_g H_{g,12,\lambda_g} M_g) \dots$$
$$\dots \quad vech(M_g H_{g,KK,\lambda_g} M_g)].$$

This modification improves the computational performance but has limitations. For big samples, one needs computational algebra methods for matrix inversion and multiplication.

13.6 Statistical Properties

In this section, we present the statistical properties of the stratum-specific estimators, the Mean Stratum estimator and the variance-covariance estimators, using sequential asymptotic theory with the number of individuals per stratum (N_g) growing to infinity and the time dimension (T_{i_g}) fixed. This implies that the total number of observations per stratum ($n_g = \sum_{i_g}^{N_g} T_{i_g}$) grows to infinity.

As mentioned in Section 13.2, we use the indexes i_g to refer to individual i belonging to stratum g and t_{i_g} for the time observation t of individual i_g.

13.6.1 Stratum Specific GLS Estimator

Theorem 13.1 *If i) Assumptions 9 to 23 and 21 hold, ii)* $\{y_{i_g}, x_{i_g}\}_{i_g=1}^{N_g}$ *is a sequence of random vectors containing* T_{i_g} *observations* $\forall g$*, iii)* $N_g \to \infty$ *and* T_{i_g} *fixed* ($n_g \to \infty$)*, then*

a) $\hat{\theta}_{g,GLS} \xrightarrow{p} \theta_g$*,* *b)* $\sqrt{n_g}(\hat{\theta}_g - \theta_g) \xrightarrow{d} N(0, Q_g)$*.*

where $Q_g = \underset{n_g \to \infty}{plim}(n_g^{-1} Z_g' \Omega_g^{-1} Z_g)^{-1}$*.*

Proof: See Appendix.

13.6.2 Variance-Covariance Matrix Estimators

Theorem 13.2 *If i) Assumptions 9 to 23 and 21 hold, ii)* $\underset{n_g \to \infty}{plim} \sum_{i_g}^{N_g} \sum_{t_{i_g}}^{T_{i_g}} n_g^{-1} C_{i_g t_{i_g}} C_{i_g t_{i_g}}' = M_g$ *with* $||M_g||_F < \infty$*, iii)* $v_{i_g t_{i_g}} \sim iid(0, \sigma_v^2)$*, iv)* $\underset{n_g \to \infty}{lim} \sum_{i_g}^{N_g} \sum_{t_{i_g}}^{T_{i_g}} n_g^{-1} C_{i_g t_{i_g}} R_{i_g t_{i_g}} = 0$*, v)* $N_g \to \infty$ *and* T_{i_g} *fixed* ($n_g \to \infty$) *then*

a) $\hat{\Omega}_g \xrightarrow{p} \Omega_g$*, b)* $\sqrt{n_g}(\hat{\Omega}_g - \Omega_g) \xrightarrow{d} N(0, var(\hat{\Omega}_g))$*.*

Proof: See Appendix.

13.6.3 Mean Stratum GLS Estimator

Corollary If i) Assumptions of theorems 13.1 and 13.2 hold $\forall g$, then

$$\sqrt{N}(\hat{\bar{\theta}} - E[\theta_g]) \xrightarrow{d} N(0, Q),$$

where $Q = \sum_g \pi_g^2 Q_g$. □

13.7 Misspecification of the Variance-Covariance Matrix

As mentioned in Section 13.5, the FGLS estimator in step 1 is non-robust to violations of the assumptions 15 and 18 that state that λ_{git} and ϵ_{git} are not serially correlated and homoskedastic within stratum. The reason is that under contemporaneous exogeneity, the FGLS is consistent only if the variance-covariance matrix of the model is diagonal. When the variance-covariance matrix of the model is not diagonal, one requires the stronger condition of strict exogeneity of all regressors included in the model. But in model (13.1), the strict exogeneity of all right-hand side regressors does not hold because of the presence of the lag of the dependent variable (Wooldridge, 2010). Then if λ_{git} and ϵ_{git} are serially correlated or/and heteroskedastic within stratum, it is better to estimate the stratum-specific parameters using OLS with a fully robust variance estimator.

The correlation of ϵ_{git} and/or λ_{git} within stratum could be caused due to clustering within stratum. For instance, students within schools belong to the same village. In this situation, we can use a one-way cluster fully robust variance estimator per stratum. If we index by j_g the cluster in stratum g and we assume that there is no cross-correlation across clusters, we can use the following within stratum one-way fully-robust variance estimator:

$$\widehat{Var(\hat{\beta}_g)} = (\sum_{j_g} X'_{j_g}\hat{\Omega}_{j_g}^{-1}X_{j_g})^{-1}(\sum_{j_g} X'_{j_g}\hat{\Omega}_{j_g}^{-1}\hat{u}_{j_g}\hat{u}'_{j_g}\hat{\Omega}_{j_g}^{-1}X_{j_g})(\sum_{j_g} X'_{j_g}\hat{\Omega}_{j_g}^{-1}X_{j_g})^{-1}.$$

The estimator is fully-robust for heteroskedasticity and serial-correlation within cluster j_g using $m_{j_g}^{-1}\sum_{j_g}\hat{u}_{j_g}\hat{u}'_{j_g}$ as an estimator of $E[u_{j_g}u'_{j_g}]$. If the number of clusters within stratum (m_{j_g}) grows and the number of observations within the clusters per stratum is fixed, the Wald t-statistic is asymptotically normal (Wooldridge, 2003, Cameron & Miller, 2015). If the number of clusters within stratum is fixed, the cluster within stratum robust variance-covariance estimator is downward-biased (Cameron & Miller, 2015, Wooldridge, 2003) and the Wald t-statistic is no longer asymptotically normal distributed (Cameron & Miller, 2015, Wooldridge, 2003). In this situation, the wild-cluster within stratum bootstrap-t method proposed by Cameron, Gelbach and Miller (2008) could be used.

Another issue is that the one-way cluster within stratum fully robust variance estimator is valid under the assumption that observations are not correlated across clusters within stratum. A solution is using a two-way cluster within stratum fully-robust variance estimator, but this estimator requires that the number of clusters within stratum and the number of time observations per individual within cluster per stratum grow to infinity. If this is not the case, we can use the two-way wild-cluster within stratum bootstrap-t method.

13.8 Relationship of the Mean Stratum Estimator with other Estimators

Here we give some comparison results omitting the proofs for the sake of space.

- The FGLS estimator of two-level panel data containing interactions of stratum dummies with one-hot encoding $(s_i^{(g)})$ with the regressors in the model is not equivalent to the MS-FGLS estimator.
- The first-difference GMM (Arellano & Bond, 1991) estimator of model (13.1) is an inconsistent estimator of $E[\rho_g]$ and $E[\beta_{git}]$ and it is equal to a weighted average of the stratum-specific parameters.
- The Mean Group estimator is infeasible when T is 3. When T is big, the Mean-Stratum estimator and the Mean Group estimator are consistent estimators of the average partial effects of model 13.1 under stratified sampling.

13.9 Relationship between the Baseline Model and Two-level Panel Data

Model (13.1) is related to a heterogeneous dynamic model for two-dimensional panel data under special conditions. To see this, let us consider the following model:

$$y_{it} = \alpha_i + \rho_i y_{it-1} + x'_{it}\beta_{it} + \varepsilon_{it}, \qquad i = 1, 2, ..., N, \quad t = 1, 2, ..., T_i. \tag{13.18}$$

The above specification needs further structuring of the β_{it}s before proceeding to the estimation stage. Below we show that by making the following assumption on the β_{it} parameters, along with some additional assumptions on α_i and ρ_i, we get to our baseline model (13.1) from the above two-level dynamic model.

Assumption 15 : The slope coefficients are conditional mean dependent on stratum belonging

$$E[\beta_{it}|s_i^{(g)}, x_{i1}, x_{i2}, ..., x_{iT}] = \beta_g.$$

This assumption is equivalent to Assumption 15.

Assumption 16 : The individual additive unobserved effect is homogeneous within stratum

$$\alpha_i = \alpha_g \quad \forall i \in g.$$

Under this assumption, the correlation of the additive unobserved individual heterogeneity with the regressors is equal within strata. For instance, the innate ability of workers is equal within city. This is feasible if workers self-select into a city based on their ability.

Assumption 17 : The individual persistence parameter is homogeneous within stratum

$$\rho_i = \rho_g \quad \forall i \in g.$$

This means that the persistence of the dynamic process is equal within strata. An example of homogeneous persistence is equal consumption persistence within village. The homogeneity of consumption persistence within village could happen if village characteristics drive consumption habits.

The use of three-level or multi-dimensional panel data models surged due to the increasing availability of big data (Mátyás, 2017, Sarafidis & Wansbeek, 2021). They are appealing because they can 1) control for unobserved heterogeneity that is not only individual and/or time specific (Sarafidis & Wansbeek, 2021), 2) accommodate the classification of each individual into strata or groups (Sarafidis & Wansbeek, 2021), 3) deal with incidental parameter bias, 4) develop appropriate inference that take into account sampling uncertainty.

13.10 Specification Tests

In order to test the assumption of strata-specific heterogeneity, we propose two specification tests that are extensions of the Hausman test (Hausman & Taylor, 1981).

First, testing the null hypothesis of stratum additive and multiplicative heterogeneity versus stratum-individual additive and multiplicative heterogeneity is not feasible when the time dimension is as short as 3.

Second, testing the null hypothesis of complete homogeneity versus stratum additive and multiplicative heterogeneity is possible. In this case, we propose to compare the Mean Stratum estimator with the Pooled OLS estimator. More specifically, the null and alternative hypothesis are the following:

$H_o : \hat{\beta}_{MC}$ consistent and inefficient, $\hat{\beta}_{POLS}$ consistent and efficient.
$H_1 : \hat{\beta}_{POLS}$ inconsistent and $\hat{\beta}_{MC}$ consistent and most efficient.
The statistic is given by:

$$Q = (\hat{\beta}_{MC} - \hat{\beta}_{POLS})' Var(\hat{\beta}_{MC} - \hat{\beta}_{POLS})^{-1} (\hat{\beta}_{MC} - \hat{\beta}_{POLS}),$$

follows a $\chi^2_{df=K}$.

In addition, testing the null hypothesis of stratum additive and multiplicative effects versus stratum-individual additive and stratum multiplicative heterogeneity is also feasible. In this case, we propose to use a Hausman-type test that compares Mean Stratum estimator s vs. a Mean Stratum First-difference GMM estimator or the Mean Stratum estimator using a Mundlak approach .

The study of the statistical properties of these tests is left for further research.

13.11 Relaxing the Assumption of Stratum Additive Specific Effects

13.11.1 Initial Conditions Generated from a Stationary Distribution

In this subsection, we relax the assumption of additive stratum specific effects and allow for the presence of additive stratum-individual correlated random effects. Therefore, Assumption 13 is replaced by the following one:

Assumption 18 : Correlated stratum-individual additive specific random effects α_{gi}. □

The inclusion of stratum-individual additive effects allows to control for endogeneity of the regressors that might not be captured by the stratum additive fixed effects. In particular, we consider the following extension of model (13.1):

$$y_{git} = \alpha_{gi} + \rho_g y_{git-1} + x'_{git}\beta_{git} + \varepsilon_{git}, \quad t = 1, ..., T_{i_g}, \tag{13.19}$$

where α_{gi} is a stratum-individual specific correlated random effect.

The estimation of model (13.19) with short time dimension has two main problems: i) the incidental parameter bias caused by the presence of the stratum-individual specific effects and ii) the impact of unobserved initial values (y_{gi0}) on the estimation.

In order to deal with the incidental parameter bias, we use a mean conditional approach instead of a linear difference approach. We choose the mean conditional approach because it is appropriate for heterogenous dynamic panel data models. As explained by Hsiao (2020), in this approach it is needed to use a linear approximation of $E(\alpha_{gi}|x_{it})$ to model the correlation of the regressors with the stratum-individual unobserved effects (This was a suggestion of Mundlak (1961) and Chamberlain (1979)). Following this suggestion, we re-express α_{gi} as a linear projection on the individual means of the regressors :

$$\alpha_{gi} = \bar{x}'_{gi.}\varphi_g + \upsilon_{gi}, \tag{13.20}$$

where $\bar{x}_{gi.} = T^{-1}\sum_{t=1}^{T} x_{git}$, υ_{gi} is an orthogonal error term such that $E(\upsilon_{gi}|\bar{x}_{gi.}) = 0$, and φ_g is a vector of unobserved parameters.

This linear projection can be replaced in model (13.19) obtaining:

$$y_{git} = \bar{x}'_{gi.}\varphi_g + \rho_g y_{git-1} + x'_{git}\beta_{git} + \upsilon_{gi} + \varepsilon_{git}, \quad t = 1, ..., T_{i_g}. \tag{13.21}$$

Now, we are only left with the problem of unobserved initial conditions dependency. Modifying Assumption 19 to allow for the presence of stratum-individual additive effects yields:

$$y_{gi0} = \rho_g^{h_{ig}} y_{gi,-h_{ig}} + \alpha_{gi}\frac{1-\rho_g^{h_{ig}}}{1-\rho_g} + \sum_{l=0}^{h_{gi}} \rho_g^l x'_{gi-l}\beta_{gi-l} + \sum_{l=0}^{h_{ig}} \rho_g^l \varepsilon_{gi-l}. \tag{13.22}$$

If we assume that $h_{g_i} \to \infty$, we can re-write the initial conditions as follows:

$$y_{gi0} = \frac{\alpha_{gi}}{1-\rho_g} + \sum_{l=0}^{\infty} \rho_g^l x'_{gi-l}\beta_{gi-l} + \sum_{l=0}^{\infty} \rho_g^l \varepsilon_{gi-l}. \tag{13.23}$$

Following Hsiao (2020), we re-call the terms of equation (13.23) such that the equation of the initial values is:

$$y_{gi0} = \frac{\alpha_{gi}}{1-\rho_g} + \psi_{gi0} + \varepsilon_{0i}. \tag{13.24}$$

Replacing the linear projection of the individual effects on the individual mean of the regressors to obtain:

$$y_{gi0} = \frac{\bar{x}'_{gi.}\varphi_g}{1-\rho_g} + \psi_{gi0} + \frac{v_{gi}}{1-\rho_g} + \varepsilon_{0i}. \tag{13.25}$$

In this equation, it is clear that we still have the problem of incidental parameters due to the presence of ψ_{gi0}. In order to deal with this issue, we follow Hsiao (2020) and assume that $E(\psi_{gi0}|x_{gi}) = \bar{x}'_{gi}\phi^*_g$. This is possible under Assumptions 19 and 20.

The combination of (13.21), (13.25), and $E(\psi_{gi0}|x_{gi}) = \bar{x}'_{gi}\phi^*_g$ leads to the system of equations:

$$\begin{aligned} y_{git} &= \bar{x}'_{gi.}\varphi_g + \rho_g y_{git-1} + x'_{git}\beta_{git} + \varepsilon^*_{git}, \quad t = 1, \ldots, T_{i_g}, \\ y_{gi0} &= \frac{\bar{x}'_{gi.}\varphi_g}{1-\rho_g} + \bar{x}'_{gi}\phi^*_g + \frac{v_{gi}}{1-\rho_g} + \varepsilon_{0i}. \end{aligned} \tag{13.26}$$

where $\varepsilon^*_{git} = \varepsilon_{git} + v_{gi}$.

For estimation of the system (13.26), we propose two different methodologies. The first one is a Bayesian hierarchical estimator with a prior for the initial conditions. The second is an estimation of the equation of the dependent variable conditional on the initial conditions. These methods are described in the following subsections:

Bayesian hierarchical estimation

In order to set up the Bayesian hierarchical estimator, we define the likelihood of the observed data by:

$$L_{\zeta|y,y_{-1},X} = \prod_{g}^{m}\prod_{i}^{N_g} L(\zeta_g|y_{gi}, X_{gi}), \tag{13.27}$$

where $\zeta_g = [\rho_g \quad \beta_g \quad \phi_g \quad \sigma^2_{\varepsilon^*}]'$, $\zeta = [\zeta_1 \quad \zeta_2 \quad \ldots \quad \zeta_m]'$, $L(\zeta_g|y_{gi}, X_{gi}) = f(y_{gi}|X_{gi};\zeta_g)$ with $f(y_{gi}|X_{gi};\zeta_g)$ representing the multivariate normal distribution with variance equal to $\sigma^2_\varepsilon I_T + \sigma^2_v \iota_T \iota'_T$ and with expectation equal to $\mu_{y,gi} = \rho_g y_{gi-1} + diag(x_{gi})\beta_{gi} + \iota_{T_{ig}}\bar{x}'_{gi.}\varphi_g$. The prior distributions for the fixed parameters are:

$$(\beta_g|\beta) \sim N(\beta, H_{\Delta_{\alpha,2}}H'_{\Delta_{\alpha,2}});\ (\rho_g|\rho) \sim N(\rho, \sigma^2_\rho)\ (\varphi_g|\varphi) \sim N(\varphi, FF').$$

While the prior for the random effects is:

$$\lambda_{git} \sim N(0, H_{\Delta_\lambda}H'_{\Delta_\lambda});\ H_{\Delta_\lambda} \sim LKJ(2).$$

The prior distribution of the variance σ^2_ε is half-normal with a location parameter equal to 0.5 and a scale parameter equal to 0.2. The prior distribution of the lower triangular matrix H_{Δ_λ} is

Lewandowski-Kurowicka-Joe (LKJ) with parameter equal to 2. The value of the parameter of the LKJ prior means that the matrix has a low correlation.
Notice that the prior set-up imposes a non-centered parametrization on β_{git} such that:

$$\beta_{git} = \beta_g + H_{\Delta_\lambda} z_{git}, \tag{13.28}$$

where z_{git} is a standard multivariate normal variable and H_{Δ_λ} is the Cholesky factor of the variance-covariance matrix of λ_{git}.
This non-centered parameterization improves the convergence of the Hamiltonian Monte Carlo (HMC) algorithm because it reduces the correlation of the parameters (Frühwirth-Schnatter & Tüchler, 2008, Betancourt & Girolami, 2015). This reduction of the correlation permits the exploration of the whole parameter space improving the mixing of the chains.

Remark 13.4 According to Rossi and Allenby (2009) and Rendon (2013), imposing prior distributions only for the parameters of the model leads to a fixed effects specification. Thus, there is not any prior specification for the hyper-parameters of the priors. Therefore, a Bayesian model for a fixed effects specification has only first-stage priors while a Bayesian model for a random effects specification includes second-stage or hyper-priors.

Under the simplifying assumption that y_{gi0} is known, we could just set up a naive Bayesian estimator. But the assumption that y_{gi0} is fixed is not plausible. Its failure leads to inconsistent estimates. This is why, we relax it and set up the following prior distribution for the initial conditions:

$$f_{y_{gi0}} \sim N(\mu_{0,gi}, \sigma^2_{y_0}), \tag{13.29}$$

where $\mu_{0,gi} = \frac{\alpha_{gi}}{1-\rho_g} + \bar{x}'_{gi}\phi^*_g$, and the prior distribution of the variance $\sigma^2_{y_0}$ is half-normal with location parameter equal to 0.5 and scale parameter equal to 0.2.

Remark 13.5 Assuming that y_{gi0} comes from the stationary distribution means that the initialization of the process happened a long time ago ($h_{i_g} \to \infty$). This implies that the parameter b_g is equal to 0.

Conditioning on the initial value
Another option for consistent estimation of the parameters of interest is the estimation of model 13.19 after conditioning on the initial value. For this purpose, we follow Hsiao (2020) and condition the first equation of the system 13.26 on the initial value y_{gi0} leading to:

$$y_{git} = \bar{x}'_{gi.}\varphi^*_g + \rho_g y_{git-1} + x'_{git}\beta_{git} + y_{gi0}\tilde{b}_g + \varepsilon^*_{git}, \quad t = 1, \ldots, T_{i_g} \tag{13.30}$$

Estimation of this model can be done using the Mean Stratum estimator with FGLS in the first stage.

When the initial conditions are not generated from a stationary distribution, one can also propose appropriate Bayesian hierarchical estimators or condition on the initial value. We leave this out for the sake of space.

13.12 Cross-sectional Dependence

13.12.1 A Model Including Common Factors

The models (13.1) and (13.19) do not consider cross-sectional correlation even though cross-sectional dependence is a common problem in panel data.

Cross-sectional dependence is caused by spatial dependence or common shocks (Bai & Li, 2021) and it can be modeled either using spatial or factor models or a combination of both.

In this section, we extend model (13.1) in order to allow for cross-sectional dependence using a factor model. For this purpose, we include a stratum-time-specific fixed effect since it represents a stratum common factor. This is possible because the stratum-time specific effect τ_{gt} can be rewritten as $\sum_i^m s_i^{(g)} f_{gt}$ with $s_i^{(g)}$ equal to 1 and 0 (Bonhomme & Manresa, 2015, Kapetanios, Mastromarco, Serlenga & Shin, 2017, Bai & Li, 2021). Additionally, we include time-specific effects that capture common global factors across strata.

The extended model includes stratum-time additive effects as well as time-fixed effects as common factors for individual i in stratum g as follows:

$$y_{git} = \alpha_g + \gamma_t + \tau_{gt} + \rho_g y_{git-1} + x'_{git}\beta_{git} + \varepsilon_{git}, \quad t = 1, ..., T_{i_g}. \tag{13.31}$$

In this setting, Assumption 20 is relaxed to allow for regressors that present common factors.

Assumption 19 : x_{git} are generated from:

$$x_{git} = \mu_g + \gamma_t + \tau_{gt} + \rho_x x_{git-l} + \omega_{git}, \qquad |\rho_x| < 1.$$

13.12.2 Identification and Estimation

The Mean Stratum estimator presented in Section 13.5 consistently estimates the parameters of interest of model 13.31, which includes time and stratum-time dummies, by exploiting the different moment conditions derived in this subsection.

We obtain moment conditions using the deviations with respect to stratum-time specific averages:

$$\begin{aligned} y_{git} - y_{g.t} &= \rho_g (y_{git-1} - y_{g.t-1}) + (x_{git} - x_{g.t})'\beta_g \\ &+ x'_{git}\lambda_{git} - x'_{g.t}\lambda_{g.t} + \varepsilon_{git} - \varepsilon_{g.t}. \end{aligned} \tag{13.32}$$

The stratum-time specific averages are equal to:

$$\frac{\sum_i y_{git}}{N_g} = \alpha_g + \gamma_t + \tau_{gt} + \rho_g \frac{\sum_i y_{git-1}}{N_g} + \frac{\sum_i x_{git}}{N_g}\beta_g + \frac{\sum_i x'_{git}\lambda_{git}}{N_g} + \frac{\sum_i \varepsilon_{git}}{N_g}. \tag{13.33}$$

We can just rename the transformed variables as:

$$\tilde{y}_{git} = \rho_g \tilde{y}_{git-1} + \tilde{x}'_{git}\beta_g + \widetilde{x'_{git}\lambda_{git}} + \tilde{\varepsilon}_{git}. \tag{13.34}$$

Thus, after this transformation we obtain the following moment conditions:

$$E(\tilde{u}_{git}\tilde{x}_{gis}) = 0, \quad s = 1, 2, ..., T, \quad i = 1, 2, ..., N_g, \quad g = 1, 2, ..., m, \tag{13.35}$$

$$E(\tilde{u}_{git}\tilde{y}_{git-1}) = 0, \quad t = 1, 2, ..., T, \quad i = 1, 2, ..., N_g, \quad g = 1, 2, ..., m. \tag{13.36}$$

In addition, we need to add the full-rank condition for the transformed regressors.

Assumption 20 :
For OLS: The matrix $E(\tilde{z}_{git}\tilde{z}'_{git})$ is full rank.
For GLS: $E[\tilde{u}_g\tilde{u}'_g]$ is positive definite and $E(\tilde{Z}'_g E[\tilde{u}_g\tilde{u}'_g]^{-1}\tilde{Z}_g)$ is nonsingular. □

13.13 Long Time Dimension

Until now, we have focused on a dynamic panel data model with stratification and short-time dimension. As mentioned, the problems in this setting are incidental parameters and initial conditions dependency. When the time dimension is long, one does not run into the problem of initial conditions dependency but the issues of non-stationarity and incidental parameter bias remain.

More specifically, if the time dimension is long the assumption that the initial conditions are generated from a stationary distribution is no longer needed (Assumption 19). The reason is that the influence of the initial conditions becomes negligible as $T \longrightarrow \infty$.

However, the assumptions of stationary regressors and stationary dependent variables are necessary for the consistency and asymptotical normality of the Mean Stratum estimator . The reason is that the stationarity of the regressors guarantees that the error term of the model is integrated of order 0. In the case of non-stationary regressors, we have two options: 1. transform the regressors to obtain stationarity or 2. estimate the model in levels if there is co-integration between the dependent variable and the regressors after including the lag of the dependent variable (Hamilton, 1994). In the last case, the assumption of stratum-additive effects is crucial to obtain asymptotically normal estimates within stratum (Choi, 2015). While in the presence of stratum-individual additive effects, it is necessary to use the fully-modified OLS estimator proposed by Phillips and Moon (1999) stratum per stratum or one can use OLS estimation with the Mundlak approach. If the dependent variable and the regressors are not cointegrated and the model presents stratum-specific additive effects, it is unclear if stratum OLS estimation and the Mean-Stratum estimator are consistent. The reason is that Phillips and Moon (1999) show that pooled OLS is a consistent estimator of the long-run average regression coefficient if the regressors are nonstationary and there is no cointegration for a model without intercept and lagged dependent variable. Thus, further research is needed to verify the consistency of the Mean Stratum estimator when there is no co-integration and the model presents stratum additive specific effects. However, the MS-OLS estimator is consistent if the model presents additive stratum-individual specific effects, and there is no cointegration if we use the Mundlak approach (Phillips & Moon, 1999). Finally, in order to test for co-integration one can extend the test proposed by Im et al. (2003) such that the model presents stratum-specific parameters instead of individual-specific parameters. Concluding that there is co-integration would entail that $u_{git} = x'_{git}\lambda_{git} + \varepsilon_{git}$ is stationary, meaning that λ_{git} could be considered as a random co-integrating vector. A study of a co-integration test and the properties of the Mean-stratum estimator when there is no co-integration is out of the scope of this paper and both issues are left for further research.

On the other hand, the problem of incidental parameter bias requires careful analysis. First, the problem of incidental parameter bias is not present in model 13.1. The intuitive explanation is that we have increasing observations to estimate stratum-specific parameters. But if we allow for stratum-individual specific effects as in model 13.19, we need to be more careful. In this setting, the estimated stratum-specific parameters using the within estimator are consistent and asymptotically normal if $lim\frac{N_g}{T_{ig}} = 0$ and the regressors are not stationary (Phillips & Moon, 1999). If the regressors are stationary, we must debias the within estimator per stratum (Hahn & Newey, 2004). A workaround to avoid the condition $lim\frac{N_g}{T_{ig}} = 0$ or debiasing is to use the Mundlak approach and project the stratum-individual specific effects onto the column space of the regressors. Finally, model (13.31) suffers from the problem of incidental parameter bias due to the presence of stratum-individual specific effects and stratum-time specific effects. But the transformation proposed in subsection 13.12.2 eliminates the incidental parameter problem.

Finally, the Mean Stratum estimator and the Mean Group estimator are consistent estimators of the mean coefficients of model (13.1) when the time dimension is long. In addition, the MS estimator remains consistent even when T is short.

13.14 Monte Carlo Simulation Experiment: Stratified Sampling

In this section, we present a Monte Carlo simulation experiment to test the proposed estimators for the baseline model and the extensions of the baseline model under stratified sampling.

For this purpose, we generate 100 datasets from two different data-generating processes called DGP 1 and DGP 2.

In the following subsections, we describe the different designs in more detail as well as the results.

13.14.1 The Design

13.14.1.1 DGP 1

In order to test the Mean Stratum estimator proposed for a model with unobserved stratum heterogeneity and mixed coefficients (model 13.1), we conduct a simulation experiment using a data-generating process that is similar to the DGP used by Arellano and Bond (1991). We use a modification of the DGP proposed by Arellano and Bond (1991) to illustrate that the first-differenced GMM estimator breaks down in the presence of multiplicative unobserved stratum heterogeneity.

The main differences with the DGP of Arellano and Bond (1991) are: 1. inclusion of stratum additive effects instead of individual-specific effects that are correlated with the regressors, 2. inclusion of multiplicative stratum-individual-time specific effects, 3. the variance and variance-covariance are stratum-specific and they are generated from Gamma and Wishart distributions. More specifically, we generate 100 samples from the following model for individual i in stratum g at period t:

$$y_{git} = \alpha_g + \rho_g y_{git-1} + x'_{git}\beta_{git} + \varepsilon_{git},$$

with $\rho_g = \bar{\rho} + \alpha_{2,g}, \bar{\rho}, \beta_{git} = \bar{\beta} + \alpha_{3,g} + \lambda_{git}$ and $\bar{\beta} = \begin{pmatrix} 1 \\ 1 \end{pmatrix}$.

The number of strata is equal to 4, the number of individuals within stratum is equal to 100, and the number of time observations is equal to 3.

The stratum additive effects α_g are generated from a normal distribution centered at 0 with heteroskedastic variance across strata ($\sigma^2_{\alpha_g} \in \{1.01, 1.01, 0.9, 0.9\}$).

The stratum effects ($\alpha_{2,g}$) added to the persistence parameter ($\bar{\rho}$) are centered at 0, and equal to $\alpha_{2,g} \in \{-0.5, -0.5, 0.5, 0.5\}$.

The stratum effects ($\alpha_{3,g}$) added to the mean coefficient vector ($\bar{\beta}$) are centered at 0, and equal to $\alpha_{3,g} \in \{-0.5, -0.5, 0.5, 0.5\}$.

The regressors $x_{git} \in \mathbb{R}^2$ follow stationary autoregressive processes similar to the process used by Arellano and Bond (1991). The key difference is that we allow for correlation with the stratum effects:

$$x_{git} = \alpha_g + \alpha_{3,g} + \phi x_{git-1} + \omega_{git},$$

with ϕ equal to 0.8.

The disturbance term of the regressors (ω_{git}) equation is sampled from the normal distribution centered at 0 with variance that is stratum specific ($\sigma^2_{\omega_g} \in \{0.9, 0.9, 1.01, 1.01\}$).

The stratum-individual-time specific effects (λ_{git}) added to the mean coefficient vector ($\bar{\beta}$) are generated from a multivariate normal distribution centered at 0 with heteroskedastic variance-

covariance matrix across strata ($\Delta_{\lambda,1} = \begin{pmatrix} 0.1 & 0.05 \\ 0.05 & 0.1 \end{pmatrix}, \Delta_{\lambda,2} = \begin{pmatrix} 0.11 & 0.05 \\ 0.05 & 0.11 \end{pmatrix}, \Delta_{\lambda,3} = \begin{pmatrix} 0.12 & 0.05 \\ 0.05 & 0.12 \end{pmatrix}, \Delta_{\lambda,4} = \begin{pmatrix} 0.13 & 0.05 \\ 0.05 & 0.13 \end{pmatrix}$).
The disturbance term (ε_{git}) is generated from a normal distribution centered at 0 with a stratum heteroskedastic variance ($\sigma^2_{\varepsilon_g} \in \{0.9, 0.9, 1.01, 1.01\}$).

13.14.1.2 DGP 2

In order to test the estimator proposed in subsection 13.11.1, we conduct a simulation experiment using a data-generating process similar to DGP 1.

DGP 2 is the same as as DGP 1 except for the following :
1. Inclusion of correlated stratum-individual effects instead of individual-specific effects,
2. The variance and variance-covariance of the stratum-individual-time specific effects is equal across strata.
More specifically, the stratum-individual additive effects $\alpha_{1,gi}$ are generated from a normal distribution centered at 0 with variance equal to 1.
3. The stratum-individual-time specific effects (λ_{git}) added to the mean coefficient vector ($\bar{\beta}$) are generated from a multivariate normal distribution centered at 0 with with variance-covariance matrix equal across strata $\Delta_{\lambda,1} = \begin{pmatrix} 0.1 & 0.05 \\ 0.05 & 0.1 \end{pmatrix}$.
The disturbance term (ε_{git}) is generated from a normal distribution centered at 0 with a stratum heteroskedastic variance ($\sigma^2_{\varepsilon_g} \in \{0.9, 0.9, 1.01, 1.01\}$).

13.14.2 The Results

13.14.2.1 DGP 1

In Table 13.1, we present the bias and RMSE of the estimated mean parameters of interest for different values of the persistence parameter. The estimates are obtained for 100 simulations for a sample with 4 strata, 100 individuals per group, and a time dimension equal to 3.

The results show that the proposed Mean Stratum FGLS estimators have lower bias and RMSE than the first-differenced GMM estimators.

13.14.2.2 DGP 2

In Table 13.2, we present the bias and RMSE of the estimated mean parameters of interest for different values of the persistence parameter. The estimates are obtained for 100 simulations for a sample with 4 strata, 100 individuals per group and a time dimension equal to 3. The results show that the proposed Mean Stratum FGLS estimators have lower bias and RMSE than the first-differenced GMM estimators and the system GMM estimators.

Table 13.1: DGP 1

	Mean	Bias	RMSE	Mean	Bias	RMSE	Mean	Bias	RMSE
	$\rho = 0.1$			$\beta_1 = 1$			$\beta_2 = 1$		
MS-OLS	0.0935	-0.0065	0.0036	1.0055	0.0055	0.0172	1.0025	0.0025	0.0161
MS-FGLS	*0.1030*	*0.0030*	*0.0036*	*1.0046*	*0.0046*	*0.0089*	1.0046	0.0046	0.0093
MS-JIFDGMM	0.5649	0.4649	85.1471	1.4842	0.4842	14.9840	0.6957	-0.3043	8.0302
JIFDGMM	-0.4678	-0.5678	21.8730	1.1253	0.1253	1.2624	1.0540	0.0540	0.7525
MC-OIFDGMM	0.5649	0.4649	85.1471	1.4842	0.4842	14.9840	0.6957	-0.3043	8.0302
OIFDGMM	0.0749	-0.0251	0.0168	0.9938	-0.0062	1.0178	0.9364	-0.0636	1.2441
MS-SYSGMM	0.0723	-0.0277	0.0074	1.1414	0.1414	0.0400	1.1130	0.1130	0.0407
SYSGMM	0.0949	-0.0051	0.0088	1.1264	0.1264	0.0429	1.0925	0.0925	0.0404
	$\rho = 0.5$			$\beta_1 = 1$			$\beta_2 = 1$		
MS-OLS	0.4921	-0.0079	0.0039	1.0191	0.0191	0.0171	1.0171	0.0171	0.0162
MS-FGLS	*0.4966*	*-0.0034*	*0.0288*	*1.0761*	*0.0761*	*0.2015*	*1.0190*	*0.0190*	*0.0357*
MC-FDGMM	-0.1210	-0.6210	108.3416	1.1141	0.1141	0.5770	0.8036	-0.1964	1.8841
FDGMM	0.5566	0.0566	0.1178	0.9929	-0.0071	0.0319	0.9725	-0.0275	0.0373
MS-OIGMM	-0.1210	-0.6210	108.3416	1.1141	0.1141	0.5770	0.8036	-0.1964	1.8841
OIGMM	0.4702	-0.0298	0.0217	1.1090	0.1090	0.8645	1.0439	0.0439	0.8128
MS-SYSGMM	0.4773	-0.0227	0.0056	1.1533	0.1533	0.0470	1.1282	0.1282	0.0526
SYSGMM	0.5159	0.0159	0.0062	1.1095	0.1095	0.0397	1.0724	0.0724	0.0503
	$\rho = 0.9$			$\beta_1 = 1$			$\beta_2 = 1$		
MS-OLS	0.8976	-0.0024	0.0019	1.0264	0.0264	0.0167	1.0247	0.0247	0.0161
MS-FGLS	*0.8975*	*-0.0025*	*0.0009*	*1.0297*	*0.0297*	*0.0090*	*1.0233*	*0.0233*	*0.0082*
MS-JIFDGMM	0.9020	0.0020	0.0033	0.9972	-0.0028	0.0319	0.9825	-0.0175	0.0337
JIFDGMM	0.9088	0.0088	0.0028	0.9883	-0.0117	0.0306	0.9736	-0.0264	0.0359
MS-OIGMM	0.9020	0.0020	0.0033	0.9972	-0.0028	0.0319	0.9825	-0.0175	0.0337
OIGMM	0.8809	-0.0191	0.0074	1.4201	0.4201	1.6271	1.3970	0.3970	1.7727
MS-SYSGMM	0.9012	0.0012	0.0010	1.1236	0.1236	0.0381	1.0906	0.0906	0.0422

Note: MS-OLS: Mean-stratum OLS estimator, MS-FGLS: Mean-stratum FGLS estimator, MS-JIFDGMM: Mean-stratum just-identified fist-differenced GMM estimator, JIFDGMM: Just-identified fist-differenced GMM estimator, MC-OIFDGMM: Mean-stratum over-identified fist-differenced GMM estimator, OIFDGMM: Over-identified fist-differenced GMM estimator, MS-SYSGMM: System GMM estimator, SYSGMM: System GMM estimator.

Table 13.2: DGP 2

	Mean	Bias	RMSE	Mean	Bias	RMSE	Mean	Bias	RMSE
	$\rho = 0.1$			$\beta_1 = 1$			$\beta_2 = 1$		
MS-OLS	0.0931	-0.0069	0.001	1.0055	0.0055	0.0057	1.0055	0.0055	0.0062
MS-OLSy0	0.0931	-0.0075	0.002	1.0066	0.0066	0.0056	1.0066	0.0066	0.0064
MS-FGLS	0.0908	-0.0092	0.0123	1.0314	0.0314	0.0782	1.0314	0.0314	0.0241
MS-FGLSy0	*0.0988*	*-0.0012*	*0.0103*	*1.0073*	*0.0073*	*0.0246*	*1.0073*	*0.0073*	*0.9594*
MS-JIFDGMM	0.5545	0.4545	82.4716	0.8048	-0.1952	15.9684	0.8048	-0.1952	8.6445
JIFDGMM	0.1727	0.0727	3.4936	0.9846	-0.1952	0.0729	0.9846	-0.1952	0.3024
MS-OIGMM	0.0995	-0.0005	0.008	1.0919	0.0919	0.822	1.0919	0.0919	0.8789
OIGMM	0.1139	0.0139	0.0116	1.2076	0.2076	1.6085	1.2076	0.2076	1.3103
MS-SYSGMM	0.0901	-0.0099	0.0034	1.0851	0.0851	0.0303	1.0851	0.0851	0.0328
SYSGMM	0.1135	0.0135	0.0047	1.0937	0.0937	0.0356	1.0937	0.0937	0.0426
FGLS-Hsiao	-0.0999	-0.1999	0.0434	1.0156	0.0156	0.0149	1.0156	0.0156	0.017
	$\rho = 0.5$			$\beta_1 = 1$			$\beta_2 = 1$		
MS-OLS	0.4931	-0.0069	0.0005	1.0083	0.0083	0.0061	1.0083	0.0083	0.0064
MS-OLSy0	0.4931	-0.0101	0.0021	1.009	0.009	0.0059	1.009	0.009	0.0065
MS-FGLS	0.4941	-0.0059	0.0081	1.0197	0.0197	0.1547	1.0197	0.0197	0.1069
MS-FGLSy0	*0.4947*	*-0.0053*	*0.0261*	*1.0028*	*0.0028*	*0.0649*	*1.0028*	*0.0028*	*0.0924*
MS-JIFDGMM	1.4552	0.9552	144.1671	1.0147	0.0147	5.5668	1.0147	0.0147	5.7164
JIFDGMM	0.5373	0.0373	0.4727	0.9605	0.0147	0.0566	0.9605	0.0147	0.0507
MS-OIGMM	0.4861	-0.0139	0.0107	0.9989	-0.0011	0.7261	0.9989	-0.0011	0.841
OIGMM	0.4962	-0.0038	0.0193	1.0505	0.0505	1.4267	1.0505	0.0505	1.1413
MS-SYSGMM	0.4894	-0.0106	0.0025	1.0912	0.0912	0.0318	1.0912	0.0912	0.0359
SYSGMM	0.5257	0.0257	0.0043	1.0782	0.0782	0.0382	1.0782	0.0782	0.0438
FGLS-Hsiao	0.2641	-0.2359	0.0594	1.0283	0.0283	0.0156	1.0283	0.0283	0.0186
	$\rho = 0.9$			$\beta_1 = 1$			$\beta_2 = 1$		
MS-OLS	0.8974	-0.0026	0.0002	1.0053	0.0053	0.0061	1.0053	0.0053	0.0064
MS-OLSy0	0.8974	-0.0062	0.0013	1.0064	0.0064	0.006	1.0064	0.0064	0.0065
MS-FGLS	0.9036	0.0036	0.0043	1.0246	0.0246	0.2824	1.0246	0.0246	0.1272
MS-FGLSy0	*0.9285*	*0.0285*	*0.0649*	*1.0095*	*0.0095*	*0.1402*	*1.0095*	*0.0095*	*0.0832*
MS-JIFDGMM	0.9536	0.0536	0.1542	0.9805	-0.0195	0.0363	0.9805	-0.0195	0.1542
JIFDGMM	0.9122	0.0122	0.0031	0.9698	-0.0195	0.032	0.9698	-0.0195	0.0386
MS-OIGMM	0.8783	-0.0217	0.0075	1.0368	0.0368	1.1681	1.0368	0.0368	1.0338
OIGMM	0.8996	-0.0004	0.0052	1.2316	0.2316	1.5646	1.2316	0.2316	1.428
MS-SYSGMM	0.9015	0.0015	0.0005	1.0766	0.0766	0.0316	1.0766	0.0766	0.0331
SYSGMM	0.9256	0.0256	0.0014	1.0576	0.0576	0.0403	1.0576	0.0576	0.0407

MS-OLSy0: MS OLS conditioning on initial value, MS-FGLSy0: MS FGLS conditioning on initial value, MS-JIFDGMM: MS just-identified ist-differenced GMM estimator, JIFDGMM: Just-identified fist-differenced GMM estimator, MC-OIFDGMM: MS over-identified fist-differenced GMM estimator, OIFDGMM: Over-identified fist-differenced GMM estimator,MS-SYSGMM: System GMM estimator, SYSGMM: System GMM estimator.

13.15 Conclusions

In this paper, we investigate the identification and estimation of dynamic heterogeneous linear models in the presence of stratum heterogeneity when stratum structure is known and panel data is unbalanced due to randomly missing data with a short or fixed time dimension.

When the number of strata is fixed, we observe all the strata and the number of individuals grows to infinity, it is possible to consistently estimate the mean slope coefficients and the persistence parameter using a Mean Stratum estimator that is an extension of the Mean Group estimator introduced by Pesaran and Smith (1995).

As an extension of the baseline model, we consider a model with stratum-individual additive effects. In this setting, we suggest a hierarchical Bayesian estimation with a prior for the unknown initial conditions. In addition, we propose to condition on the initial values in order to avoid making assumptions about the data-generating processes of the initial conditions and the regressors.

A second extension is a model that allows for cross-sectional dependence by including a common factor for the whole population and a stratum-specific common factor. In this setting, the Mean Stratum OLS estimator using the time-demeaned regressors outperforms pooled OLS.

We can conclude from the simulation experiment, that the Mean Stratum estimators have lower Relative Bias and RMSE than the MG estimator and OLS estimator. This shows that one can exploit the underlying stratification in the data to estimate the mean coefficients and the stratum-specific parameters of a heterogeneous linear dynamic panel data models.

Further, we show that the first-difference GMM estimator is inconsistent when there is multiplicative stratum heterogeneity. In fact, the first-difference GMM estimator is equal to the weighted average of the stratum-specific marginal effects. A similar conclusion can be drawn if the marginal effects are individual-specific.

Finally, we show that the Mean Group estimator is equal to the Mean Stratum estimator when the time dimension is long and the data are obtained by means of stratified sampling.

Appendix

Proofs of Theorems 13.1, and 13.2

Proof Theorem 13.1

Proof The stratum-specific GLS estimator is given by:

$$\hat{\theta}_{g,GLS} = (Z_g'\Omega_g^{-1}Z_g)^{-1}(Z_g'\Omega_g^{-1}y_g).$$

We can re-write it as follows:

$$\hat{\theta}_{g,GLS} = \theta_g + (Z_g'\Omega_g^{-1}Z_g)^{-1}(Z_g'\Omega_g^{-1}w_g),$$

with:

$$w_g = diag(X_g)\lambda_g + \epsilon_g.$$

Applying the plim operator and using Slutksy's Theorem we obtain:

$$\underset{\substack{N_g\to\infty \\ T_{ig}\, fixed}}{plim} \hat{\theta}_{g,GLS} = \theta_g + (\underset{\substack{N_g\to\infty \\ T_{ig}\, fixed}}{plim} \frac{1}{n_g} Z_g'\Omega_g^{-1}Z_g)^{-1}(\underset{\substack{N_g\to\infty \\ T_{ig}\, fixed}}{plim} \frac{1}{n_g} Z_g'\Omega_g^{-1}w_g),$$

Now, $\underset{T_{ig}\,fixed}{\underset{N_g\to\infty}{plim}} n_g^{-1} Z_g'\Omega_g^{-1} Z_g = Q_g$ by Assumption 21, and $\underset{T_{ig}\,fixed}{\underset{N_g\to\infty}{plim}} \frac{1}{n_g} Z_g'\Omega_g^{-1} w_g = 0$ by Assumptions 16 and 17. The last conclusion is obtained as follows: $\underset{T_{ig}\,fixed}{\underset{N_g\to\infty}{plim}} \frac{1}{n_g} Z_g'\Omega_g^{-1} w_g = \underset{T_{ig}\,fixed}{\underset{N_g\to\infty}{plim}} \sum_{i_g} \sum_{t_{ig}} z_{i_g t_{ig}}\, \omega_{i_g t_{ig}}\, \sigma_{i_g t_{ig}}$ because Ω_g^{-1} is a diagonal matrix with $\sigma_{i_g t_{ig}}$ in each element of the diagonal. Then,

$$\underset{T_{ig}\,fixed}{\underset{N_g\to\infty}{plim}} \frac{1}{n_g} \sum_{i_g} \sum_{t_{ig}} z_{i_g t_{ig}}\, \omega_{i_g t_{ig}}\, \sigma_{i_g t_{ig}} = \frac{1}{T_{ig}} \sum_{t_{ig}} E_g [z_{i_g t_{ig}}\, \omega_{i_g t_{ig}}\, \sigma_{i_g t_{ig}}],$$

where $E_g [z_{i_g t_{ig}}\, \omega_{i_g t_{ig}}\, \sigma_{i_g t_{ig}}]$ represents the cross-sectional expectation. Now, under assumptions 16 and 17 $E_g [z_{i_g t_{ig}}\, \omega_{i_g t_{ig}}\, \sigma_{i_g t_{ig}}] = 0$. Then, $\hat{\theta}_{g,GLS} - \theta_g = o_p(1)$.
In order to derive the asymptotic distribution of the stratum-specific parameter, I use the stabilizing factor equal to $\sqrt{n_g}$ such that:

$$\sqrt{n_g}(\hat{\theta}_{g,GLS} - \theta_g) = (\frac{1}{n_g} Z_g'\Omega_g^{-1} Z_g)^{-1} (\frac{1}{\sqrt{n_g}} Z_g'\Omega_g^{-1} w_g).$$

By Linderberg-Levy Central Limit Theorem $\frac{1}{\sqrt{n_g}} Z_g'\Omega_g^{-1} w_g \to N(0, Q_g)$.
Then,

$$\sqrt{n_g}(\hat{\theta}_{g,GLS} - \theta_g) \xrightarrow{d} N(0, Q_g^{-1}).$$

Proof Theorem 13.2

Proof As derived in section 13.5.1, the variance-covariance components stacked up in the vector η_g are estimated by the penalized LS estimator as $\hat{\eta}_g = (C_g' C_g + \tau I)^{-1} (C_g' \hat{R}_g)$ with C_g a full rank matrix obtained following the procedure proposed there. Now, for $\tau = 0$:

$$\hat{\eta}_g = (C_g' C_g)^{-1} (C_g' \hat{R}_g)$$

That is equal to:

$$\hat{\eta}_g = \eta_g + (C_g' C_g)^{-1} (C_g' \nu_{git})$$

Now, applying the plim operator and using Slutsky's theorem:

$$plim \hat{\eta}_g = \eta_g + (plim \frac{1}{n_g} C_g' C_g)^{-1} (plim \frac{1}{n_g} C_g' \nu_g)$$

Now, $plim \frac{1}{n_g} C_g' C_g = D_g$ and $plim \frac{1}{n_g} C_g' \nu_g = 0$ because ν_g is an error capturing estimation error and it is orthogonal of C_g (Assumption 22). Thus, $\hat{\eta}_g = \eta_g$.
In order to derive the asymptotic distribution of the variance-covariance estimators, I use the stabilizing factor $\sqrt{n_g}$ such that:

$$\sqrt{n_g}(\hat{\eta}_g - \eta_g) = (\frac{1}{n_g} \sum_{i_g}^{N_g} \sum_{t_{ig}}^{T_{ig}} C_{i_g t_{ig}} C_{i_g t_{ig}}')^{-1} (\frac{1}{\sqrt{n_g}} \sum_{i_g}^{N_g} \sum_{t_{ig}}^{T_{ig}} C_{i_g t_{ig}} \nu_{i_g t_{ig}}).$$

Then, by Linderberg-Levy CLT we have that $\frac{1}{\sqrt{n_g}} \sum_{i_g}^{N_g} \sum_{t_{ig}}^{T_{ig}} C_{i_g t_{ig}} \nu_{i_g t_{ig}} \to N(0, \sigma_{\nu,g}^2 D_g)$. Thus,

$$\sqrt{n_g}(\hat{\eta}_g - \eta_g) \xrightarrow{d} (0, \sigma_{\nu_g}^2 D_g^{-1}).$$

Finally, $\Omega_g = g(\triangle_{\lambda_g}, \sigma^2_{\varepsilon_g})$ and $g(.)$ is a continuous function because it is a linear decomposition. As a result, it is possible to use the Slutzky's theorem to such that:

$$\sqrt{n_g}(\hat{\Omega}_g - \Omega_g) \xrightarrow{d} N(0, var(\hat{\Omega}_g)).$$

References

Arellano, M. & Bond, S. (1991). Some tests of specification for panel data: Monte Carlo evidence and an application to employment equations. *The Review of Economic Studies*, *58*(2), 277–297.

Avila Marquez, M. (2022). Identification and estimation of dynamic panel data models with clustering. *SSRN Electronic Journal, 4282164*. https://www.researchgate.net/publication/365682596_Identification_and_Estimation_of_Dynamic_Heterogeneous_Unbalanced_Panel_Data_Models_with_Clustering.

Bai, J. (2013). Fixed-effects dynamic panel models, a factor analytical method. *Econometrica*, *81*(1), 285–314.

Bai, J. & Li, K. (2021). Dynamic spatial panel data models with common shocks. *Journal of Econometrics*, *224*(1), 134-160. Retrieved from https://www.sciencedirect.com/science/article/pii/S0304407620303961 (Annals Issue: PI Day) doi: https://doi.org/10.1016/j.jeconom.2020.12.002

Betancourt, M. J. & Girolami, M. (2015). Hamiltonian Monte Carlo for Hierarchical Models. *Current trends in Bayesian methodology with applications*, *79*(30), 2–4.

Bonhomme, S. & Manresa, E. (2015). Grouped patterns of heterogeneity in panel data. *Econometrica*, *83*(3), 1147-1184. Retrieved from https://onlinelibrary.wiley.com/doi/abs/10.3982/ECTA11319 doi: https://doi.org/10.3982/ECTA11319

Bun, M. J. & Windmeijer, F. (2010). The weak instrument problem of the system GMM estimator in dynamic panel data models. *The Econometrics Journal*, *13*(1), 95–126.

Cameron, A. C., Gelbach, J. B. & Miller, D. L. (2008). Bootstrap-based improvements for inference with clustered errors. *The Review of Economics and Statistics*, *90*(3), 414–427.

Cameron, A. C. & Miller, D. L. (2015). A practitioner's guide to cluster-robust inference. *Journal of Human Resources*, *50*(2), 317–372.

Chamberlain, G. (1979). *Analysis of covariance with qualitative data.* NBER Working Paper No. w0325.

Choi, I. (2015). Panel cointegration. *The Oxford Handbook of Panel Data.*

Cule, E. & De Iorio, M. (2012). A semi-automatic method to guide the choice of ridge parameter in ridge regression. *arXiv:1205.0686*. https://arxiv.org/abs/1205.0686.

Dynan, K. E. (2000). Habit formation in consumer preferences: Evidence from panel data. *American Economic Review*, *90*(3), 391–406.

Frühwirth-Schnatter, S. & Tüchler, R. (2008). Bayesian parsimonious covariance estimation for hierarchical linear mixed models. *Statistics and Computing*, *18*(1), 1–13.

Hahn, J. & Newey, W. (2004). Jackknife and analytical bias reduction for nonlinear panel models. *Econometrica*, *72*(4), 1295–1319.

Hamilton, J. D. (1994). *Time series analysis*. Princeton University Press.

Hausman, J. A. & Taylor, W. E. (1981). Panel data and unobservable individual effects. *Econometrica: Journal of the Econometric Society*, 1377–1398.

Hoerl, A. E., Kannard, R. W. & Baldwin, K. F. (1975). Ridge regression: some simulations. *Communications in Statistics-Theory and Methods*, *4*(2), 105–123.

Hoerl, A. E. & Kennard, R. W. (1970). Ridge regression: Biased estimation for nonorthogonal problems. *Technometrics*, *12*(1), 55–67.

Hsiao, C. (2020). Estimation of fixed effects dynamic panel data models: linear differencing or conditional expectation. *Econometric Reviews*, *39*(8), 858–874.

Hsiao, C., Hashem Pesaran, M. & Kamil Tahmiscioglu, A. (2002). Maximum likelihood estimation of fixed effects dynamic panel data models covering short time periods. *Journal of Econometrics*, *109*(1), 107-150. Retrieved from https://www.sciencedirect.com/science/article/pii/S0304407601001439 doi: https://doi.org/10.1016/S0304-4076(01)00143-9

Hsiao, C., Li, Q., Liang, Z. & Xie, W. (2019). Panel data estimation for correlated random coefficients models. *Econometrics*, *7*(1). Retrieved from https://www.mdpi.com/2225-1146/7/1/7 doi: 10.3390/econometrics7010007

Hsiao, C., Pesaran, M. H. & Tahmiscioglu, A. K. (1998). *Bayes Estimation of Short-run Coefficients in Dynamic Panel Data Models* (Cambridge Working Papers in Economics No. 9804).

Im, K. S., Pesaran, M. H. & Shin, Y. (2003). Testing for unit roots in heterogeneous panels. *Journal of econometrics*, *115*(1), 53–74.

Kapetanios, G., Mastromarco, C., Serlenga, L. & Shin, Y. (2017). Modelling in the presence of cross-sectional error dependence. In *The Econometrics of Multi-dimensional Panels* (pp. 291–322). Springer.

Krishnakumar, J., Avila Márquez, M. & Balazsi, L. (2017). Random coefficients models. In L. Matyas (Ed.), *The Econometrics of Multi-dimensional Panels: Theory and Applications* (pp. 125–161). Springer International Publishing.

Moon, H. R., Shum, M. & Weidner, M. (2018). Estimation of random coefficients logit demand models with interactive fixed effects. *Journal of Econometrics*, *206*(2), 613-644. Retrieved from https://www.sciencedirect.com/science/article/pii/S030440761830112X (Special issue on Advances in Econometric Theory: Essays in honor of Takeshi Amemiya) doi: https://doi.org/10.1016/j.jeconom.2018.06.016

Mundlak, Y. (1961). Aggregation over time in distributed lag models. *International Economic Review*, *2*(2), 154-163. Retrieved from http://www.jstor.org/stable/2525442

Mátyás, L. (Ed.). (2017). *The Econometrics of Multi-dimensional Panels*. Springer.

Pesaran, M. & Smith, R. (1995). Estimating long-run relationships from dynamic heterogeneous panels. *Journal of Econometrics*, *68*(1), 79 - 113.

Phillips, P. C. & Moon, H. R. (1999). Linear regression limit theory for nonstationary panel data. *Econometrica*, *67*(5), 1057–1111.

Rendon, S. R. (2013). Fixed and random effects in classical and bayesian regression. *Oxford Bulletin of Economics and Statistics*, *75*(3), 460–476.

Rossi, P. & Allenby, G. (2009). Bayesian applications in marketing. In *The Oxford Handbook of Bayesian Econometrics.* Oxford University Press.

Sarafidis, V. & Robertson, D. (2009). On the impact of error cross-sectional dependence in short dynamic panel estimation. *The Econometrics Journal*, *12*(1), 62–81.

Sarafidis, V. & Wansbeek, T. (2021). Celebrating 40 years of panel data analysis: Past, present and future. *Journal of Econometrics*, *220*(2), 215–226.

Wooldridge, J. M. (2003). Cluster-sample methods in applied econometrics. *American Economic Review*, *93*(2), 133–138.

Wooldridge, J. M. (2005a). Fixed-effects and related estimators for correlated random-coefficient and treatment-effect panel data models. *Review of Economics and Statistics*, *87*(2), 385–390.

Wooldridge, J. M. (2005b). Simple solutions to the initial conditions problem in dynamic, nonlinear panel data models with unobserved heterogeneity. *Journal of Applied Econometrics*, *20*(1), 39–54.

Wooldridge, J. M. (2010). *Econometric analysis of cross section and panel data.* MIT press.

Chapter 14
The Basics of the Mundlak and Chamberlain Projections

Badi H. Baltagi and Tom Wansbeek

Abstract One of the best-known results in panel data econometrics, due to Mundlak (1978), is the equality of the random-effects and fixed-effects estimators when the individual effects are correlated with the means over time of the regressors. Chamberlain (1980) showed that the same result holds when the individual effects are correlated with the regressors for all moments in time separately. In this chapter, we review basic elements of the Mundlak and Chamberlain projections. We emphasize the simplicity that is often obtained when the model is transformed into the within- and between-model, following Arellano (1993). Topics that we discuss include the augmented regression model, the Hausman test, minimum-distance estimation and its link to GMM, unbalanced data, and higher-dimensional data.

14.1 Introduction

A historical moment in the development of panel data econometrics was the first conference in the field, held in Paris, August 22-24, 1977, organized by the research unit of the French national statistics bureau INSEE. Participants included Gary Chamberlain, Zvi Griliches, Andrew Harvey, Jerry Hausman, Jim Heckman, Karl Jöreskog, G. S. Maddala, Jacques Mairesse, Yair Mundlak and Marc Nerlove. Marc participated in organizing the conference and wrote the introductory paper of the conference proceedings published as a special issue of the *Annales de l'INSEE*, Nerlove (1978), later reprinted in Nerlove (2002). This first Paris conference was followed by 29 international panel data conferences, mostly in Europe. The conferences take place under the auspices of a scientific committee of which Marc Nerlove has been a member from the beginning.

One paper in the *Annales de l'INSEE* issue was Mundlak (1978b). This paper extends, to the case of varying coefficients, Mundlak (1978a), published in the same year in *Econometrica*, containing the classical result in panel data econometrics that the random-effects (RE) estimator is the fixed-effects (FE or 'within') estimator when the individual effects correlate with the means over time of all regressors. This result has often been invoked by applied researchers to justify the use of the FE estimator, although it is costly in the sense of eliminating a lot of variation from the data. By

Badi H. Baltagi ✉
Syracuse University, Syracuse, USA, e-mail: bbaltagi@syr.edu

Tom Wansbeek
University of Groningen, The Netherlands, e-mail: t.j.wansbeek@rug.nl

B. H. Baltagi and L. Mátyás (eds.), *Seven Decades of Econometrics and Beyond*, Advanced Studies in Theoretical and Applied Econometrics 57,
https://doi.org/10.1007/978-3-031-92699-0_14

September 9, 2024, the paper has amassed 2619 citations on the Web of Science (6872 citations by Google scholar).

Shortly after, Mundlak's idea was generalized by Chamberlain, see Chamberlain (1980) (1349 citations on the Web of Science, 4081 citations by Google scholar), Chamberlain (1982) (580 citations on the Web of Science, 1800 citations by Google scholar), and the *Handbook of Econometrics* chapter on panel data, Chamberlain (1984). Chamberlain credits Mundlak for the correlated random effects model, see the *ET* interview with Chamberlain by Graham, Hirano and Imbens (2023), pp.12-13. "...I guess step one is this correlated random effects setup. There you definitely want to point to Yair Mundlak, who was visiting at Harvard while I was a graduate student and we interacted a lot. I think he was very close to Zvi and had an office in that part of the building. So that notion that you might try to build a link between the so-called fixed effects and random effects [was] very clearly in Yair's '78 Econometrica paper."

In this paper we revisit, in Section 14.2, Mundlak's projection and we propose a very brief derivation of his main result by separating the model into a 'within' and a 'between' model. This is obtained by a transformation of the model that we will call 'Arellano's transformation' throughout. We compare it with other derivations, not just for historical reasons but also this entails some algebraic results that are of more general use. We also consider the augmented regression, where Mundlak substituted for the individual effects as a linear function of all the regressors averaged across time, and discuss the Hausman test for endogeneity.

In section 14.3 we discuss the alternative projection due to Chamberlain. We discuss when this projection is better than Mundlak's and we show that, also here, Arellano's transformation leads to a simple proof that the RE and FE estimators are the same. We discuss the minimum-distance estimator put forward by Chamberlain and derive the link with GMM. The discussion of GMM brings us to consider the moment conditions underlying the linear panel data model, which suggests as an aside a discussion of two ways to test for the regression coefficients to be the same over time. This is done in Section 14.4.

Section 14.5 addresses the issue of unbalanced data. We introduce a simple way to deal with unbalancedness, for both the Mundlak and the Chamberlain projection. We discuss grouping as one particular form of unbalancedness. In Section 14.6 we use Arellano's transformation for the case where also the time effects are allowed to correlate with the regressors. This directly leads to the FE estimator for two dimensions, while in Section 14.7 we discuss the extension to the three-dimensional data. We investigate whether correlated effects still produces the FE estimator also here. When the effects are one-dimensional, it does, but when the effects are two-dimensional, it does not; the GLS estimator is then more efficient than the FE estimator. In Section 14.8 we briefly consider the model with an error term containing factors, the varying coefficient model, and the spatial regression model, to see to what extent Arellano's transformation still yields equality of the RE and FE estimator.

14.2 Mundlak's Projection

Throughout we consider the simplest static random effects model for panel data, $y_{it} = \mathbf{x}'_{it}\boldsymbol{\beta} + \alpha_i + \varepsilon_{it}$, or in matrix notation

$$\mathbf{y}_i = \mathbf{X}_i\boldsymbol{\beta} + \mathbf{i}_T\alpha_i + \boldsymbol{\varepsilon}_i, \tag{14.1}$$

for $i = 1, \ldots, n$, where $\mathbf{y}_i$ and $\boldsymbol{\varepsilon}_i$ are $T \times 1$, $\mathbf{X}_i$ is $T \times k$ while $\mathbf{i}$ is the vector of ones, its index indicating the number of elements. Further, $\mathbf{x}_i \equiv \text{vec}\mathbf{X}_i$ while $\mathbf{J}_T \equiv \mathbf{i}_T\mathbf{i}'_T$ and $\bar{\mathbf{i}}_T \equiv \mathbf{i}_T/T$ and $\bar{\mathbf{J}}_T \equiv \mathbf{i}_T\mathbf{i}'_T/T$, so the average over time of $\mathbf{X}_i$ is $\bar{\mathbf{x}}_i = \mathbf{X}'_i\bar{\mathbf{i}}_T$. The centering operator is $\mathbf{A}_T \equiv \mathbf{I}_T - \bar{\mathbf{J}}_T$ where $\mathbf{I}_T$ denotes the identity matrix of dimension T. The error term $\boldsymbol{\varepsilon}_i \sim (\mathbf{0}, \sigma^2_\varepsilon\mathbf{I}_T)$ is uncorrelated with everything else while $\alpha_i \sim (0, \sigma^2_\alpha)$ can be correlated with $\mathbf{X}_i$. To concentrate on the essentials we assume that all regressors vary over time and have been demeaned. We denote by $\mathbf{y}$ the $nT \times 1$ vector stacking the $\mathbf{y}_i$ and $\mathbf{X}$ the $nT \times k$ matrix stacking the $\mathbf{X}_i$, while $\mathbf{Y}$ is the $n \times T$ matrix with rows $\mathbf{y}'_i$ and $\dot{\mathbf{X}}$ the

$n \times kT$ matrix with rows $\mathbf{x}_i'$, and $\bar{\mathbf{X}}$ the $n \times k$ matrix with rows $\bar{\mathbf{x}}_i'$. The $nT \times nT$ covariance matrix of the error terms is

$$\boldsymbol{\Omega} \equiv \mathbf{I}_n \otimes \left(\sigma_\varepsilon^2 \mathbf{I}_T + \sigma_\alpha^2 \mathbf{J}_T\right).$$

So the model is a GLS model, where the GLS estimator is best linear unbiased. Mundlak (1978) calls this the Balestra-Nerlove estimator, after Balestra and Nerlove (1966).

To handle the possible correlation between $\mathbf{X}_i$ and α_i, Mundlak (1978) proposed to make it explicit by adding the linear projection of α_i on $\bar{\mathbf{x}}_i$ to the model,

$$\alpha_i = \bar{\mathbf{x}}_i' \boldsymbol{\pi}_{\mathrm{M}} + v_i, \tag{14.2}$$

with $\boldsymbol{\pi}_{\mathrm{M}}$ of order $k \times 1$, while v_i is by construction uncorrelated with $\bar{\mathbf{x}}_i$ and by assumption homoskedastic. Substitution of (14.2) in (14.1) yields

$$\mathbf{y}_i = \mathbf{X}_i \boldsymbol{\beta} + \mathbf{i}_T \bar{\mathbf{x}}_i' \boldsymbol{\pi}_{\mathrm{M}} + \mathbf{u}_i, \tag{14.3}$$

with $\mathbf{u}_i \equiv \mathbf{i}_T v_i + \boldsymbol{\varepsilon}_i$ the composite error term. We now have an augmented regression model, sometimes denoted as the correlated-effects (CRE) model, still a GLS model in the sense of having a non-scalar covariance matrix. However, GLS yields the same result as OLS, as can be shown as follows. Writing $\bar{\mathbf{J}}_T \mathbf{X}_i$ for $\mathbf{i}_T \bar{\mathbf{x}}_i'$, the regressors for i are $(\mathbf{X}_i, \bar{\mathbf{J}}_T \mathbf{X}_i)$, so for all i together $\tilde{\mathbf{X}} \equiv (\mathbf{X}, (\mathbf{I}_n \otimes \bar{\mathbf{J}}_T)\mathbf{X})$. There holds

$$\left(\sigma_\varepsilon^2 \mathbf{I}_T + \sigma_\alpha^2 \mathbf{J}_T\right)(\mathbf{X}_i, \bar{\mathbf{J}}_T \mathbf{X}_i) = (\mathbf{X}_i, \bar{\mathbf{J}}_T \mathbf{X}_i) \begin{pmatrix} \mathbf{I}_k & \mathbf{0} \\ T\sigma_\alpha^2 \mathbf{I}_k & (\sigma_\varepsilon^2 + T\sigma_\alpha^2)\mathbf{I}_k \end{pmatrix}.$$

Grouping this result for all i yields $\boldsymbol{\Omega}\tilde{\mathbf{X}} = \tilde{\mathbf{X}}\mathbf{B}$ for some $\mathbf{B}$. According to a classical result due to Zyskind (1967), OLS and GLS then produce the same result for $\boldsymbol{\beta}$ and $\boldsymbol{\pi}_{\mathrm{M}}$, see also Baltagi (2006). However, the standard errors of $\hat{\boldsymbol{\beta}}$ and $\hat{\boldsymbol{\pi}}_{\mathrm{M}}$ are not the standard errors produced by OLS since the error structure is still GLS, see Baltagi (2023a, 2024a) for an extensive discussion of this and its effect on test of hypothesis for the correlated random effects model using OLS rather an GLS on the augmented Mundlak regression in (16.3).

In fact, applications of the Mundlak correlated random effects model use OLS on the augmented regression (14.3) rather than GLS as suggested by Mundlak. The latter is necessary because of the presence of the random individual effect v_i. Baltagi (2006) showed that OLS is equivalent to GLS for this augmented regression and both yield the fixed effects estimator for $\boldsymbol{\beta}$, invoking the result by Zyskind (1967) when OLS and GLS coincide. However, the standard errors using OLS are different from those using GLS as they assume different variance-covariance structures. This also affects test of hypotheses and in particular the test for $H_0 : \boldsymbol{\pi}_{\mathrm{M}} = \mathbf{0}$. In particular, the Mundlak GLS regression yields the Hausman (1978) test based on the fixed versus between estimators, while the test based on the OLS estimator yields a different statistic all together. Not rejecting the null in the Mundlak (1978) augmented regression yields to non-rejection of the random effects model. While, non-rejection of the null in the OLS augmented model yields pooled OLS as the efficient estimator. This is certainly not what Mundlak (1978) intended.

A simple alternative, and a natural one given the two dimensions of panel data, is by separating the within and between dimensions. Let $\mathbf{R}_T$ be a matrix of order $T \times (T-1)$ such that $\mathbf{R}_T'\mathbf{R}_T = \mathbf{I}_{T-1}$ and $\mathbf{R}_T'\mathbf{i}_T = \mathbf{0}$ and otherwise unspecified. One possible choice is first differences properly scaled. For $T = 4$ it is

$$\mathbf{R}_4' = \frac{1}{\sqrt{2}} \begin{pmatrix} 1 & -1 & 0 & 0 \\ 0 & 1 & -1 & 0 \\ 0 & 0 & 1 & -1 \end{pmatrix}. \tag{14.4}$$

There holds $\mathbf{R}_T\mathbf{R}_T' = \mathbf{A}_T$ since both are idempotent of rank $T-1$, with the same null-space $\mathbf{i}_T$. Premultiplication of (14.1) by $(\mathbf{R}_T, \bar{\mathbf{i}}_T)'$, which is a one-to-one transformation so nothing gets lost, gives an equivalent representation of (14.1),

$$\mathbf{R}_T'\mathbf{y}_i = \mathbf{R}_T'\mathbf{X}_i\boldsymbol{\beta} + \mathbf{R}_T'\boldsymbol{\varepsilon}_i \tag{14.5}$$
$$\bar{y}_i = \bar{\mathbf{x}}_i'\boldsymbol{\beta} + \alpha_i + \bar{\varepsilon}_i, \tag{14.6}$$

with $\bar{\varepsilon}_i$ the average over time of $\boldsymbol{\varepsilon}_i$. Substitution of (14.2) in (14.6) yields

$$\begin{aligned}\bar{y}_i &= \bar{\mathbf{x}}_i'(\boldsymbol{\beta} + \boldsymbol{\pi}_{\mathrm{M}}) + v_i + \bar{\boldsymbol{\varepsilon}}_i \\ &\equiv \bar{\mathbf{x}}_i'\boldsymbol{\xi}_{\mathrm{M}} + v_i + \bar{\boldsymbol{\varepsilon}}_i.\end{aligned} \tag{14.7}$$

The two error terms, $\mathbf{R}_T'\boldsymbol{\varepsilon}_i$ in (14.5) and $v_i + \bar{\boldsymbol{\varepsilon}}_i$ in (14.7) are uncorrelated, while $\mathbf{R}_T'\boldsymbol{\varepsilon}_i$ has a scalar covariance matrix $\sigma_\varepsilon^2\mathbf{I}_{T-1}$ so OLS is optimal. Since $\boldsymbol{\xi}_{\mathrm{M}}$ is uninformative about $\boldsymbol{\beta}$, only (14.5) contains information about $\boldsymbol{\beta}$. Because of $\mathbf{R}_T\mathbf{R}_T' = \mathbf{A}_T$, the OLS estimator of $\boldsymbol{\beta}$ in (14.5) is

$$\hat{\boldsymbol{\beta}} = \left(\sum_i \mathbf{X}_i'\mathbf{A}_T\mathbf{X}_i\right)^{-1} \sum_i \mathbf{X}_i'\mathbf{A}_T\mathbf{y}_i. \tag{14.8}$$

This $\hat{\boldsymbol{\beta}}$ is the FE estimator. With $\mathbf{g}$ a column of $\mathbf{X}$ and g_{it} its (i,t)th element, the FE estimator can be obtained by the transformation $\tilde{g}_{it} = g_{it} - g_{i*}$, an asterisk in the place of an index indicating the average over that index. The OLS estimator of $\boldsymbol{\xi}_{\mathrm{M}}$ in (14.7) is the between estimator,

$$\hat{\boldsymbol{\xi}}_{\mathrm{M}} = \left(\bar{\mathbf{X}}'\bar{\mathbf{X}}\right)^{-1}\bar{\mathbf{X}}'\bar{\mathbf{y}}.$$

Anyhow, with individual effects correlated with the means over time of all regressors, there is just one estimator of $\boldsymbol{\beta}$, the FE one. This equality of RE and FE when $\mathbf{X}_i$ and α_i are correlated, due to Mundlak (1978), is one of the basic results in panel data analysis that 'everybody knows'.

As far as we know, this simple derivation of Mundlak's result was first given by Arellano (1993) and further discussed by Arellano and Bover (1995), for one particular form of $\mathbf{R}_T$, forward orthogonal deviations. To give the idea, consider the case $T = 4$. Then

$$\mathbf{R}_4' = \begin{pmatrix} \sqrt{3/4} & 0 & 0 \\ 0 & \sqrt{2/3} & 0 \\ 0 & 0 & \sqrt{1/2} \end{pmatrix} \begin{pmatrix} 1 & -1/3 & -1/3 & -1/3 \\ 0 & 1 & -1/2 & -1/2 \\ 0 & 0 & 1 & -1 \end{pmatrix}.$$

According to Arellano and Bover (1995), p.42, this specific form of $\mathbf{R}_T$ has the advantage that lags of predetermined variables are valid instruments in the transformed equations. But this property is irrelevant for our purpose here, and we will use the expression 'Arellano's transformation' without reference to one particular form of $\mathbf{R}_T$.

As to proving Mundlak's incisive result, Mundlak (1978) himself does not provide a full proof but leaves the derivation of his result partly to the reader ("Utilizing [the GLS structure arising when the projection is inserted] and the expression for the inverse of a partitioned matrix, we can obtain [the result] after some simplications[.]"). The derivation using partitioned inversion is given as Exercise 7.13 in Baltagi (2009), pp. 151-154. Baltagi (2006) simplifies the derivation using system estimation stacking the between Mundlak regression on top of the within regression and performing GLS or OLS on this system. This yields Mundlak's (1978) result without partitioned inversion. Krishnakumar (2006) considers the case with time-constant variables included in the model. As far as we know, the simple proof through Arellano's transformation is rarely mentioned in the literature; Biørn (2017), chapter 6, 'Latent heterogeneity correlated with regressors,' seems to be the only place to make the point of the simple transformation.

Alternatively, given the equality of OLS and GLS augmented regression model shown above, an algebraic proof not using Arellano's transformation is of some interest because of its wider applicability. It is based on the Frisch-Waugh theorem, to show that projecting out the $\mathbf{i}_T\bar{\mathbf{x}}_i'\boldsymbol{\pi}_{\mathrm{M}}$ corresponds with the within transformation for $\mathbf{X}$, and uses the 'double projection' general result

$$\mathbf{P}_{\mathrm{C}}\mathbf{X} = \mathbf{P}_{\mathbf{P}_{\mathrm{C}}\mathbf{X}}\mathbf{X}, \tag{14.9}$$

where $\mathbf{P}_{\mathrm{Q}}$ generically denotes the projection matrix 'onto', $\mathbf{P}_{\mathrm{Q}} \equiv \mathbf{Q}(\mathbf{Q}'\mathbf{Q})^{-1}\mathbf{Q}'$ for any appropriate $\mathbf{Q}$. We will also use $\mathbf{M}_{\mathrm{Q}} \equiv \mathbf{I} - \mathbf{P}_{\mathrm{Q}}$. The result (14.9) follows immediately from letting $\mathbf{Q} = \mathbf{P}_{\mathrm{C}}$ in the definition of $\mathbf{P}_{\mathrm{Q}}$. Now, choose $\mathbf{C} = \mathbf{I}_n \otimes \mathbf{i}_T$, so the regressor $(\mathbf{I}_n \otimes \bar{\mathbf{J}}_T)\mathbf{X}$ equals $\mathbf{P}_{\mathrm{C}}\mathbf{X}$. When we premultiply $\mathbf{X}$ with the projection matrix perpendicular to $\mathbf{P}_{\mathrm{C}}\mathbf{X}$, we obtain

$$\mathbf{M}_{\mathbf{P}_{\mathrm{C}}\mathbf{X}}\mathbf{X} = \left(\mathbf{I}_{nT} - \mathbf{P}_{\mathbf{P}_{\mathrm{C}}\mathbf{X}}\right)\mathbf{X} = \left(\mathbf{I}_{nT} - \mathbf{P}_{\mathrm{C}}\right)\mathbf{X} = \mathbf{M}_{\mathrm{C}}\mathbf{X},$$

which is the within transformation of $\mathbf{X}$. According to the Frisch-Waugh theorem, this leads to the FE estimator.

Wooldridge (2013) finds it "a bit anticlimactic" that CRE is FE but feels that there are still two reasons to consider CRE. First, one can use the estimate of $\boldsymbol{\pi}_{\mathrm{M}}$ in a test for exogeneity and make the test robust to heteroskedasticity and serial correlation and, second, time-constant regressors can still be included in the regression. As a third reason one may add the case where not all regressors are considered to be potentially endogenous. One specific case arises when the researcher has doubts about the exogeneity of one regressor in particular. When the first k_1 regressors $\mathbf{X}_1$ are considered exogenous a priori and the last k_2 regressors $\mathbf{X}_2$ endogenous, the corresponding elements of $\boldsymbol{\theta}$ in (14.7) are zero, so

$$\alpha_i = \bar{\mathbf{x}}_{2i}'\boldsymbol{\pi}_{\mathrm{M}2} + v_i, \tag{14.10}$$

Then (14.7) becomes

$$\begin{aligned}\bar{y}_i &= \bar{\mathbf{x}}_{1i}'\boldsymbol{\beta}_1 + \bar{\mathbf{x}}_{2i}'(\boldsymbol{\beta}_2 + \boldsymbol{\pi}_{\mathrm{M}2}) + v_i + \bar{\varepsilon}_i \\ &\equiv \bar{\mathbf{x}}_{1i}'\boldsymbol{\beta}_1 + \bar{\mathbf{x}}_{2i}'\boldsymbol{\xi}_{\mathrm{M}2} + v_i + \bar{\varepsilon}_i.\end{aligned}$$

So now this second equation of the model contains information about $\boldsymbol{\beta}_1$, and the fixed-effects estimator is no longer optimal. An optimal estimator is easily obtained by substituting the projection (14.10) in (14.1) to obtain

$$\mathbf{y}_i = \mathbf{X}_i\boldsymbol{\beta} + \mathbf{i}_T\bar{\mathbf{x}}_{2i}'\boldsymbol{\pi}_{\mathrm{M}2} + \mathbf{i}_T v_i + \boldsymbol{\varepsilon}_i, \tag{14.11}$$

which is the random effects model with the means over time of the endogenous variables added as regressors. This way of 'breaking the correlation' due to endogeneity by adding the means of the endogenous regressors has become quite popular with practitioners.

Comparing the difference between the between estimator $\hat{\boldsymbol{\xi}}_{\mathrm{M}}$ and the within estimator $\hat{\boldsymbol{\beta}}$ provides a simple $\chi^2(k)$ test for exogeneity, $H_0 : \boldsymbol{\pi}_{\mathrm{M}} = \mathbf{0}$ so $H_0 : \boldsymbol{\beta} = \boldsymbol{\xi}_{\mathrm{M}}$. Because of the lack of correlation between the two error terms, the variance of $\hat{\boldsymbol{\xi}}_{\mathrm{M}} - \hat{\boldsymbol{\beta}}$ is the sum of their variances. So under H_0

$$q_1 = \left(\hat{\boldsymbol{\xi}}_{\mathrm{M}} - \hat{\boldsymbol{\beta}}\right)' \left[\widehat{\mathrm{Var}}\left(\hat{\boldsymbol{\xi}}_{\mathrm{M}}\right) + \widehat{\mathrm{Var}}\left(\hat{\boldsymbol{\beta}}\right)\right]^{-1} \left(\hat{\boldsymbol{\xi}}_{\mathrm{M}} - \hat{\boldsymbol{\beta}}\right) \overset{\text{as.}}{\sim} \chi_k^2. \tag{14.12}$$

This is exactly the test statistic obtained by Mundlak (1978) for testing $H_0 : \boldsymbol{\pi}_{\mathrm{M}} = \mathbf{0}$ by performing GLS on the augmented regression in (16.3). In fact, using GLS, Mundlak (1978) obtains the within estimator as the GLS estimator of $\boldsymbol{\beta}$ and the difference between the between and within estimators as the GLS estimator for $\boldsymbol{\pi}_{\mathrm{M}}$, see Baltagi (2009). As emphasized earlier, one would not get this test statistic for $H_0 : \boldsymbol{\pi}_{\mathrm{M}} = \mathbf{0}$ if one runs OLS on the augmented regression in (16.3). While it is true that one can use a robust variance-covariance matrix to test this null with the OLS regression, not rejecting the null yields the pooled OLS rather than the random effects estimator for this uncorrelated random effects model.

We could also base a test on any other linear combination of $\boldsymbol{\xi}_{\mathrm{M}}$ and $\hat{\boldsymbol{\beta}}$ but that does not affect the numerical value of the test statistic as it is all about one-to-one linear transformations, as was first argued by Hausman and Taylor (1981). One such combination is to compare the GLS estimator with the FE estimator, which is what Hausman (1978) originally proposed and still enjoys popularity among applied researchers. Since the GLS estimator is efficient under the null, this way of testing has the elegant property that the variance of the difference between the two estimators is the difference of the variances, as follows from the classical Rao-Blackwell theorem in statistics, but it requires the estimation of σ_ε^2 and σ_α^2, needed to estimate the variance of the GLS estimator.

Note that this yields different Hausman test test statistics depending on which feasible GLS estimator is used. Stata uses the Swamy and Arora (1972) feasible GLS estimator. EViews computes two other Hausman test statistics based on the feasible GLS estimators of Wallace and Hussain (1969) and Amemiya (1971). Although these test statistics should give the same decision, they may differ in small samples. Hausman and Taylor (1981) proved the equivalence of the Hausman test based on three contrasts including (1) the between and within estimators, (2) the between and GLS estimators, and (3) the within and GLS estimators. Programs that compute the Hausman test based on the contrast of any two estimators automatically subtract the two variances. One gets the wrong Hausman test if it is applied with Stata for the contrast using the between and fixed effects estimators. Stata warns the user that one estimator should be efficient, and the other consistent under the null. But in the case of the between-versus-fixed effects estimators, neither is efficient under the null, and the variance of the difference is the sum of the two variances. The program automatically computes the difference in variances, leading to an incorrect Hausman statistic.

Incidentally, Mundlak(1978) gets this result in his augmented regression (16.3), testing that $\boldsymbol{\pi}_{\mathrm{M}} = \mathbf{0}$ with a Wald test, which yields a Hausman test based on the difference between the fixed effects and the between estimators. Applying GLS rather than OLS on this augmented regression yields the correct Hausman -type test as a Wald statistic, with the variance of the difference between these two estimators being the sum of the variances. For practical tips on what to do in case you reject the Hausman test, see Baltagi (2024b). A key point is that the Hausman test is valid only under the null and does not endorse the alternative as it is signaling misspecification which is unknown. This misspecification could be due to dynamic misspecification, ignored endogeneity of some regressors, or incorrect functional form, to mention a few.

14.3 Chamberlain's Projection

Chamberlain (1982) proposed modeling α_i as a projection on $\mathbf{X}_i$, at every point in time, rather than on their average over time. Mundlak's (1978) reduced form model in (16.2) is generalized as follows:

$$\alpha_i = \mathbf{x}_i'\boldsymbol{\pi}_{\mathrm{C}} + v_i, \tag{14.13}$$

where $\mathbf{x}_i \equiv \mathrm{vec}\mathbf{X}_i$ and $\boldsymbol{\pi}_{\mathrm{C}}$ is of order $kT \times 1$. This projection has been amply discussed in the literature, see Crépon and Mairesse (2008) for an overview.

The possible advantage of this projection over the Mundlak projection is that v_i is, by construction, uncorrelated with all elements of $\mathbf{X}_i$. So, when we substitute the projection (14.13) for α_i in the model, we get an error term that does not correlate with the regressors, which is not guaranteed with the Mundlak projection. Whether there are cases where this matters empirically is not clear, but it does give an argument of the Chamberlain projection over the Mundlak one.

One is hard put to find this argument in these simple words in the literature. Chamberlain (1982), footnote 5, focuses on the case where the two projections are the same, saying: "A solution could be based on Mundlak's (1978a) proposal that $\mathbb{E}(b|\mathbf{x}) = \psi_0 + \psi_1 \sum_{t=1}^{T} x_t$. However, even if we assume that the regression function is linear in $x_1, \ldots, x_T$, it may be difficult to justify the restriction that only $\sum_t x_t$ matters, unless T is large and we have stationarity: $\mathrm{cov}(b, x_t) = \mathrm{cov}(b, x_1)$ and $V(\mathbf{x})$ band diagonal." To make this point explicit, assume momentarily the case of a single regressor, $k = 1$, to keep notation simple. The Mundlak approach can be seen as a special case of the Chamberlain one. Since $\mathbf{R}_T\mathbf{R}_T' + \bar{\mathbf{i}}_T\mathbf{i}_T' = \mathbf{I}_k$,

$$\begin{aligned}\mathbf{x}_i'\boldsymbol{\pi}_{\mathrm{C}} &= \mathbf{x}_i'\left(\mathbf{R}_T\mathbf{R}_T' + \bar{\mathbf{i}}_T\mathbf{i}_T')\right)\boldsymbol{\pi}_{\mathrm{C}} \\ &\equiv \tilde{\mathbf{x}}_i'\tilde{\boldsymbol{\pi}}_{\mathrm{C}} + \bar{x}_i\,\pi_{\mathrm{M}},\end{aligned} \tag{14.14}$$

where $\tilde{\mathbf{x}}_i \equiv \mathbf{R}_T'\mathbf{x}_i$ and $\tilde{\boldsymbol{\pi}}_{\mathrm{C}} \equiv \mathbf{R}_T'\boldsymbol{\pi}_{\mathrm{C}}$. The second term in (14.14) corresponds with the Mundlak projection. So the two projections yield the same result when $\tilde{\boldsymbol{\pi}}_{\mathrm{C}} = \mathbf{0}$. We now use the expression for the coefficient of the linear projection of a on $\mathbf{b}$, $\boldsymbol{\Sigma}_{\mathbf{b}}^{-1}\boldsymbol{\sigma}_{\mathbf{b}a}$, in self-evident notation that we will also use below. So, $\boldsymbol{\pi}_{\mathrm{C}} = \boldsymbol{\Sigma}_{\mathbf{x}}^{-1}\boldsymbol{\sigma}_{\mathbf{x}\alpha}$. Somewhat crudely stated, when $\boldsymbol{\Sigma}_{\mathbf{x}}$ is band diagonal, its inverse will be approximately band diagonal, the difference vanishing when T becomes large. When $\boldsymbol{\sigma}_{\mathbf{x}\alpha}$ is proportional to $\mathbf{i}_k$, the same will hold for $\boldsymbol{\pi}_{\mathrm{C}}$, and hence $\tilde{\boldsymbol{\pi}}_{\mathrm{C}}$ will be zero.

Also when using the Chamberlain projection, Arellano's transformation of premultiplication of (14.1) by $(\mathbf{R}_T, \bar{\mathbf{i}}_T)'$ yields the FE estimator. Premultiplication by $\mathbf{R}_T'$ yields (14.5) again. Since

$$\mathbf{X}_i\boldsymbol{\beta} = (\boldsymbol{\beta}' \otimes \mathbf{I}_T)\,\mathbf{x}_i \tag{14.15}$$

so

$$\begin{aligned}\bar{\mathbf{i}}_T'\mathbf{X}_i\boldsymbol{\beta} &= (\boldsymbol{\beta} \otimes \bar{\mathbf{i}}_T)'\,\mathbf{x}_i \\ &= \mathbf{x}_i'\,(\boldsymbol{\beta} \otimes \bar{\mathbf{i}}_T),\end{aligned}$$

premultiplication of (14.1) by $\bar{\mathbf{i}}_T'$ yields, instead of (14.7),

$$\begin{aligned}\bar{y}_i &= \mathbf{x}_i'\,(\boldsymbol{\beta} \otimes \bar{\mathbf{i}}_T + \boldsymbol{\pi}_{\mathrm{C}}) + v_i + \bar{\varepsilon}_i \\ &= \mathbf{x}_i'\boldsymbol{\xi}_{\mathrm{C}} + v_i + \bar{\varepsilon}_i,\end{aligned} \tag{14.16}$$

with

$$\boldsymbol{\xi}_{\mathrm{C}} \equiv \boldsymbol{\beta} \otimes \bar{\mathbf{i}}_T + \boldsymbol{\pi}_{\mathrm{C}} \tag{14.17}$$

of order $kT \times 1$, as compared to the order $k \times 1$ of $\boldsymbol{\xi}_{\mathrm{M}}$. Again, the model for $\bar{y}_i$ is not informative about $\boldsymbol{\beta}$ and the error terms in (14.5) and (14.16) are not correlated, so the random-effects estimator of $\boldsymbol{\beta}$ is the FE estimator once again.

The original proof of this (and the only one we are aware of) is due to Chamberlain (1980), Section 4, p.234. It is a bit hidden in an analysis of the likelihood approach to the farms example. It says: "The ML estimator of $(\boldsymbol{\beta}, \boldsymbol{\pi})$, allowing for several variables in $\mathbf{x}_{it}$ and given $\lambda = \sigma_v^2/\sigma^2$, can be obtained from the regression of $y_{it} - \gamma\bar{y}_i$ on $\mathbf{x}_{it} - \gamma\bar{\mathbf{x}}_i$ and $(1-\gamma)\mathbf{x}_i$. The resulting estimator for $\boldsymbol{\beta}$ can be obtained from the regression of $y_{it} - \gamma\bar{y}_i$ on the residual from the regression of $\mathbf{x}_{it} - \gamma\bar{\mathbf{x}}_i$ on $\mathbf{x}_i$. This residual is $\mathbf{x}_{it} - \bar{\mathbf{x}}_i$; but the regression of $y_{it} - \gamma\bar{y}_i$ on $\bar{\mathbf{x}}_{it} - \bar{\mathbf{x}}_i$ is equivalent to the regression of $y_{it} - \bar{y}_i$ on $\bar{\mathbf{x}}_{it} - \bar{\mathbf{x}}_i$." This is followed by a new paragraph starting with "We have obtained the interesting result that a random effects specification can give a ML estimator of $\boldsymbol{\beta}$ that is identical to the fixed effects estimator, if we allow the distribution of the incidental parameters to depend upon $\mathbf{x}$," with footnote 15 added saying "This result is discussed in Mundlak (1978) for the case $\boldsymbol{\pi}'\mathbf{x}_i = \boldsymbol{\delta}'\bar{\mathbf{x}}_i$."

Also here the difference between estimators leads to a test for exogeneity, which now means $\boldsymbol{\pi}_{\mathrm{C}} = \mathbf{0}$ or $\boldsymbol{\xi}_{\mathrm{C}}' - \boldsymbol{\beta} \otimes \bar{\mathbf{i}}_T = \mathbf{0}$. The test statistic, due to Ahn and Low (1996), see also Baltagi (2021), pp.93-94, is

$$q_2 = \left(\hat{\boldsymbol{\xi}}_{\mathrm{C}} - \hat{\boldsymbol{\beta}} \otimes \bar{\mathbf{i}}_T\right)'\left[\widehat{\mathrm{Var}}\left(\hat{\boldsymbol{\xi}}_{\mathrm{C}}\right) + \widehat{\mathrm{Var}}\left(\hat{\boldsymbol{\beta}}\right) \otimes \mathbf{i}_T\mathbf{i}_T'\right]^{-1}\left(\hat{\boldsymbol{\xi}}_{\mathrm{C}} - \hat{\boldsymbol{\beta}} \otimes \bar{\mathbf{i}}_T\right) \overset{\text{as.}}{\sim} \chi^2_{kT} \tag{14.18}$$

This test conceptually directly generalizes the test in (14.12). Again, the variance of the difference is the sum of the variances. The number of degrees of freedom of this test is kT, instead of k in (14.12). So the power of the new test can be expected to be much lower when $\hat{\boldsymbol{\xi}}_{\mathrm{C}}$ shows little variation over time, which may quite well be the dominant case, since then q_2 becomes close to q_1 in (14.12), but with many more degrees of freedom.

The approach to estimation and testing taken by Chamberlain (1982), later on followed by Angrist and Newey (1991), is different and is to insert the projection (14.13) in the original model (14.1) rather than in the between regression as in (14.16). With

$$\boldsymbol{\Pi} \equiv \boldsymbol{\beta}' \otimes \mathbf{i}_T + \mathbf{I}_T \boldsymbol{\pi}_{\mathrm{C}}' \tag{14.19}$$

of order $T \times kT$ this yields, using (14.15),

$$\mathbf{y}_i = \boldsymbol{\Pi}\mathbf{x}_i + \mathbf{i}_T v_i + \boldsymbol{\varepsilon}_i. \tag{14.20}$$

Chamberlain (1982) proposed to obtain estimators $\hat{\boldsymbol{\beta}}$ and $\hat{\boldsymbol{\pi}}_{\mathrm{C}}$ of $\boldsymbol{\beta}$ and $\boldsymbol{\pi}_{\mathrm{C}}$ by the minimum-distance (MD) method, that is, by minimizing $\mathbf{d}'\widetilde{\mathbf{V}}^{-1}\mathbf{d}$, where

$$\mathbf{d} \equiv \operatorname{vec}\left(\hat{\boldsymbol{\Pi}} - \boldsymbol{\beta}' \otimes \mathbf{I}_T - \mathbf{i}_T \boldsymbol{\pi}_{\mathrm{C}}'\right), \tag{14.21}$$

with

$$\hat{\boldsymbol{\Pi}} \equiv \mathbf{Y}'\dot{\mathbf{X}}(\dot{\mathbf{X}}'\dot{\mathbf{X}})^{-1},$$

with $\widetilde{\mathbf{V}}$ a consistent estimator of the variance of $\tilde{\mathbf{d}}$ based on initial consistent estimators of $\boldsymbol{\beta}$ and $\boldsymbol{\pi}_{\mathrm{C}}$. The restrictions on $\boldsymbol{\Pi}$ according to (14.19) can be tested through the value of the criterion function in the minimum, distributed as

$$q_3 = \hat{\mathbf{d}}'\widetilde{\mathbf{V}}^{-1}\hat{\mathbf{d}} \overset{\text{as.}}{\sim} \chi^2_{k(T^2-T-1)}, \tag{14.22}$$

where $\hat{\mathbf{d}}$ is $\mathbf{d}$ with the MD estimators plugged in.

Not rejecting the null using a Hausman test leads the researcher to report the random effects estimator as the efficient estimator. Rejecting the null means there is misspecification and the random effect is not efficient. However, in practice, rejecting the null is interpreted as support for the fixed effects estimator. The random effects model assumes no correlation of the random individual effects with all the regressors, whereas the Chamberlain correlated random effects model assumes that the random individual effects are correlated with all the regressors at every point in time.

Chamberlain (1982) suggested testing the fixed effects restrictions using minimum distance estimation. In this case, the test statistic is the minimand of the objective function and is a chi-squared goodness-of-fit statistic for the restrictions on the reduced form. Angrist and Newey (1991) simplified Chamberlain's test using residuals of the fixed effects regression at every point in time. They showed that this minimand can be obtained as the sum of T terms. Each term of this sum is simply the degrees of freedom times the R^2 from a regression of the Within residuals for a particular period on all leads and lags of the independent variables.

Baltagi, Bresson and Pirotte (2009) performed Monte Carlo experiments and showed that these tests yield the same decision and are in conflict at most 2.2 % of the time. One caveat is that, like the Sargan overidentification test for dynamic panels (see Arellano & Bond, 1991), the Chamberlain test tends to understate the true variance of the test statistic as T gets large. This is because as T gets large, the number of testable restrictions increase, and the variance of the test statistic is understated. They suggest careful examination of which regressors may or may not be correlated with the individual effects. In this case, one should be willing to entertain a more restricted model where only a subset of the regressors are correlated with the individual effects, as proposed by Hausman and Taylor (1981). This would impose fewer restrictions than the general Chamberlain model and is also testable with a Hausman test.

Alternatively, one could question the endogeneity of the regressors with the disturbances, not only with the individual effects. This endogeneity leads to inconsistency of the FE estimator and invalidates the Hausman test performed based on the fixed effects versus the random effects estimator, see Baltagi (2024b).

Although Balestra and Nerlove (1966) popularized the random effects model, the dominant view in the panel data literature is that its assumptions are too restrictive, and hence the common use of the fixed effects model when a Hausman test is rejected. The fixed effects model assumes that all the regressors are correlated with the individual effects, which is perceived as less restrictive than the uncorrelated random effects model. As Chamberlain pointed out, these correlated effects restrictions are testable but unfortunately not carried out in panel data except in a handful of applications.

Angrist and Newey (1991) illustrate the Chamberlain test using two examples. The first example estimates and tests a number of models for the union wage effect using five years of data from the National Longitudinal Survey of Youth (NLSY). They find that the assumption of fixed effects in an equation for union wage effects is not at odds with the data. The second example considers a conventional human capital earnings function. They find that the fixed effects estimates of the return to schooling in the NLSY are roughly twice those of ordinary least squares. However, the overidentification test suggests that the fixed effects assumption may be inappropriate for this model.

Carey (1997) applies the Chamberlain minimum chi-square method to the estimation of a multiple output hospital cost function using a panel of 1733 facilities over the period 1987–1991. OLS (year by year), fixed effects, seemingly unrelated regressions, and Chamberlain's minimum chi-square method are reported. In this application, the minimum chi-squared test rejects the restrictions imposed by the null hypothesis. Other notable applications of Chamberlain's approach of correlated random effects include Card (1996), Islam (1995), and Nevo (2001).

As was already noticed by Arellano (2003), the MD approach is basically GMM. In the words of Cameron and Trivedi (2005), p.753, 'Minimum distance estimation has been supplanted by GMM; see Arellano (2003, pp. 22–23) and Crepon and Mairesse (1995) for comparison of Chamberlain's MD estimator with GMM. However, Chamberlain's approach of obtaining moment restrictions via exogeneity assumptions and assumptions on the individual effects has had a big impact on the panel literature."

To see the link with GMM, consider the most general kT^2 moment conditions

$$\text{vec}\mathbb{E}\left(\boldsymbol{\varepsilon}_i\mathbf{x}_i'\right) = \mathbb{E}\left(\mathbf{x}_i \otimes \boldsymbol{\varepsilon}_i\right) = \mathbf{0} \tag{14.23}$$

that follow from the exogeneity of $\mathbf{x}_i$. From (14.23),

$$\begin{aligned}\mathbf{0} &= \mathbb{E}\left(\boldsymbol{\varepsilon}_i\mathbf{x}_i'\right)\\ &= \mathbb{E}\left(\mathbf{y}_i - \dot{\mathbf{X}}_i\boldsymbol{\beta} - \mathbf{i}_T\mathbf{x}_i'\boldsymbol{\pi}_{\mathrm{C}}\right)\mathbf{x}_i'\\ &= \mathbb{E}\left(\mathbf{y}_i - \left(\boldsymbol{\beta}'\otimes\mathbf{I}_T\right)\mathbf{x}_i - \mathbf{i}_T\boldsymbol{\pi}_{\mathrm{C}}'\mathbf{x}_i\right)\mathbf{x}_i'.\end{aligned}$$

Summation over i yields

$$\mathbb{E}\left(\left[\mathbf{Y}' - \left(\boldsymbol{\beta}'\otimes\mathbf{I}_T\right)\dot{\mathbf{X}}' - \mathbf{i}_T\boldsymbol{\pi}_{\mathrm{C}}'\dot{\mathbf{X}}'\right]\dot{\mathbf{X}}\right) = \mathbf{0}.$$

Postmultiplying by $(\dot{\mathbf{X}}'\dot{\mathbf{X}})^{-1}$ yields

$$\mathbb{E}\left(\hat{\boldsymbol{\Pi}} - \boldsymbol{\beta}'\otimes\mathbf{i}_T - \mathbf{I}_T\boldsymbol{\pi}_{\mathrm{C}}'\right) = \mathbf{0},$$

or $\mathbb{E}(\mathbf{d}) = \mathbf{0}$, which corresponds with (14.21), thus showing that MD and GMM, based on the largest set of exogeneity-based moment conditions, are essentially the same.

We now consider the result of Arellano's transformation in the GMM setting and take the within and between dimensions apart. From (14.15) and (14.16), splitting up the moment conditions (14.23) into the within and between components yields

$$\begin{aligned}\mathbf{h}_{i\mathrm{W}} &\equiv \mathbf{x}_i \otimes \mathbf{R}_T'\boldsymbol{\varepsilon}_i\\ &= \mathbf{x}_i \otimes \left(\mathbf{R}_T'\mathbf{y}_i - \left(\boldsymbol{\beta}\otimes\mathbf{R}_T\right)'\mathbf{x}_i\right) \qquad (14.24)\\ \mathbf{h}_{i\mathrm{B}} &\equiv \mathbf{x}_i\bar{\mathbf{i}}_T'\boldsymbol{\varepsilon}_i\\ &= \mathbf{x}_i\left(\bar{y}_i - \boldsymbol{\xi}_{\mathrm{C}}'\mathbf{x}_i - v_i\right). \qquad (14.25)\end{aligned}$$

where we can neglect the term v_i since $\mathbb{E}\left(\mathbf{x}_i \otimes v_i\right) = \mathbf{0}$. Now, $\mathbf{h}_{i\mathrm{W}}$ and $h_{i\mathrm{B}}$ are uncorrelated when the errors are uncorrelated over time, as we assumed, and (14.24) only depends on $\boldsymbol{\beta}$ while (14.25) only depends on $\boldsymbol{\xi}_{\mathrm{C}}$. So we can consider optimal GMM separately for $\boldsymbol{\beta}$ and $\boldsymbol{\xi}_{\mathrm{C}}$. The GMM estimator of

$\boldsymbol{\xi}_C$ is the OLS estimator of the $\bar{y}_i$ on the $\mathbf{x}_i$. To derive the GMM estimator of $\boldsymbol{\beta}$, let

$$\begin{aligned}\mathbf{H}_W &\equiv \sum_i \left(\mathbf{R}_T'\mathbf{y}_i - (\boldsymbol{\beta}\otimes\mathbf{R}_T)'\mathbf{x}_i\right)\mathbf{x}_i' \\ &= \mathbf{R}_T'\mathbf{Y}'\dot{\mathbf{X}} - (\boldsymbol{\beta}\otimes\mathbf{R}_T)'\dot{\mathbf{X}}'\dot{\mathbf{X}}.\end{aligned}$$

Then the optimal GMM estimator under homoskedasticity follows from minimizing

$$\begin{aligned}q_W &\equiv (\text{vec}\mathbf{H}_W)'\left(\dot{\mathbf{X}}'\dot{\mathbf{X}}\otimes\mathbf{i}_{T-1}\right)^{-1}(\text{vec}\mathbf{H}_W) \qquad (14.26)\\ &= \text{tr}\mathbf{H}_W'\mathbf{H}_W\left(\dot{\mathbf{X}}'\dot{\mathbf{X}}\right)^{-1} \\ &= c - 2\text{tr}(\boldsymbol{\beta}\otimes\mathbf{A}_T)\mathbf{Y}'\dot{\mathbf{X}} + \text{tr}(\boldsymbol{\beta}\boldsymbol{\beta}'\otimes\mathbf{A}_T)\dot{\mathbf{X}}'\dot{\mathbf{X}} \\ &= c - 2\text{tr}\dot{\mathbf{X}}(\boldsymbol{\beta}\otimes\mathbf{A}_T)\mathbf{Y}' + \text{tr}\dot{\mathbf{X}}(\boldsymbol{\beta}\boldsymbol{\beta}'\otimes\mathbf{A}_T)\dot{\mathbf{X}}' \\ &= c - 2\left(\text{vec}\dot{\mathbf{X}}'\right)'(\mathbf{I}_n\otimes\boldsymbol{\beta}\otimes\mathbf{A}_T)\text{vec}\mathbf{Y}' + \left(\text{vec}\dot{\mathbf{X}}'\right)'(\mathbf{I}_n\otimes\boldsymbol{\beta}\boldsymbol{\beta}'\otimes\mathbf{A}_T)\text{vec}\dot{\mathbf{X}}' \\ &= c - 2\boldsymbol{\beta}'\mathbf{X}'(\mathbf{I}_n\otimes\mathbf{A}_T)\mathbf{y} + \boldsymbol{\beta}'\mathbf{X}'(\mathbf{I}_n\otimes\mathbf{A}_T)\mathbf{X}\boldsymbol{\beta}, \qquad (14.27)\end{aligned}$$

yielding the FE estimator. The result is not surprising but the length of the derivation is.

14.4 Testing if the Parameters are Constant over Time

With $\boldsymbol{\pi}_C$ entering into (14.22), the χ^2-test based on it is known as a test for FE. But the test also has power against the null-hypothesis of $\boldsymbol{\beta}$ being constant over time, which has been implicitly assumed. However, a test that is explicitly designed for this hypothesis has more power.

To simplify the discussion and focus on $\boldsymbol{\beta}$ solely, we momentarily neglect the individual effects. There are two approaches. One is the J-test on overidentification in a GMM setting. The other one is to estimate $\boldsymbol{\beta}$ for each wave and do a Wald test to see if the various $\hat{\boldsymbol{\beta}}$s differ significantly. We show that the two statistics are the same when the same weight matrix is used. Let, as before, $\dot{\mathbf{X}}_t$ of order $N\times k$ contain the regressors for time t and $\mathbf{y}_t$ likewise. Then the estimator of $\boldsymbol{\beta}$ when it is based on the data for time t only is

$$\hat{\boldsymbol{\beta}}_t = \left(\dot{\mathbf{X}}_t'\dot{\mathbf{X}}_t\right)^{-1}\dot{\mathbf{X}}_t'\mathbf{y}_t. \qquad (14.28)$$

This estimator can be seen as an MM estimator following from the moment conditions $\mathbb{E}\left(\mathbf{x}_{ti}\,\varepsilon_{ti}\right) = \mathbf{0}$, where $\mathbf{x}_{ti}$ of order $k\times 1$ is transpose of the ith row of $\dot{\mathbf{X}}_t$ or of the tth column of $\mathbf{X}_i$. By way of comparison, (14.22) uses the data in the form of $\hat{\boldsymbol{\Pi}}$. The tth column of $\hat{\boldsymbol{\Pi}}'$ is

$$\tilde{\boldsymbol{\beta}}_t \equiv \left(\dot{\mathbf{X}}'\dot{\mathbf{X}}\right)^{-1}\dot{\mathbf{X}}'\mathbf{y}_t.$$

This expression depends on t in a more limited way than does (14.28) and, intuitively, a test based on the latter will have less power. We can collect these k moment conditions into one large set of Tk moment conditions, now requiring GMM to estimate $\boldsymbol{\beta}$. Thus,

$$\mathbf{h}_i \equiv \begin{pmatrix}\mathbf{x}_{1i}\varepsilon_{1i}\\ \vdots \\ \mathbf{x}_{Ti}\varepsilon_{Ti}\end{pmatrix} = \begin{pmatrix}\mathbf{x}_{1i}y_{1i}\\ \vdots \\ \mathbf{x}_{Ti}y_{Ti}\end{pmatrix} - \begin{pmatrix}\mathbf{x}_{1i}\mathbf{x}_{1i}'\\ \vdots \\ \mathbf{x}_{Ti}\mathbf{x}_{Ti}'\end{pmatrix}\boldsymbol{\beta} \equiv \mathbf{g}_i - \mathbf{G}_i\boldsymbol{\beta}.$$

These Tk moment conditions are a subset of the full set of T^2k moment conditions (14.23). In particular,

$$\mathbf{h}_i = (\mathbf{I}_k \otimes \mathbf{H}_T)' (\mathbf{x}_i \otimes \boldsymbol{\varepsilon}_i),$$

with

$$\mathbf{H}_T \equiv \sum_t \mathbf{e}_t \otimes \mathbf{e}_t \mathbf{e}_t'$$

the 'diagonalization' matrix that selects the diagonal elements of the vec of a $T \times T$ matrix, $\boldsymbol{e}_t$ being the tth unit vector of order $T \times 1$. Let the averages over i of $\mathbf{h}_i, \mathbf{g}_i$ and $\mathbf{G}_i$ be $\mathbf{h}, \mathbf{g}$ and $\mathbf{G}$. An estimator of the variance of $\mathbf{h}$ is $\hat{\boldsymbol{\Sigma}}$. The optimal GMM estimator of $\boldsymbol{\beta}$ is

$$\tilde{\boldsymbol{\beta}} = (\mathbf{G}'\hat{\boldsymbol{\Sigma}}^{-1}\mathbf{G})^{-1}\mathbf{G}'\hat{\boldsymbol{\Sigma}}^{-1}\mathbf{g}$$

so

$$\mathbf{g} - \mathbf{G}\tilde{\boldsymbol{\beta}} = \left(\mathbf{I}_{Tk} - \mathbf{G}(\mathbf{G}'\hat{\boldsymbol{\Sigma}}^{-1}\mathbf{G})^{-1}\mathbf{G}'\hat{\boldsymbol{\Sigma}}^{-1}\right)\mathbf{g}.$$

Let $\mathbf{B}$ be of order $Tk \times (T-1)k$ with $\mathbf{B}'\mathbf{G} = \mathbf{0}$ and $(\mathbf{B}, \mathbf{G})$ of full rank, and $\hat{\mathbf{W}} \equiv \hat{\boldsymbol{\Sigma}}^{-1/2}$. The J-statistic for testing whether the $\hat{\boldsymbol{\beta}}_t$s are the same is

$$\begin{aligned}
q_J &= (\mathbf{g} - \mathbf{G}\tilde{\boldsymbol{\beta}})'\hat{\boldsymbol{\Sigma}}^{-1}(\mathbf{g} - \mathbf{G}\tilde{\boldsymbol{\beta}}) \\
&= \mathbf{g}'\left(\hat{\boldsymbol{\Sigma}}^{-1} - \hat{\boldsymbol{\Sigma}}^{-1}\mathbf{G}(\mathbf{G}'\hat{\boldsymbol{\Sigma}}^{-1}\mathbf{G})^{-1}\mathbf{G}'\hat{\boldsymbol{\Sigma}}^{-1}\right)\mathbf{g} \\
&= \mathbf{g}'\hat{\mathbf{W}}\left(\mathbf{I}_{kT} - \hat{\mathbf{W}}\mathbf{G}(\mathbf{G}'\hat{\mathbf{W}}^2\mathbf{G})^{-1}\mathbf{G}'\hat{\mathbf{W}}\right)\hat{\mathbf{W}}\mathbf{g} \\
&= \mathbf{g}'\hat{\mathbf{W}}\left(\hat{\mathbf{W}}^{-1}\mathbf{B}(\mathbf{B}'\hat{\mathbf{W}}^{-2}\mathbf{B})^{-1}\mathbf{B}'\hat{\mathbf{W}}^{-1}\right)\hat{\mathbf{W}}\mathbf{g} \\
&= \mathbf{g}'\mathbf{B}(\mathbf{B}'\hat{\boldsymbol{\Sigma}}\mathbf{B})^{-1}\mathbf{B}'\mathbf{g}, \\
&\equiv q_{\mathrm{W}},
\end{aligned}$$

which is the form of a Wald statistic. The two matrices in square brackets in the third and fourth line are identical as they are both symmetric idempotent of order $p \times p$ and rank $p - k$, orthogonal to $\hat{\mathbf{W}}\mathbf{G}$ of rank k. The result is due to Newey (1985). Now consider

$$\begin{pmatrix} \mathbf{I}_k & -\mathbf{I}_k & & \\ & \ddots & \ddots & \\ & & \mathbf{I}_k & -\mathbf{I}_k \end{pmatrix}\begin{pmatrix} \dot{\mathbf{G}}_1^{-1} & -\dot{\mathbf{G}}_2^{-1} & & \\ & \ddots & \ddots & \\ & & \dot{\mathbf{G}}_{T-1}^{-1} & -\dot{\mathbf{G}}_T^{-1} \end{pmatrix}\begin{pmatrix} \dot{\mathbf{G}}_1 \\ \vdots \\ \dot{\mathbf{G}}_T \end{pmatrix} = \mathbf{0},$$

summarized as $\mathbf{D}'\boldsymbol{\Delta}^{-1}\mathbf{G} = \mathbf{0}$. Then a $\mathbf{B}$ that suites the requirements is $\mathbf{B} \equiv \mathbf{D}'\boldsymbol{\Delta}^{-1}$. Collect the $\hat{\boldsymbol{\beta}}_t$ in the Tk-vector $\hat{\boldsymbol{\beta}}$. Then $\mathbf{B}'\mathbf{g} = \mathbf{D}'\hat{\boldsymbol{\beta}}$ and the Wald statistic can be rewritten as

$$q_{\mathrm{W}} = \hat{\boldsymbol{\beta}}'\mathbf{D}(\mathbf{D}'\hat{\boldsymbol{\Psi}}\mathbf{D})^{-1}\mathbf{D}'\hat{\boldsymbol{\beta}},$$

with $\hat{\boldsymbol{\Psi}} = \boldsymbol{\Delta}^{-1}\hat{\boldsymbol{\Sigma}}\boldsymbol{\Delta}^{-1}$ an estimator of the variance of $\hat{\boldsymbol{\beta}}$. This is the Wald statistic for $H_0 : \hat{\boldsymbol{\beta}}_1 = \cdots = \hat{\boldsymbol{\beta}}_T$.

The number of degrees of freedom of the test is $k(T-1)$, as compared to $k(T^2 - T - 1)$ for the test based on (14.22). So the power of the specific test for parameter constancy over time will have much more power.

14.5 Unbalanced Panels

Here we check how unbalancedness affects the results. Following Baltagi (2021), we capture unbalancedness by letting T depend on i. With a slight abuse of notation we write $\mathbf{i}_i$ for $\mathbf{i}_{T_i}$, $\mathbf{A}_i$ for $\mathbf{A}_{T_i}$, and $\mathbf{R}_i$ for $\mathbf{R}_{T_i}$. So now, with the subscript R indicating the reduced version of the variables, taking the unbalancedness into account,

$$\mathbf{y}_{\mathrm{R}i} = \mathbf{X}_{\mathrm{R}i}\boldsymbol{\beta} + \mathbf{i}_i\alpha_i + \boldsymbol{\varepsilon}_{\mathrm{R}i}, \tag{14.29}$$

where $\mathbf{y}_{\mathrm{R}i}$ and $\boldsymbol{\varepsilon}_{\mathrm{R}i}$ are $T_i \times 1$ and $\mathbf{X}_{\mathrm{R}i}$ is $T_i \times k$. Premultiplication of (14.29) by $(\mathbf{R}_i, \bar{\mathbf{i}}_{T_i})'$ yields basically the same result for $\boldsymbol{\beta}$ as in the balanced case,

$$\hat{\boldsymbol{\beta}} = \left(\sum_i \mathbf{X}'_{\mathrm{R}i}\mathbf{A}_i\mathbf{X}_{\mathrm{R}i}\right)^{-1} \sum_i \mathbf{X}'_{\mathrm{R}i}\mathbf{A}_i\mathbf{y}_i,$$

cf. (14.8), which again is the fixed-effects estimator, adapted to the unbalanced case. This holds for both the Mundlak and Chamberlain projections.

The between regression requires some care since the projection parameters now come to depend on i since the pattern of 'missingness' differs between cross-sectional units. Moreover, in the case of the Chamberlain projection, even the number of such parameters itself comes to depend on i. We denote by $\mathbf{S}_i$ of order $T \times T_i$ the 'selection matrix' of zeros and ones indication which observations over time are available for i, and let $\mathbf{B}_i \equiv \mathbf{I}_k \otimes \mathbf{S}_i$. Then, for the Mundlak projection there holds

$$\begin{aligned}
\alpha_i &= \bar{\mathbf{x}}'_{\mathrm{R}i}\boldsymbol{\pi}_{\mathrm{M}i} + \nu_i \\
&= \bar{\mathbf{x}}'_i\mathbf{S}_i\left(\mathbf{S}'_i\boldsymbol{\Sigma}_{\bar{\mathbf{x}}}\mathbf{S}_i\right)^{-1}\mathbf{S}'_i\boldsymbol{\sigma}_{\bar{\mathbf{x}}\alpha} + \nu_i \\
&= \left(\bar{\mathbf{x}}'_i\mathbf{S}_i\left(\mathbf{S}'_i\boldsymbol{\Sigma}_{\bar{\mathbf{x}}}\mathbf{S}_i\right)^{-1}\mathbf{S}'_i\boldsymbol{\Sigma}_{\bar{\mathbf{x}}}\right)\left(\boldsymbol{\Sigma}_{\bar{\mathbf{x}}}^{-1}\boldsymbol{\sigma}_{\bar{\mathbf{x}}\alpha}\right) + \nu_i \\
&\equiv \tilde{\bar{\mathbf{x}}}'_i\boldsymbol{\pi}_{\mathrm{M}} + \nu_i.
\end{aligned}$$

while for the Chamberlain projection there holds

$$\begin{aligned}
\alpha_i &= \mathbf{x}'_{\mathrm{R}i}\boldsymbol{\pi}_{\mathrm{C}i} + \nu_i \\
&= \mathbf{x}'_i\mathbf{B}_i\left(\mathbf{B}'_i\boldsymbol{\Sigma}_{\mathbf{x}}\mathbf{B}_i\right)^{-1}\mathbf{B}'_i\boldsymbol{\sigma}_{\mathbf{x}\alpha} + \nu_i \\
&= \left(\mathbf{x}'_i\mathbf{B}_i\left(\mathbf{B}'_i\boldsymbol{\Sigma}_{\mathbf{x}}\mathbf{B}_i\right)^{-1}\mathbf{B}'_i\boldsymbol{\Sigma}_{\mathbf{x}}\right)\left(\boldsymbol{\Sigma}_{\mathbf{x}}^{-1}\boldsymbol{\sigma}_{\mathbf{x}\alpha}\right) + \nu_i \\
&\equiv \tilde{\mathbf{x}}'_i\boldsymbol{\pi}_{\mathrm{C}} + \nu_i.
\end{aligned}$$

These indicate how adapting the variables in the projections need to be adapted in order to corrrect for unbalancedness. When making $\tilde{\bar{\mathbf{x}}}_i$ and $\tilde{\mathbf{x}}_i$ operational, the population quantities $\boldsymbol{\Sigma}_{\bar{\mathbf{x}}}$ and $\boldsymbol{\Sigma}_{\mathbf{x}}$ need to be replaced by their sample counterparts. The expression for $\tilde{\mathbf{x}}_i$ shows that replacing the 'missing' observations by zeros is not adequate, as was already pointed out by Abrevaya (2013). In particluar, he introduces a modified Chamberlain approach in which the projection depends upon the form of missingness for a given individual. This leads to orthogonality conditions that depend upon the form of exogeneity assumption maintained. These orthogonality conditions are used in a GMM framework to develop estimators of the model (and projection) parameters as well as tests of strict exogeneity and random effects.

A much related and in fact formally identical model was studied by Arkhangelsky and Imbens (2024), who have groups (with their naturally different group sizes) rather than time as the second dimension in the data. The model is

$$y_i = \mathbf{x}'_i\boldsymbol{\beta} + \alpha_{g(i)} + \varepsilon_i, \tag{14.30}$$

with $g(i)$ the group indicator and $\alpha_{g(i)}$ a fixed group effect. They state that OLS in this model yields the same result for $\boldsymbol{\beta}$ as does OLS in the model where the group fixed effects are replaced by averages per group,

$$y_i = \mathbf{x}_i'\boldsymbol{\beta} + \bar{\mathbf{x}}_{g(i)}'\boldsymbol{\gamma} + \varepsilon_i^*. \tag{14.31}$$

There is no reference to an underlying RE model to which Mundlak's projection is applied. They state that the numerical equivalence was first shown by Mundlak (1978) and follows from repeated applications of textbook omitted variable bias formulas. In fact, the step from (14.31) to (14.30) follows directly from applying the double projection result in (14.9).

14.6 Two-Way Effects

Arellano's transformation of Section 14.2 is easily extended to two dimensions that add time effects $\gamma_t, t = 1, \ldots, T$ to the model in addition to the individual effects α_i that were already there. This approach is much more concise than the one by Yang (2022) and Baltagi (2023a), who independently considered this model. The kind of situations where this may be useful will not so much involve time-series of cross-sections but e.g., trade flow data but we keep the notation with i and t.

Because the model is now symmetric in both dimensions we write also in symmetric form, for all nT observations together. The extension of the notation to this case is self-evident. The model now is

$$y_{it} = \mathbf{x}_{it}'\boldsymbol{\beta} + \alpha_i + \gamma_t + \varepsilon_{it} \tag{14.32}$$

or, in matrix format,

$$\mathbf{y} = \mathbf{X}\boldsymbol{\beta} + (\mathbf{I}_n \otimes \mathbf{i}_T)\boldsymbol{\alpha} + (\mathbf{i}_n \otimes \mathbf{I}_T)\boldsymbol{\gamma} + \boldsymbol{\varepsilon}. \tag{14.33}$$

The Mundlak projections of the effects on the means of the regressors over time are

$$\begin{aligned} \boldsymbol{\alpha} &= (\mathbf{I}_n \otimes \bar{\mathbf{i}}_T)'\mathbf{X}\mathbf{a} + \mathbf{v} \\ \boldsymbol{\gamma} &= (\bar{\mathbf{i}}_n \otimes \mathbf{I}_T)'\mathbf{X}\mathbf{c} + \mathbf{w}. \end{aligned}$$

Extending Arellano's transformation to the case of two dimensions, we premultiply the model with the orthonormal $nT \times nT$ matrix $(\mathbf{R}_n \otimes \mathbf{R}_T, \mathbf{R}_n \otimes \bar{\mathbf{i}}_T, \bar{\mathbf{i}}_n \otimes \mathbf{R}_T, \bar{\mathbf{i}}_n \otimes \bar{\mathbf{i}}_T)$ and substitute for $\boldsymbol{\alpha}$ and $\boldsymbol{\gamma}$ to obtain

$$\begin{aligned} (\mathbf{R}_n \otimes \mathbf{R}_T)'\mathbf{y} &= (\mathbf{R}_n \otimes \mathbf{R}_T)'\mathbf{X}\boldsymbol{\beta} && + (\mathbf{R}_n \otimes \mathbf{R}_T)'\boldsymbol{\varepsilon} \\ (\mathbf{R}_n \otimes \bar{\mathbf{i}}_T)'\mathbf{y} &= (\mathbf{R}_n \otimes \bar{\mathbf{i}}_T)'\mathbf{X}(\boldsymbol{\beta} + \mathbf{a}) && + (\mathbf{R}_n \otimes \bar{\mathbf{i}}_T)'\boldsymbol{\varepsilon} + \mathbf{R}_n'\mathbf{v} \\ (\bar{\mathbf{i}}_n \otimes \mathbf{R}_T)'\mathbf{y} &= (\bar{\mathbf{i}}_n \otimes \mathbf{R}_T)'\mathbf{X}(\boldsymbol{\beta} + \mathbf{c}) && + (\bar{\mathbf{i}}_n \otimes \mathbf{R}_T)'\boldsymbol{\varepsilon} + \mathbf{R}_T'\mathbf{w} \\ (\bar{\mathbf{i}}_n \otimes \bar{\mathbf{i}}_T)'\mathbf{y} &= (\bar{\mathbf{i}}_n \otimes \bar{\mathbf{i}}_T)'\mathbf{X}(\boldsymbol{\beta} + \mathbf{a} + \mathbf{c}) && + (\bar{\mathbf{i}}_n \otimes \bar{\mathbf{i}}_T)'\boldsymbol{\varepsilon} + \bar{\mathbf{i}}_n'\mathbf{v} + \bar{\mathbf{i}}_T'\mathbf{w}. \end{aligned}$$

The last equation is non-informative since the variables have been demeaned. The error terms of the four equations of this model each have a scalar covariance matrix. For the first three equations they are $\sigma_\varepsilon^2\mathbf{I}_{(N-1)(T-1)}$, $(\sigma_\varepsilon^2 + \sigma_v^2)\mathbf{I}_{N-1}$, and $(\sigma_\varepsilon^2 + \sigma_w^2)\mathbf{I}_{T-1}$, respectively. They are uncorrelated with each other and the regression coefficients do not overlap. So we have only the first equation to estimate $\boldsymbol{\beta}$. With $\mathbf{R}_n\mathbf{R}_n' \otimes \mathbf{R}_T\mathbf{R}_T' = \mathbf{A}_n \otimes \mathbf{A}_T$ the estimator is

$$\hat{\boldsymbol{\beta}} = [\mathbf{X}'(\mathbf{A}_n \otimes \mathbf{A}_T)\mathbf{X}]^{-1}\mathbf{X}'(\mathbf{A}_n \otimes \mathbf{A}_T)\mathbf{y}, \tag{14.34}$$

which again is the FE estimator, but now with fixed individual and time effects, corresponding with the transformation $\tilde{g}_{it} = g_{it} - g_{n*} - g_{*t} + g_{**}$. This result was independently derived by Wooldridge (2021), Yang (2022), and Baltagi (2023a).

Wooldridge (2021) shows that the OLS estimator of $\boldsymbol{\beta}$ in (14.32) or (14.33) is the same as the OLS estimator of $\boldsymbol{\beta}$ in the model where the effects are replaced by the means over time and the means over individuals, so in the model

$$y_{it} = \mathbf{x}'_{it}\boldsymbol{\beta} + \bar{\mathbf{x}}'_{i*}\boldsymbol{\alpha} + \bar{\mathbf{x}}'_{*t}\boldsymbol{\gamma} + \varepsilon^*_{it} \tag{14.35}$$

in self-evident notation, or again

$$\mathbf{y} = \mathbf{X}\boldsymbol{\beta} + (\mathbf{I}_n \otimes \bar{\mathbf{J}}_T)\mathbf{X}\boldsymbol{\alpha}^* + (\bar{\mathbf{J}}_n \otimes \mathbf{I}_T)\mathbf{X}\boldsymbol{\gamma}^* + \boldsymbol{\varepsilon}^*. \tag{14.36}$$

Wooldridge (2021) calls this model the 'two-way Mundlak regression'. No reference is made to an underlying idea of a projection, and hence the GLS structure that remains after the projection is not considered. But also here it can be shown that OLS and GLS coincide, while the same caveat applies as in the one-way model as to the validity of test results.

To show that OLS in this model yields the FE estimator for $\boldsymbol{\beta}$ it is of some interest to consider an alternative proof to the one based on Arellano's transformation, just as we did for the one-way model in Section 14.2. The generalization to the two-way model is straightforward and is based on the observation that the two sets of regressors containing the means, $\mathbf{Z}$, say, are orthogonal to each other, again using the fact that the variables are demeaned. Hence, the projection matrix onto the space spanned by $\mathbf{Z}$ is the sum of the separate projection matrices. So we can apply the double projection result (14.9) twice and find

$$\begin{aligned}\mathbf{P_Z} &= (\mathbf{I}_n \otimes \bar{\mathbf{J}}_T + \bar{\mathbf{J}}_n \otimes \mathbf{I}_T)\mathbf{X} \\ &= (\mathbf{A}_n \otimes \bar{\mathbf{J}}_T + \bar{\mathbf{J}}_n \otimes \mathbf{A}_T)\mathbf{X}\end{aligned}$$

and hence $\mathbf{M_Z} \equiv \mathbf{I}_{nT} - \mathbf{P_Z}$, the projection orthogonal to $\mathbf{Z}$, is

$$\mathbf{M_Z} = (\mathbf{A}_n \otimes \mathbf{A}_T)\mathbf{X},$$

which leads to the two-way FE estimator. This result is "the key algebraic result in this paper" (Wooldridge, 2021, Theorem 3.1).

Baltagi (2023a) shows that the F-tests for the significance of **a** and **c** in the augmented two-way Mundlak regression generate Hausman (1978) type tests which were generalized from the one-way to the two-way error components model by Kang (1985). Once again, it is important to emphasize that even though OLS is equivalent to GLS on this two-way Mundlak augmented model, the standard errors are different, and so are tests of hypotheses. In fact, performing the F tests for the significance of **a** and **c** with OLS on the augmented two-way Mundlak model yields completely different test statistics than those using a two-way random effects GLS regression. Non-rejection of the null finds pooled OLS to be the efficient estimator, while non-rejection of the null using GLS finds that the two-way random effects estimator is the efficient estimator. This is in the spirit of what Mundlak (1978) intended.

14.7 Extension to Three Dimensions

Up till now we considered two-dimensional data, with one or two effects. Following Baltagi (2024a) and Yang (2021) we now consider three-dimensional data, where $\mathbf{y}$ now has MNT elements, and likewise for $\mathbf{X}$ and $\boldsymbol{\varepsilon}$. Balázsi, Mátyás and Wansbeek (2018, 2024) provide an overview. With effects for the new dimension added, the model becomes $y_{mnt} = \mathbf{x}'_{mnt}\boldsymbol{\beta} + \mu + \delta_m + \alpha_n + \gamma_t + \varepsilon_{mnt}$ or

$$\mathbf{y} = \mathbf{X}\boldsymbol{\beta} + \boldsymbol{\delta} \otimes \mathbf{i}_{NT} + \mathbf{i}_M \otimes \boldsymbol{\alpha} \otimes \mathbf{i}_T + \mathbf{i}_{MN} \otimes \boldsymbol{\gamma} + \boldsymbol{\varepsilon}, \tag{14.37}$$

hence the design matrix for the effects is

$$\mathbf{D} \equiv (\mathbf{I}_M \otimes \mathbf{i}_{NT}, \mathbf{i}_M \otimes \mathbf{I}_N \otimes \mathbf{i}_T, \mathbf{i}_{MN} \otimes \mathbf{I}_T).$$

Since all three submatrices in $\mathbf{D}$ span $\mathbf{i}_{MNT}$ the rank of $\mathbf{D}$ is $M+N+T-2$.

Now consider the Mundlak projections of the effects on the means of $\mathbf{X}$ in the other two dimensions as

$$\boldsymbol{\delta} = (\mathbf{I}_M \otimes \bar{\mathbf{i}}_{NT})' \mathbf{X}\mathbf{d} + \mathbf{u} \tag{14.38}$$

$$\boldsymbol{\alpha} = (\bar{\mathbf{i}}_M \otimes \mathbf{I}_N \otimes \bar{\mathbf{i}}_T)' \mathbf{X}\mathbf{a} + \mathbf{v} \tag{14.39}$$

$$\boldsymbol{\gamma} = (\bar{\mathbf{i}}_{MN} \otimes \mathbf{I}_T)' \mathbf{X}\mathbf{c} + \mathbf{w}. \tag{14.40}$$

Like before, we premultiply the model by transposed $\mathbf{R}$ matrices and $\bar{\mathbf{i}}$ vectors. We use the notation $\mathbf{C}_{111} \equiv \mathbf{R}_M \otimes \mathbf{R}_i \otimes \mathbf{R}_T$ and $\mathbf{C}_{110} \equiv \mathbf{R}_M \otimes \mathbf{R}_i \otimes \bar{\mathbf{i}}_T$ and so on, a subscript '1' to $\mathbf{C}$ indicating an $\mathbf{R}$ and a subscript '0' an $\bar{\mathbf{i}}$. We collect all the combinations in the orthonormal $MNT \times MNT$ matrix $\mathbf{C}$,

$$\mathbf{C} \equiv (\mathbf{Q}, \mathbf{C}_{100}, \mathbf{C}_{010}, \mathbf{C}_{001}, \mathbf{C}_{000}),$$

with $\mathbf{Q}$ of order $MNT \times (MNT - M - N - T + 2)$ defined as

$$\mathbf{Q} \equiv (\mathbf{C}_{111}, \mathbf{C}_{110}, \mathbf{C}_{101}, \mathbf{C}_{011}).$$

When we follow Arellano's transformation again and premultiply (14.37) by $\mathbf{C}'$ we obtain

$$\begin{aligned}
\mathbf{Q}'\mathbf{y} &= \mathbf{Q}'\mathbf{X}\boldsymbol{\beta} && +\mathbf{Q}'\boldsymbol{\varepsilon} \\
\mathbf{C}'_{100}\mathbf{y} &= \mathbf{C}'_{100}\mathbf{X}(\boldsymbol{\beta}+\mathbf{d}) && +\mathbf{C}'_{100}\boldsymbol{\varepsilon} + \mathbf{R}'_M\mathbf{u} \\
\mathbf{C}'_{010}\mathbf{y} &= \mathbf{C}'_{010}\mathbf{X}(\boldsymbol{\beta}+\mathbf{a}) && +\mathbf{C}'_{010}\boldsymbol{\varepsilon} + \mathbf{R}'_i\mathbf{v} \\
\mathbf{C}'_{001}\mathbf{y} &= \mathbf{C}'_{001}\mathbf{X}(\boldsymbol{\beta}+\mathbf{c}) && +\mathbf{C}'_{001}\boldsymbol{\varepsilon} + \mathbf{R}'_T\mathbf{w} \\
\bar{y} &= \bar{\mathbf{x}}'(\boldsymbol{\beta}+\mathbf{d}+\mathbf{a}+\mathbf{c}) && +\bar{\boldsymbol{\varepsilon}} + \bar{\mathbf{u}} + \bar{\mathbf{v}} + \bar{\mathbf{w}}.
\end{aligned} \tag{14.41}$$

Again, the last equation is non-informative due to the demeaning of the variables, and the error terms of the equations each have a scalar covariance matrix and are uncorrelated with each other. Reasoning as above, estimation of $\boldsymbol{\beta}$ rests solely on the first equation, leading to

$$\hat{\boldsymbol{\beta}} = (\mathbf{X}'\mathbf{Q}\mathbf{Q}'\mathbf{X})^{-1}\mathbf{X}'\mathbf{Q}\mathbf{Q}'\mathbf{y}. \tag{14.42}$$

Since $\mathbf{Q}'\mathbf{D} = \mathbf{0}$ and $(\mathbf{Q}, \mathbf{D})$ is nonsingular, the matrix $\mathbf{Q}\mathbf{Q}'$ is the projection matrix orthogonal to $\mathbf{D}$. Hence $\hat{\boldsymbol{\beta}}$ in (14.42) is the FE estimator. With g a generic variable, its computation can be based on the transformation

$$\tilde{g}_{mnt} = g_{mnt} - g_{m**} - g_{*n*} - g_{**t} + 2g_{***},$$

written in scalar notation.

A variant of some interest arises when the time effects $\boldsymbol{\gamma}$ are taken fixed a priori. Then the first three equations of (14.41) are unaffected while the fourth becomes

$$\mathbf{C}'_{001}\mathbf{y} = \mathbf{C}'_{001}\mathbf{X}\boldsymbol{\beta} + \boldsymbol{\gamma} + \mathbf{C}'_{001}\boldsymbol{\varepsilon} + \mathbf{R}'_T\mathbf{w}.$$

So this leaves the estimation of $\boldsymbol{\beta}$ unaffected.

An alternative formulation that has been considered is the model with effects with double indices, $y_{mnt} = \mathbf{x}'_{mnt}\boldsymbol{\beta} + \alpha_{nt} + \gamma_{mt} + \delta_{mn} + \varepsilon_{nt}$ so

$$\mathbf{y} = \mathbf{X}\boldsymbol{\beta} + \mathbf{i}_M \otimes \boldsymbol{\alpha} + (\mathbf{I}_M \otimes \mathbf{K}_{TN})(\boldsymbol{\gamma} \otimes \mathbf{i}_N) + \boldsymbol{\delta} \otimes \mathbf{i}_T + \boldsymbol{\varepsilon},$$

with $\mathbf{K}_{NT}$ the commutation matrix with property $\mathbf{K}_{NT}(\boldsymbol{p} \otimes \boldsymbol{q}) = \boldsymbol{q} \otimes \boldsymbol{p}$ for any $\boldsymbol{p}$ and $\boldsymbol{q}$ of order $N \times 1$ and $T \times 1$, respectively. The Mundlak projections of the effects on the mean of $\mathbf{X}$ in the remaining dimension are

$$\boldsymbol{\alpha} = \left(\bar{\mathbf{i}}_M \otimes \mathbf{I}_{NT}\right)' \mathbf{Xa} + \mathbf{v} \tag{14.43}$$
$$\boldsymbol{\delta} = \left(\mathbf{I}_{MN} \otimes \bar{\mathbf{i}}_T\right)' \mathbf{Xd} + \mathbf{u} \tag{14.44}$$
$$\boldsymbol{\gamma} = \left(\mathbf{I}_M \otimes \bar{\mathbf{i}}_n \otimes \mathbf{I}_T\right)' \mathbf{Xc} + \mathbf{w}. \tag{14.45}$$

so Arellano's transformation leads to

$$\begin{aligned}
\mathbf{C}'_{111}\mathbf{y} &= \mathbf{C}'_{111}\mathbf{X}\boldsymbol{\beta} && +\mathbf{C}'_{111}\boldsymbol{\varepsilon} \\
\mathbf{C}'_{110}\mathbf{y} &= \mathbf{C}'_{110}\mathbf{X}(\boldsymbol{\beta}+\mathbf{d}) && +\mathbf{C}'_{110}\boldsymbol{\varepsilon} + (\mathbf{R}_M \otimes \mathbf{R}_i)'\mathbf{u} \\
\mathbf{C}'_{101}\mathbf{y} &= \mathbf{C}'_{101}\mathbf{X}(\boldsymbol{\beta}+\mathbf{c}) && +\mathbf{C}'_{101}\boldsymbol{\varepsilon} + (\mathbf{R}_M \otimes \mathbf{R}_T)'\mathbf{w} \\
\mathbf{C}'_{011}\mathbf{y} &= \mathbf{C}'_{011}\mathbf{X}(\boldsymbol{\beta}+\mathbf{a}) && +\mathbf{C}'_{011}\boldsymbol{\varepsilon} + (\mathbf{R}_i \otimes \mathbf{R}_T)'\mathbf{u} \\
\mathbf{C}'_{100}\mathbf{y} &= \mathbf{C}'_{100}\mathbf{X}(\boldsymbol{\beta}+\mathbf{d}+\mathbf{c}) && +\mathbf{C}'_{100}\boldsymbol{\varepsilon} + (\mathbf{R}_M \otimes \bar{\mathbf{i}}_n)'\mathbf{u} + (\mathbf{R}_M \otimes \mathbf{i}_T)'\mathbf{w} \\
\mathbf{C}'_{010}\mathbf{y} &= \mathbf{C}'_{010}\mathbf{X}(\boldsymbol{\beta}+\mathbf{d}+\mathbf{a}) && +\mathbf{C}'_{010}\boldsymbol{\varepsilon} + (\mathbf{R}_i \otimes \bar{\mathbf{i}}_T)'\mathbf{v} + (\bar{\mathbf{i}}_M \otimes \mathbf{R}_i)'\mathbf{u} \\
\mathbf{C}'_{001}\mathbf{y} &= \mathbf{C}'_{001}\mathbf{X}(\boldsymbol{\beta}+\mathbf{c}+\mathbf{a}) && +\mathbf{C}'_{001}\boldsymbol{\varepsilon} + (\bar{\mathbf{i}}_n \otimes \mathbf{R}_T)'\mathbf{v} + (\bar{\mathbf{i}}_M \otimes \mathbf{R}_T)'\mathbf{w} \\
\bar{y} &= \bar{\mathbf{x}}'(\boldsymbol{\beta}+\mathbf{d}+\mathbf{c}+\mathbf{a}) && +\bar{\boldsymbol{\varepsilon}} + \bar{\mathbf{u}} + \bar{\mathbf{v}} + \bar{\mathbf{w}}.
\end{aligned}$$

This again is a system where each equation has a scalar covariance matrix, and the errors of the equations are two by two uncorrelated. However, the regression coefficients of the various equations overlap and, unlike the previous cases, cannot be separated. Hence optimal estimation of $\boldsymbol{\beta}$ requires GLS estimation of the entire system.

As was shown by Balázsi, Mátyás, and Wansbeek (2018), the FE estimator in this case follows from the transformation by $\mathbf{C}'_{111}$, so

$$\hat{\boldsymbol{\beta}} = \left[\mathbf{X}'\left(\mathbf{A}_M \otimes \mathbf{A}_N \otimes \mathbf{A}_T\right)\mathbf{X}\right]^{-1} \mathbf{X}'\left(\mathbf{A}_M \otimes \mathbf{A}_N \otimes \mathbf{A}_T\right)\mathbf{y},$$

extending the result (14.34) for the two-dimensional case. It corresponds with the transformation

$$\tilde{g}_{mnt} = g_{mnt} - g_{mn*} - g_{m*t} - g_{*nt} + g_{m**} + g_{*n*} + g_{**t} - 1.$$

So in this case the Mundlak projection does not lead to the FE estimator. The finding that the fixed-effects and random-effects estimators are not the same for double-indexed effects was first derived by Yang (2021). Baltagi (2024a), however, shows that Mundlak's result still holds for higher-dimensional panel data with single-indexed effects. Once again, GLS rather than OLS should be applied to derive this Mundlak higher dimensional panel result and also to get the correct Hausman -type tests.

14.8 Other Applications of Arellano's Transformation

We now consider how useful applying Arellano's transformation is in three specific cases: factor models, varying coefficients, and the spatial regression model.

14.8.1 Factor Models

Instead of $\mathbf{i}_T\alpha_i$ in (14.1) we now consider the more general structure $\mathbf{F}\boldsymbol{\alpha}_i$, with $\mathbf{F}$ and $\boldsymbol{\alpha}_i$ of order $T \times r$ and $r \times 1$, respectively. The matrix $\mathbf{F}$ contains factors, that is, variables that vary over time but not over individuals. We allow for the possibility that $\mathbf{i}_T$ is a column of $\mathbf{F}$. We consider directly the Chamberlain-type projection, which now is $\boldsymbol{\alpha}_i = \mathbf{G}'\mathbf{x}_i + \mathbf{v}_i$. With this projection substituted we obtain

$$\begin{aligned}\mathbf{y}_i &= \mathbf{X}_i\boldsymbol{\beta} + \mathbf{F}\boldsymbol{\alpha}_i + \boldsymbol{\varepsilon}_i \\ &= \mathbf{X}_i\boldsymbol{\beta} + \mathbf{F}(\mathbf{G}'\mathbf{x}_i + \mathbf{v}_i) + \boldsymbol{\varepsilon}_i.\end{aligned}$$

Let $\mathbf{F}^+ \equiv (\mathbf{F}'\mathbf{F})^{-1}\mathbf{F}'$. We generalize Arellano's transformation in the sense that we now premultiply the model by $\mathbf{H}'$, of order $(T-r)\times T$ and satisfying $\mathbf{H}'\mathbf{F} = \mathbf{0}$ and $\mathbf{H}'\mathbf{H} = \mathbf{I}_{T-r}$ so $\mathbf{M}_\mathbf{F} \equiv \mathbf{I}_T - \mathbf{F}\mathbf{F}^+ = \mathbf{H}\mathbf{H}'$, and by $\mathbf{F}^+$, to obtain the equivalent two-equation model

$$\mathbf{H}'\mathbf{y}_i = \mathbf{H}'\mathbf{X}_i\boldsymbol{\beta} + \mathbf{H}'\boldsymbol{\varepsilon}_i \tag{14.46}$$

$$\begin{aligned}\mathbf{F}^+\mathbf{y}_i &= \mathbf{F}^+\mathbf{X}_i\boldsymbol{\beta} + \mathbf{G}'\mathbf{x}_i + \mathbf{v}_i + \mathbf{F}^+\boldsymbol{\varepsilon}_i \\ &= \left((\mathbf{F}^+)'\otimes\boldsymbol{\beta}' + \mathbf{G}'\right)\mathbf{x}_i + \mathbf{v}_i + \mathbf{F}^+\varepsilon_i.\end{aligned} \tag{14.47}$$

The two error terms are uncorrelated. The coefficient matrix in (14.47), $(\mathbf{F}^+)'\otimes\boldsymbol{\beta}' + \mathbf{G}'$, contains no information about $\boldsymbol{\beta}$ itself. This has to come from (14.46), which yields

$$\hat{\boldsymbol{\beta}} = \left(\sum_i \mathbf{X}_i'\mathbf{M}_\mathbf{F}\mathbf{X}_i\right)^{-1} \sum_i \mathbf{X}_i'\mathbf{M}_\mathbf{F}\mathbf{y}_i,$$

which is the estimator when the $\boldsymbol{\alpha}_i$ are taken to be fixed.

Thus, with factors, we can directly generalize (14.8), where (14.46) is the analogue of the within regression and (14.47) the analogue of the between regression. We have implicitly assumed that the factors are observable, which they usually are not, and the above just serves as a potentially useful algebraic step in a more elaborate context, see e.g. Pesaran (2006), Bai (2009) and Westerlund (2019).

14.8.2 Varying Coefficients

The panel data model with individual effects can be considered as a model with an intercept that varies over individuals or, in other words, a constant term with a coefficient that varies over individuals (Wooldridge (2019)). To check the possible relevance of Arellano's transformation here, one straightforward generalization of the basic model (14.1) is the model where one regressor has a coefficient that varies over individuals. We thus consider the model

$$y_i = \mathbf{X}_i\boldsymbol{\beta} + \mathbf{z}_i\gamma_i + \boldsymbol{\varepsilon}_i.$$

The generalization of Arellano's transformation to this case entails projecting out the $\mathbf{z}_i$, through the matrix $\mathbf{M}_i \equiv \mathbf{z}_i\mathbf{z}_i^+$, with $\mathbf{z}_i \equiv \mathbf{z}_i/\mathbf{z}_i'\mathbf{z}_i$. Then

$$\mathbf{M}_i\mathbf{y}_i = \mathbf{M}_i\mathbf{X}_i\boldsymbol{\beta} + \mathbf{M}_i\boldsymbol{\varepsilon}_i \tag{14.48}$$

$$\mathbf{z}_i^+\mathbf{y}_i = \mathbf{z}_i^+\mathbf{X}_i\boldsymbol{\beta} + \gamma_i + \mathbf{z}_i^+\boldsymbol{\varepsilon}_i. \tag{14.49}$$

This generalizes (14.5) and (14.6). We now consider the case where the variation in the varying coefficients is correlated with the $\mathbf{X}_i$ and express this correlation again by a projection of the γ_i on a function of the $\mathbf{X}_i$. The Mundlak-type projection is

$$\gamma_i = \mathbf{z}_i^+\mathbf{X}_i\boldsymbol{\pi}_\mathrm{V} + v_i$$

and substitution in (14.49) leads to

$$\mathbf{z}_i^+\mathbf{y}_i = \mathbf{z}_i^+\mathbf{X}_i(\boldsymbol{\beta} + \boldsymbol{\pi}_V) + \mathbf{z}_i^+\boldsymbol{\varepsilon}_i. \tag{14.50}$$

Now, (14.48) leads to a generalized FE estimator, while (14.49) again allows for testing whether the variation in the coefficients of the z_i is endogenous indeed.

14.8.3 The Spatial Mundlak Model

The spatial Mundlak model was first considered by Debarsy (2012) in the context of a spatial Durbin panel data model (SDM). More specifically, the SDM includes a spatial lag of the dependent variable **Wy** as well as **X** and **WX**, where **W** is a spatial weight matrix that describes the interaction between the n cross-sectional units. This weight matrix is row normalized. The random individual effects are projected on the explanatory variables averaged over time as in (14.2), but additionally, Debarsy (2012) includes the spatial weighted averages of these explanatory variables also averaged over time. Maximum likelihood estimation using the normality assumption is applied and a likelihood ratio (LR) test is used to test the significance of the correlation between the regressors and their spatial weighted average (both averaged over time) and the individual effects. Here, the fixed effects estimator is inconsistent due to the presence of a spatial lagged **y**, so the equivalence between the GLS spatial random effects and the fixed effects spatial Mundlak estimators is not considered.

Debarsy (2012) applies this spatial Mundlak-Durbin model to explain housing price variations across 588 municipalities in Belgium over the period 2004 to 2007. He finds significant non-zero coefficients for the Mundlak terms indicating correlated random effects. Baltagi (2023b) also considers the Mundlak (1978) model in the context of the spatial error panel data model (SEM) of Anselin (1988). He shows that Mundlak's classic result does not extend to the spatial panel SEM model, i.e., the spatial random effects estimator does not reduce to the spatial fixed effects estimator once the average regressors over time are included in the random effects SEM regression unless one ignores the spatial correlation in the remainder error. More specifically, GLS on this Mundlak SEM model does not yield the fixed effects estimator.

As in Debarsy (2012) for the SDM, one can use maximum likelihood estimation (MLE) for the SEM (assuming normality) to test Mundlak's (1978) idea that random effects are correlated with the regressors using an LR test. This is applied to the Belotti, Hughes and Piano Mortari (2017) data set on residential demand for electricity covering the 48 contiguous United States plus the District of Columbia for the period 1990-2010. For both SAR and SDM, Baltagi (2023b) shows that the spatial Mundlak correlated random-effects estimator does not reduce to its fixed-effects counterpart, although for this empirical example the estimates are pretty close. The LR test shows that these Mundlak averages are jointly significant in the SAR and SDM for residential demand for electricity in the United States.

References

Ahn, S. C. & Low, S. (1996). A reformulation of the Hausman test for regression models with pooled cross-section time-series data. *Journal of Econometrics*, *71*, 309–319.

Amemiya, T. (1971). The estimation of the variances in a variance-components model. *International Economic Review*, *12*, 1-13.

Angrist, J. D. & Newey, W. K. (1991). Over-identification tests in earnings fuctions with fixed effects. *Journal of Business & Economic Statistics*, *9*, 317-323.

Anselin, L. (1988). *Spatial econometrics: Methods and models*. Kluwer.
Arellano, M. (1993). On testing of correlated effects with panel data. *Journal of Econometrics*, *59*, 87-97.
Arellano, M. & Bond, S. (1991). Some tests of specification for panel data: Monte Carlo evidence and an application to employment equations. *Review of Economic Studies*, *58*, 277-297.
Arellano, M. & Bover, O. (1995). Another look at the instrumental variable estimation of error-components models. *Journal of Econometrics*, *68*, 29-51.
Arkhangelsky, D. & Imbens, G. W. (2024). Fixed effects and the generalized Mundlak estimator. *Review of Economic Studies*, *91*, 2545–2571.
Bai, J. (2009). Panel data models with interactive fixed effects. *Econometrica*, *77*, 1229-1279.
Balázsi, L., Mátyás, L. & Wansbeek, T. (2018). The estimation of multi-dimensional fixed effects panel data models. *Econometric Reviews*, *37*, 212-227.
Balázsi, L., Mátyás, L. & Wansbeek, T. (2024). Fixed effects models. In L. Mátyás (Ed.), *The econometrics of multi dimensional panels* (second ed., p. 1-37). Springer.
Balestra, P. & Nerlove, M. (1966). Pooling cross-section and time series data in the estimation of a dynamic model: the demand for natural gas. *Econometrica*, *34*, 585-612.
Baltagi, B. H. (2006). An alternative derivation of Mundlak's fixed effects results using system estimation. *Econometric Theory*, *22*, 1191-1194.
Baltagi, B. H. (2009). *A companion to econometric analysis of panel data*. Wiley.
Baltagi, B. H. (2023a). The two-way Mundlak estimator. *Econometric Reviews*, *42*, 240–246.
Baltagi, B. H. (2023b). The Mundlak spatial estimator. *Journal of Spatial Econometrics*, *4:6*.
Baltagi, B. H. (2024a). The multidimensional Mundlak estimator. *Economics Letters*, *236*, 111607.
Baltagi, B. H. (2024b). Hausman's specification test for panel data: Practical tips. In C. Parmeter, M. Tsionas & H.-J. Wang (Eds.), *Essays in honor of Subal Kumbhakar (Advances in Econometrics Book 46)* (chap. 2). Emerald.
Baltagi, B. H., Bresson, G. & Pirotte, A. (2009). Testing the fixed effects restrictions? A Monte Carlo study of Chamberlain's minimum chi-squared test. *Statistics & Probability Letters*, *79*, 1358-1362. doi: https://doi.org/10.1016/j.spl.2009.03.012
Belotti, F., Hughes, G. & Piano Mortari, A. (2017). Spatial panel data models using Stata. *The Stata Journal*, *17*, 139–180.
Biørn, E. (2017). *Econometrics of panel data*. Oxford University Press.
Cameron, A. C. & Trivedi, P. K. (2005). *Microeconometrics*. Cambridge University Press.
Card, D. (1996). The effect of unions on the structure of wages: A longitudinal analysis. *Econometrica*, *64*, 957–979.
Carey, K. (1997). A panel data design for estimation of hospital cost functions. *Review of Economics and Statistics*, *79*, 443–453.

Chamberlain, G. (1980). Analysis of covariance with qualitative data. *Review of Economic Studies*, *47*, 225-238.

Chamberlain, G. (1982). Multivariate regression models for panel data. *Journal of Econometrics*, *18*, 5-46.

Chamberlain, G. (1984). Panel data. In Z. Griliches & M. Intriligator (Eds.), *Handbook of econometrics 2* (p. 1247–1318). Amsterdam: North-Holland.

Crépon, B. & Mairesse, J. (2008). The Chamberlain approach to panel data: An overview and some simulations. In L. Mátyás & P. Sevestre (Eds.), *The econometrics of panel data* (chap. 5). Springer.

Debarsy, N. (2012). The Mundlak approach in the spatial Durbin panel data model. *Spatial Economic Analysis*, *7*, 109–131.

Graham, B., Hirano, K. & Imbens, G. W. (2023). The ET interview: Professor Gary Chamberlain. *Econometric Theory*, *39*, 1–26.

Hausman, J. A. (1978). Specification tests in econometrics. *Econometrica*, *46*, 1251-1272.

Hausman, J. A. & Taylor, W. E. (1981). Panel data and unobservable individual effects. *Econometrica*, *49*, 1377-1398.

Islam, N. (1995). Growth empirics: A panel data approach. *Quarterly Journal of Economics*, *110*, 1127–1170.

Kang, S. (1985). A note on the equivalence of specification tests in the two-factor multivariate variance components model. *Journal of Econometrics*, *28*, 193-203.

Krishnakumar, J. (2006). Time invariant variables and panel data models: A generalised Frisch–Waugh theorem and its implications. In B. H. Baltagi (Ed.), *Contributions to economic analysis* (Vol. 274, p. 119-132).

Mundlak, Y. (1978a). On the pooling of time series and cross section data. *Econometrica*, *46*, 69-85.

Mundlak, Y. (1978b). Models with variable coefficients: Integration and extension. *Annales de l'INSEE*, *30-31*, 483–509.

Nerlove, M. (1978). Econometric analysis of longitudinal data: Approaches, problems and prospects. *Annales de l'INSEE*, *46*, 7–22.

Nerlove, M. (2002). *Essays in panel data econometrics*. Cambridge University Press.

Nevo, A. (2001). Measuring market power in the ready-to-eat cereal industry. *Econometrica*, *69*, 307–342.

Newey, W. K. (1985). Generalized method of moments specification testing. *Journal of Econometrics*, *29*, 229-256.

Pesaran, M. H. (2006). Estimation and inference in large heterogeneous panels with a multifactor error structure. *Econometrica*, *74*, 967-1012.

Swamy, P. A. V. B. & Arora, S. S. (1972). The exact finite sample properties of the estimators of coefficients in the error components regression models. *Econometrica*, *40*, 261-275.

Wallace, T. & Hussain, A. (1969). The use of error components models in combining cross-section with time series data. *Econometrica*, *37*, 55-72.

Westerlund, J. (2019). Testing additive versus interactive effects in fixed-*T* panels. *Economics Letters*, *174*, 5–8.

Wooldridge, J. M. (2013). *Introductory econometrics: a modern approach* (Fifth ed.). The MIT Press.

Wooldridge, J. M. (2019). Correlated random effects models with unbalanced panels. *Journal of Econometrics*, *211*, 137–150.

Wooldridge, J. M. (2021). *Two-way fixed effects, the two-way Mundlak regression, and difference-in-difference estimators.* https://papers.ssrn.com/sol3/papers.cfm?abstract_id=3906345. SSRN 3906345.

Yang, Y. (2021). Efficient estimation of multi-level models with strictly exogenous explanatory variables. *Economics Letters*, *198*, —. doi: 10.1016/j.econlet.2020.109667

Yang, Y. (2022). A correlated random effects approach to the estimation of models with multiple fixed effects. *Economics Letters*, *213*, 110408.

Zyskind, G. (1967). On canonical forms, non-negative covariance matrices and best and simple least squares linear estimators in linear models. *Annals of Mathematical Statistics*, *36*, 1092-1109.

Chapter 15
An Algebraic Equivalence between Generalized Fixed Effects and a Generalized Mundlak Regression with Applications to Heterogeneous Trends and Difference-in-Differences

Jeffrey M. Wooldridge

Abstract Using a modified version of the Frisch-Waugh-Lovell partialling out result, I show how the standard Mundlak regression for panel data can be extended to deliver 'generalized fixed effects' estimators, where coefficients on a subset of explanatory variables vary by cross-sectional unit. The characterization shows that only fitted values from individual specific regressions need to be controlled for – not the large set of interactions between unit-specific dummy variables and explanatory variables. In the special case of a single additive heterogeneity, the fitted values are simply the unit-specific time averages of the explanatory variables. The general results have practical applications, particularly when testing the standard fixed effects estimator against heterogeneous trend models. I derive a new, fully robust test that can be viewed as a regression-based Hausman test for choosing the level of heterogeneity. As an example, I reconsider some recently proposed difference-in-differences methods with staggered interventions and heterogeneous treatment effects when the parallel trends assumption is possibly violated.

15.1 Introduction

The classic paper by Balestra and Nerlove (1966), showing how dynamic economic relationships with unobserved heterogeneity can be estimated using panel data, is now almost 60 years old. Now, the vast majority of intervention and policy analysis studies rely heavily on panel data structures, and the wealth of methods that have been developed since the early work by Mundlak (1961), Balestra and Nerlove (1966), and others.

In linear panel data analysis, the algebraic equivalence of the fixed effects estimates (on time-varying covariates) and estimating the Mundlak (1978) equation – which adds the unit-specific time averages of time-varying explanatory variables – by random effects has many applications. In Wooldridge (2019), I showed not only does the equivalence carry over to unbalanced panels, but also that pooled ordinary least squares (POLS) estimation of the Mundlak equation is identical to the random effects generalized least squares (GLS) estimator. I referred to the POLS version of the estimation as the 'Mundlak regression'.

An important application of the algebraic equivalence between the Mundlak regression and the fixed effects estimator is obtaining a fully robust approach to choosing between the random effects

Jeffrey M. Wooldridge ✉
Michigan State University, Department of Economics, East Lansing, Michigan, USA, e-mail: wooldri1@msu.edu

B. H. Baltagi and L. Mátyás (eds.), *Seven Decades of Econometrics and Beyond*, Advanced Studies in Theoretical and Applied Econometrics 57,
https://doi.org/10.1007/978-3-031-92699-0_15

and fixed effects estimators: a robust, regression-based version of the Hausman (1978) test. It also suggests that the Mundlak device can be useful for modeling the relationship between heterogeneity and covariates in nonlinear panel data models where no consistent fixed effects estimators (with the number of time periods fixed in the asymptotic analysis) are available. See, for example, (Wooldridge, 2010, Chapters 13, 15, 16, and 17).

In Wooldridge (2024), I extended the equivalence between fixed effects and the Mundlak regression to explicitly include time effects. One way to characterize the so-called 'two-way fixed effects' (TWFE) estimator is that one includes dummy variables for each time period (less one) and each unit. The 'two-way Mundlak regression' adds an intercept and the cross-sectional averages of the explanatory variables for each time period t to the POLS estimation. Baltagi (2023) shows how this equivalence leads to several insights concerning estimation of average treatment effects in difference-in-differences settings with staggered interventions. Baltagi (2023) showed that a GLS estimator that extends the usual one-way random effects estimator is equivalent to the pooled OLS estimator.

In many applications, one wants to allow for additional heterogeneity, most commonly via unit-specific trends. Even in microeconometric applications with a handful of time periods it can be important to allow units to have different trends in the outcome variable and the covariates. In a difference-in-differences setting, the outcome in the untreated state might be trending differently for control units and treated units, and even among units treated for the first time in different time periods. Such a situation would arise if the intervention decision is not based merely on levels difference before the intervention but also on pre-existing differences in trends.

Many researchers use an extended fixed effects estimator that includes unit-specific dummy variables along with those same dummy variables interacted with a linear time trend. An application of the Frisch-Waugh-Lovell (FWL) Theorem shows that including the $2N$ regressors (where N is the cross-sectional sample size) is equivalent to using unit-specific detrending of the covariates (and, optionally, the response variable). See, for example, Wooldridge (2010), Section 11.7.2.

In this paper, I extend the Mundlak regression to allow for a general model where a subset of variables has unit-specific coefficients. As discussed by Chamberlain (1992), the fixed coefficients can be estimated by generalizing the within transformation used by the simplest fixed effects estimator. Here, I show that the partialling out approach proposed by Chamberlain (1992) – see also Wooldridge (2010), Section 11.7.2 – is equivalent to adding unit-specific fitted values. The proof of equivalence relies on a general result on the equivalence between adding fitted values and using FWL partialling out. Though the equivalence, which I establish in Section 15.2, is straightforward matrix algebra, it has applications to allowing lots of heterogeneity in panel data settings.

In Section 15.3, I show, in the context of panel data, how one can obtain Chamberlain's 'generalized fixed effects' (GFE) estimator using a 'generalized Mundlak regression' (GMR). The GMR consists of adding the fitted values from first-stage unit-specific regressions, where the explanatory variables with fixed coefficients are regressed on the explanatory variables with heterogeneous coefficients. The first step results in functions of the explanatory variables for each unit i. Then, the outcome variable is regression on the explanatory variables with fixed coefficients and the unit-specific fitted values.

In Section 15.4 I consider the important special case of heterogenous trends – which can themselves be general. The key is that the variables with heterogeneous coefficients change only across time, and not by unit. Again, this generalizes the usual Mundlak regression by putting in fitted values that are linear combinations of the time trend for each unit i. The result is a relatively short regression that nevertheless allows unrestricted heterogeneity on a subset of the regressors. I also show how aggregate time variables with fixed coefficients – with the leading case being a full set of time period dummies – can be accommodated. Also, for the purposes of testing whether enough heterogeneity has been allowed, I show what changes when time-constant covariates are interacted with general time trends. I then apply the characterizations to specification testing in Section 15.5. It is straightforward to use a variable addition test to determine whether less heterogeneity is sufficient against the alternative that more is needed in the context of heterogenous trends models. The leading case is when the usual two-way fixed effects (TWFE) estimator has been computed, but one may worry that unit-specific linear trends are warranted. The test I propose appears to be novel, and it is fully robust to serial correlation and heteroskedasticity.

The equivalences here also have useful implications for staggered interventions and so-called 'difference-in-differences' estimation. In Section 15.6 I show that, starting with the flexible model I proposed in Wooldridge (2024), testing the important parallel trends assumption by interacting time-constant treatment cohort dummies with linear trends is identical to choosing between the usual fixed effects (Mundlak) estimator and Chamberlain's extension to heterogeneous trends. The equivalence provides further justification for the test that I motivated using a different argument in Wooldridge (2024).

Section 15.7 provides an empirical illustration in the context of a common timing intervention. I revisit the analysis in Moser and Voena (2012), who estimate the effects on domestic patents in the U.S. chemical industry from compulsory licensing laws. I estimate a full set of dynamic effects and show that, using the simple test from Section 15.6, that the parallel trends assumption is rejected. When heterogeneous linear trends are allowed, the average effect is about 36% larger than that reported by Moser and Voena (2012).

Section 15.8 contains concluding comments, including how the methods can be modified for unbalanced panels.

15.2 A Reformulation of the Frisch-Waugh-Lovell Theorem

To derive results for panel data structures, it is useful to begin by reformulating the Frisch-Waugh-Lovell (FWL) Theorem on the algebraic equivalence of estimators from partialling out regressions. It must be emphasized that the result in this section is purely algebraic and has nothing to do with underlying assumptions or the structure of the data. Let $\mathbf{Y}$ be an $n \times 1$ vector, $\mathbf{X}$ an $n \times k$ matrix, and $\mathbf{W}$ an $n \times m$ matrix. In what follows, we need only assume the rank condition on the regressor matrix, $\text{rank}\,(\mathbf{X}|\mathbf{W}) = k + m$.

Let $\hat{\beta}$ $(k \times 1)$ and $\hat{\gamma}$ $(m \times 1)$ be the vectors of OLS coefficients from the (long) regression

$$\mathbf{Y} \text{ on } \mathbf{X}, \mathbf{W}.$$

The well-known FWL result states that $\hat{\beta}$ can be obtained as follows:

1. Regress $\mathbf{X}$ on $\mathbf{W}$ and obtain the matrix residuals,

$$\ddot{\mathbf{X}} = \mathbf{X} - \hat{\mathbf{X}} = \left[\mathbf{I}_n - \mathbf{W}\,(\mathbf{W}'\mathbf{W})^{-1}\,\mathbf{W}'\right]\mathbf{X},$$

where $\hat{\mathbf{X}}$ is the matrix of fitted values,

$$\hat{\mathbf{X}} = \mathbf{W}\,(\mathbf{W}'\mathbf{W})^{-1}\,\mathbf{W}'\mathbf{X}.$$

2. Run the regression

$$\mathbf{Y} \text{ on } \ddot{\mathbf{X}} \tag{15.1}$$

to obtain $\hat{\beta}$. One has the option of replacing $\mathbf{Y}$ in (15.1) with $\mathbf{Y} = \left[\mathbf{I}_n - \mathbf{W}\,(\mathbf{W}'\mathbf{W})^{-1}\,\mathbf{W}'\right]\mathbf{Y}$ and $\hat{\beta}$ is unchanged.

A simple modification to FWL, useful for the current paper, is to replace the second step with the following:

2′. Run the regression

$$\mathbf{Y} \text{ on } \mathbf{X}, \hat{\mathbf{X}}. \tag{15.2}$$

The following result is fairly obvious but, as we will see, particularly useful in panel data contexts.

Proposition 15.1 *Let $\tilde{\beta}$ be the coefficients on $\mathbf{X}$ in the regression (15.2). Then $\tilde{\beta} = \hat{\beta}$.* □

The proof of this proposition is itself a straightforward application of FWL, and it is included in the appendix. Its usefulness is demonstrated in the next section when applied to panel data, where the fitted values are obtained using unit-specific time-series regressions in place of including potentially many unit-specific dummy variables and interactions of those dummy variables with other covariates.

15.3 A Generalized Mundlak Regression for Panel Data

In what follows, we have in mind a panel data model with heterogeneous coefficients, $\mathbf{c}_i$, on a subset of explanatory variables:

$$y_{it} = \mathbf{x}_{it}\beta + \mathbf{w}_{it}\mathbf{c}_i + u_{it}, \; t = 1, 2, ..., T, \tag{15.3}$$

where $\mathbf{x}_{it}$ is $1 \times K$ and $\mathbf{w}_{it}$ is $1 \times J$. For now, any aggregate time variables with fixed coefficients are included in $\mathbf{x}_{it}$. Below we consider some issues that arise when $\mathbf{x}_{it}$ and $\mathbf{w}_{it}$ both include variables that only vary across t. We are thinking of micro panels, with the number of cross-sectional units, N, notably larger than T; nevertheless, the results here are purely algebraic, with no restriction on the dimensions of T and N except those necessary for full rank conditions. Moreover, the model in (15.3) is for motivational purposes, as the key results are algebraic in nature.

In the simplest case, $\mathbf{w}_{it}$ is a scalar equal to one, which gives us the usual additive unobserved effects model:

$$y_{it} = \mathbf{x}_{it}\beta + c_i + u_{it}, \; t = 1, 2, ..., T, \tag{15.4}$$

where c_i is the so-called 'unobserved effect'. When $\mathbf{x}_{it}$ includes $T - 1$ time period dummies, estimation of (15.4) by fixed effects produces the popular 'two-way fixed effects' (TWFE) estimator of the remaining coefficients in β.

Again, even though the following is purely algebraic, it is useful to write the model in (15.3) by stacking the time periods for each unit i:

$$\mathbf{y}_i = \mathbf{X}_i\beta + \mathbf{W}_i\mathbf{c}_i + \mathbf{u}_i,$$

where

$$\mathbf{X}_i = \begin{pmatrix} \mathbf{x}_{i1} \\ \mathbf{x}_{i2} \\ \vdots \\ \mathbf{x}_{iT} \end{pmatrix}, \; \mathbf{W}_i = \begin{pmatrix} \mathbf{w}_{i1} \\ \mathbf{w}_{i2} \\ \vdots \\ \mathbf{w}_{iT} \end{pmatrix}, \; \mathbf{Y}_i = \begin{pmatrix} y_{i1} \\ y_{i2} \\ \vdots \\ y_{iT} \end{pmatrix}.$$

A general version of the FE estimator of β in (15.3) proceeds as follows. We assume that $\text{rank}(\mathbf{W}_i) = J$ for all $i = 1, ..., N$, which requires, at a minimum, that T is sufficiently large.

Procedure 1 (Generalized FE Estimation of β):

1. For each i, run the (matrix) regression

$$\mathbf{X}_i \text{ on } \mathbf{W}_i$$

   and obtain the fitted values $\hat{\mathbf{X}}_i$ and residuals,

$$\ddot{\mathbf{X}}_i \equiv \mathbf{X}_i - \mathbf{W}_i\left(\mathbf{W}_i'\mathbf{W}_i\right)^{-1}\mathbf{W}_i'\mathbf{X}_i \equiv \mathbf{X}_i - \hat{\mathbf{X}}_i.$$

2. Assuming $\text{rank}(\ddot{\mathbf{X}}_i) = K$ for all i (equivalently, $\ddot{\mathbf{X}}_i'\ddot{\mathbf{X}}_i$ is nonsingular for all i), $\hat{\beta}_{GFE}$ is obtained from system OLS estimation

$$\mathbf{y}_i \text{ on } \ddot{\mathbf{X}}_i, i = 1, 2, \ldots, N,$$

which is equivalent to pooled OLS:

$$y_{it} \text{ on } \ddot{\mathbf{X}}_{it}, t = 1, \ldots, T; \ i = 1, 2, \ldots, N.$$

Therefore,

$$\hat{\beta}_{GFE} = \left(\sum_{i=1}^{N} \ddot{\mathbf{X}}_i' \ddot{\mathbf{X}}_i \right)^{-1} \sum_{i=1}^{N} \mathbf{X}_i' \mathbf{y}_i = \left(\sum_{i=1}^{N} \ddot{\mathbf{X}}_i' \ddot{\mathbf{X}}_i \right)^{-1} \sum_{i=1}^{N} \mathbf{X}_i' \mathbf{y}_i,$$

where $\mathbf{y}_i = \mathbf{y}_i - \mathbf{W}_i \left(\mathbf{W}_i' \mathbf{W}_i \right)^{-1} \mathbf{W}_i' \mathbf{y}_i$. □

The assumption that $\ddot{\mathbf{X}}_i$ has rank K rules out certain situations. For example, if $J = T$ then, under the assumption that rank$(\mathbf{W}_i) = J$, $\hat{\mathbf{X}}_i = \mathbf{W}_i \left(\mathbf{W}_i' \mathbf{W}_i \right)^{-1} \mathbf{W}_i' \mathbf{X}_i = \mathbf{W}_i \mathbf{W}_i^{-1} \left(\mathbf{W}_i' \right)^{-1} \mathbf{W}_i' \mathbf{X}_i = \mathbf{X}_i$, and so $\ddot{\mathbf{X}}_i = \mathbf{0}$. So a necessary condition is $J < T$. We come back to this in the next section when applying the this characterization to heterogeneous trend models.

A generalized Mundlak regression is the following.

Procedure 2 (Generalized Mundlak Regression):

1. Obtain the fitted values, $\hat{\mathbf{X}}_i$, as in (15.3).
2. Run the system OLS regression

$$\mathbf{y}_i \text{ on } \mathbf{X}_i, \hat{\mathbf{X}}_i, i = 1, 2, \ldots, N$$

and obtain $\hat{\beta}_{GM}$ as the coefficient on $\mathbf{X}_i$. □

Application of Proposition 15.1 yields the following.

Proposition 15.2 *(Equivalence of Generalized FE and Generalized Mundlak): In the setting described in Procedure 2,*

$$\hat{\beta}_{GM} = \hat{\beta}_{GFE}. \ \square$$

The proof, which simply requires definitions and straightforward matrix algebra, is given in the appendix.

A special case is when $w_{it} = 1$ for all i and t, in which case, for all i,

$$\begin{aligned} \hat{\mathbf{x}}_{it} &= \bar{\mathbf{x}}_i \\ \ddot{\mathbf{X}}_{it} &= \mathbf{x}_{it} - \bar{\mathbf{x}}_i, t = 1, \ldots, T \end{aligned}$$

The conclusion is that the FE estimator and the POLS version of the Mundlak estimator – as studied in Wooldridge (2019) – are the same.

Wooldridge (2019) shows that adding time-constant regressors $\mathbf{z}_i$, to the pooled Mundlak regression,

$$y_{it} \text{ on } \mathbf{x}_{it}, \bar{\mathbf{x}}_i, 1, \mathbf{z}_i, t = 1, \ldots, T; \ i = 1, \ldots, N,$$

does not change the coefficient estimator on $\mathbf{x}_{it}$ (but it does on $\bar{\mathbf{x}}_i$, which has implications for regression-based Hausman tests). We return to the issue of time-constant variables in the generalized Mundlak regression below.

15.4 Heterogeneous Trend Models

We now consider the case where, for all i,

$$\mathbf{w}_{it} = \mathbf{g}_t,\ t = 1, \ldots, T,$$

so that the heterogeneous slopes are only on aggregate variables $\mathbf{g}_t$. The simplest extension from the Mundlak regression ($g_t = 1$) allows heterogeneous linear trends:

$$\mathbf{g}_t = (1, t). \tag{15.5}$$

The choice of $\mathbf{g}_t$ in (15.5) produces a generalized FE estimator based on unit-specific detrending of $\mathbf{x}_{it}$ (and y_{it}). In the underlying model, it allows for two sources of unobserved heterogeneity that are correlated with $\mathbf{x}_{it}$.

With a general choice of $\mathbf{g}_t$ $(1 \times J)$ with $J < T$, the matrix fitted values are

$$\hat{\mathbf{X}}_i = \mathbf{G}\,(\mathbf{G}'\mathbf{G})^{-1}\,\mathbf{G}'\mathbf{X}_i, \tag{15.6}$$

where $\mathbf{G}$ is the $T \times J$ matrix with row $\mathbf{g}_t$. The projection matrix in (15.6) does not vary with i. The residuals, $\mathbf{X}_i = \mathbf{X}_i - \hat{\mathbf{X}}_i$, are a generalized version of unit-specific detrending. It could be as simple as adding higher-order polynomials in t. If $\mathbf{g}_t = (1, t, \ldots, t^q)$ then the rank condition requires $q \leq T - 2$.

15.4.1 Adding Time Effects with Constant Coefficients

As mentioned earlier, most applications of fixed effects are actually applications of TWFE, with a full set of time dummies included. Generally, let $\mathbf{f}_t$ be a $1 \times Q$ vector of time effects. In the TWFE case, we can take $\mathbf{f}_t = (f1_t, f2_t, \ldots, fT_t)$ or $\mathbf{f}_t = (1, f2_t, \ldots, fT_t)$, where fs_t is a time period dummy such that $fs_t = 1$ if and only if $s = t$.

We have in mind the model

$$y_{it} = \mathbf{x}_{it}\beta + \mathbf{g}_t\mathbf{c}_i + \mathbf{f}_t\,\delta + u_{it},$$

where now $\mathbf{x}_{it}$ omits variables that vary only across t. As before, we run unit-specific regressions $\mathbf{x}_{it}$ on $\mathbf{g}_t$, $t = 1, \ldots, T$ to obtain the $\hat{\mathbf{x}}_{it}$. What about for $\mathbf{f}_t$? Not surprisingly, if $\mathbf{G}$ is a linear combination of $\mathbf{F}$, say $\mathbf{G} = \mathbf{FA}$ for a $Q \times J$ matrix $\mathbf{A}$, then adding the fitted values is redundant once $\mathbf{f}_t$ is included. The reason is because we can write

$$\hat{\mathbf{F}} = \mathbf{G}\,(\mathbf{G}'\mathbf{G})^{-1}\,\mathbf{G}'\mathbf{F} = \mathbf{FA}\,(\mathbf{G}'\mathbf{G})^{-1}\,\mathbf{G}'\mathbf{F} = \mathbf{F}\left[\mathbf{A}\,(\mathbf{G}'\mathbf{G})^{-1}\,\mathbf{G}'\mathbf{F}\right],$$

which is a linear combination of $\mathbf{F}$. It follows that the generalized Mundlak regression,

$$\mathbf{y}_i \text{ on } \mathbf{X}_i,\ \hat{\mathbf{X}}_i,\ \mathbf{F},\ \hat{\mathbf{F}},\ i = 1, \ldots, N$$

is the same as dropping $\hat{\mathbf{F}}$.

In the leading case where we include a full set of time dummies, we can take $\mathbf{F} = \mathbf{I}_T$, and then $\mathbf{G}$ is trivially a linear combination of $\mathbf{F}$. In other words, the generalized Mundlak regression is

$$y_{it} \text{ on } \mathbf{x}_{it},\ \hat{\mathbf{x}}_{it},\ 1,\ f2_t,\ \ldots,\ fT_t,\ t = 1, \ldots, T; i = 1, \ldots, N. \tag{15.7}$$

As mentioned previously, the usual Mundlak regression with $\hat{\mathbf{x}}_{it} = \bar{\mathbf{x}}_i$ has been very useful in obtaining specification tests, gaining new insights in difference-in-differences settings, and extending methods to nonlinear models.

15.4.2 Including Time-Constant Variables

Another useful algebraic result involves adding interactions between the elements of $\mathbf{g}_t$ and observed time-constant variables, say $\mathbf{z}_i$ $(1 \times L)$. Again, for motivational purposes only, consider the equation

$$y_{it} = \mathbf{x}_{it}\beta + \mathbf{g}_t\mathbf{c}_i + \mathbf{f}_t\delta + (\mathbf{g}_t \otimes \mathbf{z}_i)\lambda + u_{it}, \; t = 1, 2, ..., T,$$

where λ is $JL \times 1$. The coefficients $\mathbf{c}_i$ on $\mathbf{g}_t$ are dealt with as before, by projecting $\mathbf{x}_{it}$ onto $\mathbf{g}_t$ separately for each i. Remember, this approach allows the linear relationship between $\mathbf{x}_{it}$ and $\mathbf{g}_t$ to be different for each i. Therefore, intuitively, adding additional heterogeneous coefficients through $\mathbf{g}_t \otimes \mathbf{z}_i$ should not affect estimation of β. This intuition carries over to algebraic equivalence. Therefore, the pooled OLS regression

$$y_{it} \text{ on } \mathbf{x}_{it}, \hat{\mathbf{x}}_{it}, 1, f2_t, ..., fT_t, \mathbf{g}_t \otimes \mathbf{z}_i, \; t = 1, ..., T; i = 1, ..., N \tag{15.8}$$

results in the same coefficients $\hat{\beta}_{GM}$ from (15.7). Notice that the coefficients on $\hat{\mathbf{x}}_{it}$ do generally change, something that has implications for specification testing below.

That (15.8) and (15.7) produce the same coefficients on $\mathbf{x}_{it}$ is an extension of the result shown in Wooldridge (2019), Proposition 2.1 for the basic Mundlak regression obtained by taking $g_t \equiv 1$. In particular, once the time averages $\bar{\mathbf{x}}_i$ have been included, adding additional variables $\mathbf{z}_i$ has no effect: the Mundlak regression still produces the FE estimator $\hat{\beta}_{FE}$. Wooldridge (2019) further notes that, with good controls in $\mathbf{z}_i$, one may not need to include $\bar{\mathbf{x}}_i$. Testing this proposition effectively produces a robust, regression-based version of the Hausman (1978) test comparing FE and random effects. The extension of Wooldridge (2019), Proposition 2.1 is the following.

Proposition 15.3 *Assume that the $\hat{\mathbf{x}}_{it}$ are obtained from (15.6). Then the coefficients on $\mathbf{x}_{it}$ are the same with or without $\mathbf{g}_t \otimes \mathbf{z}_i$ included.* □

As in the special case in Wooldridge (2019), including $\mathbf{g}_t \otimes \mathbf{z}_i$ does generally change the estimated coefficients on $\hat{\mathbf{x}}_{it}$. In the next section explores why this is important in the context of specification testing.

15.5 Choosing the Amount of Heterogeneity via Specification Testing

The result that including $\mathbf{g}_t \otimes \mathbf{z}_i$ in (15.8) does not change the coefficient on $\mathbf{x}_{it}$ in (15.8) has implications for choosing the amount of heterogeneity in the model with heterogeneous trends, as represented by $\mathbf{g}_t\mathbf{c}_i$. Wooldridge (2019) covers the case where the decision is to include the time averages, $\bar{\mathbf{x}}_i$, in the regression

$$y_{it} \text{ on } \mathbf{x}_{it}, \bar{\mathbf{x}}_i, 1, f2_t, ..., fT_t, \mathbf{z}_i, \; t = 1, ..., T; i = 1, ..., N, \tag{15.9}$$

where $\mathbf{z}_i$ are time-constant controls. From Proposition 15.2 [or the special case in Wooldridge (2019)], the Mundlak coefficients on $\mathbf{x}_{it}$ equal the FE estimates: $\hat{\beta}_M = \hat{\beta}_{FE}$. As us well known, the inclusion of $\bar{\mathbf{x}}_i$ results in considerable multicollinearity because it is clearly correlated with $\mathbf{x}_{it}$. If some elements of $\mathbf{x}_{it}$ have little variation across t, the collinearity can make it difficult to precisely estimate elements of β. Therefore, if ξ denotes the coefficients on $\bar{\mathbf{x}}_i$, one might want to test $H_0 : \xi = \mathbf{0}$. The null hypothesis is that once $\mathbf{z}_i$ is controlled for and full time effects are allowed, $\bar{\mathbf{x}}_i$ is uncorrelated with remaining, unobserved heterogeneity. A rejection typically is taken to mean $\bar{\mathbf{x}}_i$ is retained in 15.9, in which case β is estimated using (two-way) fixed effects. As discussed in Wooldridge (2019), it does not matter whether pooled OLS or random effects is used when $\bar{\mathbf{x}}_i$ is included in the equation: they lead to the same estimates. When $\bar{\mathbf{x}}_i$ is dropped, POLS and RE are no longer the same, and one may prefer the latter on efficiency grounds (although there is no guarantee if, as is often the case, idiosyncratic errors exhibit serial correlation and heteroskedasticity).

Some econometrics software packages choose to test pooled OLS against fixed effects by replacing $\bar{\mathbf{x}}_i$ in (15.9) with unit-specific dummies $c1_i$, $c2_i$, ..., cN_i, and then testing equality of their coefficients (for $N-1$ restrictions). There are a couple of reasons this is undesirable. First, one cannot obtain a version of the test that is robust to serial correlation and heteroskedasticity because cluster-robust inference for the estimates $\hat{\alpha}_h$, $h = 1, \ldots, N$ (the coefficients on the ch_i) are not valid: in effect, we are using T time series observations to estimate each α_h. One could obtain a test under the classical linear model assumptions – which would add normality, no serial correlation, and homoskedasticity – but that is too restrictive considering the test from (15.9) is fully robust. Secondly, the test of equality of the α_h does not directly address the important question of whether c_i is correlated with the elements of $\mathbf{x}_{it}$. There could be lots of heterogeneity that is uncorrelated with $\mathbf{x}_{it}$, in which case pooled POLS would be consistent (fixed T, $N \to \infty$). In a randomized controlled trial, the treatment $\mathbf{x}_{it}$ would be independent of everything. A related issue is that, thinking of (15.9) as a regression-based Hausman test, there should be only $K = \dim(\mathbf{x}_{it})$ restrictions to test – not $N-1$.

Because most data used in economics and the social sciences are not experimental, two-way fixed effects continues to be a workhorse in empirical research, with many researchers rejecting a pure random effects analysis even with rich controls $\mathbf{z}_i$. In fact, the concern with FE often is that it allows only one source of additive heterogeneity. What if one wants to check whether allowing for a single source of additive heterogeneity is sufficient? Provided we have $T \geq 3$, a natural alternative to the usual FE estimator is a heterogeneous trends estimator where every unit is, say, allowed to have its own linear trend. Without time-constant controls, the underlying model for each unit i is

$$y_{it} = \mathbf{x}_{it}\beta + \gamma_2 f2_t + \cdots + \gamma_T fT_t + c_{i1} + c_{i2} \cdot t + u_{it},\ t = 1, \ldots, T.$$

Then the $\hat{\mathbf{x}}_{it}$ are obtained from

$$\mathbf{x}_{it} \text{ on } 1, t,\ t = 1, \ldots, T$$

and the unit-specific fitted values can be written as

$$\hat{\mathbf{x}}_{it} = \hat{\mathbf{h}}_{i1} + \hat{\mathbf{h}}_{i2} \cdot t.$$

The generalized Mundlak regression is then

$$y_{it} \text{ on } \mathbf{x}_{it}, \hat{\mathbf{h}}_{i1} + \hat{\mathbf{h}}_{i2}t, 1, f2_t, \ldots, fT_t,\ t = 1, \ldots, T; i = 1, \ldots, N.$$

From Proposition 15.2, once $\hat{\mathbf{x}}_{it}$ is included, adding $\bar{\mathbf{x}}_i$ is redundant for obtaining $\hat{\beta}_{GM}$ because $\bar{\mathbf{x}}_i$ does not change across time. In other words, the regression

$$y_{it} \text{ on } \mathbf{x}_{it}, \bar{\mathbf{x}}_i, \hat{\mathbf{x}}_{it}, 1, f2_t, \ldots, fT_t,\ t = 1, \ldots, T; i = 1, \ldots, N, \tag{15.10}$$

leads to the same $\hat{\beta}_{GM}$ coefficients on $\mathbf{x}_{it}$ as dropping $\bar{\mathbf{x}}_i$. By the usual Mundlak-FE equivalence, dropping $\hat{\mathbf{x}}_{it}$ produces $\hat{\beta}_M = \hat{\beta}_{FE}$. Consequently, the regression in (15.10) can be used to test for joint significance of the coefficients on $\hat{\mathbf{x}}_{it}$ – call these $\hat{\pi}$. The test $H_0 : \pi = \mathbf{0}$ is a test of the null of whether the usual FE estimator is sufficient of is FE rejected in favor of the heterogeneous trends model. As in the usual comparison between FE and POLS (or RE), a primary motivation for dropping $\hat{\mathbf{x}}_{it}$ from (15.10) and using FE is that adding $\hat{\mathbf{x}}_{it}$ creates even more collinearity with $\mathbf{x}_{it}$. In fact, with $T = 2$ (15.10) cannot even be carried out due to perfect collinearity. Naturally, we would use a Wald test (often reported as an F-type test) that is robust to arbitrary serial correlation and heteroskedasticity. For the reasons described earlier for the usual Hausman-type test, we do not want to base the test the $N-1$ restrictions that the coefficients on the interactions $ch_i \cdot t$ are the same across $h = 1, \ldots, N$.

When time-constant controls $\mathbf{z}_i$ are available, it is natural to allow the linear trends to vary with $\mathbf{z}_i$, in which case the regression is

$$y_{it} \text{ on } \mathbf{x}_{it}, \bar{\mathbf{x}}_i, \hat{\mathbf{x}}_{it}, 1, f2_t, \ldots, fT_t, \mathbf{z}_i, \mathbf{z}_i \cdot t,\ t = 1, \ldots, T; i = 1, \ldots, N. \tag{15.11}$$

If the linear trends are thought not change with $\mathbf{z}_i$ then $\mathbf{z}_i \cdot t$ can be dropped from (15.11). Whether including just $\mathbf{z}_i$ or $(\mathbf{z}_i, \mathbf{z}_i \cdot t)$, the estimates on $\mathbf{x}_{it}$ do not change, but those on $\hat{\mathbf{x}}_{it}$ generally do – which will affect the test statistic.

By the usual Mundlak equivalence, all coefficients except those on $\bar{\mathbf{x}}_i$, 1, $\mathbf{z}_i$ can be obtained by using fixed effects estimation. In other words, drop the time constant variables $(1, \bar{\mathbf{x}}_i, \mathbf{z}_i)$ and applying FE gives the generalized FE estimates.

Given (15.8), and the fact that $\bar{\mathbf{x}}_i$ is a linear combination of $\hat{\mathbf{h}}_{i1}$ and $\hat{\mathbf{h}}_{i2}$, the regression in (15.11) is the same as

$$y_{it} \text{ on } \mathbf{x}_{it}, \hat{\mathbf{h}}_{i1}, \hat{\mathbf{h}}_{i2} \cdot t, 1, f2_t, ..., fT_t, \mathbf{z}_i, \mathbf{z}_i \cdot t, t = 1, \ldots, T; i = 1, \ldots, N,$$

which gives another way to see that two sources of heterogeneity are being included (for each element of $\mathbf{x}_{it}$). But for specification testing, (15.10) or (15.11) should be used.

As a final comment, note that the regressions

$$y_{it} \text{ on } \mathbf{x}_{it}, \bar{\mathbf{x}}_i, 1, f2_t, ..., fT_t$$

and

$$y_{it} \text{ on } \mathbf{x}_{it}, \hat{\mathbf{x}}_{it}, 1, f2_t, ..., fT_t$$

are nonnested in the sense that $\bar{\mathbf{x}}_i$ and $\hat{\mathbf{x}}_{it}$ are different $1 \times K$ linear combinations of $\{\mathbf{x}_{is} : s = 1, \ldots, T\}$. If the goal is to choose one of the two specifications, the usual R-squareds can be used as goodness-of-fit measures. Alternatively, formal nonnested tests can be computed that are robust to serial correlation and heteroskedasticity, as in Rahmani and Wooldridge (2019).

15.6 Application to Difference-in-Differences

The previous results can be applied to recent developments in so-called "difference-in-differences" (DiD) estimation with staggered entry and heterogeneous treatment effects. Given T time periods, the first intervention period is $1 < q \leq T$, and new units enter "treatment" through period T. For simplicity, I assume here that there are some untreated units remaining in period T (the 'never treated') group. Wooldridge (2024) derives the following equation, which maintains a 'no anticipation' assumption and a 'parallel trends' assumption conditional on time-constant covariates $\mathbf{x}_i$:

$$\begin{aligned} y_{it} = & \sum_{g=q}^{T} \sum_{s=g}^{T} \tau_{gs} (dg_i \cdot fs_t) + \sum_{g=q}^{T} \sum_{s=g}^{T} (dg_i \cdot fs_t \cdot \dot{\mathbf{x}}_{ig}) \rho_{gs} \\ & + \sum_{s=2}^{T} \gamma_s fs_t + \sum_{s=2}^{T} (fs_t \cdot \mathbf{x}_i) \pi_s + c_i + u_{it}, \end{aligned} \tag{15.12}$$

where dg_i is the treatment cohort indicator, with $dg_i = 1$ if unit i is first treated in period g; fs_t is the time period dummy, as before; and $\dot{\mathbf{x}}_{ig} = \mathbf{x}_i - \bar{\mathbf{x}}_g$ is the cohort-specific demeaned row vector of covariates. The interaction $dg_i \cdot fs_t$ is a treatment indicator for cohort g in time period s for $s = g$, ..., T. The coefficients of interest, τ_{gs}, are average treatment effects on the treated.

Wooldridge (2024) suggests estimating (15.12) by fixed effects to account for c_i; with the inclusion of the time dummies, it is 'extended' TWFE, where 'extended' refers to the model beyond the simple constant (or single) effect model. The presence of $fs_t \cdot \mathbf{x}_i$ allows differential trends by observed covariates. The Mundlak equation corresponding to (15.12) is

$$
\begin{aligned}
y_{it} = & \sum_{g=q}^{T}\sum_{s=g}^{T} \tau_{gs}\left(dg_i \cdot fs_t\right) + \sum_{g=q}^{T}\sum_{s=g}^{T}\left(dg_i \cdot fs_t \cdot \dot{\mathbf{x}}_{ig}\right)\rho_{gs} \\
& + \sum_{s=2}^{T}\gamma_s fs_t + \sum_{s=2}^{T}\left(fs_t \cdot \mathbf{x}_i\right)\pi_s + \alpha + \sum_{g=q}^{T}\beta_g dg_i \\
& + \mathbf{x}_i\kappa + \sum_{g=q}^{T}\left(dg_i \cdot \dot{\mathbf{x}}_{ig}\right)\xi_g + a_i + u_{it}.
\end{aligned}
\tag{15.13}
$$

When this equation is estimated by pooled OLS, the estimates in the first line are identical to the FE estimates from (15.12).

As discussed in Wooldridge (2024), (15.12) does not account for differential trends that can depend on unobserved heterogeneity, which can cause systematic bias in estimates of the ATTs. Given at least two pre-treatment periods, we can expand the equation to

$$
\begin{aligned}
y_{it} = & \sum_{g=q}^{T}\sum_{s=g}^{T} \tau_{gs}\left(dg_i \cdot fs_t\right) + \sum_{g=q}^{T}\sum_{s=g}^{T}\left(dg_i \cdot fs_t \cdot \dot{\mathbf{x}}_{ig}\right)\rho_{gs} \\
& + \sum_{s=2}^{T}\gamma_s fs_t + \sum_{s=2}^{T}\left(fs_t \cdot \mathbf{x}_i\right)\pi_s + c_{i1} + c_{i2} \cdot t + u_{it},
\end{aligned}
$$

where $c_{i2} \cdot t$ is a unit-specific linear trend. What happens of we project $dg_i \cdot fs_t$ onto $(1, t)$, $t = 1, \ldots, T$, for each i? The fitted values are easily seen to be the same linear combinations of $(dg_i, dg_i \cdot t)$. Likewise, projecting $dg_i \cdot fs_t \cdot \dot{\mathbf{x}}_{ig}$ onto $(1, t)$ gives linear combinations of $(dg_i \cdot \mathbf{x}_{ig}, dg_i \cdot t \cdot \dot{\mathbf{x}}_{ig})$. It follows that the generalized Mundlak equation is

$$
\begin{aligned}
y_{it} = & \sum_{g=q}^{T}\sum_{s=g}^{T} \tau_{gs}\left(dg_i \cdot fs_t\right) + \sum_{g=q}^{T}\sum_{s=g}^{T}\left(dg_i \cdot fs_t \cdot \dot{\mathbf{x}}_{ig}\right)\rho_{gs} \\
& + \sum_{s=2}^{T}\gamma_s fs_t + \sum_{s=2}^{T}\left(fs_t \cdot \mathbf{x}_i\right)\pi_s + \alpha + \sum_{g=q}^{T}\beta_g dg_i + \mathbf{x}_i\kappa \\
& + \sum_{g=q}^{T}\left(dg_i \cdot \dot{\mathbf{x}}_{ig}\right)\xi_g + \sum_{g=q}^{T}\eta_g\left(dg_i \cdot t\right) + \sum_{g=q}^{T}\left(dg_i \cdot t \cdot \dot{\mathbf{x}}_{ig}\right)\lambda_g + a_i + u_{it}.
\end{aligned}
\tag{15.14}
$$

The POLS estimator using this equation was proposed by Wooldridge (2024) using different reasoning. The practical point is that allows for trends to differ linear by treatment cohort, relative to the never treated state, through the terms $dg_i \cdot t$ and $dg_i \cdot t \cdot \mathbf{x}_{ig}$. If the estimates of the τ_{gs} (and, sometimes, ρ_{gs}) are sufficiently precise, one might just got with the estimates from (15.14). However, the introduction of $dg_i \cdot t$ causes clear collinearity with the treatment indicators $dg_i \cdot fs_t$, and the estimators $\hat{\tau}_{gs}$ may be imprecise. By testing joint significance of the $dg_i \cdot t$ and, if they are included, $dg_i \cdot t \cdot \dot{\mathbf{x}}_{ig}$, one might conclude that the usual TWFE estimator applied to (15.12), or its equivalent in (15.13), are sufficient.

15.7 Empirical Example

As a simple application, I reanalyze the common timing difference-in-differences application in Moser and Voena (2012), who study the effects of a compulsory licensing law that was instituted as part of the Trading with the Enemy Act passed in October 1917 during World War I. The law allowed United States firms to violate enemy-owned patents if doing so would contribute to the war effort. The focus on subclasses of chemicals, some of which were subjected to the compulsory licensing law; the others are used as controls in a difference-in-differences analysis. Moser and Voena (2012) are interested in determining whether relaxed restrictions on foreign licensing increases or decreases domestic innovation, as measured by patents generated, in this case, in the United States.

The outcome variable, annual number of U.S. patents in a chemical class, is a count variable. Like the original authors, I use a linear model. I do not use control variables, but I do estimate a full set of dynamic effects and then average those effects to compare with the single estimated effect in Moser and Voena (2012).

With common timing and no controls, the equation for estimating dynamic treatment effects simplifies considerably:

$$y_{it} = \sum_{s=q}^{T} \tau_s (d_i \cdot fs_t) + \sum_{s=2}^{T} \gamma_s fs_t + c_{i1} + c_{i2} \cdot t + u_{it},$$

where d_i indicates whether unit i – a chemical class in this case – is subjected to the compulsory licensing law (eventually treated). With $N = 7,248$ classes, including a full set of unit-specific dummies, and especially when those are also interacted with the linear time trend, is computationally demanding (and requires a lot of memory). Without $c_{i2} \cdot t$, we can use the regular fixed effects estimator, ea. Equivalently, as follows from Wooldridge (2024) and Proposition 15.3 above, we can simply add d_i as a separate regressor and used pooled OLS – this simplifies the computation even more. To repeat, it is somewhat remarkable that even in a fully dynamic model, controlling for the single binary source of heterogeneity, d_i, is equivalent to controlling for a lot of heterogeneity by including N unit-specific dummy variables.

For the heterogeneous trend model, from the discussion in Section 15.6, we simply add $d_i \cdot t$ along with d_i. The equation is

$$y_{it} = \sum_{s=q}^{T} \tau_s (d_i \cdot fs_t) + \sum_{s=2}^{T} \gamma_s fs_t + \beta d_i + \eta (d_i \cdot t) + u_{it}. \tag{15.15}$$

If $\eta \neq 0$, selection into treatment is systematically related to the trend in the untreated state. If $\eta = 0$ but $\beta \neq 0$, selection is based on level differences before the intervention but evidently not on differing trends. This is the case that the TWFE estimator works well for. Remember, (15.15) produces estimates of the τ_s that are identical to including N unit-specific dummy variables and another N of those dummy variables interacted with t. In the current application, estimating (15.15) is essentially trivial, while including more than 14,000 regressors is much more challenging. Incidentally, Wooldridge (2024), Proposition A.1 shows that the POLS estimator using all of the data gives the same test on $d_i \cdot t$ is using only the untreated observations ($d_i \cdot fs_t = 0$), preserving the interpretation as the test detecting trends in the untreated state.

The Moser and Voena (2012) data run from 1875 through 1939, with the intervention starting in 1919 for treated chemical classes. Consequently, there are 21 dynamic treatment effects that can be identified. Out of the 7,248 total classes, 336 are treated. Across 471,120 class/year observations, the average of annual U.S. patents granted is about 0.35, with about 83.1% of the outcomes equal to zero. Here, I follow Moser and Voena (2012) and estimate linear models; Wooldridge (2023) shows how Poisson regression with an exponential mean containing exactly the same variables can be applied to nonnegative outcomes.

Table 15.1 shows the results of two models. Column (1) sets $\eta = 0$ in (15.15), and so it is identical to the TWFE estimates. The average of the coefficients, 0.255 ($se = 0.038$), reproduces the estimate Column (2) of Table 2 in Moser and Voena (2012). The pattern by exposure time differs significantly from the average. IN fact, many of the early estimates are actually negative (though not statistically significant at any reasonable level). The first statistically significant effect does not come until after eight years.

Estimating heterogeneous trends changes the exposure time estimates markedly. All coefficients are positive and all are much bigger than the corresponding TWFE estimates. Most estimates are statistically significant at the 5% level starting with one year of exposure. The average estimated effect is also notably larger, 0.349 compared with 0.255.

As a statistical matter, we can use the simple tests that are suggested by the discussion in Section 15.6. In column (1), there is clear evidence of a selection effect. The negative coefficient on d, -0.145, which is large in magnitude given the small average number of patents, is very statistically significant ($t \approx -7.70$). It suggests that compulsory licensing agreements were targeted at chemical classes that had fewer historical patents. In column (2) the coefficient is smaller in magnitude, -0.983, but it still has an absolute t statistic above five. Moreover, the coefficient on the trend $d \cdot (year - 1875)$ is negative and statistically significant ($t \approx -2.43$), implying a statistical rejection of the usual TWFE estimator. This outcome suggests selection not just based on pre-intervention levels of patents but also on trends. (Centering $year$ about 1875 ensures that the coefficient on d is the selection effect in the first year, 1875.) Assuming the difference in trends between (eventually) treated units and untreated units is roughly linear, the estimates in (2) are more reliable, and the average effects is notably larger than that reported in Moser and Voena (2012).

Table 15.1: Effects of Compulsory Licensing on Patents by Exposure Time

Exposure Time	(1) FE/Mundlak	(2) Heterogeneous Trends
0	−0.0221 (0.0340)	0.0428 (0.0335)
1	0.0064 (0.0342)	0.0742 (0.0331)
2	0.0169 (0.0432)	0.0877 (0.0351)
3	−0.0163 (0.0407)	0.0574 (0.0471)
4	0.0474 (0.0479)	0.1239 (0.0552)
5	−0.0240 (0.0453)	0.0555 (0.0429)
6	−0.0200 (0.0476)	0.0623 (0.0522)
7	0.0556 (0.0482)	0.1408 (0.0445)
8	0.1556 (0.0485)	0.2437 (0.0539)
9	0.1786 (0.0494)	0.2696 (0.0580)
10	0.0867 (0.0493)	0.1806 (0.0626)
11	0.0772 (0.0545)	0.1740 (0.0723)
12	0.1937 (0.0686)	0.2934 (0.0765)
13	0.5966 (0.1021)	0.6991 (0.1188)
14	0.5709 (0.0982)	0.6764 (0.1067)
15	0.3812 (0.0865)	0.4895 (0.0911)
16	0.5001 (0.0926)	0.6113 (0.1073)
17	0.6717 (0.1070)	0.7858 (0.1187)
18	0.5217 (0.0905)	0.6387 (0.1050)
19	0.6321 (0.1010)	0.7519 (0.1121)
20	0.718 (0.1177)	0.8745 (0.1280)
Average	0.2553 (0.0376)	0.3492 (0.0495)
d	−0.1451 (0.0188)	−0.0830 (0.0144)
$d \cdot (year - 1875)$	—	−0.0029 (0.0012)

15.8 Concluding Remarks and Extensions

Using a reformulation of the Frisch-Waugh-Lovell partialling out result, I show how the basic Mundlak (1978) – studied in the case of pooled OLS in Wooldridge (2019) – can be extended very generally when a subset of variables has heterogeneous coefficients. To estimate the fixed coefficients, the generalized Mundlak regression involves adding fitted values from unit-specific regressions. In generally, the estimators and tests do require a sequence of unit-specific regressions, which can be time consuming with a large cross-sectional sample size. Nevertheless, this is no more difficult than including many unit-specific dummy variables.

A leading application is allowing heterogenous trends, where the unit-specific fitted values are obtained as functions of trends; the leading case is a linear trend. This characterization of Chamberlain (1992) generalized fixed effects estimator leads to simple specification tests for determining whether, say, the usual two-way fixed effects is sufficient or whether unit-specific trends should be added. The test is standard because it does not involve testing hypothesis involving N (or more) coefficients on unit-specific dummies.

When applied to staggered difference-in-differences settings, the results imply that controlling for cohort dummies and cohort specific trends in a flexible regression is identical to either doing unit-specific detrending or including a full set of unit dummies along with those interacted with time trends. Practically, this is an important simplification, as illustrated in the empirical application to the Moser and Voena (2012) patents data for more than 7,000 chemical classes. The usual fixed effects specification is rejected in favor of the heterogeneous linear trend specification, and the latter produces an estimated average treatment effect more than 35% larger than that reported by Moser and Voena (2012).

The previous results can be extended to unbalanced panels, but the notation is more complicated. There are a few important considerations. First, only the complete cases are used at every step in characterizing the estimators. As in Wooldridge (2019), one can introduce a complete cases indicator, say s_{it}, where $s_{it} = 1$ if all data are observed for unit i in period t. An observation is used if and only if $s_{it} = 1$. This is true even when obtaining the $\hat{\mathbf{x}}_{it}$ even if, say, $(\mathbf{x}_{it}, \mathbf{w}_{it})$ is fully observed by y_{it} is missing. There is nothing unusual here, as the fixed effects estimator only uses the complete cases and Wooldridge (2019) shows that the Mundlak regression reproduces the FE estimator on the unbalanced sample provided only the complete cases are used. Wooldridge (2019) also emphasizes that aggregate time effects now need to be average over the complete cases – rather than the averages being constant. Naturally, the same is true here in the more general case: the fitted values $\mathbf{f}_t$ on $\mathbf{g}_t$, $t = 1, \ldots, T$ are obtained using $s_{it} = 1$. That is why these fitted values now vary by unit.

References

Balestra, P. & Nerlove, M. (1966). Pooling cross section and time series data in the estimation of a dynamic model: The demand for natural gas. *Econometrica*, *34*, 585-612.

Baltagi, B. H. (2023). The two-way Mundlak estimator. *Econometric Reviews*, *42*, 240-246.

Chamberlain, G. (1992). Efficiency bounds for semiparametric regression. *Econometrica*, *60*, 567-596.

Hausman, J. A. (1978). Specification tests in econometrics. *Econometrica*, *46*, 1251-1271.

Moser, P. & Voena, A. (2012). Compulsory licensing: Evidence from the Trading with the Enemy Act. *American Economic Review*, *102*, 396-427.

Mundlak, Y. (1961). Empirical production function free of management bias. *Journal of Farm Economics*, *43*, 44-56.

Mundlak, Y. (1978). On the pooling of cross section and time series data. *Econometrica*, *46*, 69-85.

Wooldridge, J. (2010). *Econometric analysis of cross section and panel data.* Cambridge, MA: MIT Press. (2nd edition)

Wooldridge, J. (2019). Correlated random effects models with unbalanced panels. *Journal of Econometrics*, *211*, 137-150.

Wooldridge, J. (2023). Simple approaches to nonlinear difference-in-differences with panel data. *Econometrics Journal*, *26*, C31-C66.

Wooldridge, J. (2024). *Two-way fixed effects, the two-way Mundlak regression, and difference-in-differences estimators.* (Unpublished working paper)

Chapter 16
Dimensionality and Exact Bound Tests in Simultaneous Equations

Jean-Marie Dufour, Lynda Khalaf

Abstract In linear simultaneous equations (SE), nuisance parameters and weak identification (or weak instruments) severely complicate exact and asymptotic tests. A sizable literature has proposed test procedures which aim at being robust to these difficulties. We reconsider such problems in the context of a general framework which combines statistical perspectives on multivariate linear regression, dimensionality analysis, and bound tests in econometrics. We study how hypothesis tests in a standard simultaneous equations model can be viewed as tests of rank restrictions (or dimensionality) on a multivariate linear regression (MLR). We adopt a finite-sample perspective, and show that exact tests and bounds can be obtained in this framework without relying on identification assumptions. This eschews the need for local asymptotic approximations, such as drifting sequences or local-to-nonidentification assumptions. For the problem of testing subvectors of structural parameters in a linear SE model, the bounds proposed can be viewed as a refinement of a general bound given in Dufour (1997), and the finite-sample analogue of the identification-robust asymptotic bounds proposed by Guggenberger, Kleibergen, Mavroeidis and Chen (2012) in a LIML framework. Simulation experiments illustrate the usefulness of the bounds as well as tests against alternatives which are unrestricted by the structure.

16.1 Introduction

Economic models often lead to situations where parameters of economic interest are difficult to estimate or test from observed data. This problem, which is broadly referred to as *identification* failure, occurs when the statistical objective function is not sensitive or varies little with some underlying parameters; see Dufour and Hsiao (2008) and Lewbel (2019) for formal definitions with discussion.

Identification is a key regularity condition for statistical analysis, and its failure raises serious concerns. In this regard, an active literature that may be traced back to Dufour (1997) and Staiger and Stock (1997), has emphasized the following issues: (i) identification failure occurs with methods that economists routinely use; (ii) this causes severe distortions in size and coverage of standard methods, which casts serious doubt on associated economic decisions; (iii) distortions may not

Jean-Marie Dufour ✉
McGill University, Montreal, Canada, e-mail: jean-marie.dufour@mcgill.ca

Lynda Khalaf
Carleton University, Ottawa, Canada, e-mail: Lynda_Khalaf@carleton.ca

B. H. Baltagi and L. Mátyás (eds.), *Seven Decades of Econometrics and Beyond*, Advanced Studies in Theoretical and Applied Econometrics 57,
https://doi.org/10.1007/978-3-031-92699-0_16

dissipate in large samples; (iv) alternative asymptotic and finite-sample methods are needed that do not require identification. The latter are typically referred to as *identification-robust* methods.

In this paper, we study how hypothesis tests in a standard simultaneous equations (SE) model can be viewed as tests of rank restrictions on a multivariate linear regression (MLR). We adopt a finite-sample perspective and show that finite-sample tests and bounds can be obtained in this framework without relying on identification assumptions or asymptotic arguments. The bounds are based on an approach used in our earlier work (Dufour, 1989, 1997; Dufour & Khalaf, 2002). For the problem of testing subvectors of structural parameters in a linear simultaneous model, the bounds proposed can be viewed as a refinement of the general bound given in Dufour (1997, Theorem 5.1) and the finite-sample analogue of the identification-robust asymptotic bounds proposed by Guggenberger et al. (2012) in a LIML framework.

Historically, econometricians started to comprehend the severity of identification problems with simultaneous equations (SE) models or Instrumental Variables (IV) regression, in which case identification failure stems from so-called *weak*, *i.e.* non-informative IVs; see the surveys of Stock, Wright and Yogo (2002), Dufour (2003), Mikusheva (2013) and I. Andrews, Stock and Sun (2019). This literature has experienced important developments, the most prominent build upon the Anderson-Rubin (AR) Anderson and Rubin (1949) test, which is valid regardless of whether IVs are weak or strong.

This chapter adds to this body of literature by defining a finite-sample framework which unifies several methods and allows for size and power comparisons. The proposed framework combines statistical perspectives on: (i) multivariate linear regression (MLR) (Anderson, 1984, Chapter 8; Rao, 1973, chapter 8; Berndt & Savin, 1977; Dufour & Khalaf, 2002), (ii) dimensionality tests (Rao, 1973, Section 8c.6; Saw, 1974; Schott, 1984; Gouriéroux, Monfort & Renault, 1995; Calinsky & Lejeune, 1998), and (iii) bounds tests in econometrics (Dufour, 1989, 1989, 1997; Dufour & Khalaf, 2002; Dufour & Taamouti, 2005, 2005, 2007).

The first perspective is relevant since the unrestricted reduced form of a SE is an MLR. The second perspective derives from the identification condition in IV-regression, which is a rank or dimensionality restriction. The third embeds an important property of test statistics whose null distribution depends on nuisance parameters that control identification requirements: unless the distribution in question can be bounded by another one which is invariant to these nuisance parameters, the test size can deviate arbitrarily from its nominal levels. Statistics that can be bounded along these lines are *boundedly pivotal*, using the definition of Dufour (1997); the simple existence of bounds on the distribution of a test statistic provides a *prima facie* validation of the test statistic for the null hypothesis considered. Along these lines, Dufour (1997) derives general bounds for likelihood-ratio tests in SEs. Such bounds do not depend on identification assumptions (such as rank restrictions matrices of reduced-form coefficients), irrespective of the sample size. Bounds can also lead to concrete and useful identification-robust inference procedures, which will be concretized through our analysis of SEs.

Our analysis focuses on the AR test in the context of a single SE (Dufour, 1997; Staiger & Stock, 1997; Dufour & Taamouti, 2007; D. W. K. Andrews, Moreira & Stock, 2006; Dufour & Taamouti, 2005; Doko Tchatoka & Dufour, 2020), and the related unconditional procedures that seek power improvements on this test (Wang & Zivot, 1998; Kleibergen, 2002; Guggenberger et al., 2012).[1] The AR test is by far one of the most influential identification-robust procedures; see D. W. Andrews and Marmer (2008) for non-parametric counterparts, Stock and Wright (2000) for extensions to GMM, and Beaulieu, Khalaf, Kichian and Melin (2022) for an extension to several SEs with common endogenous variables.

Our contribution is to derive exact and bounding distributions for these procedures, using the same MLR framework, which does not require local-to-zero or drifting-sequences based asymptotics, in contrast to Wang and Zivot (1998) and Guggenberger et al. (2012). Instead, bounds are validated using exact stochastic dominance arguments under the null hypothesis, based on statistics whose null distribution does not depend on identification assumptions. For the tests under consideration,

[1] On conditional tests, see Moreira (2003), Guggenberger, Kleibergen and Mavroeidis (2019) and Kleibergen (2021). The latter paper illustrates the merits of bounds in this context.

this involves criteria associated with a special case of the null hypothesis that can be tested exactly. Normality is not always needed to do this, in which case we clarify the hypotheses leading to the bound, and underlying assumptions.

The AR test is applicable when one is interested in testing the full vector of endogenous variables coefficients, denoted β, the dimension of which is m. This test is a linear least-squares based exclusion procedure, where the dimension of the excluded vector, *i.e.* the degrees-of-freedom, corresponds to the number of instruments, denoted k_2. The AR test can be inverted using tractable analytical formula (Dufour & Jasiak, 2001; Dufour & Taamouti, 2005) to derive confidence sets for inference on the components of β. We will refer to a test on a component of β based on the projected AR-based confidence set to the hypothesized value of the component in question, as the *inverted* AR (IAR) test.

Since k_2 is typically larger than m, attempts to avoid perceived degrees-of-freedom losses have driven much of the related enduring research on SEs. In particular, alternative procedures based on limited information maximum likelihood (LIML) have gained momentum. Within this class, the rank-based LIML identification-robust bound procedure of Guggenberger et al. (2012) for inference on a sub-vector of β of dimension m_1, which we denote β_1, involves $k_2 - (m - m_1)$ degrees-of-freedom, which promises to reduce the losses attributed to the IAR-test.

In this context, our contributions can be summarized as follows. *First*, we derive the AR statistic as a uniform linear (UL) restriction (Berndt & Savin, 1977; Dufour & Khalaf, 2002) test within an MLR. This provides an interesting multivariate perspective on this statistic that sets the stage for our dimensionality-based analysis in what follows. *Second*, we derive the finite-sample distribution of the AR statistic in possibly non-normal contexts. This reveals that the results of Doko Tchatoka and Dufour (2020) require separability assumptions, which we clarify. *Third*, we revisit the bound of Wang and Zivot (1998) from a finite-sample perspective. We also revisit the test proposed by Kleibergen (2002) through our context, which allows us to position the usefulness of these procedures relative the AR test. *Fourth*, we derive the IAR test and the bound procedure of Guggenberger et al. (2012) as dimensionality tests. The proposed finite-sample bounds can be viewed as the finite-sample analogue of the latter, without recourse to asymptotic arguments. The bounds given here can also be viewed as extensions of the bounds given by Saw (1974) and Schott (1984) in the more limited context of multivariate analysis-of-variance with only means (or intercepts) as explanatory variables (and no SE framework). Our analytic derivations allow us to compare the practical usefulness of these procedures on exact grounds. A simulation study further illustrates the power properties of all bounds, in contrast to the severe over-size problems that we document with IV-based Wald tests.

As demonstrated by I. Andrews et al. (2019), who examine 230 empirical analyses from 17 papers published in the American Economic Review from 2014-2018, weak IVs are a clear and evident problem in empirical work. Despite the usefulness of local-to-zero asymptotics that have facilitated improved approximations since Staiger and Stock (1997), we recommend finite-sample arguments leading to proper pivots as an alternative solid basis for robust inference. Regardless of the underlying method of proof, pivots are the key requirements for reliable inference, especially when identification may fail.

The chapter is organized as follows. Section 16.2 describes the simultaneous equations framework considered, and spells out the link with uniform linear restrictions in a multivariate linear regression model, along with LIML estimation in this context. Section 16.3 recast Anderson-Rubin-type tests as tests of dimensionality (reduced rank) on the coefficient matrix of a MLR model. Section 16.4 considers the problem of testing subvectors in a linear simultaneous equation model, and proposes finite-sample identification-robust bounds for such tests. Section 16.5 presents the results of a small simulation study. We conclude in Section 16.6.

16.2 Framework

Consider the limited-information (LI) linear simultaneous equations model:

$$
\begin{aligned}
y &= Y\beta + X_1\gamma + u = Z\delta + u, \\
Z &= [Y, X_1], \ \delta = (\beta', \gamma')', \\
Y &= X_1\Pi_1 + X_2\Pi_2 + V,
\end{aligned}
\tag{16.1}
$$

where y is a $T \times 1$ vector, Y and X_1 are $T \times m$ and $T \times k_1$ matrices which respectively contain the observations on the included endogenous and exogenous variables of the model, and X_2 refers to the $T \times k_2$ matrix of excluded exogenous variables (or instruments).

For the clarity of presentation, we will adopt the following notation:

I_s denotes an s-dimensional identity matrix;

$O_{(s,j)}$ denotes an $s \times j$ matrix of zeros;

given an $s \times j$ full-column rank matrix Z, $\quad M(Z) = I_s - Z(Z'Z)^{-1}Z'$.

$\mathcal{Z}(s)$ refers to a $T \times s$ matrix of *i.i.d.* s-dimensional standard normal variables $\mathcal{Z}_t(s)$, *i.e.*

$$
\mathcal{Z}(s) = [\mathcal{Z}_1(s), \ldots, \mathcal{Z}_T(s)]', \quad \mathcal{Z}_t(s) \overset{i.i.d.}{\sim} N[0, I_s], \quad t, \ldots, T.
$$

We denote by $F(n_1, n_2)$ the *F-distribution* with degrees of freedom n_1 and n_2, and by $F_\alpha(n_1, n_2)$ the associated α-level critical point. Similarly, $\chi^2(n)$ represents the χ^2 distribution with n degrees of freedom.

16.2.1 Distributional Assumptions

The LI reduced form associated with (16.1) is:

$$
\begin{bmatrix} y & Y \end{bmatrix} = X\Pi + \begin{bmatrix} v & V \end{bmatrix}, \ \Pi = \begin{bmatrix} \pi_1 & \Pi_1 \\ \pi_2 & \Pi_2 \end{bmatrix}, \ X = \begin{bmatrix} X_1 & X_2 \end{bmatrix}, \tag{16.2}
$$

$$
\pi_1 = \Pi_1\beta + \gamma, \ \ \pi_2 = \Pi_2\beta, \ \ v = u + V\beta,
$$

which leads to the standard condition for identification:

$$
\text{rank}(\Pi_2) = m.
$$

We suppose that the rows of $\begin{bmatrix} u & V \end{bmatrix}$ satisfy the following distributional assumptions:

$$
(u_t, V_t') \sim JW_t, \ \ t = 1, \ldots, T, \tag{16.3}
$$

where $\text{vec}(W_1, \ldots, W_T)$ has a known distribution, and J is an unknown nonsingular matrix. (16.3) implies that

$$
\begin{bmatrix} u & V \end{bmatrix} = WJ' \tag{16.4}
$$

where $W := [W_1, \ldots, W_T]'$. A special case considered below is the following mixture of normals:

$$
W = \bar{S}\mathcal{Z}(m+1) \tag{16.5}
$$

where $\bar{S}$ is $T \times T$, unknown and possibly random in which case it is independent of $\mathcal{Z}(m+1)$. As defined above, $\mathcal{Z}(m+1)$ denotes the $T \times (m+1)$ matrix of *i.i.d.* multivariate standard normal variables, so the standard Gaussian model corresponds to (16.5) with $\bar{S} = I_T$. The multivariate Student-t distribution with κ degrees-of freedom corresponds to the following structure: $\bar{S}$ is a diagonal matrix and diagonal terms denoted $\bar{S}_{t,t}$ take the form:

$$\bar{S}_{t,t} = 1/(v_t/\kappa)^{1/2}, \quad t = 1, \ldots, T,$$

where $v_t \overset{i.i.d.}{\sim} \chi^2(\kappa)$ and are independent of $\mathcal{Z}(m+1)$.

16.2.2 Background: Finite-sample Multivariate Regression Tests

The unrestricted model (16.2) is a MLR. To facilitate the analysis, this section presents some fundamental results on MLR models, based on Berndt and Savin (1977) and Dufour and Khalaf (2002). In particular, we review results from the statistical theory on dimensionality tests (Calinsky & Lejeune, 1998; Schott, 1984), which are relevant to the derivation of bounds on LIML-based tests within (16.1).

The general MLR model is given by:

$$Y = X\Pi + U$$

where Y is a $T \times n$ matrix of observations on n dependent variables, X is a $T \times k$ full-column-rank matrix of fixed regressors, and $U = [U_1, \ \ldots \ , U_T]'$ is the $T \times n$ matrix of error terms. We assume that we can condition on X for statistical analysis, and we consider the hypothesis

$$H_0 : C\Pi G = 0, \quad \text{for known } C \text{ and } G, \tag{16.6}$$

where C is $c \times k$ with rank c, G is $n \times g$ with rank g. The (16.6) form fits within the UL class, as defined by Berndt and Savin (1977). The associated trace or Wald-type statistic is given by:

$$T_0(C,G) = \mathrm{tr}[\hat{S}^{-1}\hat{S}_0] = \sum_{i=1}^{s} \theta_i,$$

$$\hat{S} = G'Y'M(X)YG, \quad \hat{S}_0 = G'Y'P_C(X)YG,$$
$$P_C(X) = X(X'X)^{-1}C'[C(X'X)^{-1}C']^{-1}C(X'X)^{-1}X',$$

where $s = \min(c,g)$ and $\theta_1 > \theta_2 > \cdots > \theta_s$ are the s positive (and distinct) eigenvalues of $\hat{S}^{-1}\hat{S}_0$. The likelihood ratio (LR) statistic corresponds to the product of these eigenvalues. Under the null hypothesis, $P_C(X)X\Pi G = 0$ because $C\Pi G = 0$, so that

$$\hat{S} = G'U'M(X)UG, \quad \hat{S}_0 = G'U'P_C(X)UG. \tag{16.7}$$

It follows that under the null hypothesis, $T_0(C,G)$ is distributed like the pivot

$$\bar{T}_0(C,G) = \mathrm{tr}[G'U'M(X)UG]^{-1}G'U'P_C(X)UG. \tag{16.8}$$

The following approximation for the null distribution of this statistic is recommended by McKeon (1974):

$$\frac{T_0(C,G)}{\varkappa_1} \sim F(cg, \varkappa_2) \tag{16.9}$$

where

$$\varkappa_1 = (cg)(\varkappa_2 - 2)/(\varkappa_2(T-k-g-1)), \quad \varkappa_2 = 4 + ((gc+2)/(\varkappa_3 - 1)),$$

$$\varkappa_3 = (T-k+c-g-1)(T-k-1)/((T-k-g-3)(T-k-g)).$$

When $\min(c, g) = 1$ and $U_t \overset{i.i.d.}{\sim} N[0, \Sigma]$, $t = 1, \ldots, T$, the distribution $F(cg, \varkappa_2)$ is valid in finite samples. In this case, the trace statistic coincides with the LR one, and (16.9) coincides with the approximation given by Rao (1973, chapter 8). For further reference, when $g = 1$ and $c = k$, (16.9) yields:

$$\frac{(T-k)}{k} \frac{G'Y'P_C(X)YG}{G'Y'M(X)YG} \sim F(k, T-k). \tag{16.10}$$

Similarly, when $c = k - (n-1) > 0$, we have:

$$\frac{(T-k)}{k-(n-1)} \frac{G'Y'P_C(X)YG}{G'Y'M(X)YG} \sim F(k-(n-1), T-k). \tag{16.11}$$

This may be verified from (16.7) by observing that UG is a vector of normal variables when $g = 1$.

Assuming $k > (n-1)$, consider now the trace statistic associated with testing whether the rank of Π is $n - 1$:

$$T_1 = \mu_n \tag{16.12}$$

where μ_n is the minimum eigenvalue of $(Y'M_XY)^{-1}Y'X(X'X)^{-1}X'Y$. By Calinsky and Lejeune (1998, equation (2.13)), we have under the null hypothesis of reduced rank:

$$\Pr\left[\frac{(T-k)}{k-(n-1)}T_1 \geq F_\alpha(k-(n-1), T-k)\right] \leq \alpha. \tag{16.13}$$

The rationale for this bound can be spelled out as follows (Dufour & Khalaf, 2002). Consider a right-tailed test statistic whose null distribution depends on nuisance parameters. Yet, it is possible to find another statistic T^* such that

$$T \leq T^* \tag{16.14}$$

for all nuisance parameters compatible with the null hypothesis, where T^* is pivotal (*i.e.*, T^* has a known nuisance-parameter free null distribution). This includes the case where this distribution can be derived by simulation. In particular, T^* may be associated with a null hypothesis which is a special case of the hypothesis under test. Let t_α refer to the α-level critical point associated with T^*. Then the dominance inequality (16.14) implies that

$$\Pr[T \geq t_\alpha] \leq \Pr[T^* \geq t_\alpha]$$

which entails that a test based on referring the observed value of T to t_α is level-correct.

These arguments lead to (16.13) as follows. The reduced rank null hypothesis can be re-expressed in the following form: there exists a $(k-(n-1)) \times k$ matrix C_* of rank $k-(n-1)$ and an $n \times 1$ non-zero vector G_* such that

$$C_*\Pi G_* = 0.$$

From standard results on matrix theory, it is also of interest to observe the following min-max property:

$$T_* = \inf_{C_*}\sup_{G_*} T_0(C_*, G_*) = \inf_{G_*}\sup_{C_*} T_0(C_*, G_*)$$

where $T_0(C_*, G_*)$ is the statistic associated with known C_* and G_*; see Calinsky and Lejeune (1998, equation (2.13) and Appendix A). Furthermore, the hypothesis with known C_* and G_* is a special case of the rank hypothesis under test, whereby (16.11) provides a known critical point which is nuisance parameter invariant.

An alternative bound can also be derived by rewriting the rank hypothesis in the following form when $k > n$: there exists an $n \times 1$ non-zero vector $G_\dagger$ such that

$$\Pi G_\dagger = 0.$$

It follows that the statistic $T_0(I_k, G_\dagger)$ associated with a known $G_\dagger$ can be used to bound the null distribution of T_1 leading to the bound

$$\Pr\left[\frac{(T-k)}{k}T_1 \geq F_\alpha(k), T-k)\right] \leq \alpha. \tag{16.15}$$

Clearly, (16.13) is a tighter bound. We use these arguments in what follows for statistical inference on β in (16.1).

16.2.3 LIML: Definitions and Multivariate Regression Perspectives

In the context of (16.1), LIML corresponds to maximizing the associated Gaussian likelihood. It is well know that the solution obtains through an eigenvalue/eigenvector problem based on the following determinantal equation

$$\det\left([\,y\ Y\,]'M(X_1)[\,y\ Y\,]-\lambda[\,y\ Y\,]'M(X)[\,y\ Y\,]\right)=0 \tag{16.16}$$

where λ refers to the eigenvalue in question. Using usual projection arguments and for presentation ease, we write (16.16) as

$$\det[\tilde{y}'\tilde{y}-\lambda\tilde{y}'M(\tilde{x})\tilde{y}]=0$$

where

$$\tilde{y}=M(X_1)[\,y\ Y\,],\quad \tilde{x}=M(X_1)X_2.$$

Indeed, it can be shown (see, for example Theil (1971, Appendix B), Wang and Zivot (1998)) that the LIML estimator of β is

$$\tilde{\beta}=\underset{\beta}{\text{ARGMIN}}\{\lambda(\beta)\}, \tag{16.17}$$

$$\lambda(\beta)=\frac{G(\beta)'\tilde{y}'\tilde{y}G(\beta)}{G(\beta)\tilde{y}'M(\tilde{x})\tilde{y}G(\beta)},\quad G(\beta)=(1,-\beta')',$$

or alternatively, the LIML estimator of δ is

$$\tilde{\delta}=\begin{bmatrix}\tilde{\beta}\\ \tilde{\gamma}\end{bmatrix}=\begin{bmatrix}Y'Y-\tilde{\lambda}Y'M(X)Y & Y'X\\ X'Y & X'X\end{bmatrix}^{-1}\begin{bmatrix}Y'-\tilde{\lambda}Y'M(X)\\ X'\end{bmatrix}y \tag{16.18}$$

where $\tilde{\lambda}$ is the smallest root of (16.16), *i.e.*

$$\tilde{\lambda}=\min_{\beta}\lambda(\beta)=\lambda(\tilde{\beta})$$

with $G(\tilde{\beta})=(1,-\tilde{\beta}')'$. Correspondingly, expressions for the estimates of the remaining parameters obtain as follows (see Theil, 1971, Appendix B):

$$[\,\tilde{\pi}_1\ \tilde{\Pi}_1\,]=(X_1'X_1)^{-1}X_1'\left([\,y\ Y\,]-X_2[\,\tilde{\pi}_2\ \tilde{\Pi}_2\,]\right), \tag{16.19}$$

$$[\,\tilde{\pi}_2\ \tilde{\Pi}_2\,]=(\tilde{x}'\tilde{x})^{-1}\tilde{x}'\tilde{y}-\frac{(\tilde{x}'\tilde{x})^{-1}\tilde{x}'y}{G(\tilde{\beta})'\hat{\Sigma}G(\tilde{\beta})}G(\tilde{\beta})G(\tilde{\beta})'\hat{\Sigma}, \tag{16.20}$$

$$\tilde{\Sigma}=\hat{\Sigma}+\frac{[\lambda(\tilde{\beta})-1]}{T}\frac{\tilde{y}'M(\tilde{x})\tilde{y}G(\tilde{\beta})G(\tilde{\beta})'\tilde{y}'M(\tilde{x})\tilde{y}}{G(\tilde{\beta})'\tilde{y}'M(\tilde{x})\tilde{y}G(\tilde{\beta})},$$

$$\hat{\Sigma}=\frac{\tilde{y}'M(\tilde{x})\tilde{y}}{T}.$$

The derivations of Theil (1971, Appendix B) also imply that $\det(\tilde{\Sigma})$ satisfies

$$\det(\tilde{\Sigma}) = \lambda(\tilde{\beta})\det(\hat{\Sigma}).$$

The above can be derived by equating the equations (7) and (9) of Berndt and Savin (1977), in the context of the regression of y on x where, in the notation of Section 16.2.2, restrictions are applied that take the form $C\left[\ \pi_2\ \Pi_2\ \right]G = 0$, with $C = I_{(k_2)}$, $E = 0$ and $G = G(\tilde{\beta})$.

For hypotheses of the form $R\delta = r$ on the coefficients of (16.1), where R is a known $q \times m$ matrix of rank q and r is known, Wald statistics are routinely applied and take the form

$$\tau_W = \frac{1}{s^2}(r - R\hat{\delta})'\{R'[ZP(P'P)^{-1}P'Z']^{-1}R\}^{-1}(r - R\hat{\delta}), \tag{16.21}$$

$$s^2 = \frac{1}{T}(y - Z\hat{\delta})'(y - Z\hat{\delta})', \quad P = \left[\ X\ X(X'X)^{-1}X'Y\ \right],$$

where $\hat{\delta}$ is a consistent asymptotically normal estimator such as (16.18) or the 2SLS

$$\hat{\delta} = [Z'P(P'P)^{-1}P'Z']^{-1}Z'P(P'P)^{-1}P'y$$

imposing identification, in which case the asymptotic null distribution of τ_W is $\chi^2(q)$. For an asymptotic theory conformable with under-identification, see Staiger and Stock (1997). Note that $\hat{\delta}$ corresponds to replacing $\tilde{\lambda}$ by 1 in (16.18). When $m = k_2$, $\tilde{\lambda} = 1$ and LIML coincides as is well known with 2SLS. We study these Wald statistics in our simulation experiment.

16.3 Anderson-Rubin-type Hypotheses

We first consider hypotheses that set the full vector β to a known value, which has generated an extensive literature since Dufour (1997) and Staiger and Stock (1997). Formally, in the context of the LI model (16.1), we consider hypotheses of the form:

$$H_{AR} : \beta = \beta_0 \tag{16.22}$$

where β_0 is a known vector. The restricted reduced form thus amounts to (16.2) with

$$\pi_1 = \Pi_1\beta_0 + \gamma, \quad \pi_2 = \Pi_2\beta_0, \quad v = u + V\beta_0. \tag{16.23}$$

16.3.1 Anderson-Rubin Test

The problem under consideration - specifically (16.23) - may be viewed in the context of the regression of y on x, namely

$$\tilde{y} = \tilde{x}\left[\ \pi_2\ \Pi_2\ \right] + M(X_1)\left[\ v\ V\ \right],$$

with restrictions of the form

$$\left[\ \pi_2\ \Pi_2\ \right]G(\beta_0) = 0, \quad G(\beta_0) = (1, -\beta_0')'.$$

As reviewed in Section 16.2.2, the associated trace statistic *against an unrestricted alternative* can be obtained as

$$\Lambda_{AR} = \frac{G_0'\tilde{y}'\tilde{x}(\tilde{x}'\tilde{x})^{-1}\tilde{x}'\tilde{y}G_0}{G_0'\tilde{y}'M(\tilde{x})\tilde{y}G_0} = \lambda(\beta_0) - 1 \tag{16.24}$$

where the function $\lambda(\cdot)$ is defined in (16.17).

Λ_{AR} corresponds to the AR statistic (up to a constant); in the context of simulation-based testing, one thus may rely on $\lambda(\beta_0)$ to obtain AR-type tests.

Theorem 16.1 *In the context of the LI model* (16.1) *with*(16.3)-(16.4), *consider the problem of testing* (16.22). *Let* Λ_{AR} *refer to the statistic defined by* (16.24). *Then, under the null hypothesis,* Λ_{AR} *is distributed like the criterion*

$$\bar{\Lambda}_{AR} = \frac{G_0'JW'M(X_1)WJ'G_0}{G_0'JW'M(X)WJ'G_0} - 1, \quad G_0 = \left(1, O_{(m,1)}'\right)'. \tag{16.25}$$

If it is further assumed that the first row of J has zeros everywhere except for the first element then

$$\bar{\Lambda}_{AR} = \frac{G_0'WM(X_1)WG_0}{G_0'W'M(X)WG_0} - 1. \tag{16.26}$$

Alternatively, under assumption (16.5),

$$\bar{\Lambda}_{AR} = \frac{G_0'\mathcal{Z}(1)'\bar{S}'M(X_1)\bar{S}\mathcal{Z}(1)G_0}{G_0'\mathcal{Z}(1)'\bar{S}'M(X)\bar{S}\mathcal{Z}(1)G_0} - 1. \tag{16.27}$$

Proof. Under the null hypothesis

$$\left[u \ V\right]G(\beta_0) = v - V\beta_0 = u$$

so that

$$\Lambda_{AR} = \frac{u'M(X_1)u}{u'M(X)u} - 1.$$

Given assumption (16.3),

$$u = \left[u \ V\right]G_0 = WJ'G_0,$$

which leads to (16.25). When the first row of J in (16.3) has zeros everywhere, except for the first element which equals (say) $\sigma \neq 0$, then $J'G_0 = \sigma G_0$ and $WJ'G_0 = \sigma WG_0$, so that

$$\Lambda_{AR} = \frac{\sigma G_0'W'M(X_1)WG_0\sigma}{\sigma G_0'W'M(X)WG_0\sigma} - 1$$

which yields (16.26). Alternatively, under assumption (16.5), $u = \left[u \ V\right]G_0 = \bar{S}\mathcal{Z}(m+1)J'G_0$. But the distribution of $\mathcal{Z}(m+1)J'G_0$ follows that of $\mathcal{Z}(1)\sigma_J$ where $\sigma_J = \left(G_0'JJ'G_0\right)^{1/2}$ which yields (16.27). □

The latter result means that an exact test can be carried out in non-normal contexts within a systems-based perspective. An assumption on the distribution of u can be formulated as in Doko Tchatoka and Dufour (2020), abstracting from the rest of the system. Our result reveals that this subsumes a block-triangular structure on J, unless the distribution is a mixture of normals. Block-triangular forms can be reasonably defended through economic theory, for example by causation arguments. Our point here is to emphasize the system perspective, in which case invariance to J should not be taken for granted. If normality is imposed, then as reviewed in Section 16.2.2 (see (16.10)) under the null hypothesis

$$\frac{T - (k_1 + k_2)}{k_2}\Lambda_{AR} \sim F[k_2, T - (k_1 + k_2)]. \tag{16.28}$$

16.3.2 Alternative Tests

Let us now turn to the LIML LR statistic associated with (16.22) (Wang & Zivot, 1998), which is a monotonic transformation of

$$\Lambda_{LI} = \lambda(\beta_0) - \lambda(\tilde{\beta}) \tag{16.29}$$

where the function $\lambda(\cdot)$ is defined in (16.17) and $\tilde{\beta}$ is the LIML estimate of β defined in (16.18). Our MLR background approach of Section 16.2.2 allows us to derive a finite-sample bound on the null distribution of this statistic, as follows.

Theorem 16.2 *In the context of the LI model* (16.1) *with assumption* (16.5) *where* $\bar{S} = I_T$, *consider the problem of testing* (16.22). *Let* Λ_{LI} *refer to the statistic defined by* (16.29). *Then, under the null hypothesis, we have:*

$$\Pr\left[\frac{T-(k_1+k_2)}{k_2}\Lambda_{LI} \geq F_\alpha\left(k_2, T-(k_1+k_2)\right)\right] \leq \alpha. \tag{16.30}$$

Proof. From (16.29), we have

$$\Lambda_{LI} = [\lambda(\beta_0) - 1] - [\lambda(\tilde{\beta}) - 1] = \Lambda_{AR} - [\lambda(\tilde{\beta}) - 1]$$

where $\lambda(\tilde{\beta}) \geq 1$. It follows that $\Lambda_{LI} \leq \Lambda_{AR}$, and the null distribution of the Λ_{AR} is given by (16.28). Then the result follows based on the argument of Section 16.2.2. □

Theorem 16.2 provides a finite-sample counterpart to the bound of Wang and Zivot (1998), which corresponds to the χ^2 approximation associated with (16.30). Theorem 16.1 further suggests that a simulation-based bound may be considered for level correct tests beyond the Gaussian special case. Yet from a power perspective, a test based on Λ_{AR} is expected to perform better. To be clear, the statistics that we consider in our simulation study below are

$$\Lambda_{AR} = \lambda(\beta_0) - 1, \tag{16.31}$$

$$\Lambda_{LI} = \lambda(\beta_0) - \lambda(\tilde{\beta}). \tag{16.32}$$

To conclude, consider the test proposed by Kleibergen (2002) in the context of (16.22). Dufour (2003) shows that the latter test corresponds to an AR-type test applied with a specific instrument choice (denoted Z_K). Specifically, equations (83) - (86) from Dufour (2003) rewritten in terms of our notation lead to the instrument

$$Z_K = X_2\overline{\Pi}_2, \quad \overline{\Pi}_2 = \hat{\Pi}_2 - \hat{\pi}_2(\beta_0)\frac{S_{\varepsilon V}(\beta_0)}{S_{\varepsilon\varepsilon}(\beta_0)},$$

$$\hat{\Pi}_2 = (\tilde{x}'\tilde{x})^{-1}\tilde{x}'\tilde{y}, \quad \hat{\pi}_2(\beta_0) = (\tilde{x}'\tilde{x})^{-1}\tilde{x}'\tilde{y}G(\beta_0),$$

$$S_{\varepsilon V}(\beta_0) = \frac{1}{T-k}G(\beta_0)'\tilde{y}'M(\tilde{x})\tilde{y}, \quad S_{\varepsilon\varepsilon}(\beta_0) = \frac{1}{T-k}G(\beta_0)'\tilde{y}'M(\tilde{x})\tilde{y}G(\beta_0).$$

Here we argue that the later expression is a constrained OLS estimator of Π_2, imposing the LIML structure. The following estimates correspond to (16.19) - (16.20) replacing $\tilde{\beta}$ by β_0:

$$\left[\,\hat{\pi}_2^0\ \hat{\Pi}_2^0\,\right] = (\tilde{x}'\tilde{x})^{-1}\tilde{x}'\tilde{y} - \frac{(\tilde{x}'\tilde{x})^{-1}\tilde{x}'\tilde{y}}{G(\beta_0)'\hat{\Sigma}G(\beta_0)}G(\beta_0)G(\beta_0)'\hat{\Sigma}$$

which in turn corresponds exactly to the estimator of Π_2, denoted $\overline{\Pi}_2$ by Kleibergen (2002). Dufour (2003) has shown that Wang and Zivot (1998)'s LM_{GMM} test obtains as an AR type test with instrument $X_2\hat{\Pi}_2$. We thus see that Kleibergen (2002) is highly related to the latter, since it is obtained in a similar way, replacing the unconstrained OLS estimator of Π_2 by a constrained OLS estimator which imposes the structure.

As mentioned in Dufour (2003), these tests are affected by the fact that estimated instruments are not independent from the error term u, and thus are not pivotal in finite samples. One way around it is to consider the split-sample AR test as proposed by Dufour and Jasiak (2001); see also Bolduc, Khalaf and Moyneur (2008) for an application of this approach.

16.4 Subset Hypotheses

Now consider hypotheses of the form

$$H_1 : \beta_1 = \beta_1^0, \tag{16.33}$$

where $\beta = (\beta_1', \beta_2')'$ and β_1 is $m_1 \times 1$ and β_2 is $m_2 \times 1$ where $m_2 = m - m_1$. Partition Y and V conformably into the $T \times m_1$ and $T \times m_2$ matrices Y_1 and Y_2, and the $T \times m_1$ and $T \times m_2$ matrices V_1 and V_2, and consider the conformable partition of Π_2 leading to the sub-system with dimension $m_2 + 1$

$$y - Y_1\beta_1^0 = Y_2\beta_2 + X_1\gamma + u,$$
$$Y_2 = X_1\Pi_{21} + X_2\Pi_{22} + V_2.$$

The restricted LIML of β_2 can be obtained as above through an eigenvalue/eigenvector problem. The determinantal equation in this case is:

$$\det\left[\tilde{y}_0'\tilde{y}_0 - \lambda_0\tilde{y}_0'M(\tilde{x})\tilde{y}_0\right]\right] = 0$$

where λ_0 refers is an eigenvalue, and

$$\tilde{y}_0 = M(X_1)\left[\, y - Y_1\beta_1^0 \; Y_2 \,\right] = M(X_1)\left[\, y \; Y \,\right]S(\beta_1^0)$$

where $S(\beta_1^0)$ is the $(m_1 + m_2 + 1) \times (m_2 + 1)$ transformation matrix. Other than zeros and ones for selection purposes, this matrix solely depends on β_1^0.

The restricted LIML estimator of β_2 is

$$\tilde{\beta}_{20} = \underset{\beta_2}{\text{ARGMIN}}\left\{\lambda_0(\beta_2)\right\},$$

$$\lambda_0(\beta_2) = \frac{G_0(\beta_2)'\tilde{y}_0'\tilde{y}_0G_0(\beta_2)}{G_0(\beta_2)\tilde{y}_0'M(\tilde{x})\tilde{y}_0G_0(\beta_2)}, \quad G_0(\beta_2) = (1, -\beta_2')',$$

or alternatively, the LIML estimator is

$$\begin{bmatrix} \tilde{\beta}_{20} \\ \tilde{\gamma}_0 \end{bmatrix} = \begin{bmatrix} Y_2'Y_2 - \tilde{\lambda}_0Y_2'M(X)Y_2 & Y_2'X \\ X'Y_2 & X'X \end{bmatrix}^{-1} \begin{bmatrix} Y_2' - \tilde{\lambda}_0Y_2'M(X) \\ X' \end{bmatrix} (y - Y_1\beta_1^0)$$

where $\tilde{\lambda}_0$ is the smallest root of the above determinantal equation, that is:

$$\tilde{\lambda}_0 = \min_{\beta_2} \lambda_0(\beta_2) = \lambda_0(\tilde{\beta}_{20}).$$

Correspondingly, expressions for the estimates of the remaining parameters obtain as follows (by replicating the above arguments based on Theil (1971, Appendix B):

$$\left[\, \tilde{\pi}_{10} \; \tilde{\Pi}_{10} \,\right] = (X_1'X_1)^{-1}X_1'(\left[\, y - Y_1\beta_1^0 \; Y_2 \,\right] - X_2\left[\, \tilde{\pi}_{20} \; \tilde{\Pi}_{20} \,\right]),$$

$$[\,\tilde{\pi}_{20}\ \tilde{\Pi}_{20}\,] = (\tilde{x}'\tilde{x})^{-1}\tilde{x}\tilde{y}_0 - \frac{(\tilde{x}'\tilde{x})^{-1}\tilde{x}\tilde{y}_0}{G_0(\tilde{\beta}_{20})'\tilde{\Sigma}_0 G_0(\tilde{\beta}_{20})} G_0(\tilde{\beta}_{20})G_0(\tilde{\beta}_{20})'\tilde{\Sigma}_0,$$

$$\tilde{\Sigma}_0 = \hat{\Sigma}_0 + \frac{(\tilde{\lambda}_0 - 1)}{T}\frac{\tilde{y}_0'M(\tilde{x})\tilde{y}_0 G_0(\tilde{\beta}_{20})G_0(\tilde{\beta}_{20})'\tilde{y}_0'M(\tilde{x})\tilde{y}_0}{G_0(\tilde{\beta}_{20})'\tilde{y}_0'M(\tilde{x})\tilde{y}_0 G_0(\tilde{\beta}_{20})},$$

$$\hat{\Sigma}_0 = \frac{\tilde{y}_0'M(\tilde{x})\tilde{y}_0}{T}.$$

16.4.1 LIML Bound Tests

In line with the previous section, we consider the following test statistics:

$$\Lambda_R = \lambda_0(\tilde{\beta}_{20}) - \lambda(\tilde{\beta}), \tag{16.34}$$

$$\Lambda_U = \lambda_0(\tilde{\beta}_{20}) - 1, \tag{16.35}$$

where the subscripts R and U aim to discern the restricted from the unrestricted alternative. $\tilde{\lambda}_0$ is the statistic of interest in Guggenberger et al. (2012).

Λ_U can be viewed as a test statistic for a rank hypothesis, based on the coefficient in the regression of y_0 on x, *i.e.* the transformed sub-system:

$$y_0 = \tilde{x}[\,\pi_2^0\ \Pi_{22}^0\,] + M(X_1)[\,v_0\ V_2\,]$$

where $[\,\pi_2^0\ \Pi_{22}^0\,] = [\,\pi_2\ \Pi_2\,]S(\beta_1^0)$ and $[\,v_0\ V_2\,] = [\,v\ V\,]S(\beta_1^0)$. In this case, we have:

$$v_0 = u + V_1(\beta_1 - \beta_{10}) + V_1\beta_2,$$

$$[\,v\ V\,] = [\,u\ V\,]B, \quad B = \begin{bmatrix} 1 & 0 \\ -\beta' & I_m \end{bmatrix}.$$

The associated hypothesis is:

$$\operatorname{rank}\left([\,\pi_2^0\ \Pi_{22}^0\,]\right) = m_2. \tag{16.36}$$

Indeed, $\tilde{\lambda}_0$ corresponds exactly to the formula of the rank test statistic in (16.12).

Theorem 16.3 *In the context of the LI model* (16.1) *with assumption* (16.5) *where* $\bar{S} = I_T$, *consider the problem of testing* (16.33) *when* $k_2 > m_2$. *Let* $\Lambda_U = \tilde{\lambda}_0$ *refer to the statistic defined by* (16.35). *Then under the null hypothesis we have*

$$\Pr\left[\frac{T-(k_1+k_2)}{k_2-m_2}\Lambda_U \geq F_\alpha\big(k_2-m_2,\, T-(k_1+k_2)\big)\right] \leq \alpha. \tag{16.37}$$

Further, the bound (16.37) *remains valid irrespective of the rank of* Π_{22}^0.

Proof. The reduced-rank null hypothesis (16.36) can be re-expressed in the following form: there exists a $(k_2 - m_2) \times k_2$ matrix C_* of rank $(k_2 - m_2)$ and an $(m_2 + 1) \times 1$ non-zero vector G_* such that

$$C_*[\,\pi_2^0\ \Pi_{22}^0\,]G_* = 0. \tag{16.38}$$

Consider the statistic denoted $T_0(C_*, G_*)$ which is associated with known C_* and G_*which is a special case of (16.38). Applying (16.7), we can write under the null hypothesis

$$T_0(C_*, G_*) = \mathrm{tr}\left[\left(G_*'U'M(X)UG_*\right)^{-1} G_*'U'M_{C_*}(X)UG_*\right],$$

$$U = M(X_1)\left[\, v_0 \; V_2 \,\right] = M(X_1)\left[\, u \; V \,\right] BS(\beta_1^0).$$

Under the null hypothesis, B depends on the unknown β_2, yet regardless of the value of β_2, the vector $\left[\, u \; V \,\right] BS(\beta_1^0)G_*$ remains normally distributed so its variance will be evacuated from the trace statistic. The rest of the proof follows from the results of Section 16.2.2 (see (16.13)) and correcting for the pre-multiplication of $\left[\, u \; V \,\right]$ by $M(X_1)$. It is important to emphasize that the null distribution of $T_0(C_*, G_*)$ does not depend on Π_{22}^0 (see (16.8) and (16.11)) the matrix that controls the identification of β_2 when $\beta_1 = \beta_1^0$. □

The fact that the bound on the finite-sample distribution of Λ_U (given by (16.37)) is valid for all matrices $\left[\, \pi_2^0 \; \Pi_{22}^0 \,\right]$ entails (by continuity) that it remains valid even if Π_{22}^0 does not have full column rank (*i.e.*, under identification failure). In other words, this bound is robust to identification failure. Guggenberger et al. (2012) derive the asymptotic χ^2 equivalent of this bound. Formally, they show that $(T - (k_1 + k_2))\Lambda_U$ can be bounded by the $\chi^2(k_2 - m_2)$ distribution without assuming identification, through drifting parameter sequences. Our proof, which builds on finite-sample dominance results, does not require such approximations. Yet interestingly, both approaches lead to similar bounds in the sense that the χ^2 bound in question is the typical limit of our F-based bound.

16.4.2 LIML Test versus Projection Anderson-Rubin Test

To begin with, observe that $H_{AR} \subseteq H_1$, so the null distribution of both Λ_U and Λ_R, as defined in the previous section, can be bounded by that of Λ_{AR} as defined by (16.24), which is pivotal under some conditions as we have shown in Theorem 16.1. This leads to the alternative bounds [see (16.15)]:

$$\Pr\left[\frac{(T - (k_1 + k_2))}{k_2}\Lambda_U \geq F_\alpha\big(k_2, T - (k_1 + k_2)\big)\right] \leq \alpha, \tag{16.39}$$

$$\Pr\left[\frac{(T - (k_1 + k_2))}{k_2}\Lambda_R \geq F_\alpha\big(k_2, T - (k_1 + k_2)\big)\right] \leq \alpha. \tag{16.40}$$

Since $\Lambda_R \leq \Lambda_U$, it makes more sense to rely on Λ_U for bound test purposes.

The bound underlying (16.39) corresponds to the one proposed by Saw (1974) to test (16.36). Calinsky and Lejeune (1998) rely instead on the improved bound of Schott (1984), which we have used to derive (16.37). The key ingredient underlying the validity of both bounds is that the null distribution of the bounding statistics are invariant to Π_{22}^0.

If m_1 is a scalar, then a bound test based on (16.39) corresponds to the above defined IAR test, which involves checking whether the projection-based confidence interval for β_1 associated with inverting the AR test based on Λ_{AR} covers β_{10}. This is because Λ_U corresponds to the minimum of the Λ_{AR} criterion over β_2 subject to $\beta_1 = \beta_{10}$. If this minimum over β_2 exceeds $F_\alpha(k_2, T - (k_1 + k_2))$, then the AR test associated with all other values of β_2 will also exceed this same cut-off, which in turn entails that the relevant projection will not cover β_{10}.

From a practical perspective, inverting the test based on (16.37) for confidence set purposes needs to be conducted numerically. In contrast, the AR test can be inverted analytically (Dufour & Taamouti, 2005). Since both tests rely on the same statistic, it makes sense to obtain the projection based confidence intervals for all model parameters, and then proceed to an analytical refinement using the tighter bound. The numerical burden may be alleviated this way, as the search set may be guided by the projections.

To conclude, note that a simulation-based alternative to the $F_\alpha(k_2, T - (k_1 + k_2))$ can be easily derived, following Theorem 16.1. A simulation-based alternative to reliance on the $F(k_2 - m_2, T - (k_1 + k_2))$ can be envisaged using Theorem 16.3, yet a supremum over C_* will be required. This said, and as emphasized by Calinsky and Lejeune (1998), the approximation due to McKeon (1974) performs well and provides a useful finite-sample alternative to the corresponding χ^2 from Guggenberger et al. (2012).

16.5 A Simulation Study

This section reports an investigation, by simulation, of the performance of the various proposed test procedures. Each experiment relies on 1000 replications. All experiments are based on the LI model (16.1). We consider three endogenous variables ($m = 2$) and $k = 4$, 5 and 6 exogenous variables. In all cases, the structural equation includes only one exogenous variable (so $k_1 = 1$), the constant regressor. In the following tables

$$d = (k - 1) - (p - 1) \tag{16.41}$$

refers to the *degree of over-identification.* We consider in turn: hypotheses which set the full vector of endogenous variables coefficients, *i.e.* of the form (16.22), and hypotheses which set a subset of endogenous variables coefficients:

$$\beta_1 = \beta_1^0 \tag{16.42}$$

where $\beta = (\beta_1', \beta_2')'$ and β_1 is $m_1 \times 1$, with $m_1 = 1$. The sample sizes are set to $T = 25, 50, 100$. These sizes are small by design, given the focus of our paper. The exogenous regressors are independently drawn from the normal distribution, with means zero and unit variances. These were drawn only once. The errors were generated according to a multinormal distribution with mean zero and covariance

$$\Sigma = \begin{bmatrix} 1 & .95 & -.95 \\ .95 & 1 & -1.91 \\ -.95 & -1.91 & 12 \end{bmatrix}.$$

The other coefficients were:

$$\gamma_1 = 1,\ \beta = (10, -1.5)',\quad \Pi_1 = (1.5, 2)',\quad \Pi_2 = \begin{bmatrix} \bar{\Pi} \\ O_{(k-3,2)} \end{bmatrix}.$$

The identification problem becomes mores serious as the determinant of $\Pi_2'\Pi_2$ gets closer to zero. In view of this, we consider various choices for $\bar{\Pi}$:

$$\bar{\Pi}_{(1)} = \begin{bmatrix} 2 & 1 \\ 1 & 2 \end{bmatrix},\quad \bar{\Pi}_{(2)} = \begin{bmatrix} 2 & 1.999 \\ 1.999 & 2 \end{bmatrix}, \tag{16.43}$$

$$\bar{\Pi}_{(3)} = \begin{bmatrix} .5 & .499 \\ .499 & .5 \end{bmatrix},\quad \bar{\Pi}_{(4)} = \begin{bmatrix} .01 & .009 \\ .009 & .01 \end{bmatrix}. \tag{16.44}$$

We first study the standard Wald statistics (see (16.21)), to document the severity of the weak IV problems in our design. Despite the two decades of research on this problem, such tests are still reported in empirical work. In each case, we consider 2SLS and LIML-based Wald tests, as defined in Section 16.2, and denoted $\tau_{w/2SLS}$ and $\tau_{w/LIML}$. We next examine the various identification-robust statistics that we have discussed in the previous sections.

The statistics are fully defined in the notes to each table with reference to the previous sections, and the identification-robust distributions considered are also fully defined. In sum, Λ_{LI} and Λ_R consider an alternative restricted by the model structure, whereas Λ_U and Λ_{AR} consider an unrestricted alternative. With the exception of Λ_{AR} which is exactly size correct, the remaining tests are exactly level correct where the underlying bounds are justified in the previous sections. We report size results for the asymptotic Wald tests under various scenarios for identification strength, and power results for the exact identification-robust tests for our scenario which reflects strong identification. We have verified that the exact tests control size regardless of identification strength in line with our analytical derivations. Power is not analyzed for the (here grossly) over-sized tests. We also do not report power for the weak IV cases; the correctly sized test have no power - as it should be - when the IVs are not informative.

Our main purpose is to document the usefulness of the bounds, to emphasize that the tests although level - but not necessarily size - correct remain powerful when power is expected. Tables 16.1 - 16.3 summarize our findings.

Table 16.1: Empirical size: testing a subset of endogenous variables coefficients, Wald tests.

			$\tau_{w/2SLS}$	$\tau_{w/LIML}$				$\tau_{w/2SLS}$	$\tau_{w/LIML}$
d	T	$\bar{\Pi}$	Asy	Asy	d	T	$\bar{\Pi}$	Asy	Asy
1	25	$\bar{\Pi}_{(1)}$	8.6	8.3	1	25	$\bar{\Pi}_{(3)}$	10.9	6.0
	50		6.4	6.2		50		7.2	4.8
	100		5.4	5.5		100		6.8	5.9
2	25		11.0	9.9	2	25		17.7	10.5
	50		8.0	8.5		50		13.3	6.7
	100		7.6	7.2		100		11.0	8.3
3	25		14.2	14.3	3	25		22.6	10.2
	50		10.4	10.9		50		18.3	10.4
	100		8.1	7.4		100		14.3	6.3
1	25	$\bar{\Pi}_{(2)}$	8.2	8.6	1	25	$\bar{\Pi}_{(4)}$	88.9	75.1
	50		4.6	5.2		50		84.9	66.8
	100		4.2	5.1		100		85.0	68.0
2	25		12.6	13.9	2	25		85.0	79.7
	50		8.3	10.4		50		55.5	76.9
	100		7.6	11.7		100		95.3	74.3
3	25		14.7	18.7	3	25		99.3	84.4
	50		13.4	18.8		50		98.9	81.6
	100		11.6	17.1		100		98.9	77.8

Note: The null hypothesis has the form (16.42), with $\beta_{10} = 10$. $\tau_{w/.}$ refers to the Wald statistic (16.21); the subscripts $2SLS$ versus $LIML$ identify the underlying estimator $\hat{\delta}$. $\bar{\Pi}_{(j)}, j = 1, \ldots, 4$ are defined in (16.43)-(16.44) and control identification strength: the quality of the instruments worsens moving from $\bar{\Pi}_{(1)}$ to $\bar{\Pi}_{(4)}$. d as defined in (16.41) is the degree of over-identification. 'Asy' refers to the asymptotic $\chi^2(1)$ approximation which requires strong identification.

Table 16.2: Power: Testing the full vector of endogenous variables coefficients

$H_0 : \beta_{11} = 10$			Λ_{LI}	Λ_{AR}	$H_0 : \beta_{11} = 10$			Λ_{LI}	Λ_{AR}
T	d	β_{11}	Bound	Exact	T	d	β_{11}	Bound	Exact
50	1	10.1	15.6	19.3	100	1	10.1	31.6	37.9
		10.2	54.0	60.2			10.2	87.0	90.3
		10.3	88.4	90.7			10.5	1.0	1.0
		11.0	1.0	1.0			11.0	1.0	1.0
	2	10.1	11.8	20.1		2	10.1	18.9	31.4
		10.2	45.8	57.5			10.2	76.6	84.3
		10.3	83.6	89.1			10.3	98.4	98.8
		11.0	1.0	1.0			10.5	1.0	1.0
	3	10.1	6.9	16.9		3	10.1	13.7	27.3
		10.2	30.7	46.2			10.2	70.1	82.0
		10.3	67.2	79.4			10.5	1.0	1.0
		11.0	1.0	1.0			11.0	1.0	1.0

Note: The null hypothesis is of the form (16.22), with $\beta_0 = (10, -1.5)'$. Λ_{AR} is the AR statistic (16.31); Λ_{LI} is the LIML-based statistic (16.32). The design imposes $\bar{\Pi}_{(1)}$ as defined in (16.43), which reflects instruments strength. d as defined in (16.41) is the degree of over-identification. 'Bound' refers to the dominance result (16.30), and 'Exact' to the null distribution (16.28).

Table 16.1 clearly reveals the dire consequences of weak IVs: identification problems severely distort the sizes of standard asymptotic tests. While the evidence of size distortions is notable even in identified models, the problem is far more severe in near-unidentified situations. The results for the Wald test are especially striking: empirical sizes exceeding 80 and 90% were observed. More importantly, increasing the sample size does not correct the problem.

Table 16.2 emphasizes the power superiority of the AR test. This said, the LIML test does not perform poorly despite its reliance on a bound, and catches up with the AR test as the DGP departs from the null hypothesis. Overall, there seems to be no advantage to imposing the structure under the alternative hypothesis. In some sense, this may be counter-intuitive. Concretely, the information advantages which may stem from imposing the restrictions implied by the structure are offset by the nuisance parameters that are introduced to do this. These nuisance parameters depend on the quality of the IVs, which calls for bounds to correct this problem. The net effect is some power loss, which although not dramatic, is worth noting. There is another advantage associated with relaxing the structure, as shown by Dufour and Taamouti (2007). The AR test can be shown to be robust to missing IVs, which brings in the broader question of model incompleteness. In line with our research question, we have kept our design within a correctly specified DGP. Issues resulting from misspecification or incomplete models are worthy research questions (beyond the scope of our paper) that may call for a different perspective on comparing the considered tests.

Table 16.3 concerns nuisance parameter dependent problems, so all tests though level correct are not size-correct. Our results clearly show the usefulness of the bounds. We have verified that power reaches one as the sample size increases or as one departs from the null. For comparison purposes, we report results for the same designs across tables. The power advantage of the bound test based on Λ_U and (16.37) is evident. Yet again, the IAR test (that is, the bound test based on Λ_U and (16.39))

Table 16.3: Power: Testing a subset of endogenous variables coefficients

$H_0 : \beta_{11} = 10$			Λ_R		Λ_U		$H_0 : \beta_{11} = 10$			Λ_R		Λ_U	
T	d	β_{11}	BD	BD$_*$	BD	BD$_*$	T	d	β_{11}	BD	BD$_*$	BD	BD$_*$
50	1	10.1	2.1	5.2	3.7	8.8	100	1	10.1	5.0	11.7	10.3	18.4
		10.3	8.1	19.6	11.3	24.7			10.3	26.2	40.3	33.3	47.9
		10.5	13.6	26.4	17.4	32.3			10.5	36.4	54.0	43.0	61.2
		11.0	18.9	32.1	22.4	37.5			11.0	47.9	64.5	54.6	71.5
	2	10.1	1.3	4.5	4.2	10.9		2	10.1	2.8	5.9	7.6	15.1
		10.3	6.0	14.1	17.1	24.8			10.3	19.1	30.4	30.3	46.8
		10.5	10.0	20.0	17.9	33.3			10.5	29.6	43.2	41.7	60.1
		11.0	14.8	28.3	24.0	41.1			11.0	39.6	55.4	53.3	69.9
	3	10.1	0.9	2.5	4.0	10.6		3	10.1	1.4	3.5	7.1	13.7
		10.3	4.9	10.4	12.0	23.9			10.3	12.2	21.2	25.3	39.4
		10.5	8.4	17.7	17.9	32.9			10.5	21.8	33.7	37.5	52.6
		11.0	13.3	22.7	24.2	41.1			11.0	31.7	44.4	47.2	63.8

Note: The null hypothesis is of the form (16.42), with $\beta_{10} = 10$. Λ_R is the LIML statistic (16.34) against an alternative restricted by the structure; Λ_U is its counterpart against an unrestricted reduced form (16.35). 'BD' refers to the dominance results (16.39) and (16.40), and 'BD$_*$' to its alternative (16.37), which is based on a tighter bound. The test based on Λ_U and the BD bound corresponds to the IAR test, which corresponds to referring the projection-based AR confidence set to the tested value (here 10). The design imposes $\bar{\Pi}_{(1)}$ as defined in (16.43), which reflects instruments strength. d as defined in (16.41) is the degree of over-identification.

has good power, despite the superiority of the improved bound. Observe that the IAR test is easier to generalize beyond normality from a simulation-based finite-sample perspective, in contrast to the improved bound as shown analytically above. We have also pointed out above that the improved bound will involve numerical searches for inversion purpose, in contrast with the AR test. All these considerations should be weighed in, to interpret the benefit/cost trade-off of the improved bound. Finally, whether the improved bound would deliver robustness to missing instruments remains an open question. We reiterate the importance of further work on incomplete models.

16.6 Conclusion

The serious inadequacy of standard asymptotic tests in finite samples is widely observed in the SE context. Here, we have proposed alternative exact and bound procedures, and demonstrated their feasibility. Particular attention was given to the identification problem. The simulation results show that relaxing the structure under the alternative hypothesis pays off power wise. While structures hold information, this comes at an important cost: imposing the structure introduces nuisance parameters that are influenced by the model's identification status. Pivotal bounds will correct this problem, yet post-correction, the unrestricted tests perform better. Overall, pivotal statistics - even when bounds-based - provide a sound basis for inference in the presence of weak IVs, provided the model is complete (no excluded instruments).

This said, identification robust procedures will only avoid spurious rejections. Weak IVs will not hold information on parameters of interest so power cannot be expected in this case. We conclude by pointing to a promising research direction aiming to capture the information content of any IV, weak or strong, through a reparametrized setting which is always identified; see Doko Tchatoka and Dufour (2014), Beaulieu et al. (2022) and Beaulieu, Dufour, Khalaf and Melin (2023). The parameter that one can identify is different from the original inference target, in the sense that the former embeds the extent of endogeneity. Its interpretation is thus problem dependent. As the discipline pursues the analysis of incomplete models, such alternative parameters which embed the effects of unobservables will gain credibility in economics. Finally, it is important to remember that tests based on the complete specification are not robust.

References

Anderson, T. W. (1984). *An introduction to multivariate statistical analysis* (Second ed.). New York: John Wiley & Sons.

Anderson, T. W. & Rubin, H. (1949). Estimation of the parameters of a single equation in a complete system of stochastic equations. *Annals of Mathematical Statistics*, *20*, 46-63.

Andrews, D. W. & Marmer, V. (2008). Exactly distribution-free inference in instrumental variables regression with possibly weak instruments. *Journal of Econometrics*, *142*(1), 183–200.

Andrews, D. W. K., Moreira, M. J. & Stock, J. H. (2006). Optimal two-sided invariant similar tests for instrumental variables regression. *Econometrica*, *74*, 715-752.

Andrews, I., Stock, J. H. & Sun, L. (2019). Weak instruments in IV regression : theory and practice. *Annual Review of Economics*, *11*, 727-753.

Beaulieu, M.-C., Dufour, J.-M., Khalaf, L. & Melin, O. (2023). Identification-robust beta pricing, spanning, mimicking portfolios, and the benchmark neutrality of catastrophe bonds. *Journal of Econometrics*, *236*(1), 105464.

Beaulieu, M.-C., Khalaf, L., Kichian, M. & Melin, O. (2022). Finite sample inference in multivariate instrumental regressions with an application to catastrophe bonds. *Econometric Reviews*, *41*(10), 1205-1242.

Berndt, E. R. & Savin, N. E. (1977). Conflict among criteria for testing hypotheses in the multivariate linear regression model. *Econometrica*, *45*, 1263-1277.

Bolduc, D., Khalaf, L. & Moyneur, E. (2008). Identification-robust simulation-based inference in joint discrete/continuous models for energy markets. *Computational Statistics and Data Analysis*, *52*, 3148-3161.

Calinsky, T. & Lejeune, M. (1998). Dimensionality in Manova tested by a closed testing procedure. *Journal of Multivariate Analysis*, *65*, 181-194.

Doko Tchatoka, F. & Dufour, J.-M. (2014). Identification-robust inference for endogeneity parameters in linear structural models. *The Econometrics Journal*, *17*(1), 165-187.

Doko Tchatoka, F. & Dufour, J.-M. (2020). Exogeneity tests, incomplete models, weak identification and non-gaussian distributions: Invariance and finite-sample distributional theory. *Journal of Econometrics*, *218*(2), 390-418.

Dufour, J.-M. (1989). Nonlinear hypotheses, inequality restrictions, and non-nested hypotheses: Exact simultaneous tests in linear regressions. *Econometrica*, *57*, 335-355.

Dufour, J.-M. (1997). Some impossibility theorems in econometrics, with applications to structural and dynamic models. *Econometrica*, *65*, 1365-1389.

Dufour, J.-M. (2003). Identification, weak instruments and statistical inference in econometrics. *Canadian Journal of Economics*, *36*, 767-808.

Dufour, J.-M. & Hsiao, C. (2008). Identification. In L. Blume & S. Durlauf (Eds.), *The new Palgrave dictionary of economics* (Second ed.). Basingstoke, Hampshire, England.: Palgrave MacMillan.

Dufour, J.-M. & Jasiak, J. (2001). Finite sample limited information inference methods for structural equations and models with generated regressors. *International Economic Review*, *42*, 815-843.

Dufour, J.-M. & Khalaf, L. (2002). Simulation based finite and large sample tests in multivariate regressions. *Journal of Econometrics*, *111*, 303-322.

Dufour, J.-M. & Taamouti, M. (2005). Projection-based statistical inference in linear structural models with possibly weak instruments. *Econometrica*, *73*, 1351-1365.

Dufour, J.-M. & Taamouti, M. (2007). Further results on projection-based inference in IV regressions with weak, collinear or missing instruments. *Journal of Econometrics*, *139*, 133-153.

Gouriéroux, C., Monfort, A. & Renault, E. (1995). Inference in factor models. In G. S. Maddala, P. C. B. Phillips & T. N. Srinivasan (Eds.), *Advances in econometrics and quantitative economics* (p. 311-353). Oxford, U.K.: Blackwell.

Guggenberger, P., Kleibergen, F. & Mavroeidis, S. (2019). A more powerful subvector Anderson Rubin test in linear instrumental variables regression. *Quantitative Economics*, *10*, 487-526.

Guggenberger, P., Kleibergen, F., Mavroeidis, S. & Chen, L. (2012). On the asymptotic sizes of subset Anderson-Rubin and Lagrange Multipliers tests in linear instrumental variables regression. *Econometrica*, *80*, 2649-2666.

Kleibergen, F. (2002). Pivotal statistics for testing structural parameters in instrumental variables regression. *Econometrica*, *70*, 1781-1803.

Kleibergen, F. (2021). Efficient size correct subset inference in homoskedastic linear instrumental variables regression. *Journal of Econometrics*, *221*, 78-96.

Lewbel, A. (2019). The identification zoo: Meanings of identification in econometrics. *Journal of Economic Literature*, *57*, 835-903.

McKeon, J. J. (1974). F approximation to the distribution of Hotelling's T^2. *Journal of Business and Economic Statistics*, *15*, 74-81.

Mikusheva, A. (2013). Survey on statistical inferences in weakly-identified instrumental variable models. *Applied Econometrics*, *29*, 117-131.

Moreira, M. J. (2003). A conditional likelihood ratio test for structural models. *Econometrica, 71*(4), 1027–1048.

Rao, C. R. (1973). *Linear statistical inference and its applications* (Second ed.). New York: John Wiley & Sons.

Saw, J. G. (1974). A lower bound for the distribution of a partial product of latent roots. *Communications in Statistics, 3*, 665-669.

Schott, J. R. (1984). Optimal bounds for the distribution of some test criteria for tests of dimensionality. *Biometrika, 71*, 561-567.

Staiger, D. & Stock, J. H. (1997, May). Instrumental variables regression with weak instruments. *Econometrica, 65*, 557-586.

Stock, J. H. & Wright, J. H. (2000). GMM with weak identification. *Econometrica, 68*, 1097-1126.

Stock, J. H., Wright, J. H. & Yogo, M. (2002). A survey of weak instruments and weak identification in generalized method of moments. *Journal of Business and Economic Statistics, 20*, 518-529.

Theil, H. (1971). *Principles of econometrics.* New York: John Wiley & Sons.

Wang, J. & Zivot, E. (1998). Inference on structural parameters in instrumental variable regression with weak instruments. *Econometrica, 66*, 1389-1404.

Chapter 17
Dynamic Log-Linear Probability Model with Interactions

Christian Gouriéroux and Nour Meddahi

Abstract The log-linear probability model has been initially introduced by Nerlove and Press (1973) for the analysis of contingency tables constructed from business survey data. We extend this modelling approach to the dynamic analysis of multivariate qualitative processes with the application to technical analysis of financial returns in mind. We develop the dynamic qualitative models with pairwise and/or three-wise interactions, discuss the interpretations of the interaction parameters, study the filtering and prediction algorithms, and compare the approach to machine learning models as the restricted Boltzmann machine and the normalizing flows.

17.1 Introduction

The chartism approach (or technical analysis) in finance provides empirical charts to predict the future path of returns (or prices, or exchange rates) based on the observed past paths. This is an example of chart (pattern) recognition technique with specific informative patterns known as Elliot waves or Dow Theory (Rhea, 1932; Russell, 2012), Head and Shoulder (Lo, Mamaysky & Wang, 2000), Bull Market versus Bear Market (Lofton, 1986), and for a general presentations of technical analysis see Frankel and Froot (1990); Taylor and Allen (1992); Archer and Bickford (2007); Neely and Weller (2012). They lead to portfolio management strategies known as momentum or reversal (Menkhoff, Sarno, Schmeling & Schrimpf, 2012). These empirical methods were largely applied on commodity markets and foreign exchange rates (FOREX). They are pure univariate time series approaches that were frequently compared with fundamental analysis (Vigfusson, 1997; Lui & Mole, 1998; Oberlechner, 2001; Dick & Menkhoff, 2013). They highlight the importance of nonlinear dynamics features to predict future returns (prices, exchange rates) in contrast with the random walk hypothesis defended in Fama (1995).

These approaches had the drawback to be mainly descriptive without the support of a stochastic dynamic model for the observed series. In particular, they do not provide the associated inference as the possibility to get term structure of pointwise predictions, prediction intervals, or do not provide any diagnostic tools. This explains why these methods, less the portfolio strategies themselves, have

Christian Gouriéroux ✉
University of Toronto, Canada, Toulouse School of Economics and CREST, France, e-mail: christian.gourieroux@ensae.fr

Nour Meddahi
Toulouse School of Economics, Toulouse, France, e-mail: nour.meddahi@tse-fr.eu

B. H. Baltagi and L. Mátyás (eds.), *Seven Decades of Econometrics and Beyond*, Advanced Studies in Theoretical and Applied Econometrics 57,
https://doi.org/10.1007/978-3-031-92699-0_17

been partly abandoned, especially with the apparition of the ARCH models and their extensions (Engle, 1982; Bollerslev, 1986) able to capture some non-linear dynamic features as the volatility persistence, or the fact that a higher volatility can be an advanced indicator of a turning point in the evolution of the series in the ARCH-in-Mean models (Engle, Lilien & Robins, 1987).

Nonlinear dynamic features, however, such as the leverage effects (Black, 1976), the different dynamics of left and right extreme returns, the waves and more generally the threshold effects are not well-captured by the extended family of ARCH models (see Gourieroux & Monfort, 1992; Zakoian, 1994; and Section 5.4 of Gourieroux, 1997 for the first introductions of thresholds in the ARCH modelling).

To capture these additional dynamic features, we propose to first transform the series into a multivariate qualitative series. For instance into a bivariate qualitative series with alternatives: positive returns versus negative ones, or standard returns versus extreme ones, or in a trivariate qualitative series with alternatives: large negative returns, standard returns, and large positive returns, and then to apply a flexible nonlinear dynamic model to such qualitative time series.

More precisely we adjust the log-linear probability models with interactions (Nerlove & Press, 1973; Bishop, Fienberg & Holland, 1975; Liang & Zeger, 1989) to the case of qualitative Markov processes of order H, taking into account the fact that the alternatives are exclusive (this is a competitor to the (multiple) Threshold Autoregressive (TAR) model recently applied to stock returns by Zhang, Li and Tong (2023), whose implementation seems to impose arbitrarily a rather small lag H in practice). This leads to transition distributions that are conditional logit models with lagged endogenous variables that can include pairwise or three-wise interactions of the past. These conditional models are easily estimated by composite maximum likelihood, or by maximum likelihood with LASSO (Least Absolute Shrinkage and Selection Operator) penalties on the interactions even with rather large lag H. These dynamic models have the advantage of providing closed form expressions of the prediction and filtering distributions. They also underlie the Deep Boltzmann Machine (DM), that treat complex interactions in (static) neural networks and deep learning (see Ackley, Hinton & Sejnowski, 1985; Salakhutdinov & Hinton, 2009). In other words the autoregressive logit model with interactions introduced in this chapter can also be used for deep learning in a dynamic framework of neural network (i.e., for deep recurrent neural network).

The plan of the chapter is the following. Section 17.2 introduces the log-linear probability model with interactions adjusted to dynamic qualitative observations and to exclusive alternatives. First we adopt a progressive presentation for binary processes, then for qualitative processes with three alternatives. These qualitative Markov processes can encounter the curse of dimensionality if they are let unconstrained (saturated models). We explain how this dimensionality can be significantly reduced by limiting the interactions to pairwise and/or three-wise interactions. In the three exclusive alternatives case, we also distinguish the autoregressive logit model with interactions from the autoregressive recursive model. Section 17.3 considers statistical inference. We review the properties of the maximum likelihood estimator, explain how to use the estimated pairwise or three-wise interactions as diagnostic tool. We also discuss the introduction of LASSO penalties, and the possibility to also estimate the threshold to separate 'standard' and 'extreme' returns. Section 17.4 provides prediction and filtering distributions, while Section 17.5 concludes. Different technical issues are gathered in the Appendices: a discussion of the ergodicity conditions in Appendix 1, the static log-linear probability model with interactions and its properties are recalled in Appendix 2, and the binary AR(1) process is discussed in Appendix 3.

17.2 The Dynamic Log-Linear and Logit Models with Interactions

This section explains how the static log-linear/logit modelling (see also Appendix 2) can be adjusted to define the dynamics of a stationary univariate or multivariate series of binary variables. We first consider univariate series, such as $Y_t = 1$, if the daily return is positive, $Y_t = 0$, if it is negative. Then we extend the model to the multivariate case, with in mind a qualitative variable Y_t with three exclusive alternatives corresponding to extreme negative returns, standard ones, and extreme positive ones, say.

In this section, we assume that the process $\{Y_t\}$ is a Markov process of order H and that this process is ergodic with a unique stationary distribution which is not degenerate, that is: $P(Y_t = y_t \mid Y_{t-1} = y_{t-1}, \dots, Y_{t-H} = y_{t-H}) > 0$, for any possible values of $(y_t, y_{t-1}, \dots, y_{t-H})$. Under this assumption, the distribution of the process is characterized by the positive transition distribution:

$$P(Y_t = y_t \mid Y_{t-1} = y_{t-1}, \dots, Y_{t-H} = y_{t-H}), \quad \forall y_t, y_{t-1}, \dots, y_{t-H} \in \{0, 1\},$$

or equivalently by the joint distribution:

$$P(Y_t = y_t, Y_{t-1} = y_{t-1}, \dots, Y_{t-H} = y_{t-H}), \quad \forall y_t, y_{t-1}, \dots, y_{t-H} \in \{0, 1\},$$

since the stationary distribution of the path $(Y_t, Y_{t-1}, \dots, Y_{t-H})$ is uniquely defined from the positive transition distribution under the ergodicity condition (see the discussion of the ergodicity condition in Appendix 1). This is especially important in the log-linear/logit modelling, where the log-linear probability models with interactions introduced for the joint distributions are associated with logit models with interactions for the conditional distributions.

17.2.1 Univariate Binary Process

17.2.1.1 Model with Pairwise Interactions

Let us first consider a log-linear probability model with marginal effects and pairwise interactions. In the time series framework the binary variables are indexed by $t - h$, $h = 0, \dots, H$. Then the log-linear model involves:

$\alpha_h, h = 0, \dots, H$, marginal effects and $\beta_{hk}, h, k = 0, \dots, H, h < k$, pairwise interaction effects, to get:

$$P(Y_t = y_t, Y_{t-1} = y_{t-1}, \dots, Y_{t-H} = y_{t-H}) \equiv p(y_t, y_{t-1}, \dots, y_{t-H}),$$

such that:

$$\log p(y_t, y_{t-1}, \dots, y_{t-H}) = \mu + \sum_{h=0}^{H} \alpha_h y_{t-h} + \sum_{h=0}^{H} \sum_{k=h+1}^{H} \beta_{hk} y_{t-h} y_{t-k}, \tag{17.1}$$

where μ is fixed by the unit mass restriction (see Appendix 2). However the specification above does not account for the stationarity of the binary process (Y_t). Under the strict stationarity assumption, the parameters are not asymptotically identified. Intuitively, we can impose that the marginal effects α_h are independent of h and the pairwise interaction effects β_{hk} depend on $h, k, k > h$, by the difference $k - h$ only. Under this identification restriction, the number of parameters is reduced to H+1 with new parameters $\alpha, \beta_k, k = 1, \dots, H$, such that:

$$\alpha_k \equiv \alpha, \ \forall k = 1, \dots, H, \ \beta_{h,h-k} \equiv \beta_k, \ k = 1, \dots, H, \quad \text{for any } h. \tag{17.2}$$

Then we deduce the conditional distribution of Y_t given $(Y_{t-1}, \dots, Y_{t-H})$ as:

$$\begin{aligned} \log p(y_t \mid y_{t-1}, \dots, y_{t-H}) &= \tilde{\mu}_t + \alpha y_t + \sum_{k=1}^{H} \beta_k y_t y_{t-k} \\ &= \tilde{\mu}_t + y_t \left(\alpha + \sum_{k=1}^{H} \beta_k y_{t-k} \right), \end{aligned} \tag{17.3}$$

where $\tilde{\mu}_t$ depends on the parameters and $y_{t-1}, \dots, y_{t-H}$, by the unit mass restriction. Equivalently, one has an autoregressive dichotomous logit model of order H with:

$$\log p(1 \mid y_{t-1}, \dots, y_{t-H}) = \exp\left(\alpha + \sum_{k=1}^{H} \beta_k y_{t-k}\right)\left[1 + \exp\left(\alpha + \sum_{k=1}^{H} \beta_k y_{t-k}\right)\right]^{-1},$$

$$\log p(0 \mid y_{t-1}, \dots, y_{t-H}) = \left[1 + \exp\left(\alpha + \sum_{k=1}^{H} \beta_k y_{t-k}\right)\right]^{-1}.$$

Remark 1: In the static log-linear probability model (see Appendix 2), the binary variables are often indexed by individual, or localization (in spatial models). The marginal and interaction effects are not jointly constrained ex-ante and the identification and estimation of the parameters are performed by composite conditional likelihood that combines all possible conditional distributions, such as the conditional distribution of Y_{t-1} given $(Y_t, Y_{t-2}, \dots, Y_{t-H})$ for instance. In the Markov framework, this is not necessary since the transition distribution contains all the sufficient information on the (restricted) set of parameters.

Remark 2: To understand the identification restriction (2.2), we can consider the average $\frac{1}{T} \log p(y_t, y_{t-1}, \dots, y_{t-H})$ under (2.1). We get an exponential family of distributions, where the sufficient statistic associated with β_{hk}, say, is $\frac{1}{T} \sum_{t=1}^{T} y_{t-h} y_{t-k}$. Asymptotically, this statistic converges to $E[y_{t-h} y_{t-k}] = m_2(h-k)$ by stationarity. Therefore this exponential family is asymptotically degenerate, that explains the identification restriction (2.2).

Remark 3: Log-linear probability models with pairwise interactions are used in the machine learning literature to define the architecture of Boltzmann machines (Salakhutdinov & Hinton, 2009). The underlying stochastic model assumes a sample drawing of two groups of binary variables $(X_i', Y_i'), i = 1, \dots, n$, where $X_i = (X_{1i}, \dots, X_{Ki})', Y_i = (Y_{1i}, \dots, Y_{Li})'$. The log-linear probability model with pairwise interactions is used to define the distribution of $(X_i', Y_i')'$. This Boltzmann approach differs from our approach in two respects:

i) It is usually applied on individual data, and not to time series. In particular there is no parameter restriction for stationarity purpose.

ii) Other restrictions are introduced in order to interpret the Boltzmann machine architecture as a neural network with two layers, that are an entry layer with neurons the components of X and the exit layer with neurons the components of Y. Typically the Restricted Boltzmann Machine (RBM) does not allow for intra-layer connections, that is there is no interaction between the neurons of a given group.

The knowledge of the transition is equivalent to the knowledge of the joint distribution by applying formula (17.3) and taking into account the stationarity constraints on the parameters. We deduce:

$$\log p(y_t, y_{t-1}, \dots, y_{t-H}) = \mu + \alpha \sum_{h=0}^{H} y_{t-h} + \sum_{k=1}^{H} \left[\beta_k \sum_{h=0}^{H-k} y_{t-h} y_{t-h-k}\right], \tag{17.4}$$

where μ is a function of the parameters fixed by the unit mass restriction. This expression can be used to derive the other conditional logit model for providing the distribution of Y_{t-k} given Y_{t-h}, $h = 0, \dots, H, h \neq k$, that is the nonlinear filtering of Y_{t-k} that can be compared to the observed value to construct diagnostic tools appropriate for binary variables (see Section 17.4).

By summing over t formula (17.4), we can also note that we get an exponential family with the sufficient statistics:

$$\frac{1}{H+1} \frac{1}{T} \sum_{t=1}^{T} \sum_{k=0}^{H} y_{t-k} \approx \frac{1}{T}(y_1 + y_2 + \dots + y_T),$$

and

$$\frac{1}{H-K+1}\frac{1}{T}\sum_{t=1}^{T}\sum_{h=0}^{H-k} y_{t-h}y_{t-h-k} \approx \frac{1}{T}\sum_{t=1}^{T} y_t y_{t-k}, k = 1,\ldots,H,$$

that are the first and second-order moments.[1] These are the expected summary statistics for linear analysis of a binary time series.

The interpretations of the pairwise interactions and of the associated model (17.4) are deduced from the following proposition.

Proposition 1: *Under the log-linear probability model with pairwise interactions, the pairwise interaction parameter β_k measures the partial dependence between Y_t and Y_{t-k} given Y_{t-h}, $h \neq 0, k$. This measure is independent of the conditioning values. Moreover $\beta_k = 0$ if and only if Y_t and Y_{t-k} are conditionally independent.*

Proof. Let us consider $k = 1$ for exposition purpose, the proof being similar for any k. From (2.3), we get

$$\beta_1 = \log(p(1 \mid 1, y_{t-2},\ldots,y_{t-H})/p(1 \mid 0, y_{t-2},\ldots,y_{t-H})).$$

In particular $\beta_1 = 0$ if and only if Y_t and Y_{t-1} are independent conditional on $(Y_{t-2},\ldots,Y_{t-H})$. □

The sign of β_1 is also informative on the sign of the conditional dependence.

Proposition 2: *Under the log-linear probability model with pairwise interactions, the parameter β_k is positive if and only if the (partial) correlation between Y_t and Y_{t-k} given Y_{t-h}, $h \neq 0, k$, is positive (for any conditioning values).*

Proof. Let us consider $k = 1$. We get:

$$\begin{aligned}
\beta_1 > 0 &\Leftrightarrow P[Y_t = 1 \mid Y_{t-1} = 1, \underline{Y_{t-2}}] > P[Y_t = 1 \mid Y_{t-1} = 0, \underline{Y_{t-2}}]\\
&\Leftrightarrow P[Y_t = 1, Y_{t-1} = 1 \mid \underline{Y_{t-2}}]P[Y_{t-1} = 0 \mid \underline{Y_{t-2}}]\\
&\quad > P[Y_t = 1, Y_{t-1} = 0 \mid \underline{Y_{t-2}}]P[Y_{t-1} = 1 \mid \underline{Y_{t-2}}]\\
&\Leftrightarrow P[Y_t = 1, Y_{t-1} = 1 \mid \underline{Y_{t-2}}]P[Y_t = 0, Y_{t-1} = 0 \mid \underline{Y_{t-2}}]\\
&\quad > P[Y_t = 1, Y_{t-1} = 0 \mid \underline{Y_{t-2}}]P[Y_t = 0, Y_{t-1} = 1 \mid \underline{Y_{t-2}}]\\
&\Leftrightarrow E[Y_t Y_{t-1} \mid \underline{Y_{t-2}}]E[(1-Y_t)(1-Y_{t-1}) \mid \underline{Y_{t-2}}]\\
&\quad > E[Y_t(1-Y_{t-1}) \mid \underline{Y_{t-2}}]E[(1-Y_t)Y_{t-1} \mid \underline{Y_{t-2}}]\\
&\Leftrightarrow E[Y_t Y_{t-1} \mid \underline{Y_{t-2}}] > E[Y_t \mid \underline{Y_{t-2}}]E[Y_{t-1} \mid \underline{Y_{t-2}}]\\
&\Leftrightarrow Cov[Y_t, Y_{t-1} \mid \underline{Y_{t-2}}] > 0.
\end{aligned}$$

The result follows.□

As shown in Appendix 3 that β_1 is a Kullback-Leilbler (KL) measure of dependence between Y_t and Y_{t-1} given $\underline{Y_{t-2}}$, β_2 of dependence between Y_t and Y_{t-2} given $Y_{t-1}, \underline{Y_{t-3}}, \ldots$. Properties and interpretations of the conditional Kullback-Leibler measure of dependence are discussed in Appendix 2.

Remark 4: If β_1 is positive, all the conditional covariances $Cov[Y_t, Y_{t-1} \mid \underline{Y_{t-2}}]$ are positive, but this does not necessarily implies that the unconditional covariance $Cov[Y_t, Y_{t-1}]$ is positive. This is a consequence of the covariance decomposition:

$$Cov[Y_t, Y_{t-1}] = E[Cov[Y_t, Y_{t-1} \mid \underline{Y_{t-2}}]] + Cov[E[Y_t \mid \underline{Y_{t-2}}], E[Y_{t-1} \mid \underline{Y_{t-2}}]],$$

where the second term of the right hand side can be of any sign.

[1] Note that $\frac{1}{T}\sum_{t=1}^{T} y_t^2 = \frac{1}{T}\sum_{t=1}^{T} y_t$, since $y_t \in \{0,1\}$.

Then the next result easily follows:

Proposition 3: *In the log-linear probability model with pairwise interactions, the following properties are equivalent:*
i) (Y_t) *is a strong white noise, that is the* Y_t'*s are i.i.d.*
ii) The Y_t'*s are uncorrelated.*
iii) All pairwise interaction parameters are equal to zero.

Remark 5: There exist equivalent parametrizations of the conditional logit model (17.3). For instance,

$$\log p(y_t \mid y_{t-1}, \ldots, y_{t-H}) = \tilde{\mu}_t + (1 - y_t)\left[\tilde{\alpha} + \sum_{k=1}^{H} \tilde{\beta}_k (1 - y_{t-k})\right],$$

where the code $(0, 1)$ is implicitly replaced by the code $(1, 0)$. There also exists a more symmetric parametrization, where:

$$\log p(y_t \mid y_{t-1}, \ldots, y_{t-H}) = \tilde{\mu}_t + y_t\left[\tilde{\alpha}^* + \sum_{k=1}^{H} \beta^*_{1k} y_{t-k} + \sum_{k=1}^{H} \beta^*_{0k}(1 - y_{t-k})\right], \tag{17.5}$$

where $\beta^*_{1k} + \beta^*_{0k} = 0$, $k = 1, \ldots, H$, in order to treat the identification issue due to the collinearity restrictions $y_{t-k} + (1 - y_{t-k}) = 1$, $\forall t, k$.

17.2.1.2 Model with Three-wise Interactions

Similar computations can be developed for log-linear probability models with three-wise interactions. Sufficient stationarity restrictions on the parameters lead to a new parametrization $\alpha, \beta_k, \gamma_{k,l}$ such that:

$$\alpha \equiv \alpha_k,\ \forall k,\ \beta_k \equiv \beta_{h,h+k},\ k = 1, \ldots, H \quad \text{for any } h, \tag{17.6}$$

and

$$\gamma_{k,l} \equiv \gamma_{h,h+k,h+l},\ k, l = 1, \ldots, H,\ l > k, \quad \text{for any } h. \tag{17.7}$$

Then the transition distribution becomes a dichotomous logit model, with a score function which is linear in lagged and crossed lagged observations:

$$\begin{aligned}\log p(y_t \mid y_{t-1}, \ldots, y_{t-H}) &= \tilde{\mu}_t + \alpha y_t + \sum_{k=1}^{H} \beta_k y_t y_{t-k} + \sum_{k=1}^{H}\sum_{l=k+1}^{H} \gamma_{kl} y_t y_{t-k} y_{t-l} \\ &= \tilde{\mu}_t + y_t\left(\alpha + \sum_{k=1}^{H} \beta_k y_{t-k} + \sum_{k=1}^{H}\sum_{l=k+1}^{H} \gamma_{kl} y_{t-k} y_{t-l}\right).\end{aligned} \tag{17.8}$$

Therefore the pointwise prediction of Y_t, that is:

$$E[Y_t \mid y_{t-1}, \ldots, y_{t-H}] = p(1 \mid y_{t-1}, \ldots, y_{t-H}),$$

has a logit form with "explanatory" variables the lagged and cross lagged values.

To get an interpretation of a three-wise interaction, let us consider the case $H = 2$. We have:

$$y_t \gamma_{12} = \log\left(\frac{p(y_t \mid y_{t-1} = y_{t-2} = 1)\, p(y_t \mid y_{t-1} = y_{t-2} = 0)}{p(y_t \mid y_{t-1} = 1, y_{t-2} = 0)\, p(y_t \mid y_{t-1} = 0, y_{t-2} = 1)}\right),$$

and then

$$\gamma_{12} = 0 \Longleftrightarrow$$

$$p(y_t \mid y_{t-1} = y_{t-2} = 1)p(y_t \mid y_{t-1} = y_{t-2} = 0) =$$
$$= p(y_t \mid y_{t-1} = 1, y_{t-2} = 0)p(y_t \mid y_{t-1} = 0, y_{t-2} = 1).$$

It is shown in Appendix 2 that γ_{12} measures the effect of a shock on Y_{t-2} on the Kullback-Leibler measure of dependence between Y_t and Y_{t-1} given $\underline{Y_{t-2}}$ (see Proposition A.2).

17.2.1.3 Dimensionality

The objective of limiting the type of interactions is to reduce the curse of dimensionality of the saturated conditional logit model that involves 2^H parameters. Despite this reduction, the number of parameters under the stationarity assumption equals to $1 + H(H+1)/2$ in the model with three-wise interactions can still be large. These dimensions, and the degrees of overidentification $2^H - 1 - H(H+1)/2$ are provided in Table 17.1 below for different lags: $H = 5-7$ corresponding to a week of opening days for daily data, $H = 12$ for the year and monthly data. Then it can be useful to continue to reduce this dimension by looking for zero pairwise and three-wise interactions by an automatic approach as LASSO (Hastie, Tibshirani & Wainwright, 2015). Observe that three-wise interactions are possible when $H \geq 3$.

Table 17.1: Number of Parameters and Degree of Overidentification
Log-linear probability model with three-wise interactions

H	3	4	5	6	7	12
Number of parameters	7	14	15	21	28	66
Degree of overidentification	1	5	16	47	99	4029

17.2.2 Multivariate Qualitative Process

The conditional autoregressive logit model of Section 17.2.1.1 can be extended to qualitative processes with more than two exclusive alternatives. For the application to chartism, we focus below on the case of three exclusive alternatives corresponding to 'large negative returns', 'standard returns', and 'large positive returns'. It is usual to code the three alternatives as $(1,0,0)$, $(0,1,0)$, and $(0,0,1)$ and to consider the process (Y_t) as a 3-dimensional process taking these values, or equivalently as the 2-dimensional process $(\tilde{Y}_t)$ with values $(1,0)$, $(0,1)$ and $(0,0)$ corresponding to the two first components of Y_t, since the third component is then uniquely defined.

However, there exists another way for coding these observations. We can first introduce a binary process X_t such that $X_t = 1$, if the returns are extreme, and $X_t = 0$, otherwise. Then, if $X_t = 1$, a second binary variable Z_t characterizes the sign of the return. Whereas X_t is observed at all dates, Z_t is only observed when $X_t = 1$.

Remark 6: The case of three alternatives is more complex than the case of four alternatives. Indeed with four alternatives the second code will lead to $\tilde{Y}_t$ with values $(1,1), (1,0), (0,1), (0,0)$, where $\tilde{Y}_t$ can be written as a bivariate vector $(\tilde{Y}_{1t}, \tilde{Y}_{2t})'$ of binary variables. Then it will be possible to consider a joint log-linear probability model with pairwise interactions on the multivariate variables $(\tilde{Y}_{1,t}, \tilde{Y}_{2,t}, \tilde{Y}_{1,t-1}, \tilde{Y}_{2,t-1}, \dots, \tilde{Y}_{1,t-H}, \tilde{Y}_{2,t-H})$, and derive the associated transitions.

The different interpretations above and their associated codes lead to two extensions of the autoregressive dichotomous logit model that are the autoregressive polytomous logit model and the autoregressive recursive logit model, respectively.

17.2.2.1 The Autoregressive Polytomous Logit Model with Interactions

Let us consider the first code and denote $Y_t = (Y_{1,t}, Y_{2,t}, Y_{3,t})'$. The transition distribution of Y_t is characterized by the conditional probability that $Y_{1,t} = 1$ (then $Y_{2,t} = Y_{3,t} = 0$) and the conditional probability that $Y_{2,t} = 1$ (then $Y_{1,t} = Y_{3,t} = 0$), since the alternatives are exclusive.

Model with pairwise interactions

The model is an extension of the autoregressive dichotomous logit model introduced for the univariate binary process. It can be written with different equivalent parametrizations (see Remark 5). The direct extension of model (17.3) takes into account all possible pairwise interactions, but considers only the components $Y_{1,t}, Y_{2,t}$ (i.e., the process $\tilde{Y}_t$). The model is defined by:

$$\begin{aligned}
\log p(Y_{1,t} = 1 \mid y_{t-1}, \dots, y_{t-H}) &= \tilde{\mu}_t + \alpha_1 + \sum_{k=1}^{H} \beta_{1k} \tilde{y}_{t-k} \\
&= \tilde{\mu}_t + \alpha_1 + \sum_{k=1}^{H} \beta_{11,k} y_{1,t-k} + \sum_{k=1}^{H} \beta_{12,k} y_{2,t-k}, \\
\log p(Y_{2,t} = 1 \mid y_{t-1}, \dots, y_{t-H}) &= \tilde{\mu}_t + \alpha_2 + \sum_{k=1}^{H} \beta_{2k} \tilde{y}_{t-k} \\
&= \tilde{\mu}_t + \alpha_2 + \sum_{k=1}^{H} \beta_{21,k} y_{1,t-k} + \sum_{k=1}^{H} \beta_{22,k} y_{2,t-k}, \\
\log p(Y_{3,t} = 1 \mid y_{t-1}, \dots, y_{t-H}) &= \tilde{\mu}_t,
\end{aligned}$$

where $\beta_{1k} = (\beta_{11,k}, \beta_{12,k})$, $\beta_{2k} = (\beta_{21,k}, \beta_{22,k})$.

This definition of the conditional polytomous logit model does not highlight the 'symmetry' among the alternatives. In particular the interpretation of the parameters depend on the ordering of the alternatives, that is the choice of the alternative where the dependence from the past is entirely captured by the term $\tilde{\mu}_t$.

A more symmetric formulation extends the specification (17.5) in Remark 5 and corresponds to another parametrization. We can write:

$$\begin{aligned}
\log p(Y_{1,t} = 1 \mid y_{t-1}, \dots, y_{t-H}) &= \tilde{\mu}_t + \alpha_1^* + \sum_{k=1}^{H} \beta_{1k}^* \tilde{y}_{t-k} \\
&= \tilde{\mu}_t + \alpha_1^* + \sum_{k=1}^{H} \beta_{11,k}^* y_{1,t-k} + \sum_{k=1}^{H} \beta_{12,k}^* y_{2,t-k} \\
&\quad + \sum_{k=1}^{H} \beta_{13,k}^* y_{3,t-k},
\end{aligned}$$

$$\log p(Y_{2,t} = 1 \mid y_{t-1}, \dots, y_{t-H}) = \tilde{\mu}_t + \alpha_2^* + \sum_{k=1}^{H} \beta_{2k}^* \tilde{y}_{t-k}$$

$$= \tilde{\mu}_t + \alpha_2^* + \sum_{k=1}^{H} \beta_{21,k}^* y_{1,t-k} + \sum_{k=1}^{H} \beta_{22,k}^* y_{2,t-k} + \sum_{k=1}^{H} \beta_{23,k}^* y_{3,t-k},$$

$$\log p(Y_{3,t} = 1 \mid y_{t-1}, \ldots, y_{t-H}) = \tilde{\mu}_t.$$

However, to avoid the collinearity issue among the regressors:

$$y_{1,t-k} + y_{2,t-k} + y_{3,t-k} = 1, \forall t, k,$$

identification restrictions have to be introduced on the interactions. The pairwise interactions $\beta_k^+ = \begin{pmatrix} \beta_{1k}^* \\ \beta_{2k}^* \end{pmatrix}$ can be constrained by

$$\beta_k^+ \begin{pmatrix} 1 \\ 1 \\ 1 \end{pmatrix} = \begin{pmatrix} 0 \\ 0 \\ 0 \end{pmatrix}, \ \forall k = 1, \ldots, H \Leftrightarrow \begin{cases} \beta_{11,k}^* + \beta_{12,k}^* + \beta_{13,k}^* = 0 \\ \beta_{21,k}^* + \beta_{22,k}^* + \beta_{23,k}^* = 0 \end{cases}, \ \forall k = 1, \ldots, H$$

(see Nerlove and Press (1973) for such a parametrization).

The unconstrained (i.e., saturated) model depends on 2×3^H parameters, that are the conditional elementary probabilities. The autoregressive polytomous logit model with pairwise interactions depends on a number of independent parameters equal to 2 (for α^*s)+$6H$ (for β^*s)-2H (for identification restrictions)=2+4H.

The parameter dimensions and the degrees of overidentification are summarized in Table 17.2.

Table 17.2: Number of Parameters and Degree of Overidentification (3 exclusive alternatives, pairwise interactions)

H	2	3	4	5	6	7	12
Number of parameters	10	14	18	22	26	30	50
Degree of overidentification	8	40	144	464	1432	4344	1062832

Model with three-wise interactions

This extension will also include cross-terms for the conditioning variables appearing in the score function of the polytomous logit models. The conditional logit model will admit as regressors the quadratic functions of y_{t-h} and y_{t-k}. Since $y_{1t} + y_{2t} + y_{3t} = 1$, $\forall t$, it is equivalent to use only the quadratic functions of $\tilde{y}_{t-h}$ and $\tilde{y}_{t-k}$. This leads to the following specification (since $\tilde{y}_{t-k}\tilde{y}'_{t-k} = diag\tilde{y}_{t-k}$, the cross-effects from the values at a same lag are already included in the marginal effects):

$$\log p(Y_t = 1 \mid y_{t-1}, \ldots, y_{t-H}) = \tilde{\mu} + \alpha_1 + \sum_{k=1}^{H} \beta_{1k} \tilde{y}_{t-k} + \sum_{k=1}^{H} \sum_{l=k+1}^{H} \tilde{y}'_{t-k} C_{1kl} \tilde{y}_{t-l},$$

$$\log p(Y_t = 2 \mid y_{t-1}, \ldots, y_{t-H}) = \tilde{\mu} + \alpha_2 + \sum_{k=1}^{H} \beta_{2k} \tilde{y}_{t-k} + \sum_{k=1}^{H} \sum_{l=k+1}^{H} \tilde{y}'_{t-k} C_{2kl} \tilde{y}_{t-l},$$

where the matrices C_{1kl}, C_{2kl} define the three-wise interactions. The parameter dimension is increased by $4H(H-1)$ parameters.

Table 17.3: Number of Parameters and Degree of Overidentification (3 alternatives, three-wise interactions)

H	3	4	5	6	7	12
Number of parameters	38	66	102	146	198	578
Degree of overidentification	16	96	384	1312	4176	1062304

Tables 17.2 and 17.3 have been introduced to show the inflation of the number of parameters with the increase of lag H and of the orders of interactions. Then we can encounter the curse of dimensionality if the number of observations is not large enough with both effects on the estimator's accuracy (statistical efficiency) and the cost of estimation (numerical efficiency).

17.2.2.2 The Recursive Autoregressive Logit Model

The recursive model is based on two latent binary processes X_t, Z_t. In our application they define the alternatives "standard" versus "extreme": $X_t = 1$, if extreme, $X_t = 0$, if standard. Then Z_t defines the sign of the potential extreme: $Z_t = 1$ if positive extreme, $Z_t = 0$, if negative extreme. The model has the form of a state-space model:

Measurement equation:

We have $Y_t = \begin{pmatrix} 1 \\ 0 \\ 0 \end{pmatrix}$, if $X_t = 0$, $Y_t = \begin{pmatrix} 0 \\ 1 \\ 0 \end{pmatrix}$, if $X_t = 1$, $Z_t = 0$, $Y_t = \begin{pmatrix} 0 \\ 0 \\ 1 \end{pmatrix}$, if $X_t = 1$, $Z_t = 1$.

When Y_t is observed, X_t is observed too, but Z_t is not observed when $X_t = 0$. Therefore Z_t is partially a latent variable.

Transition equation:

The model is completed by autoregressive logit models with interactions, that can include the effects of X_{t-k}, Z_{t-k}, not only the effects of the lagged observed qualitative variables y_{t-k}. For instance the transitions can be defined as:

$$\log p(X_t = 1 \mid y_{t-1}, \ldots, y_{t-H}) = \mu_t + \alpha + \sum_{k=1}^{H} \beta_k x_{t-k} + \sum_{k=1}^{H} \delta_k x_{t-k} z_{t-k},$$

$$\log p(Z_t = 1 \mid y_{t-1}, \ldots, y_{t-H}) = \tilde{\mu}_t + \tilde{\alpha} + \sum_{k=1}^{H} \tilde{\beta}_k x_{t-k} + \sum_{k=1}^{H} \tilde{\delta}_k x_{t-k} z_{t-k},$$

assuming that the variables X_t, Z_t are independent conditional on the lagged observations. Note that it is equivalent to observe y_t, or to observe x_t and $x_t z_t$. This independence assumption means that the variable sign, i.e., Z_t, is latent when the return is standard, but can be filtered by using the same conditional distribution as if the returns was extreme, that is when $X_t = 1$.

It has been observed in empirical studies that the signs of the return were close to a strong white noise. Thus we can expect that the interactions parameters $\tilde{\beta}_k$, $\tilde{\delta}_k$ are often non significant.

We can also expect that the dynamics of extreme features is exogenous, that is that the $\tilde{\delta}_k$'s parameters are zero. Indeed, this recursive model with latent variables is an analogue for qualitative Y_t of the stochastic volatility model introduced for quantitative returns, with stochastic volatility as the latent process. Such a recursive logit model with latent qualitative variables appears also in the (static) Boltzmann machine literature (see Ackley et al., 1985). In our framework it is extended to dynamic models.

17.3 Statistical Inference

Let us focus on the conditional autoregressive logit models with interactions introduced in Sections 17.2.1.1 and 17.2.2. These models depend on the interaction parameters $a = vec(\alpha),\ b = vec(\beta), c = vec(C)$, where the different types of interactions are stacked in a vector. The models can also depend on additional parameters δ that define the alternatives. For instance in the three exclusive alternative specification, the alternatives can be defined as

$$\begin{aligned} &- \text{extreme negative return if} \quad r_t < -\delta, \\ &- \text{standard return if} \quad | r_t | \leq \delta, \\ &- \text{extreme positive return if} \quad r_t > \delta. \end{aligned} \tag{17.9}$$

Then, the transition at date t can be written as $p(y_t \mid \underline{y_{t-1}}; a, b, c, \delta)$ and the log-likelihood as:

$$L_T(y; a, b, c, \delta) = \sum_{t=H+1}^{T} \log p(y_t \mid \underline{y_{t-1}}; a, b, c, \delta). \tag{17.10}$$

17.3.1 Maximum Likelihood Estimation

Different maximum likelihood approaches can be implemented.[2] In general the maximum likelihood estimators have no closed form expressions and they are unfeasible. The parameters can be estimated online by applying the Average Stochastic Gradient Descent (ASGD) algorithm (Ruppert, 1988; Polyak & Juditsky, 1992).

17.3.1.1 Maximum Likelihood with Known δ

When δ is known, the conditional autoregressive logit model is a generalized linear model for which special optimization softwares are available and the expression of the estimated information matrix is greatly simplified (see Nelder & Wedderburn, 1972). The ML estimators are the solution of the maximization below:

$$[\hat{a}_t(\delta), \hat{b}_t(\delta), \hat{c}_t(\delta)] = \underset{a,b,c}{Argmax} L_T(y; a, b, c, \delta). \tag{17.11}$$

[2] Likewise, composite likelihood approaches could be implemented; see Appendix 2. These methods can be used sequentially with first the estimation of the α's by marginal composite likelihood, second the estimation of the β's by pairwise conditional composite likelihood, then of the γ's by conditional three-wise composite likelihood. This approach is numerically efficient.

Note that the online ASGD estimator has the same asymptotic properties as the (unfeasible) maximum likelihood estimator. If the model is well specified and the observed process is stationary ergodic, this estimator is consistent, asymptotically normal and asymptotically efficient. Moreover its asymptotic variance, that is the inverse of the information matrix, can be also estimated online.

17.3.1.2 Maximum Likelihood with Unknown δ

When δ is unknown, the log-likelihood can also be maximized with respect to the threshold parameter δ. This can be done in two steps by first concentrating the log-likelihood with respect to the interaction parameters. More precisely, let us denote:

$$\hat{L}_T(y;\delta) = L_T(y;\hat{a}_t(\delta),\hat{b}_t(\delta),\hat{c}_t(\delta),\delta), \tag{17.12}$$

the maximum value of the concentrated objective function. Then the ML estimator $\hat{\hat{a}}_T, \hat{\hat{b}}_T, \hat{\hat{c}}_T, \hat{\delta}_T$ is defined as:

$$\hat{\delta}_T = \underset{\delta}{Argmax}\hat{L}_T(y;\delta),\ \hat{\hat{a}}_T = \hat{a}_T(\hat{\delta}_T),\ \hat{\hat{b}}_T = \hat{b}_T(\hat{\delta}_T),\ \hat{\hat{c}}_T = \hat{c}_T(\hat{\delta}_T). \tag{17.13}$$

This method is the analogue in the pure qualitative framework of the ML approach used for instance in Zhang et al. (2023) with threshold autoregressive model.

The interpretation of δ as a threshold implies that the ML estimators and their online analogues are consistent, but no longer satisfy the standard asymptotic properties of normality and efficiency.

17.3.1.3 Maximum Likelihood with LASSO

To reduce the number of parameters, the objective function of the ML estimation can also be penalized by LASSO with penalties written on the different interactions (see Hastie et al. (2015) for a general introduction to LASSO). Let us for instance consider the case with known δ. The objective function is replaced by:

$$L_T(y;a,b,c,\delta) - \lambda_1 \parallel a \parallel_1 - \lambda_2 \parallel b \parallel_1 - \lambda_3 \parallel c \parallel_1,$$

where $\parallel a \parallel_1$ denotes the l_1-norm of the vector a, that is the sum of the absolute values of the its components and $\lambda_1, \lambda_2, \lambda_3$ are positive tuning scalars.

The solution of the penalized optimization depends on the tuning parameters $\lambda_1, \lambda_2, \lambda_3$. Larger the λ's, smaller the number of nonzero interactions. Note also that the penalization is the same for all pairwise interactions (resp. three-wise interactions). This choice is justified by the behavior of the ML estimates in a neighborhood of the strong white noise hypothesis.

17.3.2 Close to Independence Behavior (Binary Process)

Let us consider the case of a binary process. By Proposition 1, the binary process is a strong white noise if and only if the interaction parameters β_k, γ_{kl} are all equal to zero. Under the independence hypothesis, the ML estimator $\hat{\beta}_{k,T}$, $\hat{\gamma}_{kl,T}$ have simplified asymptotic properties, similar to the properties of estimated ACF. In particular, the asymptotic variance of the $\hat{\beta}_{k,T}$ (resp. $\hat{\gamma}_{kl,T}$) are independent of the index k (resp. of the pairwise index kl).

Therefore the estimates can be provided with their (fixed) confidence bound under the null, in a plot of $\hat{\beta}_{k,T}$ function of k and a heat plot of the $\hat{\gamma}_{kl,T}$ function of k, l. These plots can also be given setting to 0, the nonsignificant values, as a diagnostic tool of the real interactions, before introducing LASSO penalties.

17.4 Prediction and Filtering

Once the models are estimated, they can be used for prediction and filtering. Since the processes are qualitative, the standard pointwise analysis based on conditional expectations has no real meaning and the complete predictive distributions have to be considered. In our framework we get closed form expressions of the prediction and filtering distributions due to the conditional logit models corresponding to the joint log-linear probability model (see Appendix 2).

17.4.1 Prediction of Future Patterns

With in mind the technical analysis, we are interested in the future pattern of returns. More precisely, let us fix a horizon H^*, we are interested at date T in the prediction of $(Y_{T+1}, \dots, Y_{T+H^*})$ given $(Y_T, \dots, Y_{T-H+1})$. By the Markov property, this predictive distribution takes the form:

$$p_{H^*}(y_{T+1}, \dots, y_{T+H^*} \mid y_T, \dots, y_{t-H+1}) = \prod_{k=1}^{H^*} p(y_{T+k} \mid y_{T+k-1}, \dots, y_{t+k-H}).$$

Remark 7: This prediction of future patterns differ from the prediction at date T of Y_{T+H^*} only. In terms of derivative products, the technical analysis is more focused on American derivatives than on European ones.

17.4.2 Filtering

As mentioned in the introduction, an advantage of the log-linear probability model with interactions is to satisfy invariance properties by conditioning and to provide closed form expressions of the predictive distributions. As simple example, we can consider the problem of backward forecast and the determination of the filtering distributions for a binary time series.

Backward forecasting

This concerns the predictive distribution of Y_t given $\overline{Y_{t+1}} = (Y_{t+1}, Y_{t+2}, \dots)$.

Proposition 4:

i) The conditional logit autoregressive process with interactions is both Markov in calendar and reversed times

ii) The process with pairwise interactions is reversible.

Proof:

i) Let us consider a process of order 1 for expository purpose and compute the backward transition. We have:

$$\begin{aligned} p(y_t \mid y_{t+1}, \dots, y_{t+H}) &= \frac{p(y_t, y_{t+1}, \dots, y_{t+H})}{p(y_{t+1}, \dots, y_{t+H})} \\ &= \frac{\pi(y_t) p(y_{t+1} \mid y_t) \dots p(y_{t+H} \mid y_{t+H-1})}{\pi(y_{t+1}) p(y_{t+2} \mid y_{t+1}) \dots p(y_{t+H} \mid y_{t+H-1})} \end{aligned}$$

$$= \frac{\pi(y_t)p(y_{t+1} \mid y_t)}{\pi(y_{t+1})},$$

where $\pi(\cdot)$ denotes the stationary distribution. We deduce that:

$$p(y_t \mid y_{t+1}, \ldots, y_{t+H}) = p(y_t \mid y_{t+1}), \ \forall H,$$

and then by taking the limit where H tends to infinity, we deduce that the process satisfies the Markov property in reversed time (see Cambanis and Fakhre-Zakeri (1995) for a similar property for real valued processes).

ii) More generally for a logit autoregressive process of any order H, we deduce that the process satisfies also a conditional logit autoregressive process of the type (see (2.4)):

$$\log p(y_t, y_{t+1}, \ldots, y_{t+H}) = \mu^* + \alpha^* \sum_{h=0}^{H} y_{t+h} + \sum_{k=1}^{H} (\beta_k^* \sum_{h=0}^{H-k} y_{t+h} y_{t+h+k}).$$

We deduce that $\alpha^* = \alpha$, $\beta_k^* = \beta_k$, $\forall k$, $k = 1, \ldots, H$, by the interpretation of the interactions parameters. For instance β_k^* is the Kullback-Leibler (KL) measure of independence between Y_t and Y_{t-k} and coincides with the KL measure between Y_t and Y_{t+k}.□

Whereas the Markov conditions in calendar and reversed times are still equivalent for processes with three alternatives, the reversibility can require constraints on the three-wise interaction parameters.

Filtering distributions

This concerns the prediction of Y_t given $\underline{Y_{t-1}}$ and $\overline{Y_{t+1}}$. The result below extends to this qualitative framework the closed form expression known for the univariate Gaussian AR(1) model, $Y_t \mid \underline{y_{t-1}} \sim N(\rho y_{t-1}, 1-\rho^2)$, where the filtering distribution for Y_t is Gaussian, with a conditional mean equal to $\frac{\rho}{1+\rho^2}(y_{t-1} + y_{t+1})$ and a variance equal to $\frac{1-\rho^2}{1+\rho^2}$.

Proposition 5:

i) The filtering distribution of Y_t given $\underline{Y_{t-1}}$ and $\overline{Y_{t+1}}$ is equal to the filtering distribution of Y_t given $Y_{t-1}, \ldots, Y_{t-H}, Y_{t+1}, \ldots, Y_{t+H}$. It has the form of a logit model with interactions.

ii) When $H = 1$, the filtering distribution is symmetric in Y_{t-1} and Y_{t+1}.

Proof: Let us consider $H = 1$.

i) This is a consequence of the Markov property that the past and future are independent given the present. In general the result can be derived following the same approach as in the proof of Proposition 4 i).

ii) It is easily checked that:

$$P(Y_t = 1 \mid y_{t-1}, y_{t+1}) = \exp(\alpha^* + \beta_1(y_{t-1} + y_{t+1}))/(1 + \exp(\alpha^* + \beta_1(y_{t-1} + y_{t+1}))),$$

with

$$\alpha^* = \alpha - \log\left(\frac{1+\exp(\alpha+\beta)}{1+\exp(\alpha)}\right).$$

The result follows.□

17.5 Concluding Remarks

The aim of this paper was to extend the static log-linear probability model considered by Nerlove and Press (1973) to the dynamic analysis of qualitative processes with two or three alternatives. The introduction of interactions of small orders allows for solving the curse of dimensionality, whereas the structure (architecture) of the model facilitates the statistical inference and the numerical cost of training.

For a stationary (ergodic) Markov process (Y_t) of order H, it is equivalent to specify the conditional distribution of Y_t given $Y_{t-1}, \dots, Y_{t-H}$, or the joint distribution of $Y_t, Y_{t-1}, \dots, Y_{t-H}$. Recursive machine learning techniques have been introduced in the literature to approximate recursively such joint distributions. They are known under the name of Normalizing Flows (see, e.g., Papamakarios, Nalisnick, Rezende, Mohamed & Lakshminarayanan, 2021 for a survey). However they have some drawbacks in our framework: i) They are not able to introduce the constraint of stationarity; ii) They are mainly defined for continuous multivariate variables (Y_t), with no real analogue for discrete or qualitative processes (see Papamakarios et al., 2021, Section 5.3, and Kobyzev, Prince & Brubaker, 2020, Section 5.2.2).

Acknowledgements We thank the Agence Nationale de la Recherche (ANR), grants *ANR-24-CE26-3681-01* and *ANR-17-EURE-0010*, for financial support. The second author would also like to thank for financial support the Institut Universitaire de France (IUF).

Appendix 1: Ergodicity

Let us consider the log-linear probability model with interactions for a binary process (the proof for process with three alternatives is similar), and the process $Y_t^* = (Y_t, Y_{t-1}, \dots, Y_{t-H+1})'$ obtained by stacking H consecutive observations of Y_t. The process Y_t^* is also a Markov chain and due to the conditional logit expression all the transition probabilities for Y_t^* are strictly positive. Thus this Markov chain is irreducible aperiodic and then necessarily ergodic (Norris, 1998, Chapter 1). In particular, there is no transient state and a unique class of recurrence. In addition, there exists a unique stationary distribution for Y_t^* (resp. Y_t) obtained as the limit of the predictive distribution at infinite horizon.

Appendix 2: Log-Linear Probability Models with Interactions

We provide a brief review of the (static) log-linear probability model with interactions (Nerlove & Press, 1973, Bishop et al., 1975, Schmidt & Strauss, 1975, Lee, 1981, Liang & Zeger, 1989), with special attention to the case of binary variables.

2.1 The Models

Let us consider n binary variables $X_1, \dots, X_n$, where X_i can take values in $\{0, 1\}$. The log-linear probability models with interactions provide parametric specifications for their joint distribution. We will essentially consider models with interactions up to order 2, or up to order 3.

The model with interactions up to order 2 assumes that the elementary probabilities $P[X = x] = P[X_1 = x_1, \dots, X_n = x_n] \equiv p(x_1, \dots, x_n)$ are of the form:

$$\log p(x_1, \dots, x_n) = \mu + \sum_{i=1}^{n} \alpha_i x_i + \sum_{i=1}^{n} \sum_{j=i+1}^{n} \beta_{ij} x_i x_j.$$

The coefficients α_i (resp. β_{ij}) define the marginal effects (resp. the interaction effects at order 2). The parameter μ is fixed by the unit mass restriction and then is a (complicated function) of the α_i's and β_{ij}'s. Whereas the unconstrained joint distribution depends on $2^n - 1$ parameters, the log-linear probability model with pairwise interactions depends on $n + n(n-1)/2 = n(n+1)/2$ parameters, that are the α_i's and β_{ij}'s.

The model above can be extended to allow for interactions of higher order, such as three-wise interactions. Then, it is defined as:

$$\log p(x_1, \ldots, x_n) = \mu + \sum_{i=1}^{n} \alpha_i x_i + \sum_{i=1}^{n} \sum_{j=i+1}^{n} \beta_{ij} x_i x_j + \sum_{i=1}^{n} \sum_{j=i+1}^{n} \sum_{k=j+1}^{n} \gamma_{ijk} x_i x_j x_k,$$

where the γ_{ijk} are the threewise interaction coefficients. This parametric specification depends on $n + n(n-1)/2 + n(n-1)(n-2)/6 = n(n^2+2)/6$ independent parameters.

When n is large, these models can still encounter the curse of dimensionality. It is usual in applications to constrain the interactions by assuming that they depend on a 'distance' between the pairs or triplets of individuals:

$$\beta_{ij} = d_2(z_i, z_j; \beta), \quad \gamma_{ijk} = d_3(z_i, z_j, z_k; \gamma),$$

where the z_i's are characteristics of the individuals and β, γ are hyperparameters.

2.2 Conditioning and Logit Models

As noted above the parameter μ has a complicated expression in terms of marginal and interaction effects. This parameter can be easily eliminated by conditioning. Moreover the conditional models are logit models with interaction effects (see e.g., Liang & Zeger, 1989, Section 2). Let us for instance consider the conditional distribution of X_1 given $X_2, \ldots, X_n$. We have

$$\begin{aligned}\log p(x_1 \mid x_2 \ldots, x_n) &= \mu_1(x_2, \ldots, x_n) + \alpha_1 x_1 + \sum_{j=2}^{n} \beta_{1j} x_1 x_j \\ &= \mu_1(x_2, \ldots, x_n) + x_1 \left(\alpha_1 + \sum_{j=2}^{n} \beta_{1j} x_j \right),\end{aligned}$$

for the model with pairwise interactions, and

$$\log p(x_1 \mid x_2 \ldots, x_n) = \mu_1(x_2, \ldots, x_n) + x_1 \left(\alpha_1 + \sum_{j=2}^{n} \beta_{1j} x_j + \sum_{j=2}^{n} \sum_{k=j+1}^{n} \gamma_{1jk} x_j x_k \right),$$

for the model with threewise interactions.

These conditional models are simply dichotomous logit models with $x_j, j = 2, \ldots, n$ (resp. x_j, $x_j x_k$) as explanatory variables introduced in a linear score function for the model with pairwise interactions (resp. threewise interactions).

2.3 Aggregation of Interaction Effects

Pairwise interactions

Let us consider a model with pairwise interactions and denote I, J, a partition of $\{1,\dots,n\}$. The joint distribution can be written as:

$$p(x_I, x_J) = \exp(\mu)\times$$
$$\exp\left[\sum_I \alpha_i x_i + \sum_J \alpha_j x_j + \sum_I\sum_I \beta_{ij} x_i x_j + \sum_J\sum_J \beta_{ij} x_i x_j + \sum_I\sum_J \beta_{ij} x_i x_j\right].$$

We deduce the conditional distribution:

$$p(x_I \mid x_J) = \exp(\mu(x_J))\exp\left[\sum_I x_i\left(\alpha_i + \sum_J \beta_{ij} x_j\right) + \sum_I\sum_I \beta_{ij} x_i x_j\right],$$

which is also a log linear probability model with pairwise interactions that depend on the conditioning variables. In particular, if the partition is $I = \{1,2\}$, $J = \{3,\dots,n\}$, we get

$$\log p(x_1, x_2 \mid x_J) = \mu(x_J) + x_1\alpha_1(x_J) + x_2\alpha_2(x_J) + \beta_{12}x_1x_2,$$

with clear notations. This is the conditional distribution of a log-linear probability model with pairwise interactions. The we can apply the result derived for a bivariate binary variable in Appendix 3).

Proposition 6: *β_{12} is the Kullback measure of dependence between X_1 and X_2 conditional on $\{X_3,\dots,X_n\}$. This measure does not depend on the values of the conditional variables.*

We have a similar interpretation of any β_{ij} parameter as a Kullback measure of partial dependence. We deduce the following Corollaries:

Corollary 1: In a pairwise log-linear probability model, the binary variables are independent if and only if X_i and X_j are independent given $\{x_k, k = 1,\dots,n, k \neq i, k \neq j\}$, for any pair (i,j), $i \neq j$.

Next consider a saturated log-linear probability model with $I = \{1,2\}$, $J = \{3,\dots,n\}$. We easily deduce that:

$$\log p(x_1, x_2 \mid x_J) = \mu(x_J) + x_1\alpha_1(x_J) + x_2\alpha_2(x_J) + \beta_{12}(x_J)x_1x_2.$$

Corollary 2: A saturated log-linear probability model reduces to a log-linear probability model with pairwise interactions if and only if $\beta_{ij}(x_J)$, $J = \{k = 1,\dots,n, k \neq i, k \neq j\} \equiv \{i,j\}^c$ is independent of x_J for any pair (i,j).

Three-wise interactions

Let us now consider a model with three-wise interactions and the partition $I^* = \{1,2,3\}$ and $J^* = \{1,2,3\}^c$. By aggregation we get:

$$\log p(x_1, x_2 \mid x_3, x_{J^*}) = \mu(x_3, x_{J^*}) + x_1\alpha_1(x_3, x_{J^*}) + x_2\alpha_2(x_3, x_{J^*})$$
$$+ x_1x_2(\beta_{12}(x_{J^*}) + \gamma_{123}x_3).$$

Let us denote $K_{12}(x_3, x_{J^*})$ the Kullback measure of dependence between X_1 and X_2 given x_3, x_{J^*}. We have:

$$K_{12}(x_3, x_{J^*}) = \beta_{12}(x_{J^*}) + \gamma_{123} x_3,$$

and then:

$$\gamma_{123} = K_{12}(1, x_{J^*}) - K_{12}(0, x_{J^*}).$$

We get the following interpretation of the three-wise interaction parameter.

Proposition 7: *In the log-linear probability model with threewise interactions, γ_{123} measures the effect on the conditional measure of dependence between X_1 and X_2 of a shock on x_3 (going from $x_3 = 0$ to $x_1 = 1$).*

Such interpretations in term of shock and impulse responses are related to different notions of causality between binary variables (see Mosconi & Seri, 2006 for non-causality in binary time series).

2.4 Composite Conditional Likelihood

The implementation of the standard maximum likelihood approach to estimate the α's and the β's can be numerically cumbersome due to the complicated expression for μ and a number of parameters of order n^2 in the model with pairwise interactions (and even more with three-wise interactions). However, it is seen from the expressions of the conditional distributions that the parameters can be identified and estimated by considering the conditional distributions. More precisely, we can apply the conditional maximum likelihood to the distribution of X_1 given $X_2, \ldots, X_n$, i.e., estimate the conditional logit model to get consistent estimators of $\alpha_1, \beta_{1j}, j = 2, \ldots, n$. Similarly we can consider the conditional model of X_2 given $X_3, \ldots, X_n$ to estimate $\alpha_2, \beta_{2j}, j = 3, \ldots, n$, and so on. We can also put together these different conditional log-likelihood functions into a composite conditional log-likelihood function (Besag, 1974, Varin, Reid & Firth, 2011).

A similar approach can be applied for the model with threewise interactions, based on the joint distribution of X_1, X_2 given $X_3, \ldots, X_n$. The conditional model is a logit polytomous model with 4 alternatives, in which the number of parameters is of order n instead of n^3 if the joint ML were used.

Appendix 3: The Binary AR(1) Process

3.1 Markov Chain

This is the simplest case of a Markov chain. The joint distribution of (Y_t, Y_{t-1}) is given by:

$$\begin{pmatrix} p_{11} & p_{01} \\ p_{10} & p_{00} \end{pmatrix} = \frac{1}{1 + 2\exp(\alpha) + \exp(2\alpha + \beta)} \begin{pmatrix} \exp(2\alpha + \beta) & \exp(\alpha) \\ \exp(\alpha) & 1 \end{pmatrix}.$$

The transition probabilities are:

$$P = \begin{pmatrix} p_{1|1} & p_{0|1} \\ p_{1|0} & p_{0|0} \end{pmatrix} = \begin{pmatrix} \frac{\exp(\alpha+\beta)}{1+\exp(\alpha+\beta)} & \frac{1}{1+\exp(\alpha+\beta)} \\ \frac{\exp(\alpha)}{1+\exp(\alpha)} & \frac{1}{1+\exp(\alpha)} \end{pmatrix}.$$

The transition matrix has the eigenvalues 1 and $p_{1|1} + p_{0|0} - 1 \equiv \lambda$. We see that $2 > p_{1|1} + p_{0|0} \Leftrightarrow \lambda < 1$. Therefore the Markov chain is ergodic. Its stationary distribution is deduced from the joint distribution as:

$$\pi = P(Y_t = 1) = P(Y_{t-1} = 1) = \frac{\exp(2\alpha + \beta) + \exp(\alpha)}{1 + \exp(2\alpha + \beta) + 2\exp(\alpha)}.$$

For $H = 1$, the conditional logit model is just identified and provides an equivalent parametrization of the transition matrix assuming that all transition probabilities are strictly positive.

3.2 Linear AR(1) Representation

An alternative parametrization is π, λ, where $0 < \pi < 1$, $-1 < \lambda < 1$. Since

$$P(Y_t = 1 \mid Y_{t-1}) = E[Y_t \mid Y_{t-1}] = \pi + \lambda(Y_{t-1} - \pi), \tag{17.14}$$

the transition matrix can also be written as

$$P = \begin{pmatrix} \pi + (1-\pi)\lambda & (1-\pi)(1-\lambda) \\ \pi(1-\lambda) & (1-\pi) + \pi\lambda \end{pmatrix} = \begin{pmatrix} 1 \\ 1 \end{pmatrix} (\pi, 1-\pi) + \lambda \begin{pmatrix} 1-\pi \\ -\pi \end{pmatrix} (1, -1).$$

The recursive equation (17.14) provides the interpretation of parameters π and λ. π is the marginal expectation of Y_t and has a long run interpretation, whereas λ measures the serial correlation and has a short run interpretation. In fact, (17.14) is a linear AR(1) representation of the binary process.

This alternative parametrization is appropriate for deriving the term structure of predictions. Indeed the autoregressive equation (17.14) can be iterated to get:

$$P(Y_{t+k-1} = 1 \mid Y_{t-1}) = \pi + \lambda^k (Y_{t-1} - \pi),$$

and

$$P^k = \begin{pmatrix} \pi + (1-\pi)\lambda^k & (1-\pi)(1-\lambda^k) \\ \pi(1-\lambda^k) & (1-\pi) + \pi\lambda^k \end{pmatrix}.$$

3.3 Measure of Dependence

Let us now compute the first-order autocorrelation. We directly deduce from (17.14) and the stationarity that:

$$\rho(1) = Corr(Y_t, Y_{t-1}) = \lambda = \frac{\exp(\alpha)(\exp(\beta) - 1)}{(1 + \exp(\alpha))(1 + \exp(\alpha + \beta))}.$$

Even if $\rho(1) = 0$ if and only if $\beta = 0$, the pairwise interaction β cannot be interpreted as an autocorrelation. Instead, it is a Kulback-Leibler measure of dependence:

$$\beta = \log\left(\frac{p_{1|1} p_{0|0}}{p_{0|1} p_{1|0}}\right) = \log\left(\frac{p_{11} p_{00}}{p_{01} p_{10}}\right).$$

This measure is more appropriate in a nonlinear dynamic framework where the qualitative feature of the observed series has to be taken into account.

References

Ackley, D., Hinton, G. & Sejnowski, T. (1985). A learning algorithm for Boltzmann machines. *Cognitive Science*, *9*, 147-169.

Archer, M. & Bickford, J. (2007). *The FOREX chartist companion: a visual approach to technical analysis*. John Wiley and Sons.

Besag, J. (1974). Spatial interaction and the statistical analysis of lattice systems. *Journal of the Royal Statistical Society: Series B (Methodological)*, *36*, 192-225.

Bishop, Y. M. M., Fienberg, S. E. & Holland, P. W. (1975). *Discrete multivariate analysis: theory and practice*. MIT Press. Retrieved from https://archive.org/details/discretemultivar00bish

Black, F. (1976). Studies of stock market volatility changes. In *Proceedings of the American Statistical Association, Business and Economic Statistics Section* (p. 177-181).

Bollerslev, T. (1986). Generalized autoregressive conditional heteroskedasticity. *Journal of Econometrics*, *31*, 307-327.

Dick, C. & Menkhoff, L. (2013). Exchange rate expectations of chartists and fundamentalists. *Journal of Economic Dynamics and Control*, *37*, 1362-1383.

Engle, R. (1982). Autoregressive conditional heteroscedasticity with estimates of the variance of United Kingdom inflation. *Econometrica*, *50*, 987-1007.

Engle, R., Lilien, D. & Robins, R. (1987). Estimating time varying risk premia in the term structure: The ARCH-M model. *Econometrica*, *55*, 391-407.

Fama, E. (1995). Random walks in stock market prices. *Financial Analysts Journal*, *51*, 75-80.

Frankel, J. & Froot, K. (1990). The rationality of the foreign exchange rate: chartists, fundamentalists and trading in the foreign exchange market. *American Economic Review*, *80*, 181-185.

Gourieroux, C. (1997). *ARCH models and financial applications*. Springer Science and Business Media.

Gourieroux, C. & Monfort, A. (1992). Qualitative threshold ARCH models. *Journal of Econometrics*, *52*, 159-199.

Hastie, T., Tibshirani, R. & Wainwright, M. (2015). *Statistical learning with sparsity: the LASSO and generalizations*. CRC press.

Kobyzev, I., Prince, S. & Brubaker, M. (2020). Normalizing flows: An introduction and review of current methods. *IEEE Transactions on Pattern Analysis and Machine Intelligence*, *43*, 3964-3979.

Lee, L. (1981). Fully recursive probability models and multivariate log-linear probability models for the analysis of qualitative data. *Journal of Econometrics*, *16*, 51-69.

Liang, K. & Zeger, S. (1989). A class of logistic regression models for multivariate binary time series. *Journal of the American Statistical Association*, *84*, 447-451.

Lo, A., Mamaysky, H. & Wang, J. (2000). Foundations of technical analysis: Computational algorithms, statistical inference, and empirical implementation. *The Journal of Finance*, *55*, 1705-1765.

Lofton, T. E. (1986). *Trading Tactics: A Livestock Futures Anthology*. Chicago Mercantile Exchange.

Lui, Y. & Mole, D. (1998). The use of fundamental and technical analyses by foreign exchange dealers: Hong Kong evidence. *Journal of International Money and Finance*, *17*, 535-545.

Menkhoff, L., Sarno, L., Schmeling, M. & Schrimpf, A. (2012). Currency momentum strategies. *Journal of Financial Economics*, *106*, 660-684.

Mosconi, R. & Seri, R. (2006). Non-causality in bivariate binary time series. *Journal of Econometrics*, *132*, 379-407.

Neely, C. & Weller, P. e. (2012). *Technical analysis in the foreign exchange market*. Wiley, Handbook of exchange rates.

Nelder, J. & Wedderburn, R. (1972). Generalized linear models. *Journal of the Royal Statistical Society Series A: Statistics in Society*, *135*, 370-384.

Nerlove, M. & Press, S. (1973). *Univariate and multivariate log-linear and logistic models*. Rand Corporation, https://www.rand.org/content/dam/rand/pubs/reports/2006/R1306.pdf.

Norris, J. (1998). *Markov chains*. Cambridge University Press.

Oberlechner, T. (2001). Importance of technical and fundamental analysis in the European foreign exchange market. *International Journal of Finance and Economics*, *6*, 81-93.

Papamakarios, G., Nalisnick, E., Rezende, D., Mohamed, S. & Lakshminarayanan, B. (2021). Normalizing flows for probabilistic modeling and inference. *Journal of Machine Learning Research*, *22*, 1-64.

Polyak, B. & Juditsky, A. (1992). Acceleration of stochastic approximation by averaging. *SIAM Journal on Control and Optimization*, *30*, 838-855.

Rhea, R. (1932). *The Dow theory: An explanation of its development and an attempt to define its usefulness as an aid in speculation*. Fraser Publishing Company.

Ruppert, D. (1988). *Efficient estimations from a slowly convergent Robbins-Monro process*. Cornell University Operations Research and Industrial Engineering.

Russell, R. (2012). *The Dow Theory Today*. Snowball Publishing.

Salakhutdinov, R. & Hinton, G. (2009). Deep Boltzmann Machines. *Artificial Intelligence and Statistics*, *5*, 448-455.

Schmidt, P. & Strauss, R. (1975). Estimation of models with jointly dependent qualitative variables: A simultaneous logit approach. *Econometrica*, *43*, 745-755.

Taylor, M. & Allen, H. (1992). The use of technical analysis in the foreign exchange market. *Journal of International Money and Finance*, *11*, 304-314.

Varin, C., Reid, N. & Firth, D. (2011). An overview of composite likelihood methods. *Statistica Sinica*, *21*, 5-42.

Vigfusson, R. (1997). Switching between chartists and fundamentalists: a Markov regime-switching approach. *International Journal of Finance and Economics*, *2*, 291-305.

Zakoian, J. (1994). Threshold heteroskedastic models. *Journal of Economic Dynamics and Control*, *18*, 931-955.

Zhang, X., Li, D. & Tong, H. (2023). On the least squares estimation of multiple-threshold-variable autoregressive models. *Journal of Business and Economic Statistics*, *42*, 1-29.

Correction to: Dynamic Heterogeneous Linear Models for Three-level Panel Data with Short Time Dimension and Stratification

Monika Avila Márquez and Jaya Krishnakumar

Correction to:
Chapter 13 in: B. H. Baltagi and L. Mátyás (eds.), *Seven Decades of Econometrics and Beyond*, Advanced Studies in Theoretical and Applied Econometrics 57, https://doi.org/10.1007/978-3-031-92699-0_13

The original version of this book was inadvertently published with incorrect author name "Márquez, M.A." in Chapter 13, which has now been changed to "Avila Márquez, M.". The book and the chapter have been updated with the changes.

The updated version of this chapter can be found at
https://doi.org/10.1007/978-3-031-92699-0_13

B. H. Baltagi and L. Mátyás (eds.), *Seven Decades of Econometrics and Beyond*, Advanced Studies in Theoretical and Applied Econometrics 57,
https://doi.org/10.1007/978-3-031-92699-0_18

Index

B. H. Baltagi and L. Mátyás (eds.), *Seven Decades of Econometrics and Beyond*, Advanced Studies in Theoretical and Applied Econometrics 57, https://doi.org/10.1007/978-3-031-92699-0

The manufacturer's authorised representative in the EU is Springer Nature Customer Service Centre GmbH, Europaplatz 3, 69115 Heidelberg, Germany. If you have any concerns regarding our products, please contact ProductSafety@springernature.com

Printed and bound by CPI Group (UK) Ltd, Croydon, CR0 4YY
01/10/2026
02225954-0019